Ronald Tetzlaff
Editor

Memristors and Memristive Systems

Editor
Ronald Tetzlaff
Faculty of Electrical
and Computer Engineering
Institute of Circuits and Systems
Technische Universität Dresden
Dresden, Germany

ISBN 978-1-4614-9067-8 ISBN 978-1-4614-9068-5 (eBook)
DOI 10.1007/978-1-4614-9068-5
Springer New York Heidelberg Dordrecht London

Library of Congress Control Number: 2013953316

Printed on acid-free paper

Springer is part of Springer Science+Business Media (www.springer.com)

Memristors and Memristive Systems

Preface

Since 2007, when Stanley Williams and his team at HP Labs constructed a nanoelectronic device [1] showing certain fingerprints of a memristor, there has been a resurgence of research activities in developing elements of electronic circuits and analyzing physical systems exhibiting those fingerprints. The existence of a memristor (for MEMory ResISTOR) as a fourth basic circuit element relating flux-linkage to charge had been postulated by Chua [2] in 1971. A memristor behaves like a nonlinear resistor with memory depending on the past history of the current or voltage in the device. An extension of the notion to memristive systems [3] was given by Chua and Kang (1976) allowing that these systems depend additionally on a state. In 2009, Ventra et. al. [4] claimed that these resistive elements belong to a larger class of memory circuit elements including capacitive and inductive systems. Recently, Chua proposed to denote memristive systems as memristors as well, while a circuit element classified according to the original definition can be regarded as an ideal memristor. Since circuits and systems with memristors exhibit several different nonlinear phenomena, these elements are considered not only for the development of memory technology but moreover for playing an important role in the development of new nonlinear information processing methods and their implementation in hardware embodiments. Especially, major efforts will be made to develop neuromorphic memristor technology to build complex brain-like computing structures in the future.

This book aims at providing a comprehensive overview of major memristor aspects. It includes memristor fundamentals, models, and their simulation, the theory of memristor circuits, an overview of devices developed recently, logic gates, neuromorphic systems, and applications. I am pleased to have contributions from renowned technical and scientific experts who ensure to make this volume a useful source for readers outside the field as well as for those inside. The book should be helpful in getting a broad state-of-the-art overview, a deeper insight into the theory of memristors and into the technology. It will be interesting for graduate students and specialists in engineering, physics, neuroscience, biology, and applied mathematics. The content of this book has been organized around four major parts. While fundamentals of memristor theory and the discovery of the so-called *HP* memristor are summarized in Part I, computational models, simulation, and an application of the Volterra series approach to multi memristor circuits are discussed in Part II. Part III gives a state-of-the-art overview of memristor devices including applications and, finally, Part IV addresses logic circuits and neuromorphic systems. I hope this book will contribute to future technology.

I would like to express my appreciation and sincere thanks to all authors and especially to Leon Chua for his valuable comments.

Dresden, Germany Ronald Tetzlaff

References

1. D.B. Strukov, G.S. Snider, D.R. Stewart, R.S. Williams, The missing memristor found. Nature **453**, 80–83 (2008)
2. L.O. Chua, Memristor: the missing circuit element. IEEE Trans. Circuit Theory **18**(5), 507–519 (1971)
3. L.O. Chua, S.-M. Kang, Memristive devices and systems. Proc. IEEE **64**(2), 209–223 (1976)
4. M. Di Ventra, Yu.V. Pershin, L.O. Chua, Circuit elements with memory: memristors, memcapacitors and meminductors. Proc. IEEE **97**, 1717–1724 (2009)

Contents

Contributors

Alon Ascoli Institute of Circuits and Systems, Technische Universität Dresden, Dresden Germany

Dalibor Biolek Department of Electrical Engineering, University of Defence, Brno Czech Republic

Department of Microelectronics, Brno University of Technology, Brno Czech Republic

Zdenek Biolek Department of Microelectronics, Brno University of Technology, Brno Czech Republic

Ting Chang Department of Electrical Engineering and Computer Science, University of Michigan, Ann Arbor MI USA

Yiran Chen Swanson School of Engineering, University of Pittsburgh, Pittsburgh PA USA

Leon Chua Department of Electrical Engineering and Computer Sciences, University of California, Berkeley CA USA

Fernando Corinto Dipartimento di Elettronica e Telecomunicazioni, Politecnico di Torino, Torino Italy

Giovanni De Micheli Integrated Systems Laboratory, École polytechnique fédérale de Lausanne, Lausanne Switzerland

Marco Gilli Faculty of Engineering Politecnico di Torino, Torino Italy

Yusuf Leblebici Microelectronic Systems Laboratory, École polytechnique fédérale de Lausanne, Lausanne Switzerland

Hai Li Swanson School of Engineering, University of Pittsburgh, Pittsburgh PA USA

Bernabé Linares-Barranco Instituto de Microelectrónica de Sevilla (IMSE-CNM-CSIC), Seville Spain

Wei Lu Department of Electrical Engineering and Computer Science, University of Michigan, Ann Arbor MI USA

Matthew D. Pickett Hewlett-Packard Laboratories, Palo Alto CA USA

Vikas Rana Forschungszentrum Jülich, Jülich Germany

Davide Sacchetto Microelectronic Systems Laboratory, École polytechnique fédérale de Lausanne, Lausanne Switzerland

Torsten Schmidt Hochschule Ansbach, Ansbach Germany

Teresa Serrano-Gotarredona Instituto de Microelectrónica de Sevilla (IMSE-CNM-CSIC), Seville Spain

Sangho Shin Department of Electrical Engineering, Jack Baskin School of Engineering, University of California, Santa Cruz CA USA

Sung Mo Steve Kang Korea Advanced Institute of Science and Technology, Yuseong-gu Daejeon Republic of Korea

Zhenyu Sun Swanson School of Engineering, University of Pittsburgh, Pittsburgh PA USA

Ronald Tetzlaff Institute of Circuits and Systems, Technische Universität Dresden, Dresden Germany

Rainer Waser Department of Electronic Materials II and Institute of Electronic Materials. RWTH Aachen University, Aachen Germany

Forschungszentrum Jülich, Jülich Germany

R. Stanley Williams Memristor Research Group, Hewlett-Packard, Palo Alto CA USA

Qiangfei Xia Nanodevices and Integrated Systems Laboratory, Department of Electrical and Computer Engineering, University of Massachusetts, Amherst MA USA

Yuchao Yang Department of Electrical Engineering and Computer Science, University of Michigan, Ann Arbor MI USA

Part I
Introduction

Chapter 1
How We Found the Missing Memristor

R. Stanley Williams

1.1 The Memristor—the Functional Equivalent of a Synapse—Could Revolutionize Circuit Design

It's time to stop shrinking. Moore's Law, the semiconductor industry's obsession with the shrinking of transistors and their commensurate steady doubling on a chip about every 2 years, has been the source of a 50-year technical and economic revolution. Whether this scaling paradigm lasts for 5 more years or 15, it will eventually come to an end. The emphasis in electronics design will have to shift to devices that are not just increasingly infinitesimal but increasingly capable.

Earlier this year, I and my colleagues at Hewlett-Packard Labs, in Palo Alto, Calif., surprised the electronics community with a fascinating candidate for such a device: the memristor. It had been theorized nearly 40 years ago, but because no one had managed to build one, it had long since become an esoteric curiosity. That all changed on 1 May, when my group published the details of the memristor in *Nature*.

Combined with transistors in a hybrid chip, memristors could radically improve the performance of digital circuits without shrinking transistors. Using transistors more efficiently could in turn give us another decade, at least, of Moore's Law performance improvement, without requiring the costly and increasingly difficult doublings of transistor density on chips. In the end, memristors might even become the cornerstone of new analog circuits that compute using an architecture much like that of the brain (Fig. 1.1).

For nearly 150 years, the known fundamental passive circuit elements were limited to the capacitor (discovered in 1745), the resistor (1827), and the inductor (1831). Then, in a brilliant but underappreciated 1971 paper, Leon Chua, a professor of electrical engineering at the University of California, Berkeley, predicted the

R.S. Williams (✉)
Hewlett-Packard Laboratories, Palo Alto, CA, USA
e-mail: stan.williams@hp.com

R. Tetzlaff (ed.), *Memristors and Memristive Systems*,
DOI 10.1007/978-1-4614-9068-5_1,

Fig. 1.1 Thinking Machine: This artist's conception shows a stack of multiple crossbar arrays, of memristors. Because memristors behave functionally like synapses, replacing a few transistors in a circuit with memristors could lead to analog circuits that can think like a human brain. *Image: bryan christie design*

existence of a fourth fundamental device, which he called a memristor. He proved that memristor behavior could not be duplicated by any circuit built using only the other three elements, which is why the memristor is truly fundamental.

Memristor is a contraction of "memory resistor," because that is exactly its function: to remember its history. A memristor is a two-terminal device whose resistance depends on the magnitude and polarity of the voltage applied to it and the length of time that voltage has been applied. When you turn off the voltage, the memristor remembers its most recent resistance until the next time you turn it on, whether that happens a day later or a year later (Fig. 1.2).

Think of a resistor as a pipe through which water flows. The water is electric charge. The resistor's obstruction of the flow of charge is comparable to the diameter of the pipe: the narrower the pipe, the greater the resistance. For the history of circuit design, resistors have had a fixed pipe diameter. But a memristor is a pipe that changes diameter with the amount and direction of water that flows through it. If water flows through this pipe in one direction, it expands (becoming less resistive). But send the water in the opposite direction and the pipe shrinks (becoming more resistive). Further, the memristor remembers its diameter when water last went through. Turn off the flow and the diameter of the pipe "freezes" until the water is turned back on.

That freezing property suits memristors brilliantly for computer memory. The ability to indefinitely store resistance values means that a memristor can be used as a nonvolatile memory. That might not sound like very much, but go ahead and pop the battery out of your laptop, right now—no saving, no quitting, nothing. You'd lose your work, of course. But if your laptop were built using a memory based on memristors, when you popped the battery back in, your screen would return to life with everything exactly as you left it: no lengthy reboot, no half-dozen auto-recovered files.

But the memristor's potential goes far beyond instant-on computers to embrace one of the grandest technology challenges: mimicking the functions of a brain. Within a decade, memristors could let us emulate, instead of merely simulate, networks of neurons and synapses. Many research groups have been working toward a brain in silico: IBM's Blue Brain project, Howard Hughes Medical Institute's Janelia Farm, and Harvard's Center for Brain Science are just three. However, even a mouse brain simulation in real time involves solving an astronomical number of coupled partial differential equations. A digital computer capable of coping with this staggering workload would need to be the size of a small city, and powering it would require several dedicated nuclear power plants.

Memristors can be made extremely small, and they function like synapses. Using them, we will be able to build analog electronic circuits that could fit in a shoebox and function according to the same physical principles as a brain.

A hybrid circuit—containing many connected memristors and transistors—could help us research actual brain function and disorders. Such a circuit might even lead to machines that can recognize patterns the way humans can, in those critical ways computers can't—for example, picking a particular face out of a crowd even if it has changed significantly since our last memory of it.

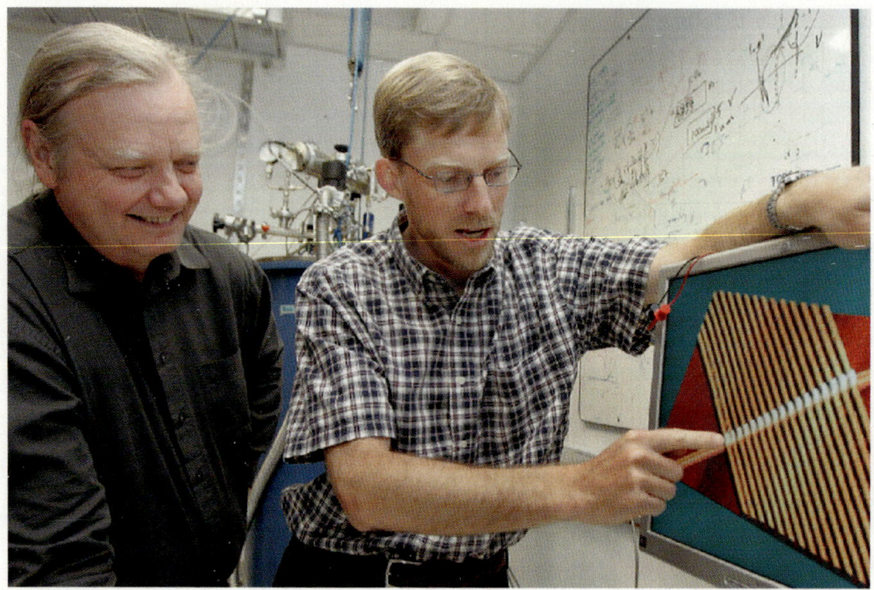

Fig. 1.2 Picturing Memristance: HP Labs senior fellow R. Stanley Williams [*left*] and research physicist Duncan Stewart [*right*] explain the fourth fundamental circuit element. Williams worked with nearly 100 scientists and engineers to find the memristor. *Photo: Paul Sakuma/AP Photo*

The story of the memristor is truly one for the history books. When Leon Chua, now an IEEE Fellow, wrote his seminal paper predicting the memristor, he was a newly minted and rapidly rising professor at UC Berkeley. Chua had been fighting for years against what he considered the arbitrary restriction of electronic circuit theory to linear systems. He was convinced that nonlinear electronics had much more potential than the linear circuits that dominate electronics technology to this day.

Chua discovered a missing link in the pairwise mathematical equations that relate the four circuit quantities—charge, current, voltage, and magnetic flux—to one another. These can be related in six ways. Two are connected through the basic physical laws of electricity and magnetism, and three are related by the known circuit elements: resistors connect voltage and current, inductors connect flux and current, and capacitors connect voltage and charge. But one equation is missing from this group: the relationship between charge moving through a circuit and the magnetic flux surrounded by that circuit—or more subtly, a mathematical doppelgänger defined by Faradays Law as the time integral of the voltage across the circuit. This distinction is the crux of a raging Internet debate about the legitimacy of our memristor (see Sect. 1.2, "Resistance to Memristance").

Chua's memristor was a purely mathematical construct that had more than one physical realization. What does that mean? Consider a battery and a transformer. Both provide identical voltages—for example, 12 volts of direct current—but they do so by entirely different mechanisms: the battery by a chemical reaction going on inside the cell and the transformer by taking a 110 V ac input, stepping that down to 12 V ac, and then transforming that into 12 V dc. The end result is mathematically identical—both will run an electric shaver or a cellphone, but the physical source of that 12 V is completely different.

Conceptually, it was easy to grasp how electric charge could couple to magnetic flux, but there was no obvious physical interaction between charge and the integral over the voltage.

Chua demonstrated mathematically that his hypothetical device would provide a relationship between flux and charge similar to what a nonlinear resistor provides between voltage and current. In practice, that would mean the device's resistance would vary according to the amount of charge that passed through it. And it would remember that resistance value even after the current was turned off. He also noticed something else—that this behavior reminded him of the way synapses function in a brain.

Even before Chua had his eureka moment, however, many researchers were reporting what they called "anomalous" current-voltage behavior in the micrometer-scale devices they had built out of unconventional materials, like polymers and metal oxides. But the idiosyncrasies were usually ascribed to some mystery electrochemical reaction, electrical breakdown, or other spurious phenomenon attributed to the high voltages that researchers were applying to their devices.

As it turns out, a great many of these reports were unrecognized examples of memristance. After Chua theorized the memristor out of the mathematical ether, it took another 35 years for us to intentionally build the device at HP Labs (Fig. 1.3), and we only really understood the device about 2 years ago. So what took us so long?

It's all about scale. We now know that memristance is an intrinsic property of any electronic circuit. Its existence could have been deduced by Gustav Kirchhoff or by James Clerk Maxwell, if either had considered nonlinear circuits in the 1800s. But the scales at which electronic devices have been built for most of the past two centuries have prevented experimental observation of the effect. It turns out that the influence of memristance obeys an inverse square law: memristance is a million times as important at the nanometer scale as it is at the micrometer scale, and it's essentially unobservable at the millimeter scale and larger. As we build smaller and smaller devices, memristance is becoming more noticeable and in some cases dominant. That's what accounts for all those strange results researchers have described. Memristance has been hidden in plain sight all along. But in spite of all the clues, our finding the memristor was completely serendipitous.

In 1995, I was recruited to HP Labs to start up a fundamental research group that had been proposed by David Packard. He decided that the company had become large enough to dedicate a research group to long-term projects that would be protected from the immediate needs of the business units. Packard had an altruistic vision that HP should "return knowledge to the well of fundamental science from which HP had been withdrawing for so long." At the same time, he understood that long-term research could be the strategic basis for technologies and inventions that

Fig. 1.3 Crossbar Architecture: A memristor's structure, shown here in a scanning tunneling microscope image, will enable dense, stable computer memories. *Image: R. Stanley Williams/HP Labs*

would directly benefit HP in the future. HP gave me a budget and four researchers. But beyond the comment that "molecular-scale electronics" would be interesting and that we should try to have something useful in about 10 years, I was given carte blanche to pursue any topic we wanted. We decided to take on Moore's Law.

At the time, the dot-com bubble was still rapidly inflating its way toward a resounding pop, and the existing semiconductor road map didn't extend past 2010. The critical feature size for the transistors on an integrated circuit was 350 nanometers; we had a long way to go before atomic sizes would become a limitation. And yet, the eventual end of Moore's Law was obvious. Someday semiconductor researchers would have to confront physics-based limits to their relentless descent into the infinitesimal, if for no other reason than that a transistor cannot be smaller than an atom. (Today the smallest components of transistors on integrated circuits are roughly 45 nm wide, or about 220 silicon atoms.)

That's when we started to hang out with Phil Kuekes, the creative force behind the Teramac (tera-operation-per-second multiarchitecture computer)—an experimental supercomputer built at HP Labs primarily from defective parts, just to show it could be done. He gave us the idea to build an architecture that would work even if a substantial number of the individual devices in the circuit were dead on arrival. We didn't know what those devices would be, but our goal was electronics that would keep improving even after the devices got so small that defective ones would become common. We ate a lot of pizza washed down with appropriate amounts of beer and speculated about what this mystery nanodevice would be.

We were designing something that wouldn't even be relevant for another 10–15 years. It was possible that by then devices would have shrunk down to the molecular scale envisioned by David Packard or perhaps even be molecules. We could think of no better way to anticipate this than by mimicking the Teramac at the nanoscale. We decided that the simplest abstraction of the Teramac architecture was the crossbar, which has since become the de facto standard for nanoscale circuits because of its simplicity, adaptability, and redundancy.

The crossbar is an array of perpendicular wires. Anywhere two wires cross, they are connected by a switch. To connect a horizontal wire to a vertical wire at any point on the grid, you must close the switch between them. Our idea was to open and close these switches by applying voltages to the ends of the wires. Note that a crossbar array is basically a storage system, with an open switch representing a zero and a closed switch representing a one. You read the data by probing the switch with a small voltage.

Like everything else at the nanoscale, the switches and wires of a crossbar are bound to be plagued by at least some nonfunctional components. These components will be only a few atoms wide, and the second law of thermodynamics ensures that we will not be able to completely specify the position of every atom. However, a crossbar architecture builds in redundancy by allowing you to route around any parts of the circuit that don't work. Because of their simplicity, crossbar arrays have a much higher density of switches than a comparable integrated circuit based on transistors.

But implementing such a storage system was easier said than done. Many research groups were working on such a cross-point memory—and had been since the 1950s. Even after 40 years of research, they had no product on the market. Still, that didn't stop them from trying. That's because the potential for a truly nanoscale crossbar memory is staggering; picture carrying around the entire Library of Congress on a thumb drive.

One of the major impediments for prior crossbar memory research was the small off-to-on resistance ratio of the switches (40 years of research had never produced anything surpassing a factor of 2 or 3). By comparison, modern transistors have an off-to-on resistance ratio of 10,000 to 1. We calculated that to get a high-performance memory, we had to make switches with a resistance ratio of at least 1,000 to 1. In other words, in its off state, a switch had to be 1,000 times as resistive to the flow of current as it was in its on state. What mechanism could possibly give a nanometer-scale device a three-orders-of-magnitude resistance ratio?

We found the answer in scanning tunneling microscopy (STM), an area of research I had been pursuing for a decade. A tunneling microscope generates atomic-resolution images by scanning a very sharp needle across a surface and measuring the electric current that flows between the atoms at the tip of the needle and the surface the needle is probing. The general rule of thumb in STM is that moving that tip 0.1 nm closer to a surface increases the tunneling current by one order of magnitude.

We needed some similar mechanism by which we could change the effective spacing between two wires in our crossbar by 0.3 nm. If we could do that, we would have the 1,000:1 electrical switching ratio we needed.

Our constraints were getting ridiculous. Where would we find a material that could change its physical dimensions like that? That is how we found ourselves in the realm of molecular electronics.

Conceptually, our device was like a tiny sandwich. Two platinum electrodes (the intersecting wires of the crossbar junction) functioned as the "bread" on either end of the device. We oxidized the surface of the bottom platinum wire to make an extremely thin layer of platinum dioxide, which is highly conducting. Next, we assembled a dense film, only one molecule thick, of specially designed switching molecules. Over this "monolayer" we deposited a 2–3 nm layer of titanium metal, which bonds strongly to the molecules and was intended to glue them together. The final layer was the top platinum electrode.

The molecules were supposed to be the actual switches. We built an enormous number of these devices, experimenting with a wide variety of exotic molecules and configurations, including rotaxanes, special switching molecules designed by James Heath and Fraser Stoddart at the University of California, Los Angeles. The rotaxane is like a bead on a string, and with the right voltage, the bead slides from one end of the string to the other, causing the electrical resistance of the molecule to rise or fall, depending on the direction it moves. Heath and Stoddart's devices used silicon electrodes, and they worked, but not well enough for technological applications: the off-to-on resistance ratio was only a factor of 10, the switching was slow, and the devices tended to switch themselves off after 15 min.

Our platinum devices yielded results that were nothing less than frustrating. When a switch worked, it was spectacular: our off-to-on resistance ratios shot past the 1,000 mark, the devices switched too fast for us to even measure, and having switched, the device's resistance state remained stable for years (we still have some early devices we test every now and then, and we have never seen a significant change in resistance). But our fantastic results were inconsistent. Worse yet, the success or failure of a device never seemed to depend on the same thing.

We had no physical model for how these devices worked. Instead of rational engineering, we were reduced to performing huge numbers of Edisonian experiments, varying one parameter at a time and attempting to hold all the rest constant. Even our switching molecules were betraying us; it seemed like we could use anything at all. In our desperation, we even turned to long-chain fatty acids—essentially soap—as the molecules in our devices. There's nothing in soap that should switch, and

Bow Ties: Leon Chua's original graph of the hypothetical memristor's behavior is shown at the top right; the graph of R. Stanley Williams's experimental result in the *Nature* paper is shown below. The loops map the switching behavior of the device: it begins with a high resistance, and as the voltage increases, the current slowly increases. As charge flows through the device, the resistance drops, and the current increases more rapidly with increasing voltage until the maximum is reached. Then, as the voltage decreases, the current decreases but more slowly, because charge is flowing through the device and the resistance is still dropping. The result is an on-switching loop. When the voltage turns negative, the resistance of the device increases, resulting in an off-switching loop. *– R.S.W.*

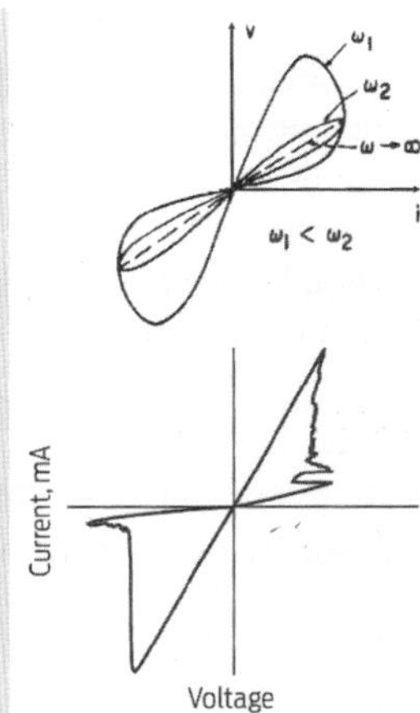

Fig. 1.4 Bow ties

yet some of the soap devices switched phenomenally. We also made control devices with no molecule mono-layers at all. None of them switched.

We were frustrated and burned out. Here we were, in late 2002, 6 years into our research. We had something that worked, but we couldn't figure out why, we couldn't model it, and we sure couldn't engineer it. That's when Greg Snider, who had worked with Kuekes on the Teramac, brought me the Chua memristor paper. "I don't know what you guys are building," he told me, "but this is what I want."

To this day, I have no idea how Greg happened to come across that paper. Few people had read it, fewer had understood it, and fewer still had cited it. At that point, the paper was 31 years old and apparently headed for the proverbial dustbin of history. I wish I could say I took one look and yelled, "Eureka!" But in fact, the paper sat on my desk for months before I even tried to read it. When I did study it, I found the concepts and the equations unfamiliar and hard to follow. But I kept at it because something had caught my eye, as it had Greg's: Chua had included a graph that looked suspiciously similar to the experimental data we were collecting.

The graph described the current–voltage (I–V) characteristics that Chua had plotted for his memristor. Chua had called them "pinched-hysteresis loops"; we called our I–V characteristics "bow ties." A pinched hysteresis loop looks like a diagonal infinity symbol with the center at the zero axis, when plotted on a graph of current against voltage. The voltage is first increased from zero to a positive maximum value, then decreased to a minimum negative value, and finally returned to zero. The bow ties on our graphs were nearly identical (see Fig. 1.4).

That's not all. The total change in the resistance we had measured in our devices also depended on how long we applied the voltage: the longer we applied a positive voltage, the lower the resistance until it reached a minimum value. And the longer we applied a negative voltage, the higher the resistance became until it reached a maximum limiting value. When we stopped applying the voltage, whatever resistance characterized the device was frozen in place, until we reset it by once again applying a voltage. The loop in the I–V curve is called hysteresis, and this behavior is startlingly similar to how synapses operate: synaptic connections between neurons can be made stronger or weaker depending on the polarity, strength, and length of a chemical or electrical signal. That's not the kind of behavior you find in today's circuits.

Looking at Chua's graphs was maddening. We now had a big clue that memristance had something to do with our switches. But how? Why should our molecular junctions have anything to do with the relationship between charge and magnetic flux? I couldn't make the connection.

Two years went by. Every once in a while I would idly pick up Chua's paper, read it, and each time I understood the concepts a little more. But our experiments were still pretty much trial and error. The best we could do was to make a lot of devices and find the ones that worked.

But our frustration wasn't for nothing: by 2004, we had figured out how to do a little surgery on our little sandwiches. We built a gadget that ripped the tiny devices open so that we could peer inside them and do some forensics. When we pried them apart, the little sandwiches separated at their weakest point: the molecule layer. For the first time, we could get a good look at what was going on inside. We were in for a shock.

What we had was not what we had built. Recall that we had built a sandwich with two platinum electrodes as the bread and filled with three layers: the platinum dioxide, the monolayer film of switching molecules, and the film of titanium.

But that's not what we found. Under the molecular layer, instead of platinum dioxide, there was only pure platinum. Above the molecular layer, instead of titanium, we found an unexpected and unusual layer of titanium dioxide. The titanium had sucked the oxygen right out of the platinum dioxide! The oxygen atoms had somehow migrated through the molecules and been consumed by the titanium. This was especially surprising because the switching molecules had not been significantly perturbed by this event—they were intact and well ordered, which convinced us that they must be doing something important in the device.

The chemical structure of our devices was not at all what we had thought it was. The titanium dioxide—a stable compound found in sunscreen and white paint—was not just regular titanium dioxide. It had split itself up into two chemically different layers. Adjacent to the molecules, the oxide was stoichiometric TiO_2, meaning the ratio of oxygen to titanium was perfect, exactly 2 to 1. But closer to the top platinum electrode, the titanium dioxide was missing a tiny amount of its oxygen, between 2 and 3 %. We called this oxygen-deficient titanium dioxide TiO_{2-x}, where x is about 0.05.

Because of this misunderstanding, we had been performing the experiment backward. Every time I had tried to create a switching model, I had reversed the switching polarity. In other words, I had predicted that a positive voltage would switch the device off and a negative voltage would switch it on. In fact, exactly the opposite was true.

It was time to get to know titanium dioxide a lot better. They say 3 weeks in the lab will save you a day in the library every time. In August of 2006 I did a literature search and found about 300 relevant papers on titanium dioxide. I saw that each of the many different communities researching titanium dioxide had its own way of describing the compound. By the end of the month, the pieces had fallen into place. I finally knew how our device worked. I knew why we had a memristor.

The exotic molecule monolayer in the middle of our sandwich had nothing to do with the actual switching. Instead, what it did was control the flow of oxygen from the platinum dioxide into the titanium to produce the fairly uniform layers of TiO_2 and TiO_{2-x}. The key to the switching was this bilayer of the two different titanium dioxide species (see diagram, "How Memristance Works" in Fig. 1.5). The TiO_2 is electrically insulating (actually a semiconductor), but the TiO_{2-x} is conductive, because its oxygen vacancies are donors of electrons, which makes the vacancies themselves positively charged. The vacancies can be thought of like bubbles in a glass of beer, except that they don't pop—they can be pushed up and down at will in the titanium dioxide material because they are electrically charged.

Now I was able to predict the switching polarity of the device. If a positive voltage is applied to the top electrode of the device, it will repel the (also positive) oxygen vacancies in the TiO_{2-x} layer down into the pure TiO_2 layer. That turns the TiO_2 layer into TiO_{2-x} and makes it conductive, thus turning the device on. A negative voltage has the opposite effect: the vacancies are attracted upward and back out of the TiO_2, and thus the thickness of the TiO_2 layer increases and the device turns off. This switching polarity is what we had been seeing for years but had been unable to explain.

On 20 August 2006, I solved the two most important equations of my career—one equation detailing the relationship between current and voltage for this equivalent circuit, and another equation describing how the application of the voltage causes the vacancies to move—thereby writing down, for the first time, an equation for memristance in terms of the physical properties of a material. This provided a unique insight. Memristance arises in a semiconductor when both electrons and charged dopants are forced to move simultaneously by applying a voltage to the system. The memristance did not actually involve magnetism in this case; the integral over the voltage reflected how far the dopants had moved and thus how much the resistance of the device had changed.

We finally had a model we could use to engineer our switches, which we had by now positively identified as memristors. Now we could use all the theoretical machinery Chua had created to help us design new circuits with our devices.

Triumphantly, I showed the group my results and immediately declared that we had to take the molecule monolayers out of our devices. Skeptical after years of false starts and failed hypotheses, my team reminded me that we had run control samples without molecule layers for every device we had ever made and that those devices had never switched. And getting the recipe right turned out to be tricky indeed. We needed to find the exact amounts of titanium and oxygen to get the two layers to do

How Memristance Works

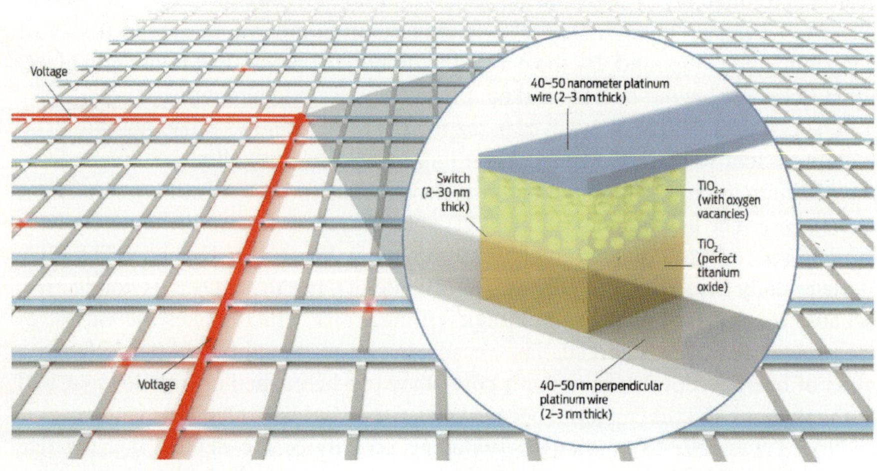

THE CROSSBAR ARCHITECTURE: The crossbar architecture is a fully connected mesh of perpendicular wires. Any two crossing wires are connected by a switch. To close the switch, a positive voltage is applied across the two wires to be connected. To open the switch, the voltage is reversed.

THE SWITCH: A switch is a 40-nanometer cube of titanium dioxide (TiO_2) in two layers: The lower TiO_2 layer has a perfect 2:1 oxygen-to-titanium ratio, making it an insulator. By contrast, the upper TiO_2 layer is missing 0.5 percent of its oxygen (TiO_{2-x}), so x is about 0.05. The vacancies make the TiO_{2-x} material metallic and conductive.

APPLIED MEMRISTANCE: The oxygen deficiencies in the TiO_{2-x} manifest as "bubbles" of oxygen vacancies scattered throughout the upper layer. A positive voltage on the switch repels the (positive) oxygen deficiencies in the metallic upper TiO_{2-x} layer, sending them into the insulating TiO_2 layer below. That causes the boundary between the two materials to move down, increasing the percentage of conducting TiO_{2-x} and thus the conductivity of the entire switch. The more positive voltage is applied, the more conductive the cube becomes.

A negative voltage on the switch attracts the positively charged oxygen bubbles, pulling them out of the TiO_2. The amount of insulating, resistive TiO_2 increases, thereby making the switch as a whole resistive. The more negative voltage is applied, the less conductive the cube becomes.

What makes this switch special—memristive—is that when the voltage is turned off, positive or negative, the oxygen bubbles do not migrate. They stay where they are, which means that the boundary between the two titanium dioxide layers is frozen. That is how the memristor "remembers" how much voltage was last applied.

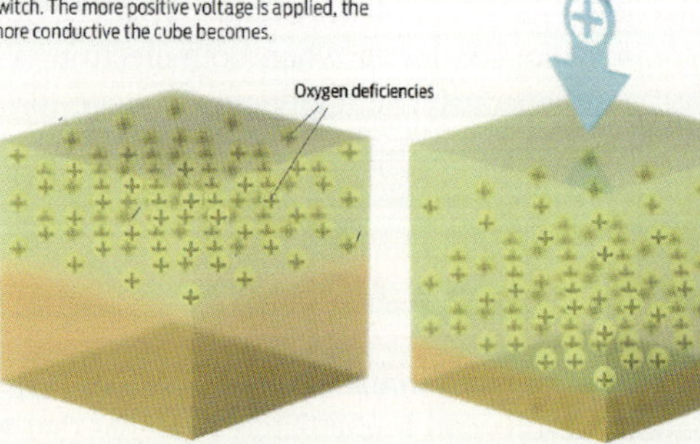

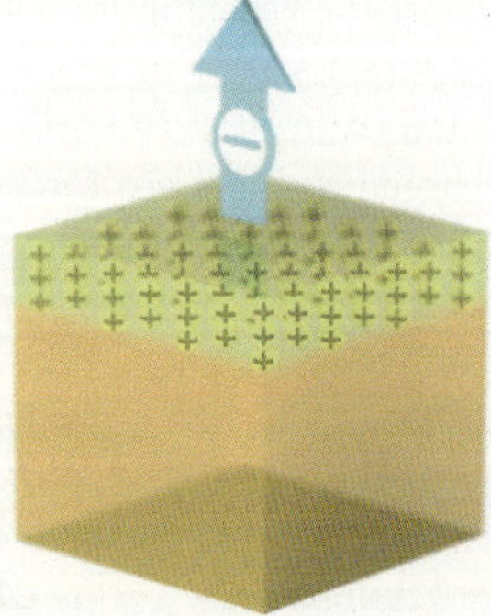

Fig. 1.5 How Memristance Works. *Image: bryan christie design*

their respective jobs. By that point we were all getting impatient. In fact, it took so long to get the first working device that in my discouragement I nearly decided to put the molecule layers back in. A month later, it worked. We not only had working devices, but we were also able to improve and change their characteristics at will.

But here is the real triumph. The resistance of these devices stayed constant whether we turned off the voltage or just read their states (interrogating them with a voltage so small it left the resistance unchanged). The oxygen vacancies didn't roam around; they remained absolutely immobile until we again applied a positive or negative voltage. That's memristance: the devices remembered their current history. We had coaxed Chua's mythical memristor off the page and into being.

Emulating the behavior of a single memristor, Chua showed, requires a circuit with at least 15 transistors and other passive elements. The implications are extraordinary: just imagine how many kinds of circuits could be supercharged by replacing a handful of transistors with one single memristor.

The most obvious benefit is to memories. In its initial state, a crossbar memory has only open switches, and no information is stored. But once you start closing switches, you can store vast amounts of information compactly and efficiently. Because memristors remember their state, they can store data indefinitely, using energy only when you toggle or read the state of a switch, unlike the capacitors in conventional DRAM, which will lose their stored charge if the power to the chip is turned off. Furthermore, the wires and switches can be made very small: we should eventually get down to a width of around 4 nm, and then multiple crossbars could be stacked on top of each other to create a ridiculously high density of stored bits.

Greg Snider and I published a paper last year showing that memristors could vastly improve one type of processing circuit, called a field-programmable gate array, or FPGA. By replacing several specific transistors with a crossbar of memristors, we showed that the circuit could be shrunk by nearly a factor of 10 in area and improved in terms of its speed relative to power-consumption performance. Right now, we are testing a prototype of this circuit in our lab.

And memristors are by no means hard to fabricate. The titanium dioxide structure can be made in any semiconductor fab currently in existence. (In fact, our hybrid circuit was built in an HP fab used for making inkjet cartridges.) The primary limitation to manufacturing hybrid chips with memristors is that today only a small number of people on Earth have any idea of how to design circuits containing memristors. I must emphasize here that memristors will never eliminate the need for transistors: passive devices and circuits require active devices like transistors to supply energy.

The potential of the memristor goes far beyond juicing a few FPGAs. I have referred several times to the similarity of memristor behavior to that of synapses. Right now, Greg is designing new circuits that mimic aspects of the brain. The neurons are implemented with transistors, the axons are the nanowires in the crossbar, and the synapses are the memristors at the cross points. A circuit like this could perform real-time data analysis for multiple sensors. Think about it: an intelligent physical infrastructure that could provide structural assessment monitoring for bridges. How much money—and how many lives—could be saved?

I'm convinced that eventually the memristor will change circuit design in the twenty-first century as radically as the transistor changed it in the twentieth. Don't forget that the transistor was lounging around as a mainly academic curiosity for a decade until 1956, when a killer app—the hearing aid—brought it into the marketplace. My guess is that the real killer app for memristors will be invented by a curious student who is now just deciding what EE courses to take next year.

1.2 Resistance to Memristance

Introducing a new fundamental circuit element earned Stanley Williams some grief along with his newfound fame. After the Nature article appeared in May, online comments pages boiled over with skepticism. "Is this a hoax?" someone asked on the Wikipedia memristor page on 30 April 2008, the day the news broke, in one of the milder statements of disbelief. Seven months later, the debate continues.

Skeptics argue that the memristor is not a fourth fundamental circuit element but an example of bad science. The crux of their argument rests on two fundamental misunderstandings: first, skeptics overlook the expanded design space that arises from working with nonlinear circuit elements. The second and more profound misunderstanding concerns Leon Chua's mathematical definition of a memristor.

At first, most people—including Williams—assumed that Chua defined memristance strictly as the relationship between electric charge and magnetic flux. However, the actual definition of memristance is more general. Linking charge and magnetic flux is one way to satisfy the definition, but it's not the only one. In fact, it turns out you can bypass magnetic interaction altogether.

Chua's general memristance definition has two parts. The first equation defines how the memristor's voltage depends on current and a "state variable"—that is, a quantity that measures some physical property of a device, like the length of a column of mercury in a thermometer. The column's length correlates with the thermometer's temperature, and adding or removing heat makes the column longer or shorter. In Williams's memristor, the state variable is the thickness of the stoichiometric titanium dioxide in the switch; increasing or decreasing that thickness causes the device's resistance to increase or decrease.

The second equation expresses how the changing state variable (the TiO_2's thickness) depends on the amount of charge flowing through the device. In Williams's memristor, the TiO_2's thickness depends on the distribution of the oxygen vacancies throughout the material.

Here is what you need to remember: one, a magnetic interaction is not necessary for memristance. Two, in nonlinear circuit elements, memristance is not the same thing as nonlinear resistance. Three, because no combination of passive devices can reproduce the properties of a memristor, memristance is a fundamental circuit quantity.

Williams himself is sanguine about the memristor's reputation. "A hundred years after Einstein proposed his theory of relativity," he says, shrugging, "you still have some people arguing against it." *Sally Adee*

Chapter 2
If It's Pinched It's a Memristor

Leon Chua

This chapter consists of two parts. Part I gives a circuit-theoretic foundation for the first four elementary nonlinear 2-terminal circuit elements, namely, the resistor, the capacitor, the inductor, and the memristor. Part II consists of a collection of colorful "Vignettes" with carefully articulated text and colorful illustrations of the rudiments of the memristor and its characteristic fingerprints and signatures. It is intended as a self-contained pedagogical primer for beginners who have not heard of memristors before.

L. Chua (✉)
Department of Electrical Engineering and Computer Sciences, University of California, Berkeley, CA 94720, USA
e-mail: chua@eecs.berkeley.edu

R. Tetzlaff (ed.), *Memristors and Memristive Systems*,
DOI 10.1007/978-1-4614-9068-5_2,

Part I

2.1 Abstract

This tutorial clarifies the axiomatic definition of $(v^{(\alpha)}, i^{(\beta)})$ circuit elements via a look-up-table dubbed an *A-pad*, of admissible (v,i) signals measured via Gedanken Probing Circuits. The $(v^{(\alpha)}, i^{(\beta)})$ elements are ordered via a complexity metric. Under this metric, the *memristor* emerges naturally as the *fourth element* [1], characterized by a state-dependent Ohm's law. A logical generalization to memristive devices reveals a common *fingerprint* consisting of a dense continuum of *pinched hysteresis loops* whose area decreases with the frequency ω and tends to a straight line as $\omega \to \infty$, for all bipolar periodic signals and for all initial conditions. This common fingerprint suggests that the term memristor be used henceforth as a moniker for memristive devices.

2.1.1 *Axiomatic Definition of Circuits Elements*

How do you *characterize* a 2-terminal "black box" B such that its response to any electrical signal can be predicted? Since you are not allowed to peek inside B your only recourse is to carry out measurements by probing B with *all possible* electrical circuits, containing arbitrary interconnections of circuit elements, such as resistors, capacitors, inductors, diodes, transistors, op amps, batteries, voltage and current sources with arbitrary time functions, etc. We will henceforth call such circuits "*Gedanken Probing Circuits*," as depicted in the *Gedanken* experimental setup shown in Fig. 2.1. Let us insert an instrument called an ammeter in series with the top wire to record a time function $i(t)$ called the *current* in Amperes entering the top terminal (labeled by a plus $(+)$ sign). Next let us connect an instrument called a *voltmeter* across B to record a time function $v(t)$ called the *voltage* in Volts across the *plus-minus* terminals of B.[1] Let us call $(v(t), i(t))$ an *admissible* (v,i) signal of *B*. The recorded list

$$B(v,i) \triangleq \{(v_1(t), i_1(t)), (v_2(t), i_2(t)), \ldots, (v_n(t), i_n(t)), \ldots\} \tag{2.1}$$

of *admissible* (v,i) *signals* (***AVIS***) from *all possible Gedanken Probing Circuits* constitutes the *complete* characterization of the 2-terminal black box B in the sense that given any voltage signal or current signal, one can search the ***AVIS*** "memory bank," henceforth called the ***AVIS****-pad* of B or just *A-pad*, and identify the unique admissible signals $(\tilde{v}(t), \tilde{i}(t))$ being sought. The *A-pad must* contain this entry in its memory bank because the signal is associated with *some* circuit connected to B, and

[1]Observe that the voltage v and the current i are defined axiomatically via two instruments called voltmeter and ammeter, without invoking any physical concepts such as electric field, magnetic field, charge, flux linkages, etc. One does not even have to know how a voltmeter, or an ammeter, works. They are just names assigned to the instruments.

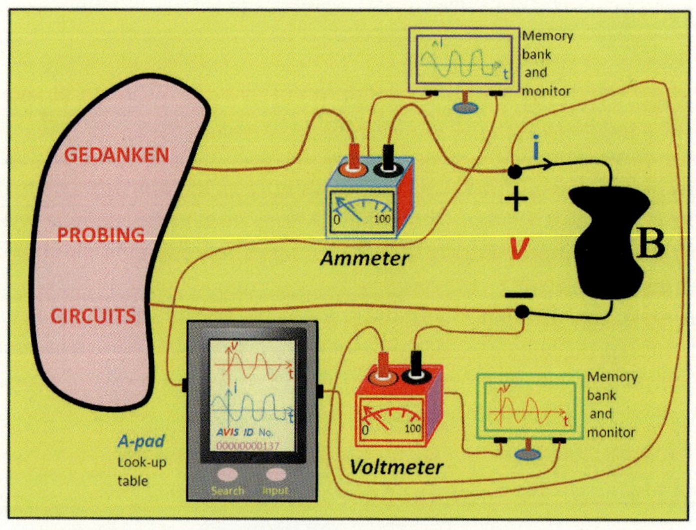

Fig. 2.1 Axiomatic definition of a 2-terminal circuit element

this circuit is a Gedanken probing circuit, *by definition*. The ***A**-pad* is just a *look-up-table* containing *all* admissible (v,i) signals of B. Observe that the ***A**-pad* is in general an infinitely long pad containing infinitely many pairs of admissible signal waveforms $(v(t), i(t))$ of B, as depicted in Fig. 2.1.

The above Gedanken experiment is only a thought experiment. However, for a large number of real-world 2-terminal devices, the ***A**-pad* for B can be generated via equations.

Example 2.1 (Ohm's Law). A very small subset of all 2-terminal black boxes are characterized by an ***A**-pad* that satisfies Ohm's Law; namely,

$$v = Ri \quad \text{or} \quad i = Gv \tag{2.2}$$

where R is called the *resistance* in Ohms (Ω) of B and G is called the *conductance* in Siemens (S) of B. In this case

$$\boldsymbol{AVIS} = \{(Ri_1(t), i_1(t)), (Ri_2(t), i_2(t)) \ldots (Ri_n(t), i_n(t)), \ldots\} \tag{2.3}$$

can be reconstructed by (2.2). When Ohm's law is written with i as the independent variable, namely; $v = Ri$, it is called *current controlled*. If it is written in the form $i = Gv$, it is called *voltage controlled*. Often it is more convenient to recast (2.2) in the *implicit form*

$$f_R(v,i) = v - Ri = 0 \tag{2.4}$$

Since (2.4) is neither a function of v, nor of i, it is called a *relation* in mathematics. In *nonlinear circuit theory*, it is called a *constitutive relation* [2–4]. Observe that the constitutive relation is just a compact formula, or algorithm, for generating the ***A**-pad* of B.

Example 2.2. Suppose the ***A**-pad* of the 2-terminal black box B in Fig. 2.1 can be written in the form

$$\boldsymbol{AVIS} = \left\{ \left(v_1, v_1 + \frac{1}{3}v_1^3\right), \left(v_2, v_2 + \frac{1}{3}v_2^3\right), \ldots, \left(v_n, v_n + \frac{1}{3}v_n^3\right) \ldots \right\} \tag{2.5}$$

for all possible voltage signals

$$v(t) = v_1(t), v(t) = v_2(t), \ldots v(t) = v_n(t) \ldots$$

then the *A-pad* of B can be generated by the much more compact constitutive relation

$$f_R(v,i) = v + \frac{1}{3}v^3 - i = 0 \tag{2.6}$$

Since both (2.4) of Example 2.1 and (2.6) of Example 2.2 involve the same pair of circuit variables (voltage, current), and since all 2-terminal devices that can be characterized by a constitutive relation

$$f_R(v,i) = 0 \tag{2.7}$$

between the variable pair (v,i) can be proved to be dissipative (or passive) if $v \times i > 0$ for all (v,i) listed in the ***A**-pad*, this class of 2-terminal elements are called *resistors* [2–4].

Example 2.3. Most 2-terminal black boxes can *not* be described by a constitutive relation between the variable pair (v,i). However, another important subclass can be expressed by a relationship between the variable pair (v,q), where

$$q(t) = \int_{-\infty}^{t} i(\tau)d\tau = q_0 + \int_{t_0}^{t} i(\tau)d\tau \tag{2.8}$$

and

$$q_0 \triangleq \int_{-\infty}^{t_0} i(\tau)d\tau \tag{2.9}$$

is called the initial state[2] of $q(t)$ at the initial time $t = t_0$. This subclass of 2-terminal black boxes can be characterized by a collection of admissible signals between the variable pair (v, q), namely,

$$B(v,q) = \{(v_1(t), q_1(t)), (v_2(t), q_2(t)), \ldots, (v_n(t), q_n(t)), \ldots\} \tag{2.10}$$

where

$$q = Cv \tag{2.11}$$

and C is a constant called the *Capacitance* of B. Equation (2.11) is the constitutive relation of B because we can generate the corresponding ***AVIS*** $(v(t), i(t))$ via (2.8); namely

$$i(t) = \frac{dq(t)}{dt} \tag{2.12}$$

Indeed, any relationship

$$q = f_C(v) \tag{2.13}$$

is a valid constitutive relation and this class of 2-terminal devices are called *capacitors.*

By the same reasoning, the constitutive relation

$$\varphi = f_L(i) \tag{2.14}$$

involving the variable pair (i, φ) defines a third subclass of 2-terminal devices called *inductors*, where

$$\varphi(t) = \int_{-\infty}^{t} v(\tau)d\tau = \varphi_0 + \int_{t_0}^{t} v(\tau)d\tau \tag{2.15}$$

Observe that the above three classes of basic circuit elements, called resistors, capacitors, and inductors, are defined *axiomatically*, via a constitutive relation between a pair of variables chosen from $\{v, i, q, \varphi\}$. There are six different pairs that can be formed from these four variables; namely

$$\{(v,\varphi), (i,q), (v,i), (v,q), (i,\varphi), (\varphi,q)\} \tag{2.16}$$

---

[2] In practice one can never know the precise signal $i(t)$ over the infinite past. Rather we can only set up our measurements to begin at some initial time $t = t_0$. Consequently, the initial condition q_0 in Eq. (2.8) represents a summary of the past memory of $q(t)$ measured at $t = t_0$.

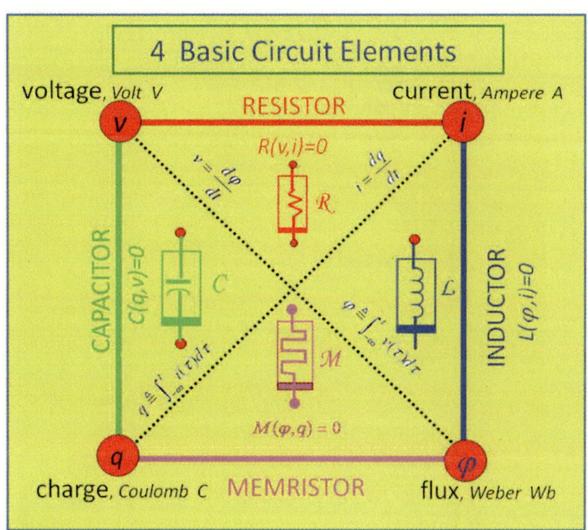

Fig. 2.2 Four axiomatically defined circuit elements

The first two pairs (v,φ) and (i,q) are already related via (2.15) and (2.8), respectively, and are *not* constitutive relations because they cannot predict the corresponding current $i(t)$ and voltage $v(t)$. However, the last pair (φ,q) defines yet another constitutive relation since given any *admissible* signals $(\varphi(t),q(t))$, one can recover the corresponding $(v(t),i(t))$ via (2.15) and (2.8). For logical consistency, and symmetry considerations, it is necessary to define a 4*th circuit element* [1] via the constitutive relation

$$f_M(\varphi,q) = 0 \tag{2.17}$$

between the variables φ and q. This element was postulated and named the *memristor* (acronym for *mem*ory *resistor* in [5]). A physical approximation of such an element has been fabricated in 2008 as a TiO_2 nano device by Dr. Stanley Williams group at hp [6]. The above axiomatic definition of the four basic circuit elements is summarized in Fig. 2.2, along with their respective symbols [7]. Note that the standard symbols for resistor, capacitor, and inductor are enclosed by a thin rectangle with a dark band at the bottom because it is essential to distinguish the reference polarity of each nonlinear element if its constitutive relation is *not* odd-symmetric.

We wish to stress that although the symbols of q and φ in Fig. 2.2 are given the names *charge* and *flux*, respectively, *they need not* be associated with a real physical *charge* as in the case of a classical *capacitor* built by sandwiching a pair of parallel metal plates between an insulator, or a real physical *flux* as in the case of a classical *inductor* built by winding a copper wire around an iron core.

2.1.2 $(v^{(\alpha)} - i^{(\beta)})$ *Circuit Elements*

Let us introduce the notations [4]

$$v^{(\alpha)}(t) \triangleq \begin{cases} \frac{d^{\alpha}v(t)}{dt^{\alpha}}, & \text{if } \alpha = 1,2,\ldots,\infty \\ v(t), & \text{if } \alpha = 0 \\ \int_{-\infty}^{t} v(\tau)d\tau, & \text{if } \alpha = -1 \\ \int_{-\infty}^{t}\int_{-\infty}^{\tau_{|\alpha|}} \cdots \int_{-\infty}^{\tau_2} v(\tau_1)d\tau_1 d\tau_2 \cdots d\tau_{|\alpha|}, & \text{if } \alpha = -2,-3,\ldots,\infty \end{cases} \tag{2.18}$$

and

$$i^{(\beta)}(t) \triangleq \begin{cases} \frac{d^{\beta}i(t)}{dt^{\beta}}, & \text{if } \beta = 1,2,\ldots,\infty \\ i(t), & \text{if } \beta = 0 \\ \int_{-\infty}^{t} i(\tau)d\tau, & \text{if } \beta = -1 \\ \int_{-\infty}^{t}\int_{-\infty}^{\tau_{|\beta|}} \cdots \int_{-\infty}^{\tau_2} i(\tau_1)d\tau_1 d\tau_2 \cdots d\tau_{|\beta|}, & \text{if } \beta = -2,-3,\ldots,\infty \end{cases} \tag{2.19}$$

where $|\alpha|$ and $|\beta|$ are integers. Let us identify a $(v^{(0)}, i^{(0)})$ element as a *resistor*, a $(v^{(0)}, i^{(-1)})$ element as a *capacitor*, a $(v^{(-1)}, i^{(0)})$ element as an *inductor*, and a $(v^{(-1)}, i^{(-1)})$ element as a *memristor*. Using this notation, we can define an infinite family of circuit elements, each one identified by its element code $(v^{(\alpha)} - i^{(\beta)})$ and referred to simply as an (α, β) element.

The first 25 (α, β) elements are listed in Fig. 2.3, each coded by an integer pair (α, β), and identified by a rectangular box where "α" and "β" are printed on the "top," and at the "bottom," respectively. Each (α, β) element is located at the intersection between a vertical line through α, and a horizontal line through β. The four circuit element symbols shown in Fig. 2.2 are printed in their corresponding locations in Fig. 2.3. The two elements $(\alpha, \beta) = (-1, -2)$ and $(\alpha, \beta) = (-2, -1)$ are called *memcapacitor* and *meminductor*, respectively [8], and are identified by their corresponding symbols.

The above infinite family of circuit elements are defined *not* for the sake of generality. Rather, they are *essential* for developing a rigorous mathematical theory of nonlinear circuits in the sense that if one excludes all elements with $|\alpha| > k$ and $|\beta| > k$, for any finite integer k, then one can construct hypothetical circuits whose solutions do *not* exist after certain finite times $t \geq T_k$ due to the presence of a "singularity" called an *impasse point* [2, 3, 9]. It is unlikely, however, that (α, β) elements with $|\alpha| > 2$ and $|\beta| > 2$ will be needed in modeling most real-world devices.

It can be proved that any (α, β) element with $|\alpha| + |\beta| > 2$ is *active* in the sense that it can be built only with active components, such as transistors and op amps, which requires a power supply. Finally, we remark that every (α, β) element can be built by the same procedure illustrated in [2, 5, 10] using a family of linear active

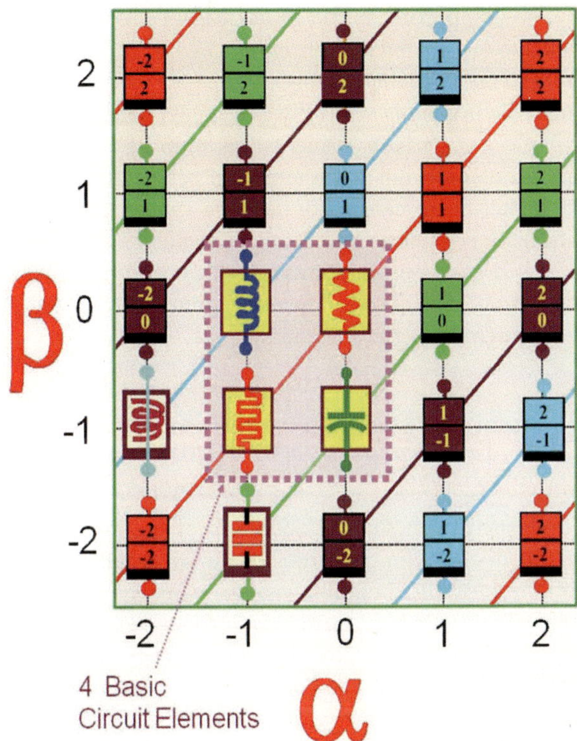

Fig. 2.3 The first 25 (α, β) circuit elements, $-2 \leq \alpha \leq 2$, $-2 \leq \beta \leq 2$

2-ports called *mutators*. They can also be *emulated* via various off-the-shelf digital components [11], or by programmable softwares interfaced with analog-to-digital (A/D) and digital-to-analog (D/A) converters.

2.1.3 Complexity Metric of Circuit Elements

For each (α, β) element, let

$$\chi \triangleq |\alpha| + |\beta| \tag{2.20}$$

be its associated *complexity metric* [12]. For example, $\chi(0,0) = 0$ for a *resistor*, $\chi(0,-1) = 1$ for a *capacitor*, $\chi(-1,0) = 1$ for an *inductor*, $\chi(-1,-1) = 2$ for a *memristor*, $\chi(-1,-2) = 3$ for a *memcapacitor*, and $\chi(-2,-1) = 3$ for a *meminductor*. If one associates the vertical and horizontal lines passing through the elements in Fig. 2.3 as streets of Manhattan, New York city, then the complexity metric χ of an (α, β) element gives a measure of its distance from the resistor $(\alpha, \beta) = (0,0)$. The larger the metric $\chi(\alpha, \beta)$, the farther it is from the resistor.

The complexity metric measures not just only the distance of (α,β) element from the resistor but also the *minimum number* of capacitors (or inductors) needed to build an (α,β) element using off-the-shelf components. For example, a minimum of one capacitor along with active elements such as transistors and op amps is needed to build a *memristor* while a minimum of two capacitors are needed to build a meminductor. From a mathematical perspective, the larger the complexity metric, the higher the dimension of the *state space* and the larger the number of nonlinear differential equations and exotic dynamical phenomena that can emerge.

Based on any of the above measures of complexity, the four elements depicted in Fig. 2.3 are indeed the simplest circuit elements, with the memristor ranked as the 4th element in increasing complexity.

2.1.4 Fingerprint of Memristors

The formal mathematical definition of the memristor is given in [5], along with its circuit-theoretic properties. Here we recall that the memristor is defined by a collection of all admissible signals, namely, an *A-pad* listing all signals measured from all admissible "Gedanken Probing Circuits" (Fig. 2.1) and which can be completely reproduced by the constitutive relation (2.17).

For example, a charge-controlled memristor can be defined by

$$\varphi = f_M(q) \tag{2.21}$$

where f_M is a *piecewise-differentiable* function [12]. In this case, we can generate all $(v(t), i(t))$ from the *A-pad* via the following q-dependent Ohm's law:

$$\boxed{\begin{aligned} v &= R(q)i & \text{(2.22a)} \\ R(q) &\triangleq \frac{df_M(q)}{dq} & \text{(2.22b)} \end{aligned}}$$

The function $R(q)$ is called the memristance (acronym for Memory Resistance) where

$$R(q) \geq 0 \tag{2.23}$$

for all passive *memristors* [2].

Now observe from (2.8) that since

$$\frac{dq}{dt} = 0 \quad \text{when} \quad i = 0 \tag{2.24}$$

the memristor can assume a *continuous* range of distinct equilibrium states

$$q = q(t_0), \quad t \geq t_0 \tag{2.25}$$

when the power is switched off at any time $t = t_0$. It follows that the *memristor* can be used as a *nonvolatile analog memory*. In particular, it can be used as a nonvolatile *binary* memory where two sufficiently different values of resistance are chosen to code the binary states "0" and "1," respectively. Because the hp *memristor* reported in [6] as well as in many other nano devices [13] can be scaled down to atomic dimensions, the *memristor* offers immense potentials for an ultra low-power and ultra dense nonvolatile memory technology that could replace flash memories and DRAMS.

An incisive analysis of (2.22) reveals that the *nonvolatile* memory property possessed by the memristor is a direct consequence of its *state-dependent* Ohm's law. Moreover, all circuit-theoretic properties possessed by the *memristor* are preserved if we generalize (2.22) to the form [14].

$$v = R(x,i)i \tag{2.26a}$$

$$dx/dt = \mathbf{f}(x,i) \tag{2.26b}$$

The generalized *memristor* defined in (2.26) is dubbed a memristive *device* in [14] where $x = (x_1, x_2, \ldots, x_n)$ denotes *n states variables*, which do not depend on any external voltages or currents. However, since both (2.22) and (2.26) are endowed with the same circuit-theoretic properties, it is more convenient and logical to refer to both equations as defining a *memristor*. In the rare events where a distinction may be desirable, one can refer to (2.22) as defining an "ideal memristor."

The most important common property of (2.22) and (2.26) is that the loci (i.e., Lissajous figure) of $(v(t), i(t))$ due to *any* periodic current source, or periodic voltage source, which assumes both positive and negative values, must always be *pinched* at the origin in the sense that $(v,i) = (0,0)$ must always lie on the (v,i)-loci, called a *pinched hysteresis loop* in the literature [13]. We wish to stress that (2.22) and (2.26) imply that the pinched hysteresis loop phenomenon of the memristor must hold for *any periodic signal*, $v(t)$ or $i(t)$, that assumes both positive and negative values, as well as for any intial condition used to integrate the differential equations to obtain the corresponding steady state $i(t)$ and $v(t)$, respectively.

Another unique property shared by all memristor hysteresis loops is that for every given periodic function $i = f(t)$ (where $f(\bullet)$ assumes both positive and negative values), and for any initial state $\mathbf{x}(0)$ the area enclosed within the part of the pinched hysteresis loop in the first quadrant, and the third quadrant, of the $v - i$ plane shrinks continuously as the frequency ω increases, and the hysteresis loop tends to a *single-valued function* through the origin as ω tends to ∞.

The above dense continuum of pinched hysteresis loops, as well as their *single-valued function* limiting phenomenon as $\omega \to \infty$ must hold for *all memristors*. Any purported system which may exhibit a pinched hysteresis loop but which violates the above continuum and frequency-dependent limiting *memristor fingerprint* is *not* a memristor, the reader is referred to [15] for several contrived examples which fails the above "*memristor fingerprint* test."

We end this tutorial by pointing out that not all memristors are nonvolatile memories. In fact there is an even *larger class* of *locally active* memristors [2, 4, 9] which exhibit many exotic nonlinear dynamical phenomena. A very interesting and scientifically significant example is the classic Hodgkin–Huxley Axon circuit model of the squid giant axon.[3] Notwithstanding the immense importance of their circuit model, Hodgkin and Huxley had erroneously named two circuit elements in their model associated with the potassium ion, and the sodium ion, respectively, as *time-varying conductances*. This mistaken identity has led to numerous confusions and paradoxes ever since the publications of their classic axon circuit model [16]. Well-known physiologists were puzzled by experimentally observed *rectification phenomenon* as well as *gigantic inductances* that could not exist within the soft tissues of the brain. The following quotation from Cole (see page 78 of [17]), an eminent physiologist and the recipient of the 1967 USA *National Medal of Science*, is a case in point:

"The suggestion of an inductive reactance anywhere in the system was shocking to the point of being unbelievable"

We have solved the above conundrum, and many other hitherto unresolved paradoxes associated with the Hodgkin–Huxley Axon, by showing the Hodgkin–Huxley *time-varying* potassium conductance is in fact a 1st-order memristor, and the Hodgkin–Huxley *time-varying* sodium conductance is in fact a 2nd-order memristor, as defined in Fig. 2.4b, c, respectively [18]. Also depicted in Fig. 2.4 are the pinched hysteresis loops associated with each memristor. Observe that they are all pinched at the origin, and that the lobe area in the first and third quadrants shrinks continuously to a straight line as ω increases, both being the fingerprint of memristors.

We conclude this tutorial by stressing that memristors are not inventions. They are *discoveries* and are ubiquitous. Indeed, many devices, including the "*electric arc*" dating back to 1801, have now been identified as memristors [19, 20]. Aside from serving as nonvolatile memories [21], *locally passive* memristors, have been used for switching electromagnetic devices [22], for field programmable logic arrays [23–27], for synaptic memories [28–30], for learning [31–33], etc.

In addition, *locally active* memristors have been found to exhibit many exotic dynamical phenomena, such as *oscillations* [34], *chaos* [35, 36], *Hamiltonian vortices* [37] and *autowaves* [38], etc.

---

[3]Hodgkin and Huxley were awarded the 1965 Nobel Prize in physiology for their derivation of the circuit shown in Fig. 2.4a, where the two memristors were drawn as time-varying resistors in Fig. 1 (page 501) of [16].

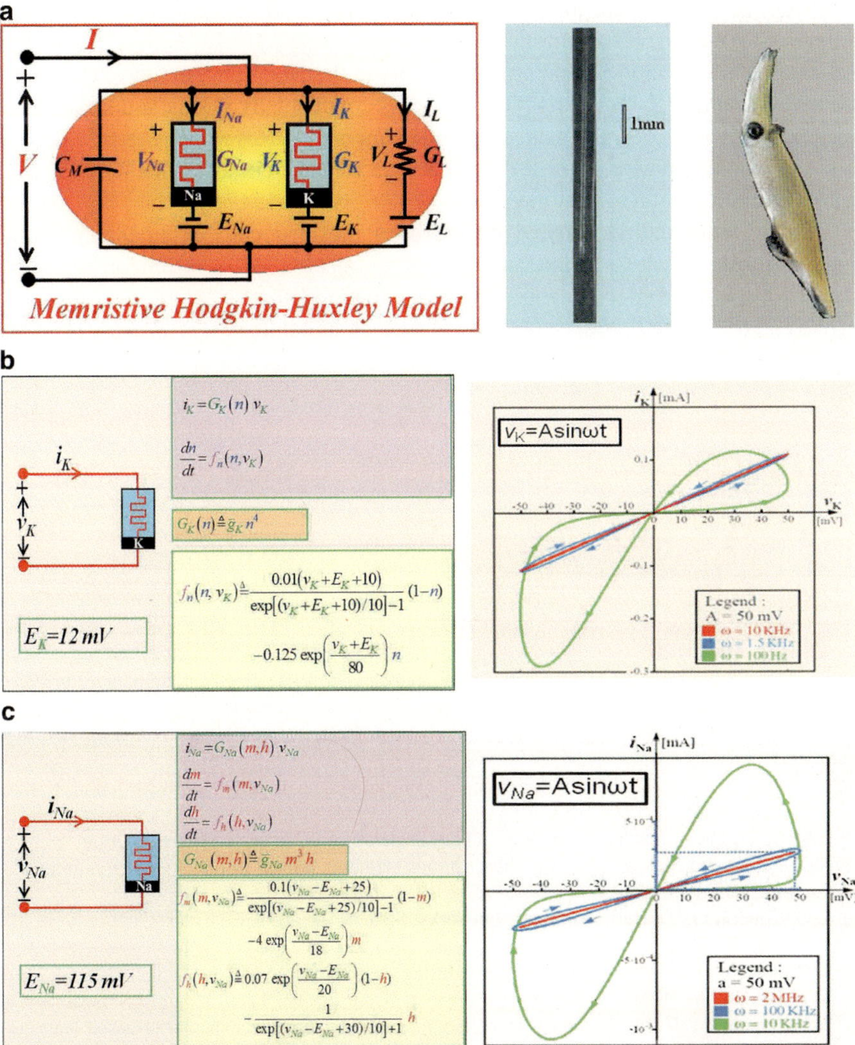

Fig. 2.4 Hodgkin–Huxley Axon. (**a**) Memristive Hodgkin–Huxley Circuit model of giant axon (*center*) of North Atlantic squid Loligo (*right*). (**b**) Postassium ion-channel memristor and its pinched hysteresis loops. (**c**) Sodium ion-channel memristor and its pinched hysteresis loops [18]

2.2 Concluding Remarks

Any 2-terminal device which exhibits a pinched hysteresis loop in the v-i plane when driven by *any* bipolar periodic voltage or current waveform, for *any* initial conditions, is a *memristor*. In the case where the *memristance* $R(x_1, x_2, \ldots, x_n)$ does not depend on the current i, the loop shrinks to a straight line whose slope depends on the excitation waveform, as the excitation frequency tends to infinity.

Except in ideal cases, memristors, memcapacitors, and meminductors do *not* behave like resistors, capacitors, and inductors, respectively. For example, the potassium and sodium ion channel memristors in the Hodgkin–Huxley axon circuit model behave like R-L circuits ([18, 39]). It is conceptually wrong and misleading to identify memristors, memcapacitors, and meminductors with resistors, capacitors, and inductors. Each (α, β) element is a distinct circuit element because it cannot be built from the other elements.

Readers who may have been misled by some erroneous commentary in the popular press which associates an earlier *gadget* called a *memistor* with the *memristor* are referred to a technical clarification in [40].

We end this tutorial with the following succint signature of a *memristor* [13]: *If it's pinched it's a memristor.*

References

1. J.M. Tour, T. He, The fourth element. Nature **453**, 42–43 (1 May 2008)
2. L.O. Chua, *Introduction to Nonlinear Network Theory* (McGraw Hill Book Co., New York, 1969)
3. L.O. Chua, Device modeling via basic nonlinear circuit elements. IEEE Trans. Circuit Syst. **27**, 1014–1044 (1980)
4. L.O. Chua, Nonlinear circuit foundations for nano devices, Part I: The four-element torus. Proc. IEEE **91**, 1830–1859 (2003)
5. L.O. Chua, Memristor: The missing circuit element. IEEE Trans. Circuit Theory **18**, 507–519 (1971)
6. D.B. Strukov, G.S. Snider, D.R. Stewart, R.S. Williams, The missing memristor found. Nature **453**, 80–83 (2008)
7. L.O. Chua, Introduction to memristors. IEEE Expert Now, Educational Course, 2009
8. M. Di Ventra, Y.V. Pershin, L.O. Chua, Circuit elements with memory: memristors, memcapacitors, and meminductors. Proc. IEEE **97**, 1717–1723 (2009)
9. L.O. Chua, L.O.: Dynamic nonlinear networks: State of the art. IEEE Trans. Circuits Syst. **27**, 1059–1087 (1980)
10. D. Biolek, V. Bioleova, Mutator for transforming memristor into memcapacitor. Electron. Lett. **46**, 1428–1429 (2010)
11. Y.V. Pershin, M. Di Ventra, Teaching memory circuit elements via experiment-based learning. arXiv: 1112.5427v1 [physics.ins-det]
12. R.G. Bartle, *The Elements of Real Analysis*, 2nd edn. (Wiley, New York, 1976)
13. L.O. Chua, Resistance switching memories are memristors. Appl. Phys. A **102**, 765–783 (2011)
14. L.O. Chua, S.M. Kang, Memristive devices and systems. Proc. IEEE **64**, 209–223 (1976)
15. H. Kim, M.P. Sah, S.P. Adhikari, Pinched hysteresis loop is the fingerprint of memristive devices. arXiv:1202.2437v1 [cond-mat.mes-hall]
16. A.L. Hodgkin, A.F. Huxley, A quantitative description of membrane current and its application to the conduction and excitation in nerve. J. Physiol. **117**, 500–544 (1952)
17. K.S. Cole, *Membranes, Ions and Impulse* (University of California Press, Berkeley, 1972)
18. L. Chua, V. Sbitnev, H. Kim, Hodgkin Huxley axon is made of memristors. Int. J. Bifurcat. Chaos **22**(3), 1230011 (2012)
19. H. Ayrton, *The Electric Arc* (D.Van Nostrand Co., London, 1902)

20. T. Prodromakis, C. Toumazou, L. Chua, Two centuries of memristors. Nat. Mater. **11**, 478–481 (June 2012)
21. I. Valov, R. Waser, J.R. Jameson, M.N. Kozicki, Electrochemical metallization memories: Fundamentals, applications, prospects. Nanotechnol. **2**, 254003 (2011)
22. M.G. Bray, D.H. Werner, Passive switching of electromagnetic devices with memristors. Appl. Phys. Lett. **96**, 0735041-3 (2010)
23. J. Borghetti, G.S. Snider, P.J. Kukes, J.J. Yang, D.R. Stewart, R.S. Williams, Williams memristive' switches enable 'stateful' logic operations via material implication. Nature **464**, 873–876 (2010)
24. D.B. Strukov, R.S. Williams, Four-dimensional address topology for circuits with stacked multilayer crossbar arrays. Proc. Natl. Acad. Sci. **106**, 20155–20158 (2009)
25. W. Lehtonen, M. Laiho, Stateful implication logic with memristors. Proc. IEEE/ACM International Symposium on Architectures, pp. 33–36, San Francisco, CA 2009
26. J. Borghetti, Z. Li, X. Straznicky, X. Li, A. Ohlberg, W. Wu, D.R. Stewart, R.S. Williams, A hybrid nanomemristor/transistor logic circuit capable of self-programming. Proc. Natl. Acad. Sci. **106**, 1699–1703 (2009)
27. K. Kim, S. Shin, S.-M. Kang, Field programmable stateful logic array. IEEE Trans. Comput. Aided Des. Integr. Circuits Syst. **30**, 1800–1813 (2011)
28. B. Linares-Barranco, T. Serrano-Gotarredona, Memristance can explain spike-time-dependent-plasticity in neural synapses. Nature Precedings, hdl:10101/npre. 2009.3010.1 : 31 March 2009
29. S.H. Jo, T. Chang, I. Ebong, B.B. Bhadviya, P.Lu.W. Mazumder, Nanoscale memristor device as synapse in neuromorphic systems. Nano Lett. **10**, 1297–1301
30. Pershin, Y.V., Di Ventra, M.: Neuromorphic, digital and quantum computation with memory circuit elements. arXiv:1009.6025v3 [cond-mat.mes-hall]
31. G.S. Snider, Spike-timing dependent learning in memristive nanodevices, IEEE/ACM International Symposium on Nanoscale Architecture, Anaheim, CA, 85–92, 2008
32. T. Liu, Y. Kang, M. Verma, M. Orlowski, Novel Highly Nonlinear Memristive Circuit Elements for Neural Networks. Proceedings, 2012 IJCNN International Joint Conference on Neural Networks, in Brisbane, Australia, June 2012, doi 10.1109/IJCNN.2012.6252460
33. T. Chang, S.-H. Jo, W. Lu, Short-term memory to long-term memory transition in a nanoscale memristor. Am. Chem. Soc. (ACS) Nano **5**, 7669–7676 (2011)
34. M. Itoh, L.O. Chua, Memristor oscillators. Int. J. Bifurcat. Chaos **18**, 3183–3206 (2008)
35. B. Muthuswamy, L.O. Chua, Simplest chaotic circuit. Int. J. Bifurcat. Chaos **20**, 1567–1580 (2010)
36. J.-M. Ginoux, C. Letellier, L.O. Chua, Topological analysis of chaotic solution of a three-element memristive circuit. Int. J. Bifurcat. Chaos **20**, 3819–3827 (2010)
37. M. Itoh, L.O. Chua, Memristor Hamiltonian circuits. Int. J. Bifurcat. Chaos **21**, 2395–2425 (2011)
38. V.-T. Pham, A. Buscarino, M. Frasca, L. Fortuna, Autowaves in memristive cellular neural networks. Int. J. Bifurcat. Chaos **22**, 12300271-9 (2012)
39. L. Chua, V. Sbitnev, H. Kim, Neurons are poised near the edge of chaos. Int. J. Bifurcat. Chaos **22**(4), 1250098 (2012)
40. H. Kim, S.P. Adhikari, Memistor is mot memristor. IEEE Circuits and Systems Magazine, Fist Quarter 2012, pp. 75–78

Part II

Vignettes

From

Memristor

Leon Chua

Professor of University of California, Berkeley

and

Honorary Fellow of Institute of Advanced Study

Technische Universität München

What is this Device ?

A 4.5 billion-year
old *meteorite*
(chondrite) from
the *milky way*

What is a ***resistor, inductor,*** and ***capacitor*** ?

Shocking Fact !

Before the publication of my book *Introduction to Nonlinear Network Theory in 1969, there was no* scientific definition of *basic circuit elements.*

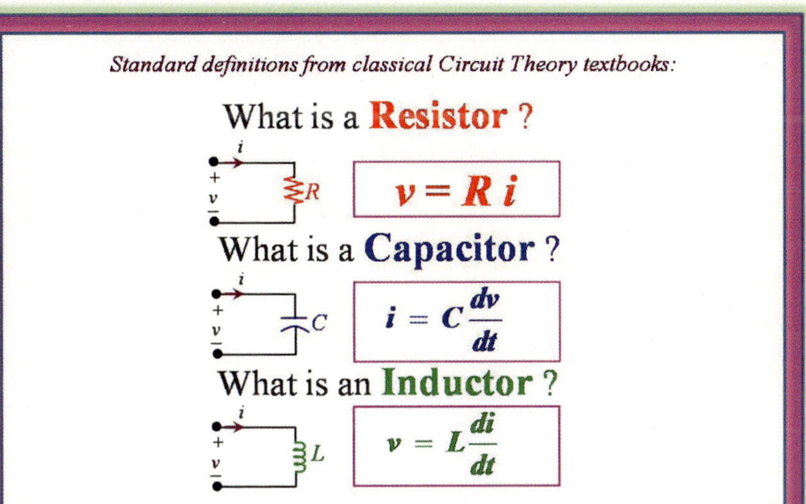
Standard definitions from classical Circuit Theory textbooks:
What is a Resistor ?
$v = R\,i$
What is a Capacitor ?
$i = C\frac{dv}{dt}$
What is an Inductor ?
$v = L\frac{di}{dt}$

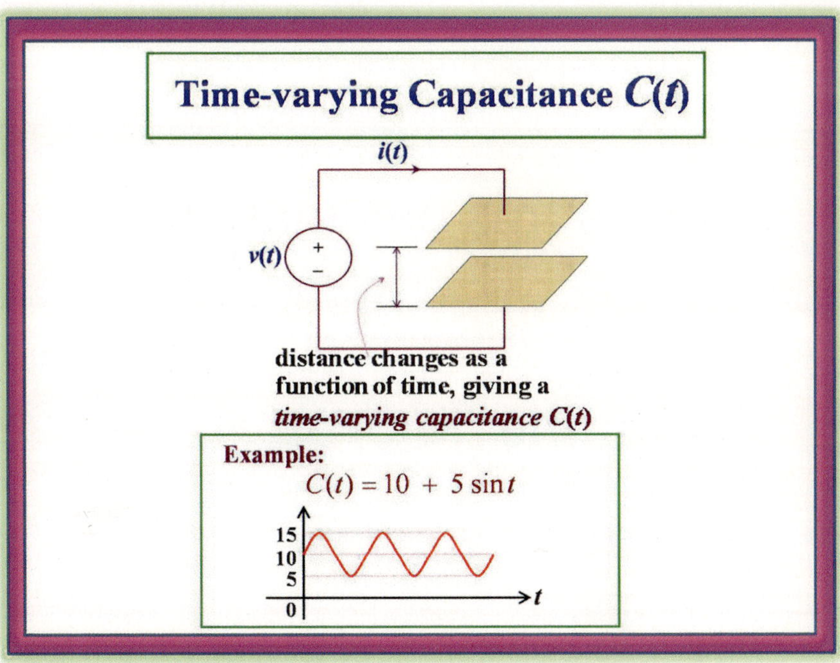
Time-varying Capacitance $C(t)$
$i(t)$
$v(t)$
distance changes as a
function of time, giving a
time-varying capacitance $C(t)$
Example:
$C(t) = 10 + 5\sin t$
15
10
5
0
t

Example : Time-varying Capacitance

$$i(t) = \underbrace{\left(10+5\sin t\right)}_{C(t)} \frac{dv(t)}{dt}$$

Does this obvious generalization of the formula

$$i = C\,\frac{dv}{dt}$$

give the correct current $i(t)$ for any applied voltage $v(t)$?

Does this calculated *capacitor current* agree with the *laboratory measurement*?

No !

Reason :
Our classical definition of the *capacitor* is wrong.

The *classical* definition

$$i(t) = C\,\frac{dv(t)}{dt}$$

uses an *incorrect pair* of variables

$i(t)$ and $v'(t) \triangleq \frac{dv(t)}{dt}$

Let us *integrate* both sides to obtain

$$q(t) = C\,v(t)$$

This equation defines a *relationship* between a *different pair* of variables $q(t)$ and $v(t)$, and gives the *correct* answer.

Correct formula:

$$q(t) = C(t)\, v(t)$$

$$q(t) = \underbrace{(10 + 5 \sin t)}_{C(t)}\, v(t)$$

$$i = \frac{dq(t)}{dt} = \underbrace{(10+5 \sin t)}_{C(t)} \frac{dv(t)}{dt} + (5 \cos t)\, v(t)$$

$$i = \frac{dq(t)}{dt} = C(t) \frac{dv(t)}{dt} + \underbrace{v(t) \frac{dC(t)}{dt}}$$

extra term is needed !

An analogy of a

similar mistake

from **Mechanics**

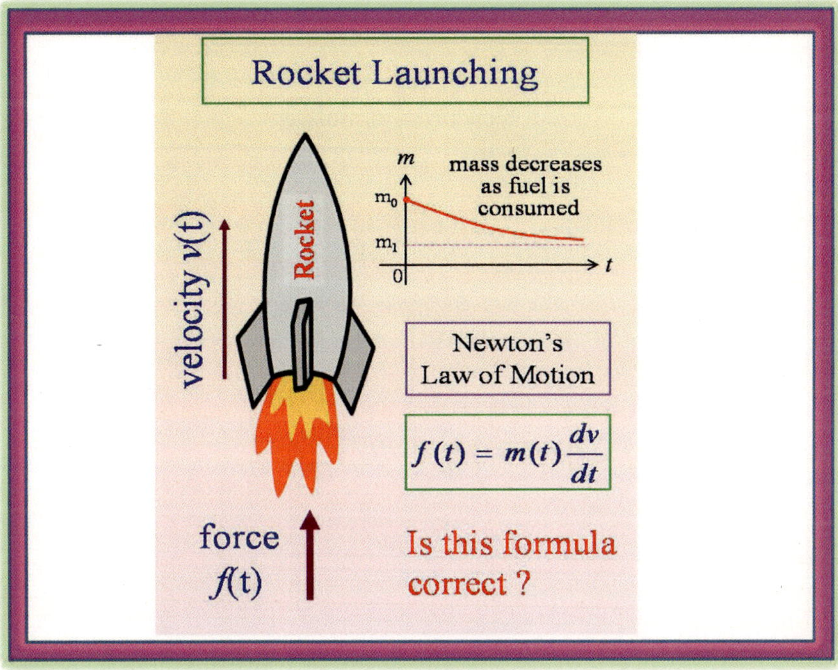
Rocket Launching
m
mass decreases
as fuel is
consumed
m_0
m_1
0
t
Rocket
velocity v(t)
Newton's
Law of Motion
$f(t) = m(t)\frac{dv}{dt}$
force
f(t)
Is this formula
correct ?

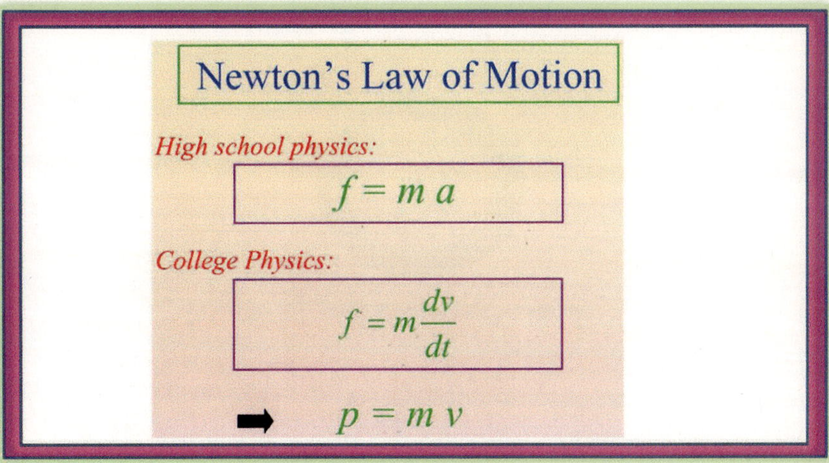
Newton's Law of Motion
High school physics:
$f = m\,a$
College Physics:
$f = m\frac{dv}{dt}$
$p = m\,v$

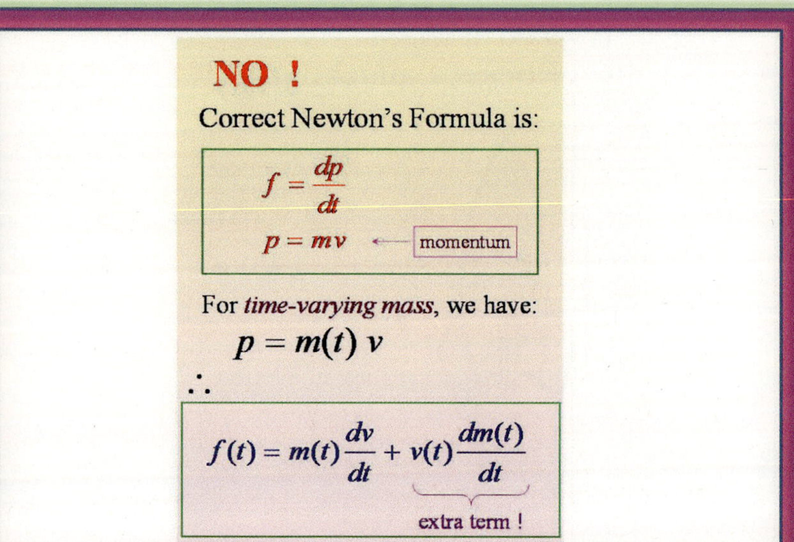
NO !
Correct Newton's Formula is:
$f = \frac{dp}{dt}$
$p = mv$
momentum
For *time-varying mass*, we have:
$p = m(t)\, v$
$\therefore$
$f(t) = m(t)\frac{dv}{dt} + v(t)\frac{dm(t)}{dt}$
extra term !

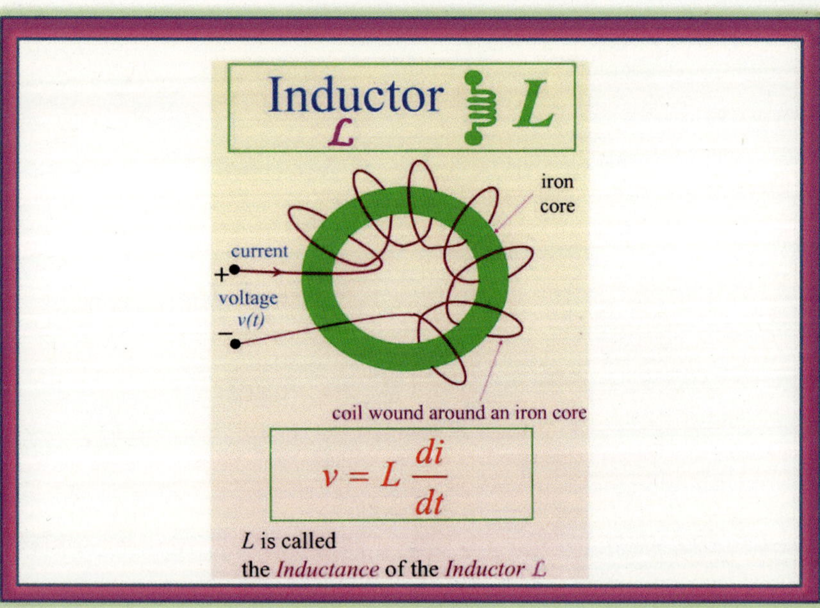
Inductor
L
L
iron
core
current
+
voltage
v(t)
−
coil wound around an iron core
$v = L\frac{di}{dt}$
L is called
the Inductance of the Inductor L

Inductor

The *classical definition*

$$v = L\,\frac{di}{dt}$$

which relates the 2 variables *voltage* v and

the *derivative* $i' \triangleq \dfrac{di}{dt}$ gives the ***wrong*** answer !

The correct definition

$$\varphi = L\,i$$

relates the 2 variables *flux* φ and *current* i

Correct formula:

$$\varphi(t) = L(t)\,i(t)$$

$$\varphi(t) = \underbrace{(10 + 5\sin t)}_{L(t)}\,i(t)$$

$$v = \frac{d\varphi(t)}{dt} = \underbrace{(10+5\sin t)}_{L(t)}\frac{di(t)}{dt} + (5\cos t)\,i(t)$$

$$v = \frac{d\varphi(t)}{dt} = L(t)\frac{di(t)}{dt} + \underbrace{i(t)\frac{dL(t)}{dt}}$$

extra term is needed !

Conclusion:

A *time-varying inductance* $L(t)$ must be defined by a relationship between the *flux* $\varphi(t)$ and the *current* $i(t)$, and *not* between $v(t)$ and $\dfrac{di(t)}{dt}$, as in the *incorrect* formula $v = L(t)\,\dfrac{di}{dt}$

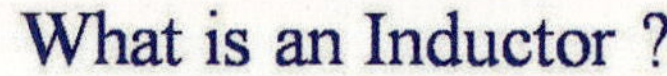

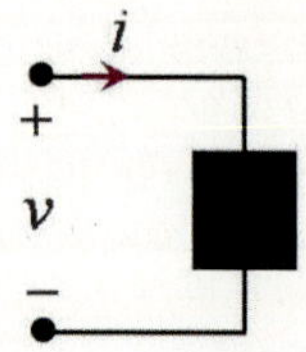

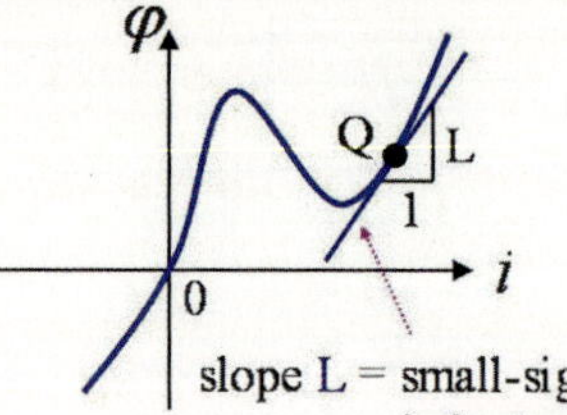

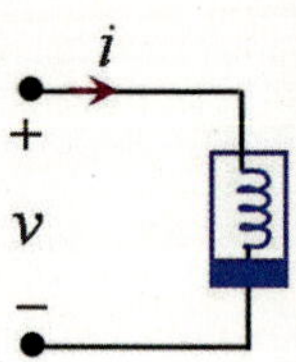

Definition

Any device which imposes a *relationship* between the **flux** $\boldsymbol{\varphi} \triangleq \int_{-\infty}^{t} \boldsymbol{v(t)dt}$ and the **current *i*** is an **inductor**. The *slope* at any point Q is called the *small-signal inductance L* at Q

Axiomatic Definitions

To obtain the correct definitions of *elementary circuit elements*, it is necessary to introduce an *axiomatic approach* involving the 4 basic circuit variables *voltage v*(t), *current i*(t), *flux* φ(t), and *charge q(t)*.

Axiomatic Definition of Circuit Elements

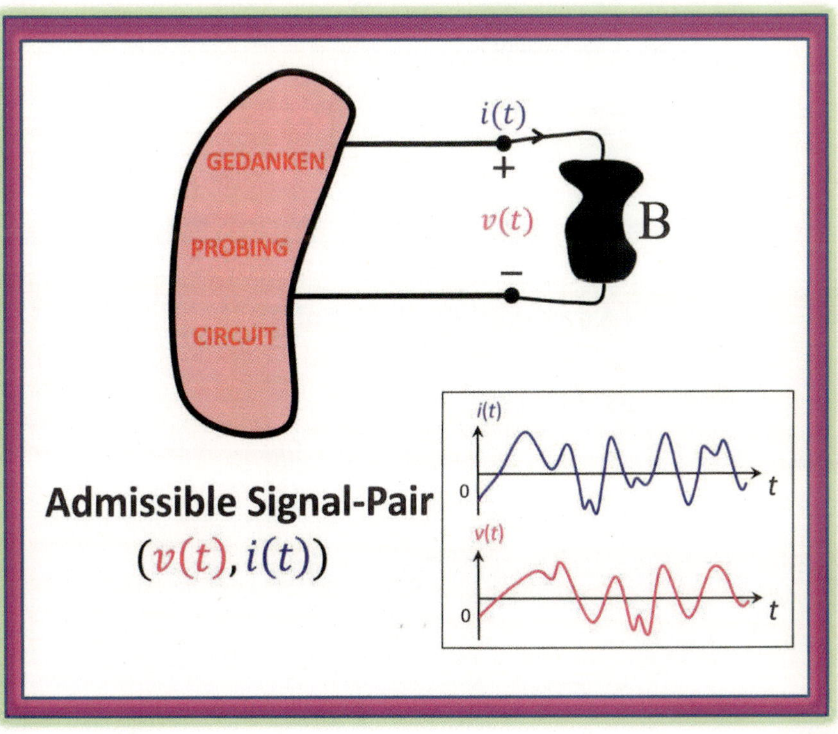

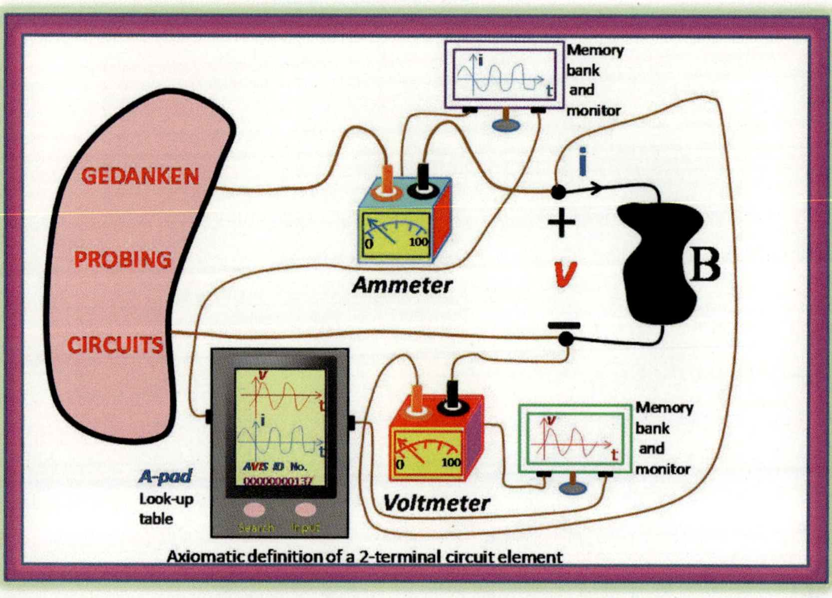
Memory
bank
and
monitor
GEDANKEN
PROBING
CIRCUITS
Ammeter
i
+
v
−
B
A-pad
Look-up
table
AVIS ID No.
0000000137
Voltmeter
Memory
bank
and
monitor
Axiomatic definition of a 2-terminal circuit element

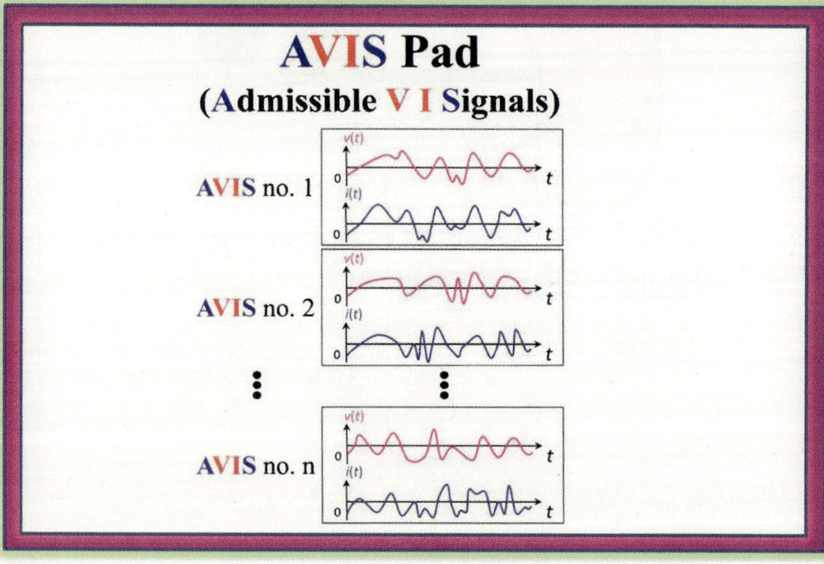
AVIS Pad
(Admissible V I Signals)
AVIS no. 1
AVIS no. 2
AVIS no. n
v(t)
i(t)
t

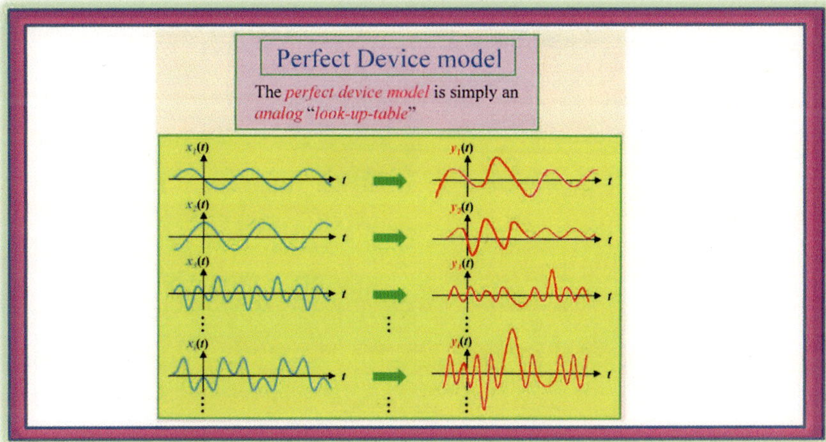

Perfect Device model
The *perfect device model* is simply an *analog* "*look-up-table*"

No device model is Perfect

A useful device model must reproduce the input-output behaviors of a physical device to acceptable engineering accuracy.

Model

must

Predict !

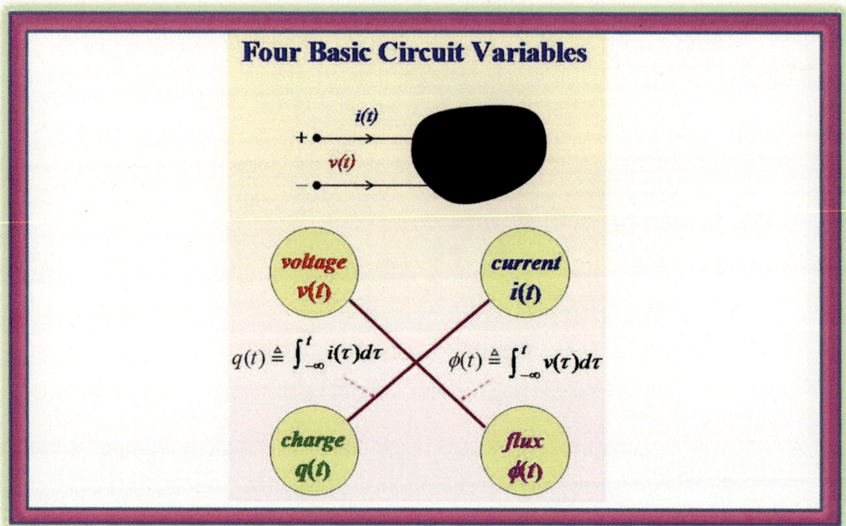

FOUR

ELEMENTARY

CIRCUIT ELEMENTS

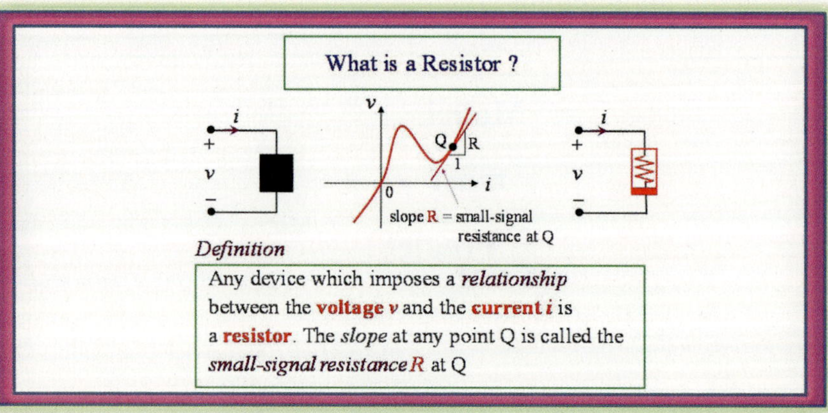

What is a Resistor ?
i
+
v
−
v
Q
R
1
0
i
slope R = small-signal
resistance at Q
i
+
v
−
Definition
Any device which imposes a relationship
between the voltage v and the current i is
a resistor. The slope at any point Q is called the
small-signal resistance R at Q

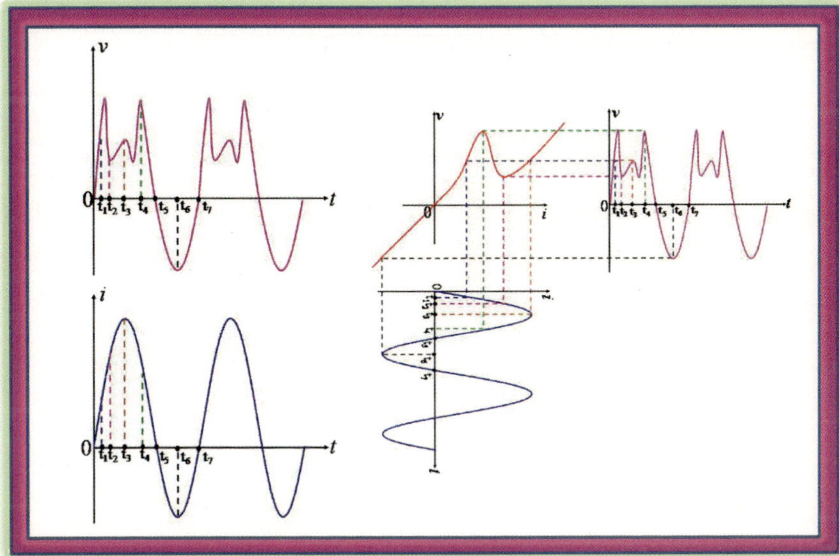

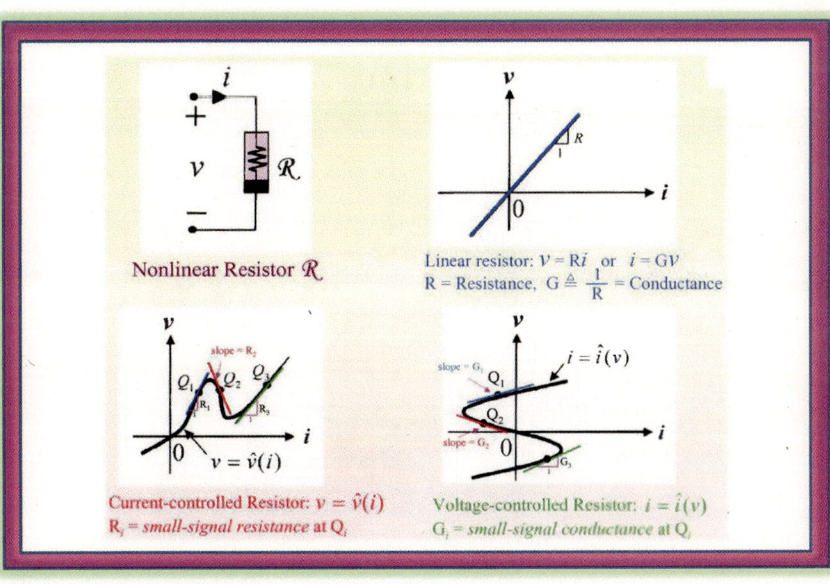
Nonlinear Resistor $\mathcal{R}$
Linear resistor: $v = Ri$ or $i = Gv$
R = Resistance, $G \triangleq \frac{1}{R}$ = Conductance
Current-controlled Resistor: $v = \hat{v}(i)$
R_i = *small-signal resistance* at Q_i
Voltage-controlled Resistor: $i = \hat{i}(v)$
G_i = *small-signal conductance* at Q_i

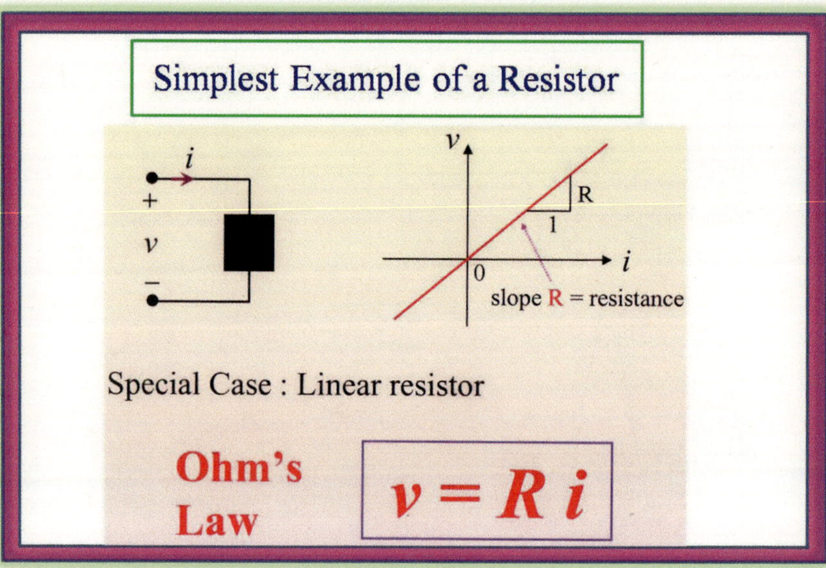
Simplest Example of a Resistor
i
+
v
−
v
R
1
i
0
slope R = resistance
Special Case : Linear resistor
Ohm's Law
v = R i

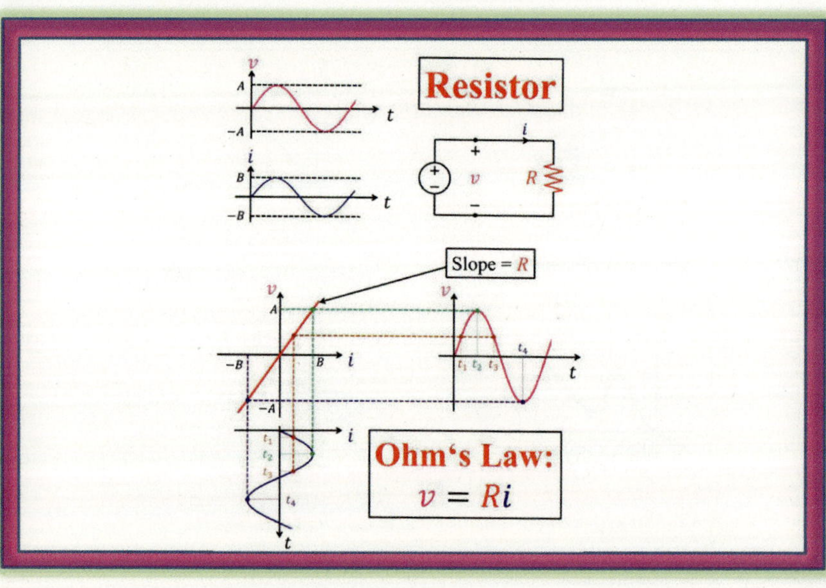
Resistor
Slope = R
Ohm's Law:
v = Ri

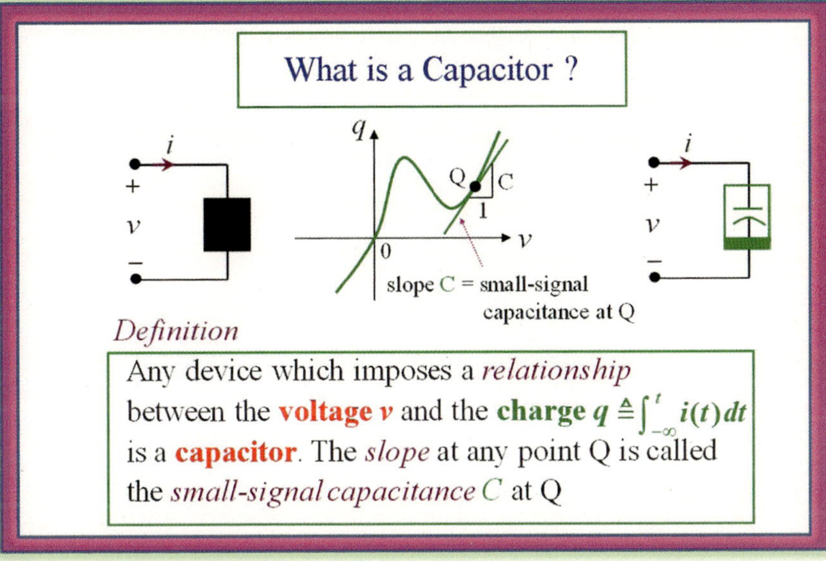
What is a Capacitor ?
i
+
v
−
q
Q
C
1
0
v
slope C = small-signal capacitance at Q
i
+
v
−
Definition
Any device which imposes a relationship between the voltage v and the charge $q \triangleq \int_{-\infty}^{t} i(t)dt$ is a capacitor. The slope at any point Q is called the small-signal capacitance C at Q

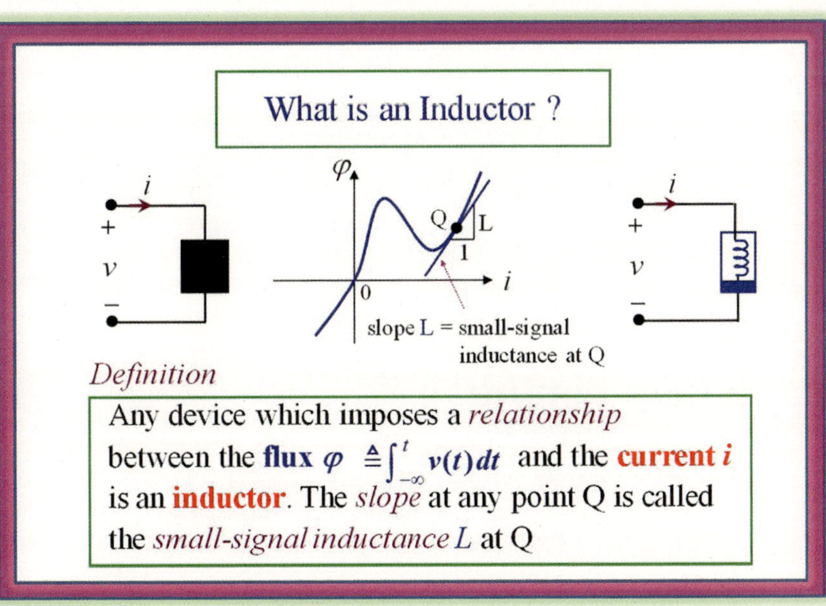
What is an Inductor ?
i
+
v
−
φ
Q
L
1
0
i
slope L = small-signal inductance at Q
i
+
v
−
Definition
Any device which imposes a relationship between the flux $\varphi \triangleq \int_{-\infty}^{t} v(t)dt$ and the current i is an inductor. The slope at any point Q is called the small-signal inductance L at Q

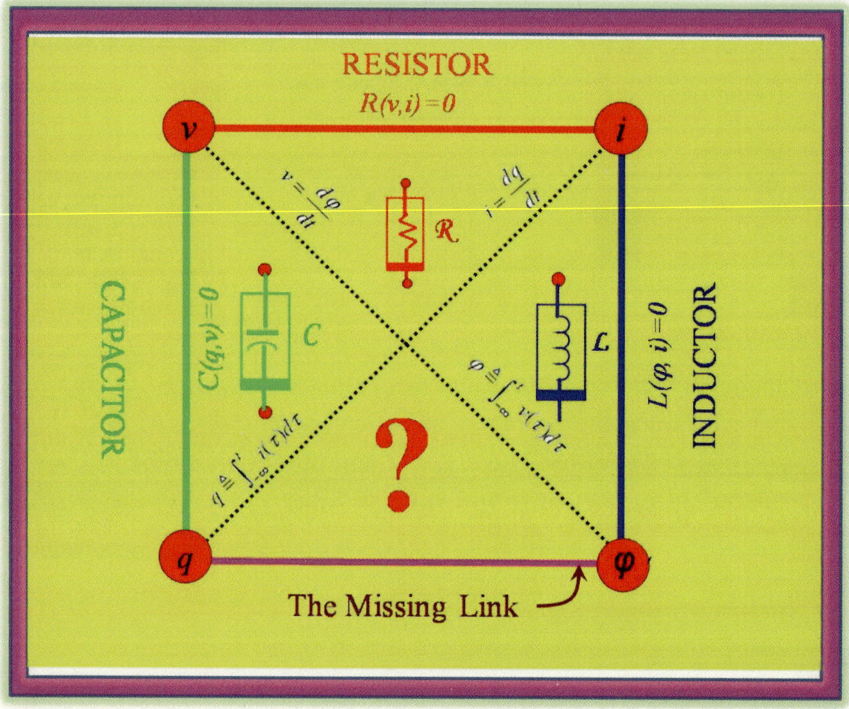

The missing
circuit element
is the
memristor !

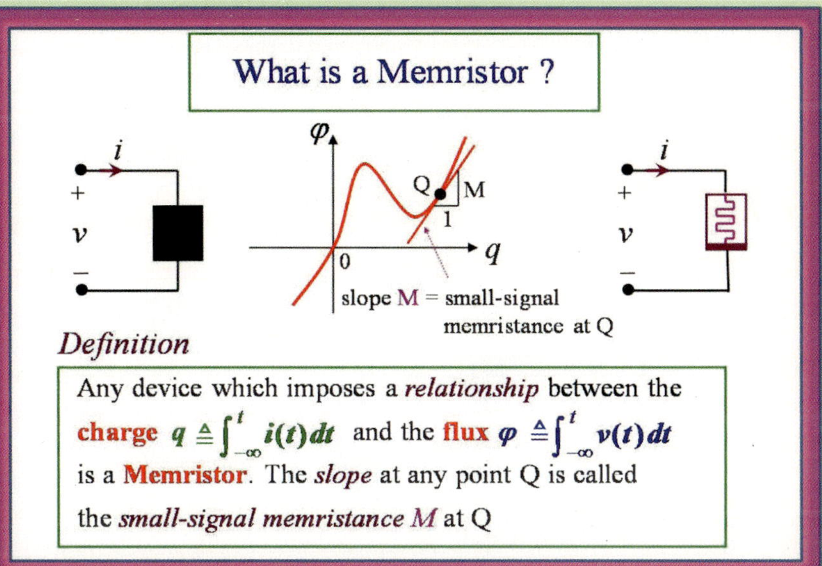
What is a Memristor ?
i
+
v
−
φ
Q
M
1
0
q
slope M = small-signal
memristance at Q
Definition
Any device which imposes a relationship between the
charge $q \triangleq \int_{-\infty}^{t} i(t)dt$ and the flux $\varphi \triangleq \int_{-\infty}^{t} v(t)dt$
is a Memristor. The slope at any point Q is called
the small-signal memristance M at Q

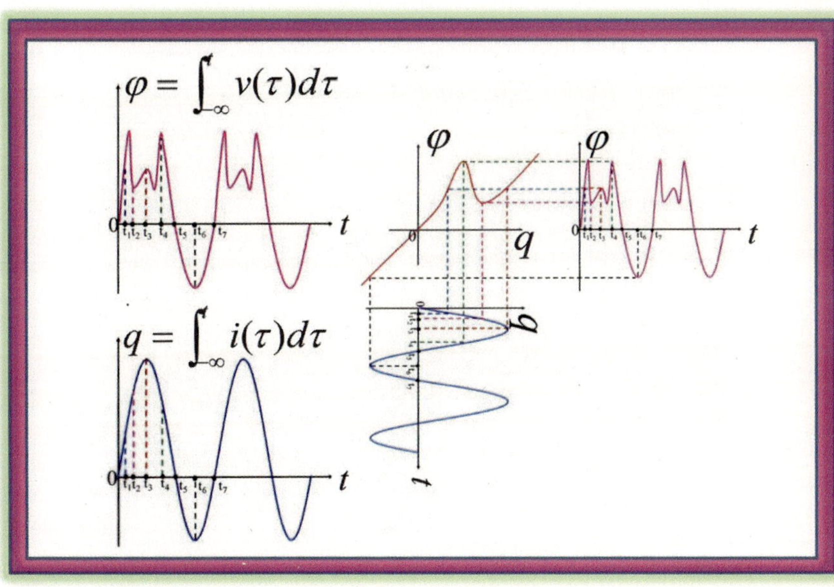
$\varphi = \int_{-\infty}^{t} v(\tau)d\tau$
$q = \int_{-\infty}^{t} i(\tau)d\tau$
φ
q
t

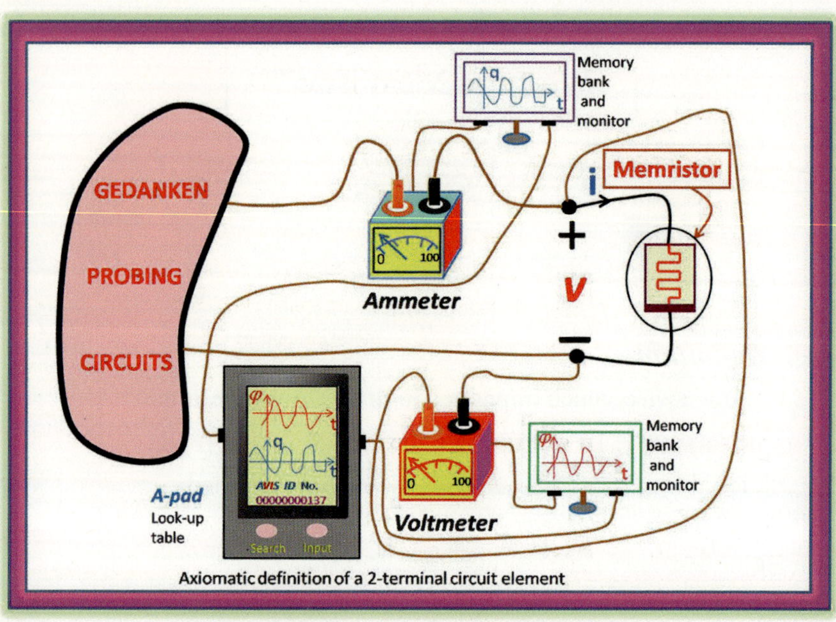
Memory
bank
and
monitor
GEDANKEN
PROBING
CIRCUITS
Memristor
Ammeter
A-pad
Look-up
table
AVIS ID No.
00000000137
Search
Input
Voltmeter
Memory
bank
and
monitor
Axiomatic definition of a 2-terminal circuit element

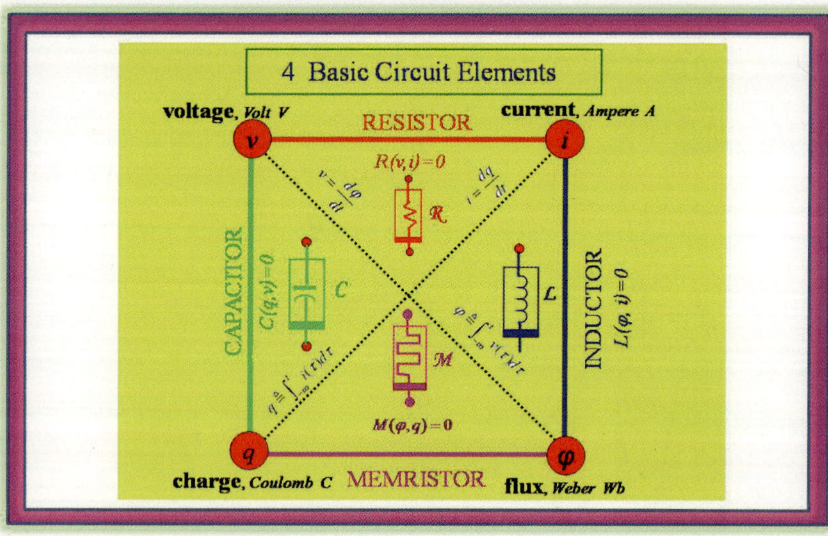
4 Basic Circuit Elements
voltage, Volt V
current, Ampere A
RESISTOR
R(v,i)=0
CAPACITOR
C(q,v)=0
INDUCTOR
L(φ, i)=0
M(φ,q)=0
charge, Coulomb C
MEMRISTOR
flux, Weber Wb

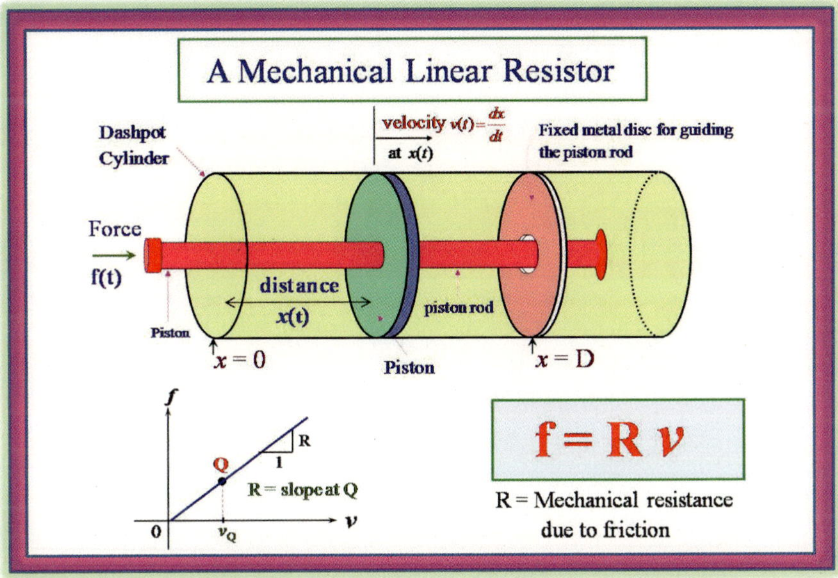
A Mechanical Linear Resistor
Dashpot Cylinder
velocity $v(t)=\frac{dx}{dt}$
at $x(t)$
Fixed metal disc for guiding the piston rod
Force
f(t)
distance
$x(t)$
piston rod
Piston
$x=0$
Piston
$x=D$
f
R
1
Q
R = slope at Q
0
v_Q
v
f = R v
R = Mechanical resistance due to friction

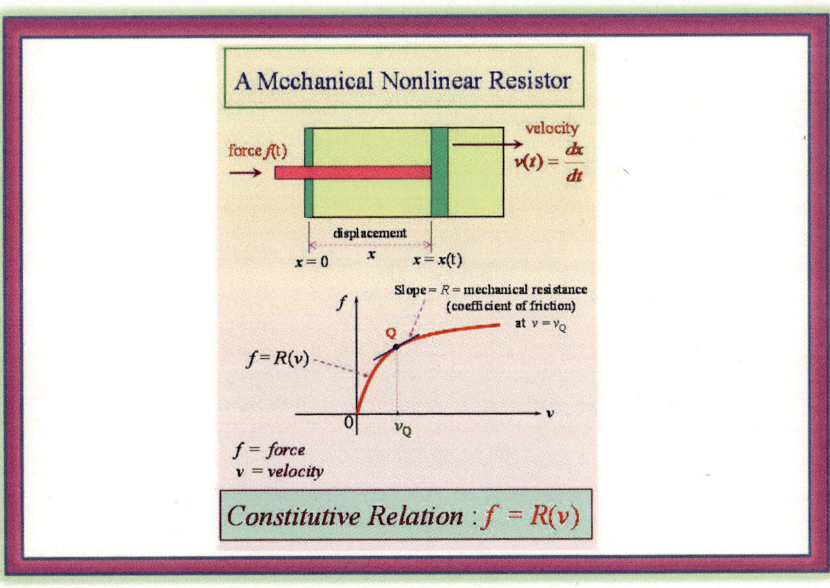
A Mechanical Nonlinear Resistor
force $f(t)$
velocity
$v(t)=\frac{dx}{dt}$
displacement
$x=0$
x
$x=x(t)$
Slope = R = mechanical resistance
(coefficient of friction)
at $v=v_Q$
f
Q
$f=R(v)$
0
v_Q
v
f = force
v = velocity
Constitutive Relation : $f = R(v)$

Newton's Equation

High School version:

force = mass × acceleration

$$f = m\frac{dv}{dt}$$

Integrating both sides:

$$\underbrace{\int_{-\infty}^{t} f(t)dt}_{\text{Momentum}} = m\underbrace{\int_{-\infty}^{t} \frac{dv}{dt}dt}_{\text{velocity}}$$

Correct Newton's Equation:

$$p = mv$$

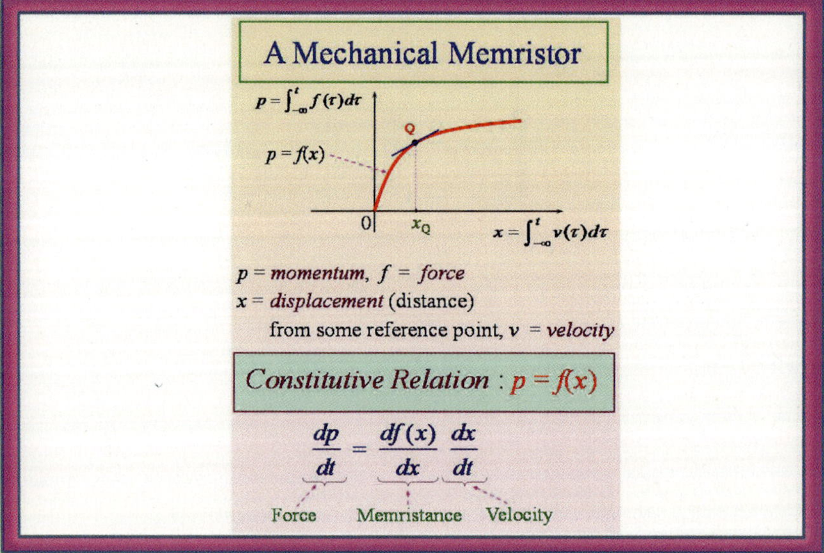

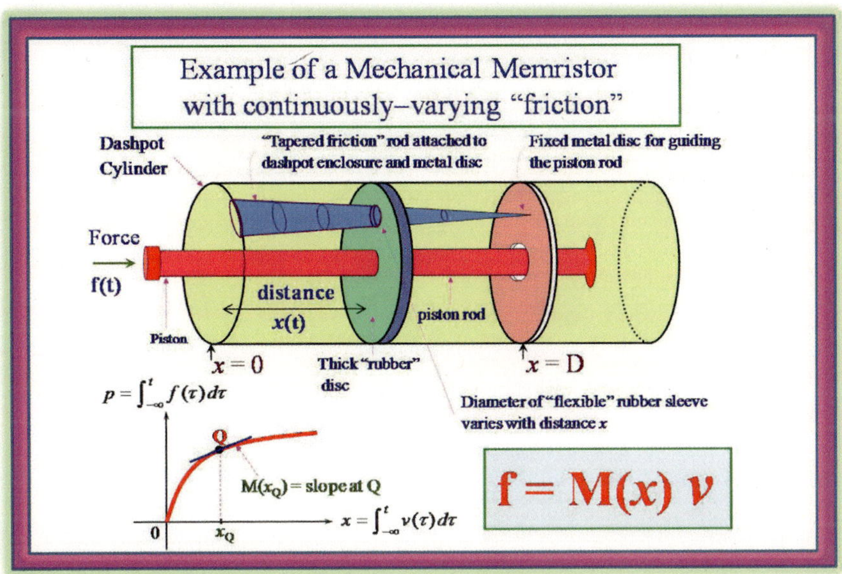
Example of a Mechanical Memristor
with continuously–varying "friction"
Dashpot
Cylinder
"Tapered friction" rod attached to
dashpot enclosure and metal disc
Fixed metal disc for guiding
the piston rod
Force
f(t)
distance
x(t)
piston rod
Piston
x = 0
Thick "rubber"
disc
x = D
$p = \int_{-\infty}^{t} f(\tau)d\tau$
Diameter of "flexible" rubber sleeve
varies with distance x
Q
$M(x_Q)$ = slope at Q
0
x_Q
$x = \int_{-\infty}^{t} v(\tau)d\tau$
$f = M(x)\,v$

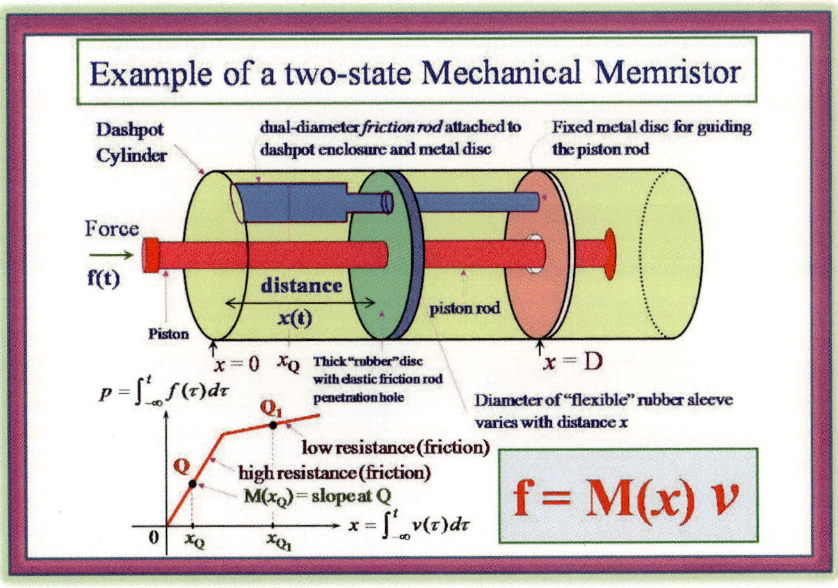
Example of a two-state Mechanical Memristor
Dashpot
Cylinder
dual-diameter friction rod attached to
dashpot enclosure and metal disc
Fixed metal disc for guiding
the piston rod
Force
f(t)
distance
x(t)
piston rod
Piston
x = 0
x_Q
Thick "rubber" disc
with elastic friction rod
penetration hole
x = D
$p = \int_{-\infty}^{t} f(\tau)d\tau$
Diameter of "flexible" rubber sleeve
varies with distance x
Q_1
low resistance (friction)
Q
high resistance (friction)
$M(x_Q)$ = slope at Q
0
x_Q
x_{Q_1}
$x = \int_{-\infty}^{t} v(\tau)d\tau$
$f = M(x)\,v$

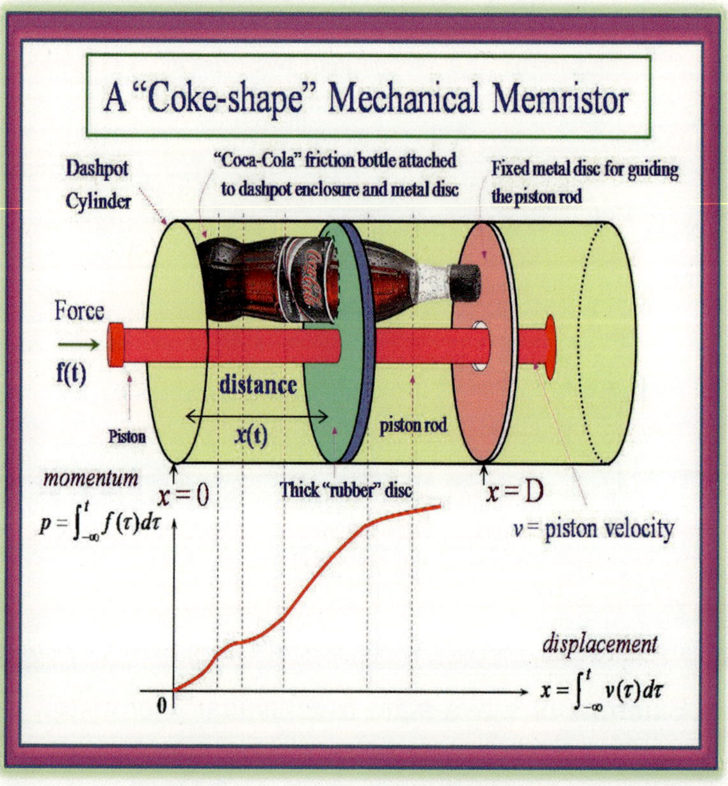
A "Coke-shape" Mechanical Memristor
Dashpot Cylinder
"Coca-Cola" friction bottle attached to dashpot enclosure and metal disc
Fixed metal disc for guiding the piston rod
Force
f(t)
distance
x(t)
Piston
piston rod
momentum
$p = \int_{-\infty}^{t} f(\tau)d\tau$
x = 0
Thick "rubber" disc
x = D
v = piston velocity
displacement
$x = \int_{-\infty}^{t} v(\tau)d\tau$
0

Our
4 Circuit-Element
Axiomatic
Approach
dates back to
Aristotle !
Aristotle, 350 BC

Our *axiomatic definitions* of the *4 elementary circuit elements* is analogous to *Aristotle's definitions* of the 4 *building blocks of matter*.

Aristotle's Theory of Matter

All *matter* consisted of *four elements*:

1. EARTH
2. WATER
3. AIR
4. FIRE

Each of these elements exhibited two of *four fundamental properties* :

- Moistness
- Dryness
- Coldness
- Hotness

Aristotle's 4 Pairs of Complementary Variables
• (Dryness, Hotness) ⇨ FIRE
• (Moistness, Hotness) ⇨ AIR
• (Moistness, Coldness) ⇨ WATER
• (Dryness, Coldness) ⇨ EARTH

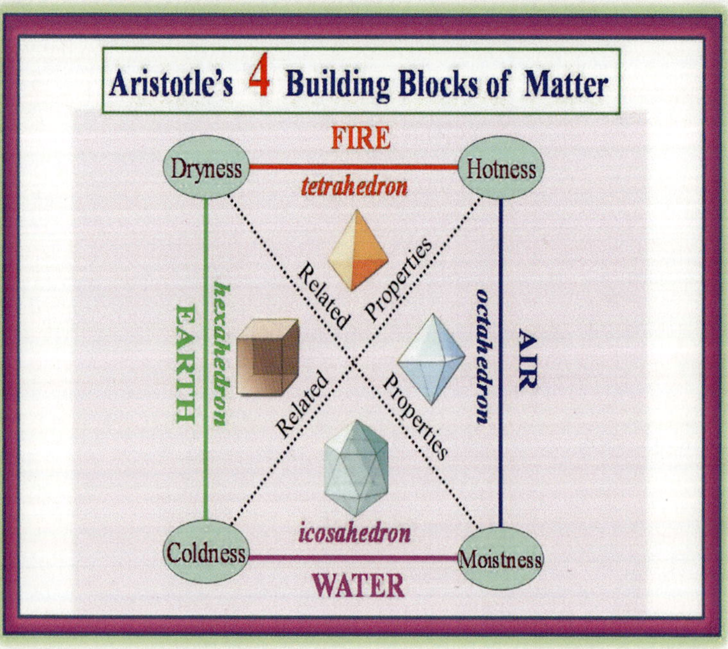
Aristotle's 4 Building Blocks of Matter
FIRE
Dryness
Hotness
tetrahedron
Related
Properties
EARTH
hexahedron
AIR
octahedron
Related
Properties
icosahedron
Coldness
Moistness
WATER

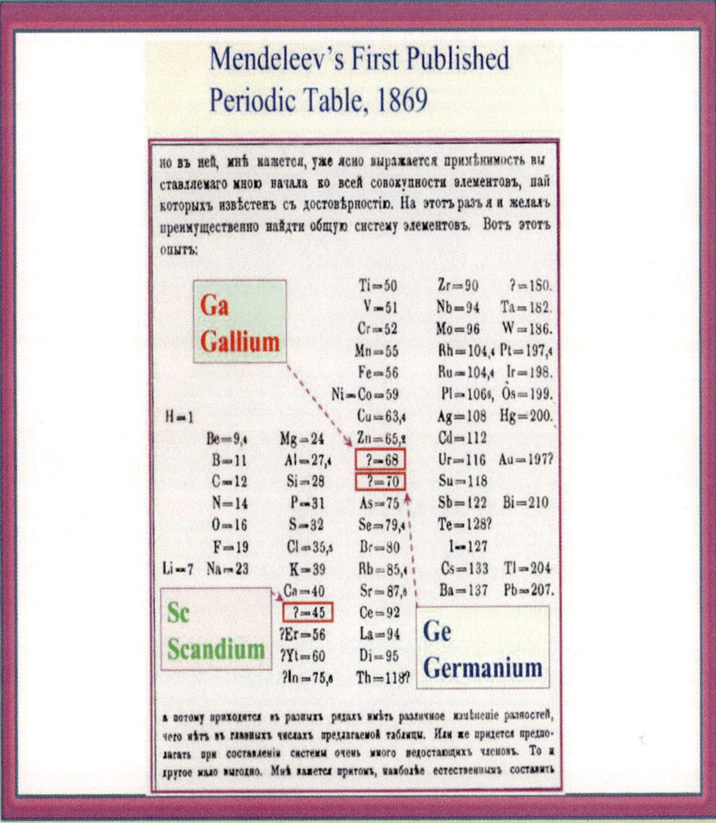

но въ ней, мнѣ кажется, уже ясно выражается примѣнимость выставляемаго мною начала ко всей совокупности элементовъ, пай которыхъ извѣстенъ съ достовѣрностію. На этотъ разъ я и желалъ преимущественно найдти общую систему элементовъ. Вотъ этотъ опытъ:

			Ti=50	Zr=90	?=180.
			V=51	Nb=94	Ta=182.
			Cr=52	Mo=96	W=186.
			Mn=55	Rh=104,4	Pt=197,4
			Fe=56	Ru=104,4	Ir=198.
			Ni=Co=59	Pl=106,6	Os=199.
H=1			Cu=63,4	Ag=108	Hg=200.
	Be=9,4	Mg=24	Zn=65,2	Cd=112	
	B=11	Al=27,4	?=68	Ur=116	Au=197?
	C=12	Si=28	?=70	Sn=118	
	N=14	P=31	As=75	Sb=122	Bi=210
	O=16	S=32	Se=79,4	Te=128?	
	F=19	Cl=35,5	Br=80	I=127	
Li=7	Na=23	K=39	Rb=85,4	Cs=133	Tl=204
		Ca=40	Sr=87,6	Ba=137	Pb=207.
		?=45	Ce=92		
		?Er=56	La=94		
		?Yt=60	Di=95		
		?In=75,6	Th=118?		

а потому приходится въ разныхъ рядахъ имѣть различное измѣненіе разностей, чего нѣтъ въ главныхъ числахъ предлагаемой таблицы. Или же придется предполагать при составленіи системы очень много недостающихъ членовъ. То и другое мало выгодно. Мнѣ кажется притомъ, наиболѣе естественнымъ составить

Dmitri Mendeleev's idea of the *periodic table* of elements originated from a *dream*.

Nature pages 42–43 **May 1, 2008**

ELECTRONICS

The fourth element

James M. Tour and Tao He

Almost four decades since its existence was first proposed, a fourth basic circuit element joins the canonical three. The 'memristor' might herald a step-change in the march towards ever more powerful circuitry.

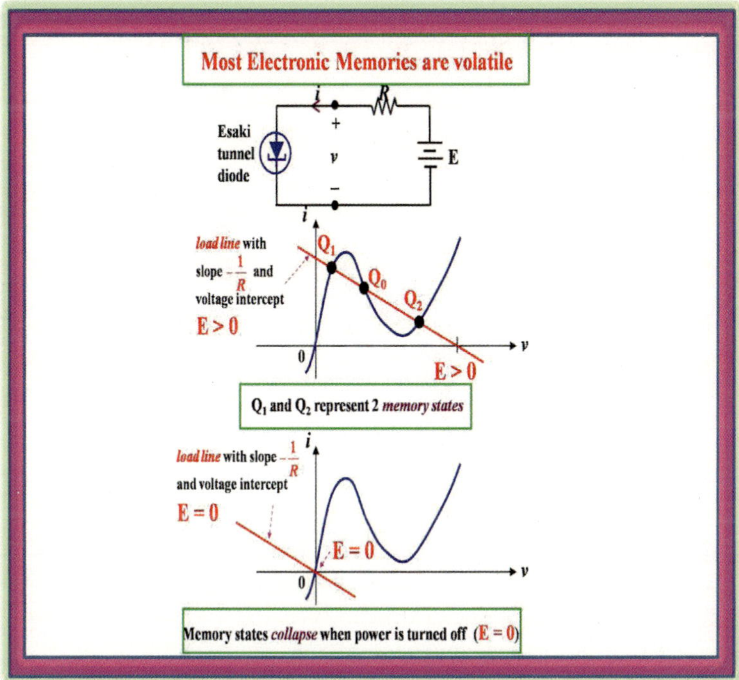
Most Electronic Memories are volatile
Esaki tunnel diode
load line with slope $-\frac{1}{R}$ and voltage intercept
E > 0
Q1 and Q2 represent 2 memory states
load line with slope $-\frac{1}{R}$ and voltage intercept
E = 0
Memory states collapse when power is turned off (E = 0)

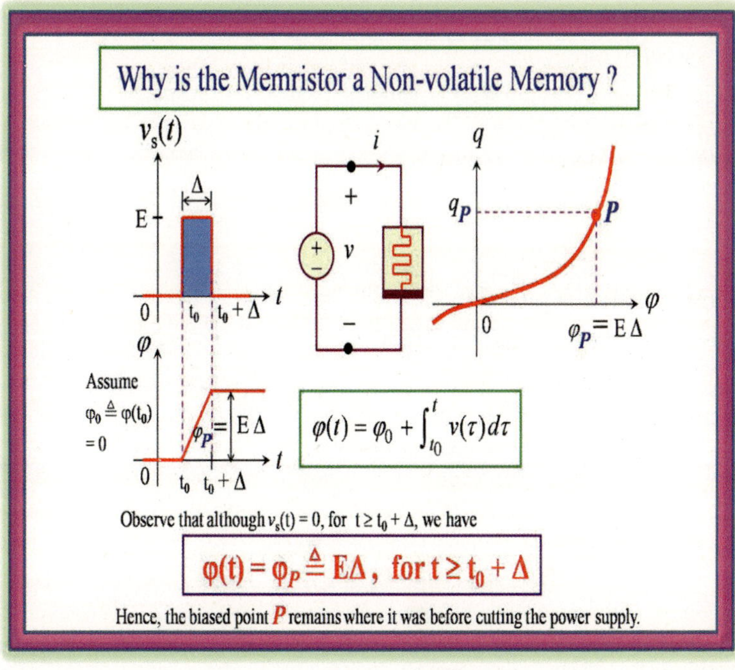
Why is the Memristor a Non-volatile Memory ?
Assume $\varphi_0 \triangleq \varphi(t_0) = 0$
$\varphi(t) = \varphi_0 + \int_{t_0}^{t} v(\tau)d\tau$
$\varphi_P = E\Delta$
Observe that although $v_s(t) = 0$, for $t \geq t_0 + \Delta$, we have
$\varphi(t) = \varphi_P \triangleq E\Delta$, for $t \geq t_0 + \Delta$
Hence, the biased point P remains where it was before cutting the power supply.

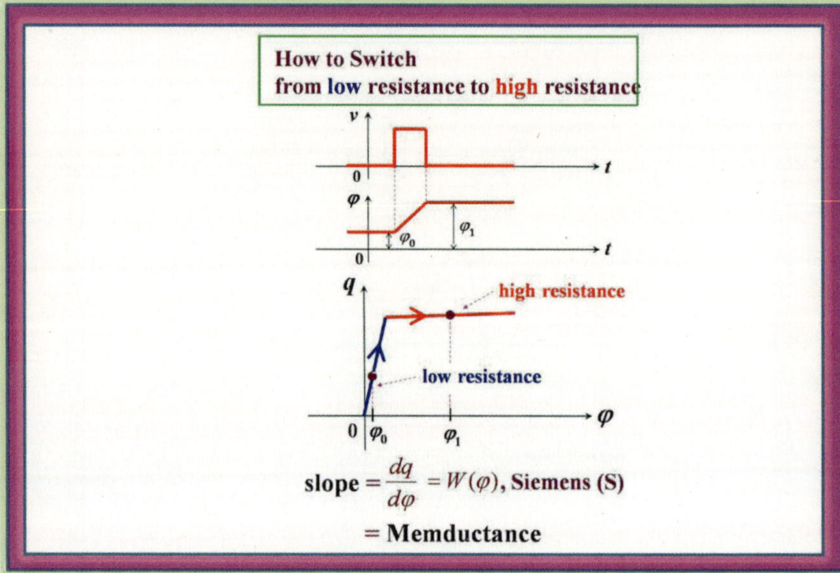
How to Switch
from low resistance to high resistance
v
t
0
φ
φ1
φ0
t
0
q
high resistance
low resistance
0
φ0
φ1
φ
slope = dq/dφ = W(φ), Siemens (S)
= Memductance

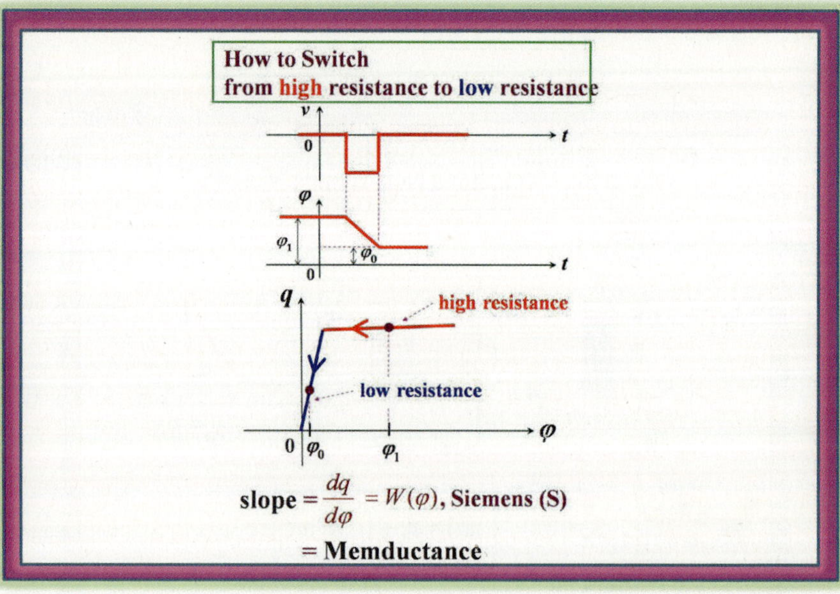
How to Switch
from high resistance to low resistance
v
0
t
φ
φ1
φ0
0
t
q
high resistance
low resistance
0
φ0
φ1
φ
slope = dq/dφ = W(φ), Siemens (S)
= Memductance

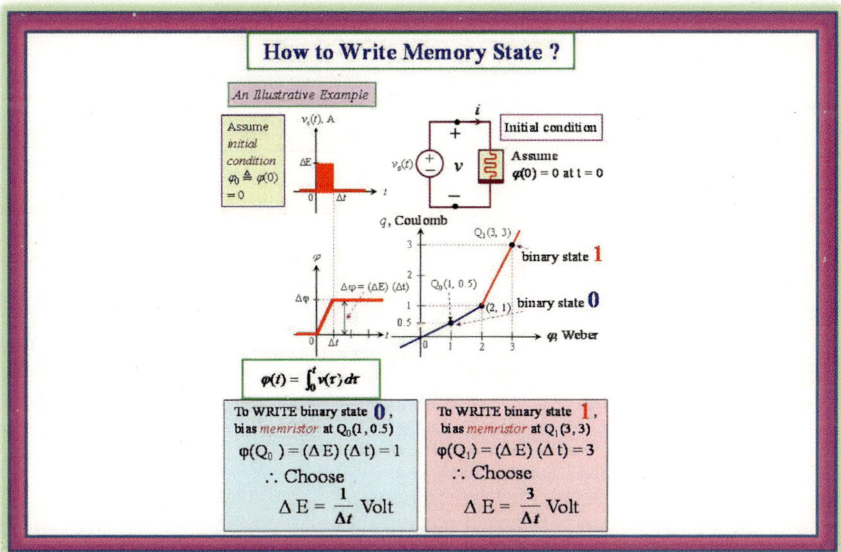
How to Write Memory State ?
An Illustrative Example
Assume initial condition $\varphi_0 \triangleq \varphi(0) = 0$
Initial condition
Assume $\varphi(0) = 0$ at $t = 0$
q, Coulomb
$Q_1(3, 3)$
binary state 1
$Q_0(1, 0.5)$
binary state 0
$\Delta\varphi = (\Delta E)(\Delta t)$
φ, Weber
$\varphi(t) = \int_0^t v(\tau)d\tau$
To WRITE binary state 0, bias memristor at $Q_0(1, 0.5)$
$\varphi(Q_0) = (\Delta E)(\Delta t) = 1$
∴ Choose
$\Delta E = \frac{1}{\Delta t}$ Volt
To WRITE binary state 1, bias memristor at $Q_1(3, 3)$
$\varphi(Q_1) = (\Delta E)(\Delta t) = 3$
∴ Choose
$\Delta E = \frac{3}{\Delta t}$ Volt

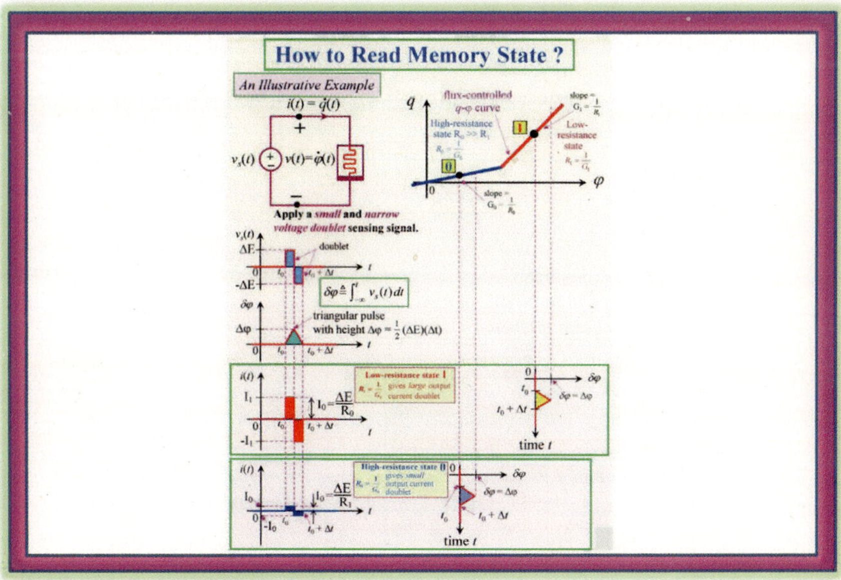
How to Read Memory State ?
An Illustrative Example
$i(t) = \dot{q}(t)$
$v_s(t)$
$v(t) = \dot{\varphi}(t)$
flux-controlled q-φ curve
High-resistance state $R_0 >> R_1$
Low-resistance state
Apply a small and narrow voltage doublet sensing signal.
doublet
$\delta\varphi \triangleq \int_{-\infty}^{t} v_s(t)dt$
triangular pulse with height $\Delta\varphi = \frac{1}{2}(\Delta E)(\Delta t)$
Low-resistance state 1
$I_0 = \frac{\Delta E}{R_0}$
time t
High-resistance state 0
$I_0 = \frac{\Delta E}{R_1}$
time t

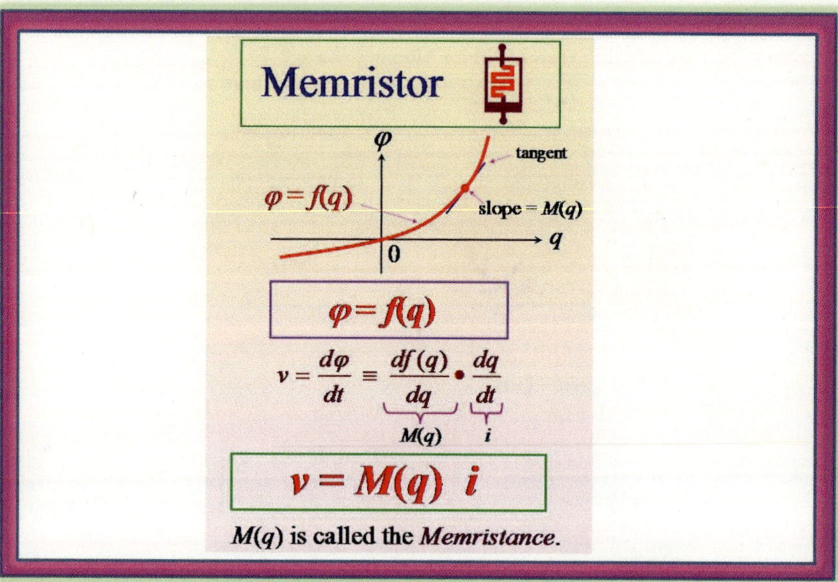
Memristor
φ
tangent
$\varphi = f(q)$
slope = $M(q)$
0
q
$\varphi = f(q)$
$v = \frac{d\varphi}{dt} \equiv \underbrace{\frac{df(q)}{dq}}_{M(q)} \bullet \underbrace{\frac{dq}{dt}}_{i}$
$v = M(q)\ i$
$M(q)$ is called the *Memristance*.

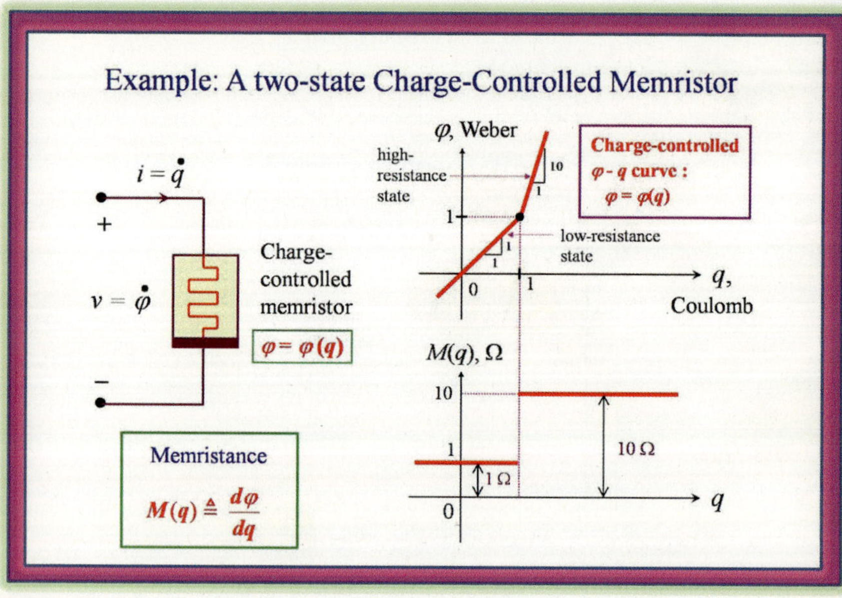
Example: A two-state Charge-Controlled Memristor
$i = \dot{q}$
+
$v = \dot{\varphi}$
−
Charge-controlled memristor
$\varphi = \varphi(q)$
Memristance
$M(q) \triangleq \frac{d\varphi}{dq}$
φ, Weber
high-resistance state
10
1
1
low-resistance state
0
1
q, Coulomb
Charge-controlled
φ - q curve :
$\varphi = \varphi(q)$
$M(q)$, Ω
10
1
1 Ω
10 Ω
0
q

Memristor is an Analog Non-volatile memory

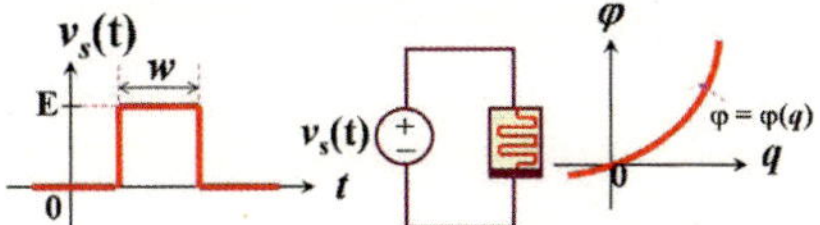

For a *memristor* with a *smooth* function $\varphi = \varphi(q)$, we can obtain a *continuous range* of *resistances*, *not* just binary states, by simply choosing the *pulse height* E, or the *pulse width* w of a biasing voltage pulse.

***Application* : Ideal for neural network learning via *synaptic* tunings. With the tiny HP memristor, which can be scaled down to 2 nanometers, it is possible to mimic biological neurons with more than 20,000 synapses per neuron.**

How do you know your device may be a Memristor ?

Examine:

$$v = M(q)\, i$$

or

$$i = G(\varphi)\, v$$

⇨

$$v(t) = 0 \quad \Leftrightarrow \quad i(t) = 0$$

Memristor Fingerprint

Both $v(t)$ and $i(t)$ of a *memristor* must have *identical zero crossings*, i.e., they must be *in phase*.

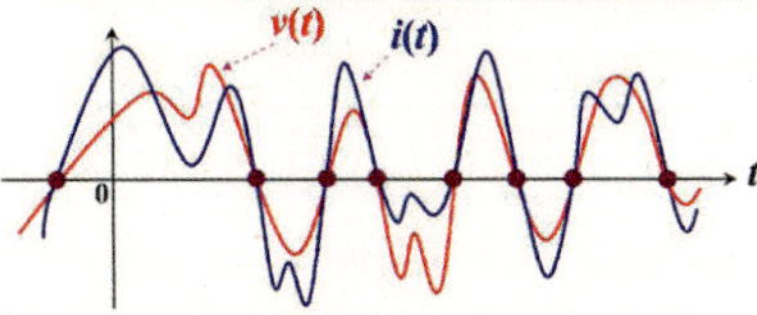

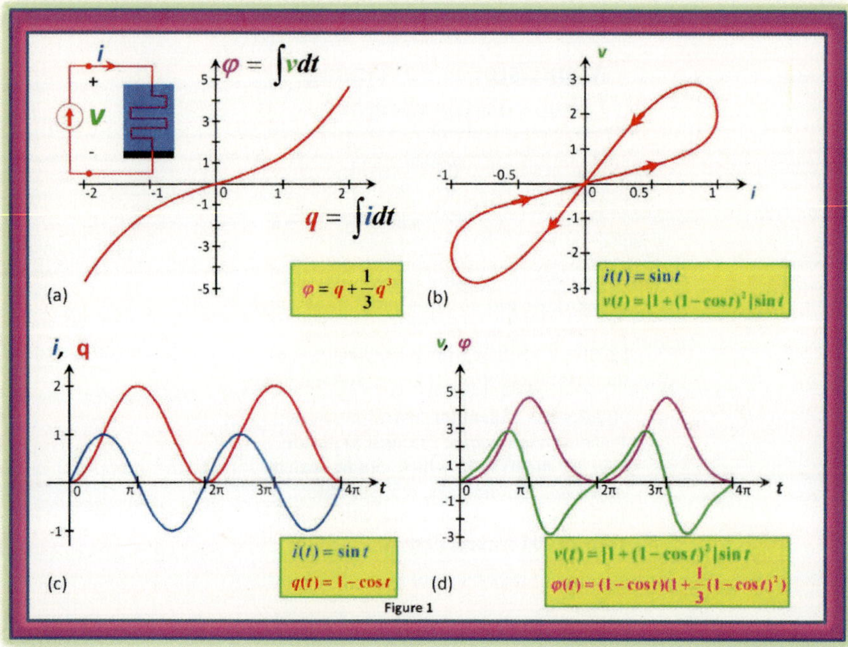

$\varphi = \int v dt$
$q = \int i dt$
$\varphi = q + \frac{1}{3}q^3$
(a)
(b)
$i(t) = \sin t$
$v(t) = [1 + (1 - \cos t)^2] \sin t$
i, q
v, φ
(c)
$i(t) = \sin t$
$q(t) = 1 - \cos t$
(d)
$v(t) = [1 + (1 - \cos t)^2] \sin t$
$\varphi(t) = (1 - \cos t)(1 + \frac{1}{3}(1 - \cos t)^2)$

Figure 1

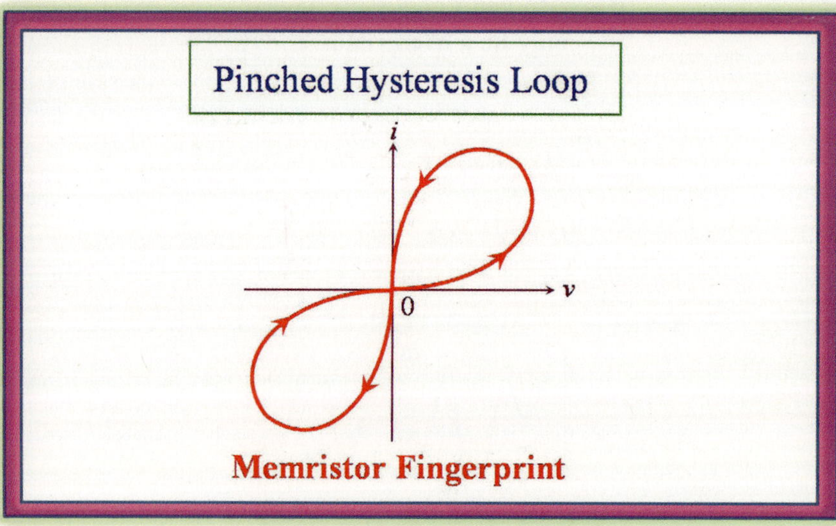

Pinched Hysteresis Loop
i
v
0
Memristor Fingerprint

Theorem

Memristor pinched hysteresis loop shrinks continuously as frequency increases.

Example

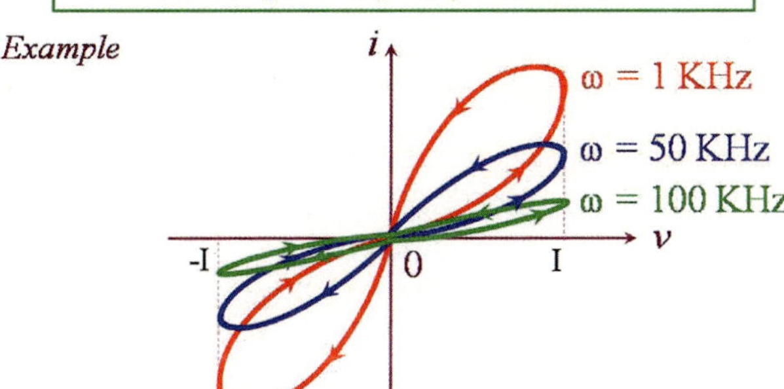

The *pinched hysteresis loop* of a memristor becomes *thinner* as frequency ω increases, and tends to a *straight line* as ω → ∞

High-Frequency Memristor Behavior

$$\varphi(t) = \varphi_0 + \int_{t_0}^{t} A\sin(\omega\tau)\,d\tau$$

$$= \varphi_0 + \frac{A}{\omega}\left(\cos\omega t - \cos\omega t_0\right) \rightarrow \varphi_0, \quad \text{as } \omega \rightarrow \infty$$

Memductance **at ω → ∞**

$$W(\varphi(t)) \rightarrow W(\varphi_0)$$

Memristance **at ω → ∞**

$$M(q(t)) \rightarrow M(q_0)$$

where

$$M(q_0) = \frac{1}{W(\varphi_0)}$$

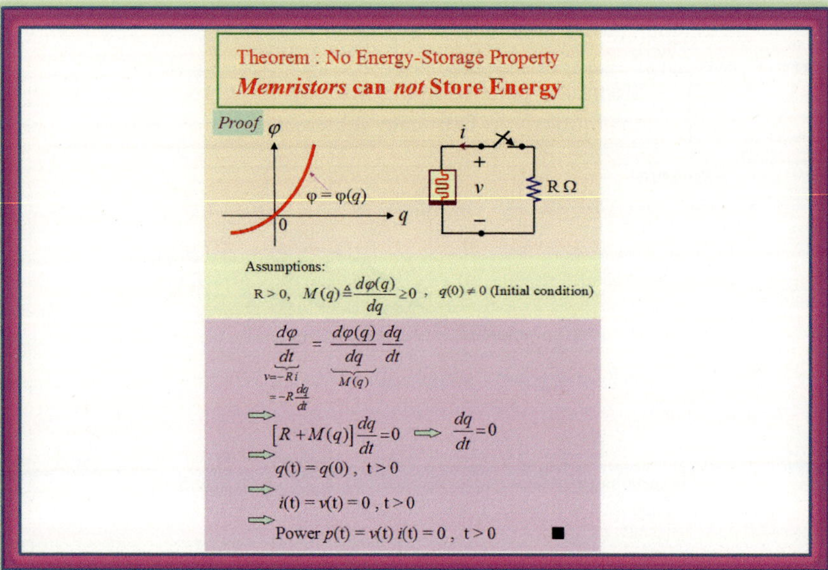
Theorem : No Energy-Storage Property
Memristors can not Store Energy
Proof
φ = φ(q)
R Ω
Assumptions:
R > 0, $M(q) \triangleq \frac{d\varphi(q)}{dq} \geq 0$, $q(0) \neq 0$ (Initial condition)
$\underbrace{\frac{d\varphi}{dt}}_{v=-Ri = -R\frac{dq}{dt}} = \underbrace{\frac{d\varphi(q)}{dq}}_{M(q)} \frac{dq}{dt}$
$[R + M(q)]\frac{dq}{dt} = 0 \Rightarrow \frac{dq}{dt} = 0$
$q(t) = q(0)$, t > 0
$i(t) = v(t) = 0$, t > 0
Power $p(t) = v(t)\, i(t) = 0$, t > 0

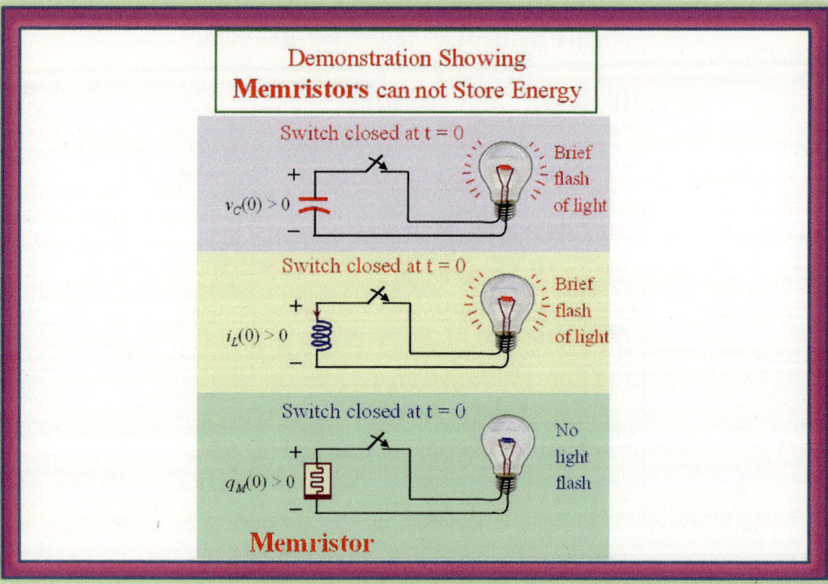
Demonstration Showing
Memristors can not Store Energy
Switch closed at t = 0
$v_C(0) > 0$
Brief flash of light
Switch closed at t = 0
$i_L(0) > 0$
Brief flash of light
Switch closed at t = 0
$q_M(0) > 0$
No light flash
Memristor

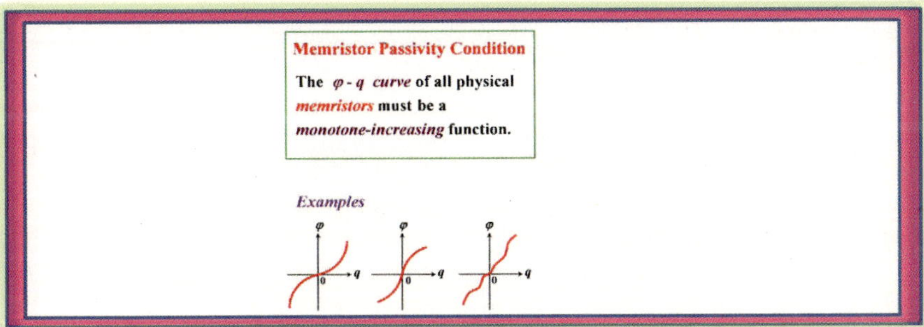

Memristor Passivity Condition
The φ - q curve of all physical memristors must be a monotone-increasing function.
Examples
φ
q
0

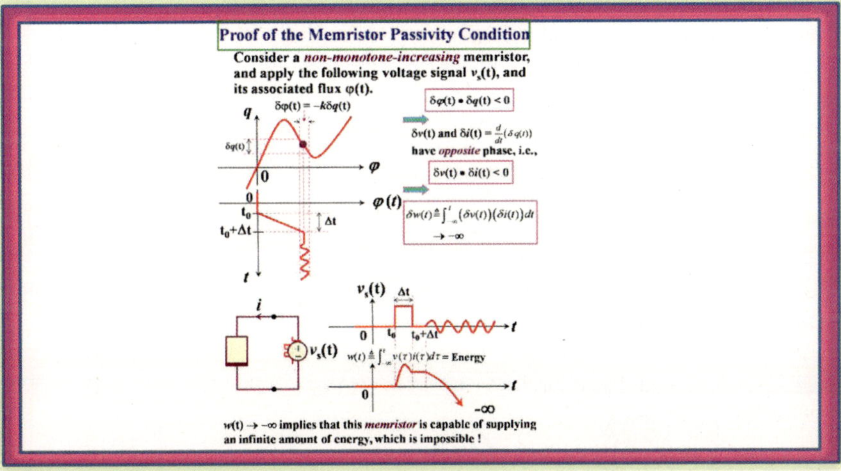

Proof of the Memristor Passivity Condition
Consider a non-monotone-increasing memristor, and apply the following voltage signal $v_s(t)$, and its associated flux φ(t).
$\delta\varphi(t) = -k\delta q(t)$
$\delta\varphi(t) \bullet \delta q(t) < 0$
$\delta v(t)$ and $\delta i(t) = \frac{d}{dt}(\delta q(t))$ have opposite phase, i.e.,
$\delta v(t) \bullet \delta i(t) < 0$
$\delta w(t) \triangleq \int_{-\infty}^{t} (\delta v(t))(\delta i(t)) dt \to -\infty$
q
δq(t)
φ
0
φ(t)
t_0
$t_0+\Delta t$
Δt
t
i
$v_s(t)$
t_0
$w(t) \triangleq \int_{-\infty}^{t} v(\tau) i(\tau) d\tau$ = Energy
-∞
$w(t) \to -\infty$ implies that this memristor is capable of supplying an infinite amount of energy, which is impossible !

Is it possible to build a *passive solid state memristor* ?

Answer :

Yes, provided the *memristor* satisfies the *memristor passivity condition*.

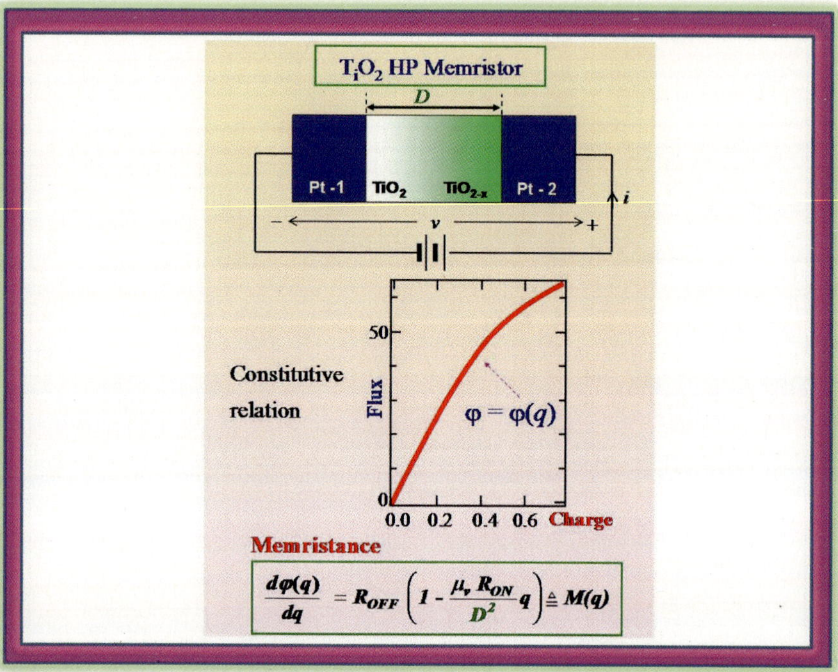
T_iO_2 HP Memristor
D
Pt - 1
TiO_2
TiO_{2-x}
Pt - 2
i
v
Constitutive relation
Flux
50
0
0.0 0.2 0.4 0.6 Charge
$\varphi = \varphi(q)$
Memristance
$\frac{d\varphi(q)}{dq} = R_{OFF}\left(1 - \frac{\mu_v R_{ON}}{D^2} q\right) \triangleq M(q)$

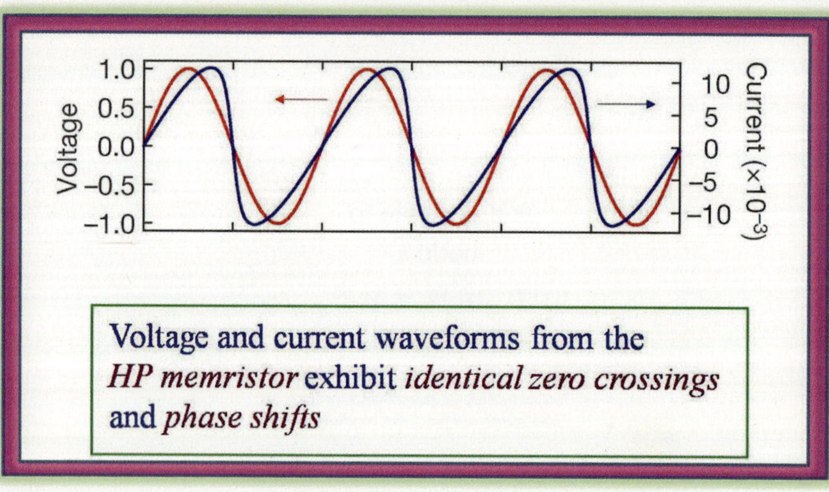
Voltage
1.0
0.5
0.0
−0.5
−1.0
Current (×10^{-3})
10
5
0
−5
−10
Voltage and current waveforms from the *HP memristor* exhibit *identical zero crossings* and *phase shifts*

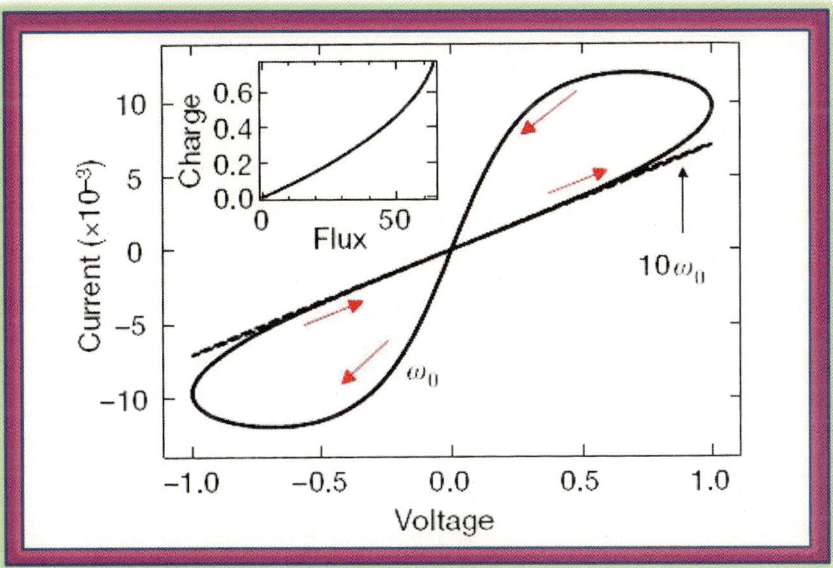

Memristor Model of an Experimental Nano Device

The following slide shows a nano device, reported from Professor Lieber's Harvard Nano-device Laboratory, whose experimentally measured ***v-i characteristic*** is a ***pinched hysteresis loop***. The conductance of the device can be switched from "0" nS (*off* state), to 800 nS (*on* state), by applying a ***square wave***.

Professor Lieber had confirmed (private communication) that the loop "shrinks" with increasing frequency. This device can therefore be modeled as a ***memristor***.

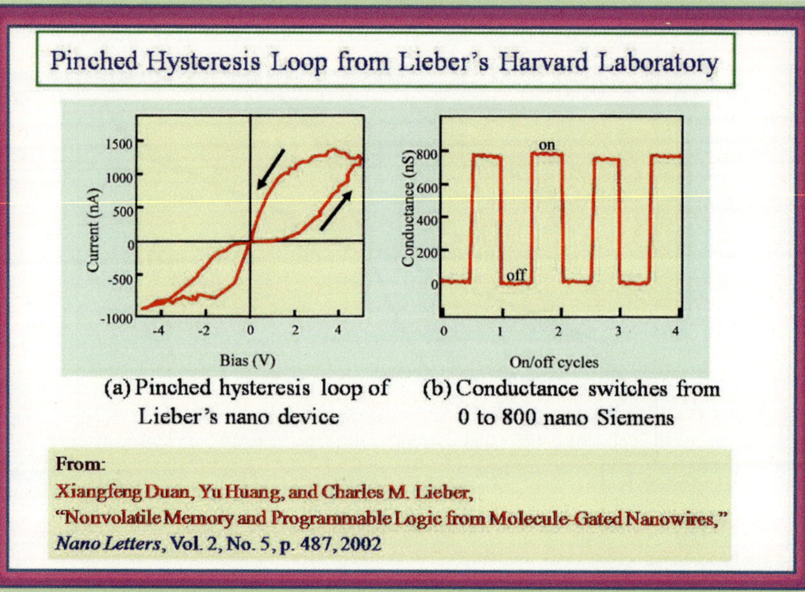
Pinched Hysteresis Loop from Lieber's Harvard Laboratory
Current (nA)
1500
1000
500
0
-500
-1000
-4
-2
0
2
4
Bias (V)
Conductance (nS)
800
600
400
200
0
on
off
0
1
2
3
4
On/off cycles
(a) Pinched hysteresis loop of Lieber's nano device
(b) Conductance switches from 0 to 800 nano Siemens
From:
Xiangfeng Duan, Yu Huang, and Charles M. Lieber,
"Nonvolatile Memory and Programmable Logic from Molecule-Gated Nanowires,"
Nano Letters, Vol. 2, No. 5, p. 487, 2002

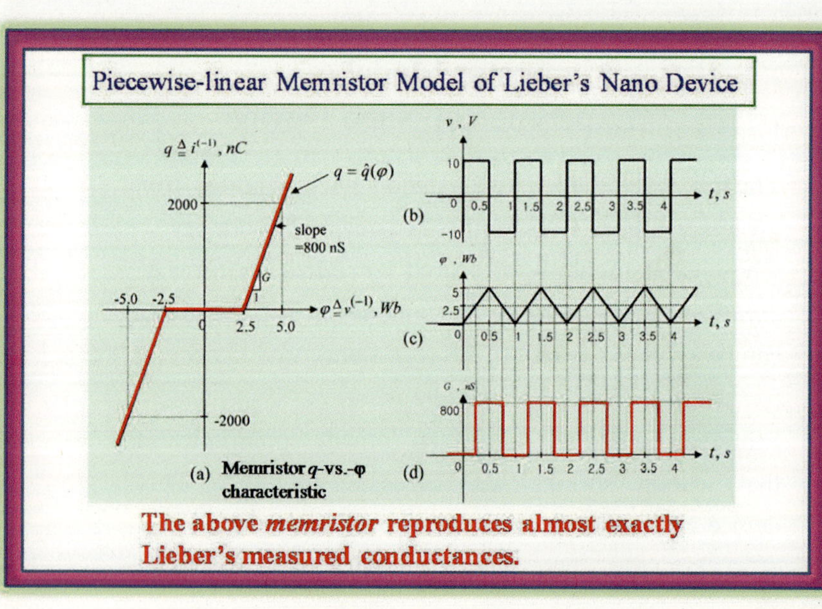
Piecewise-linear Memristor Model of Lieber's Nano Device
q ≜ i^(-1), nC
q = q̂(φ)
2000
slope
=800 nS
G
1
-5.0
-2.5
0
2.5
5.0
φ ≜ v^(-1), Wb
-2000
(a) Memristor q-vs.-φ characteristic
v, V
10
0
-10
0.5
1
1.5
2
2.5
3
3.5
4
t, s
(b)
φ, Wb
5
2.5
(c)
G, nS
800
(d)
The above memristor reproduces almost exactly Lieber's measured conductances.

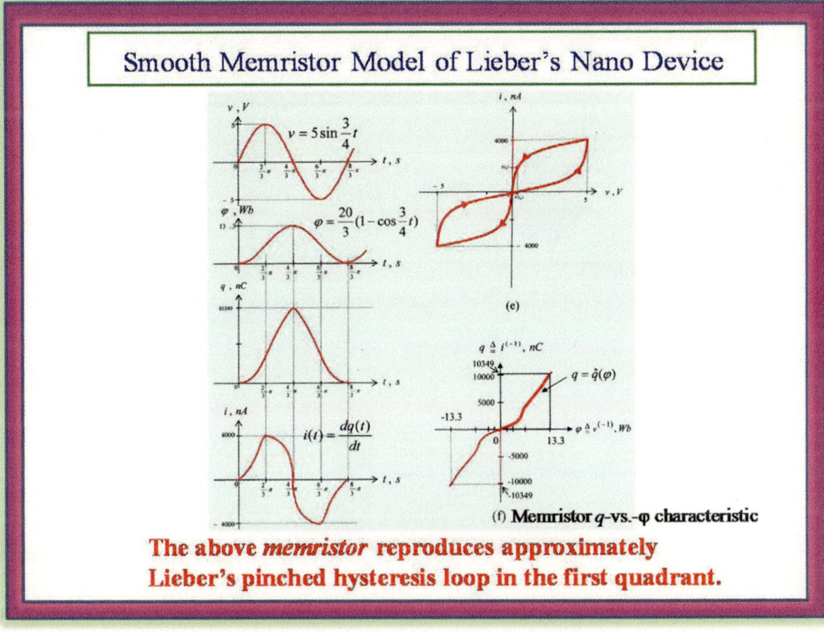
Smooth Memristor Model of Lieber's Nano Device
$v = 5\sin\frac{3}{4}t$
$\varphi = \frac{20}{3}(1-\cos\frac{3}{4}t)$
$i(t) = \frac{dq(t)}{dt}$
(e)
$q = \hat{q}(\varphi)$
(f) Memristor q-vs.-φ characteristic
The above *memristor* reproduces approximately
Lieber's pinched hysteresis loop in the first quadrant.

Ideal and Generalized Memristors

$$v = M(q)i$$
$$\frac{dq}{dt} = i$$

Ideal Memristor
($x = q =$ charge)

1. Change "charge" q to "n" state variables $x = (x_1, x_2, \ldots, x_n)$ and current i:
$$M(q) \rightarrow M(x, i)$$
2. Generalize to a nonlinear differential equation:
$$i \rightarrow f(x, i)$$

$$v = M(x, i)i$$
$$\frac{dx}{dt} = f(x, i)$$

Memristor

What is a Memristor ?

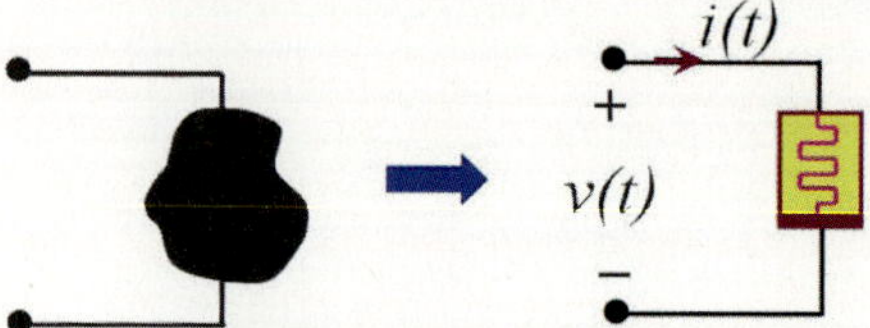

1. It has 2 electrical terminals
2. It obeys *Ohm's law*

$$v = R\,i$$

3. R is not a constant:

$$R = g(x_1, x_2, \ldots x_n; i)$$
$$\dot{x}_1 = f_1(x_1, x_2, \ldots x_n; i)$$
$$\dot{x}_2 = f_2(x_1, x_2, \ldots x_n; i)$$
$$\vdots \qquad \vdots$$
$$\dot{x}_n = f_n(x_1, x_2, \ldots x_n; i)$$

Physical Definition of Memristor

Any 2-terminal device defined by a *state-dependent* Ohm's Law is a *memristor*.

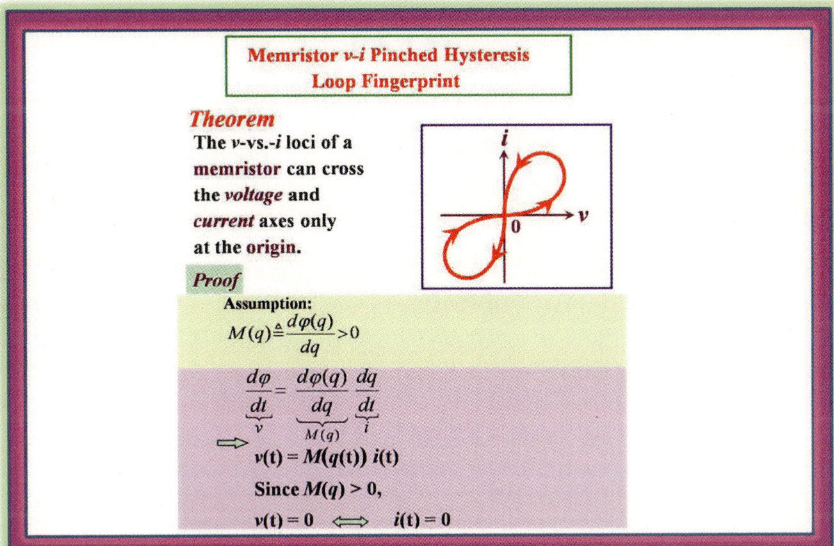
Memristor v-i Pinched Hysteresis Loop Fingerprint
Theorem
The v-vs.-i loci of a memristor can cross the voltage and current axes only at the origin.
i
v
0
Proof
Assumption:
$M(q) \triangleq \frac{d\varphi(q)}{dq} > 0$
$\underbrace{\frac{d\varphi}{dt}}_{v} = \underbrace{\frac{d\varphi(q)}{dq}}_{M(q)} \underbrace{\frac{dq}{dt}}_{i}$
$\Rightarrow v(t) = M(q(t))\, i(t)$
Since $M(q) > 0$,
$v(t) = 0 \iff i(t) = 0$

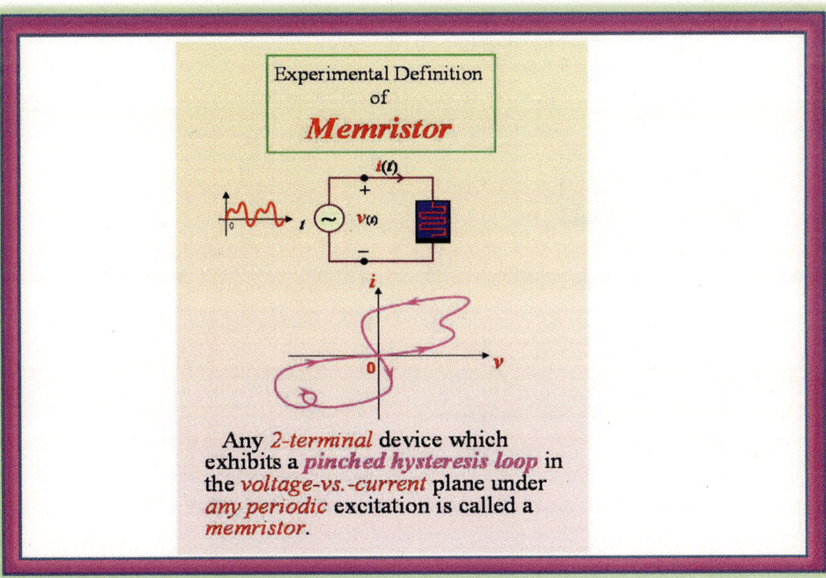
Experimental Definition of Memristor
i(t)
+
v(t)
−
t
0
i
v
0
Any 2-terminal device which exhibits a pinched hysteresis loop in the voltage-vs.-current plane under any periodic excitation is called a memristor.

Concise Definition of Memristor

A memristor is any *state-dependent linear 2-terminal* resistor.

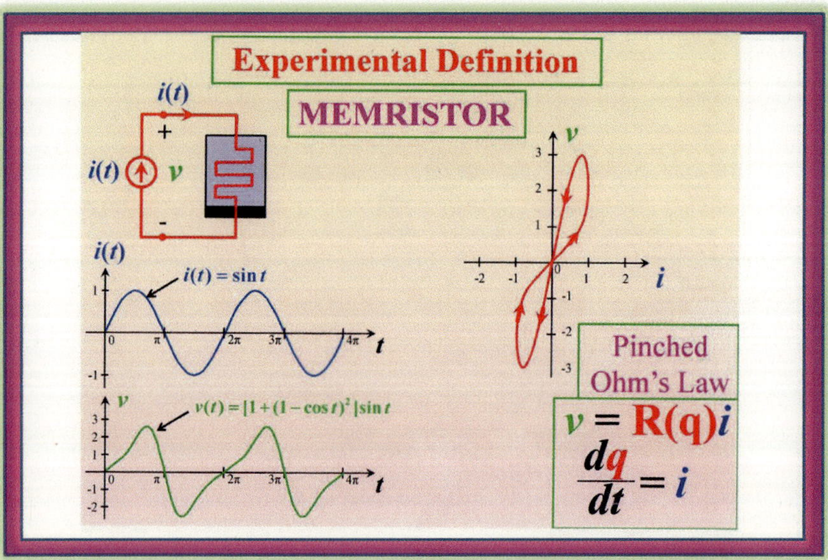

Experimental Test for Memristors

Any 2-terminal device exhibiting the following fingerprint characteristics is a ***memristor***:

1. The **Lissajoux figure** in the voltage-current plane is a ***pinched hysteresis loop*** when driven by ***any*** bipolar periodic voltage $v(t)$, or current $i(t)$, and under ***any*** initial conditions.

2. The area of each lobe of the ***pinched hysteresis loop*** shrinks as the frequency ω of the forcing signal increases.

3. As the frequency ω tends to infinity, the ***pinched hysteresis loop*** degenerates to a straight line through the origin, whose slope depends on the ***amplitude*** and ***shape*** of the forcing signal.

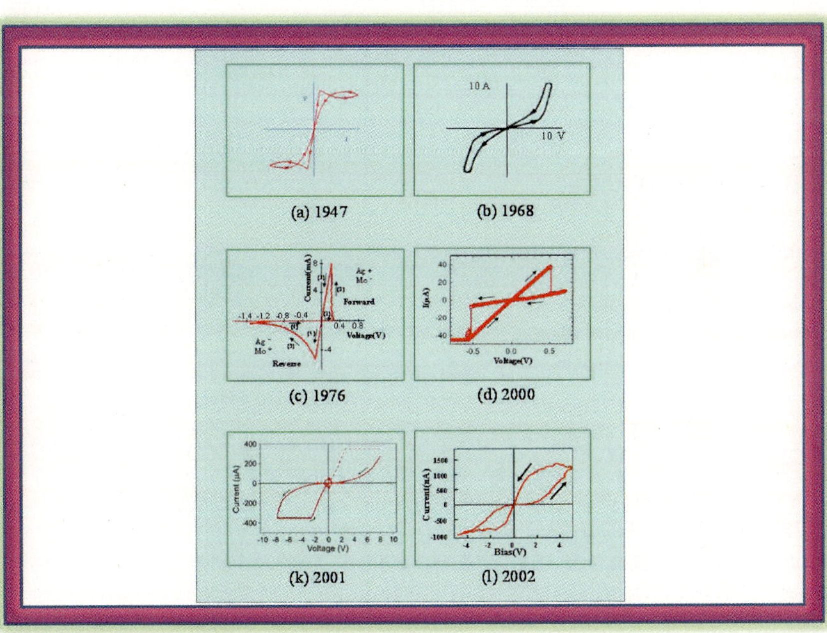

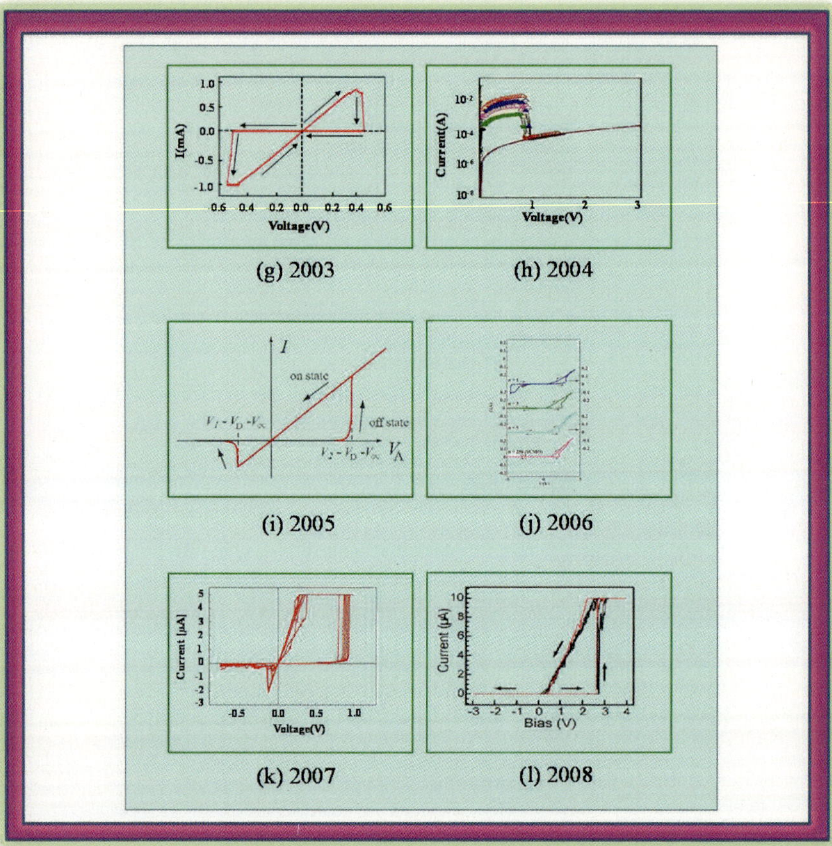

If it's pinched,
It's a memristor

Resistance Switching

implies

Memristor

RRAM

and

Phase Change

Memory

are

Memristors

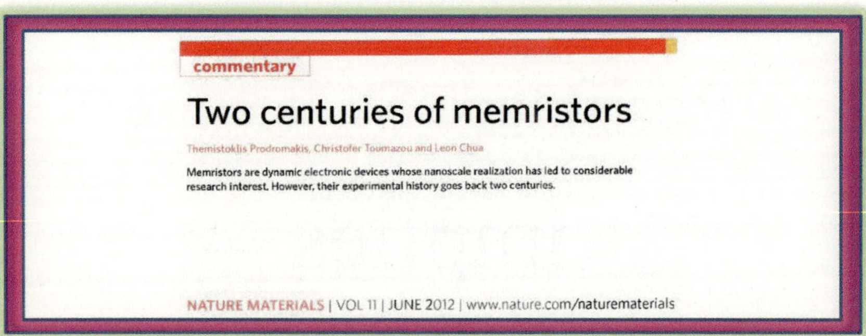
commentary

Two centuries of memristors

Themistoklis Prodromakis, Christofer Toumazou and Leon Chua

Memristors are dynamic electronic devices whose nanoscale realization has led to considerable research interest. However, their experimental history goes back two centuries.

NATURE MATERIALS | VOL 11 | JUNE 2012 | www.nature.com/naturematerials

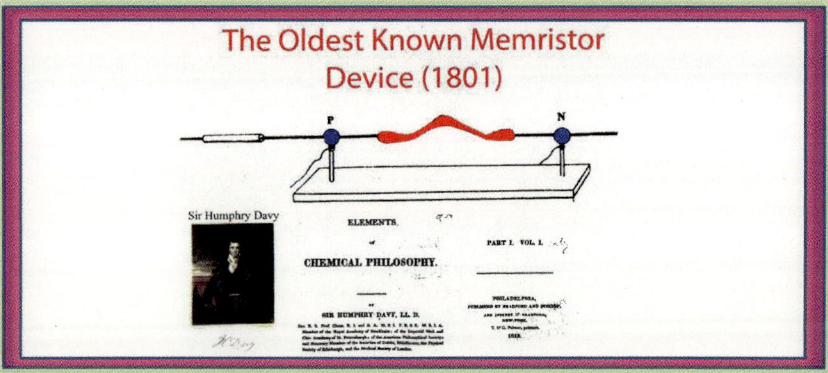

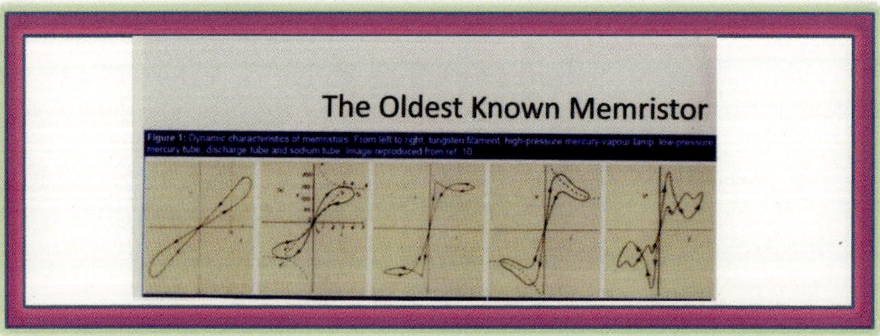

Breaking News!

A memristor has been built with only one molecule.

July 3, 2012

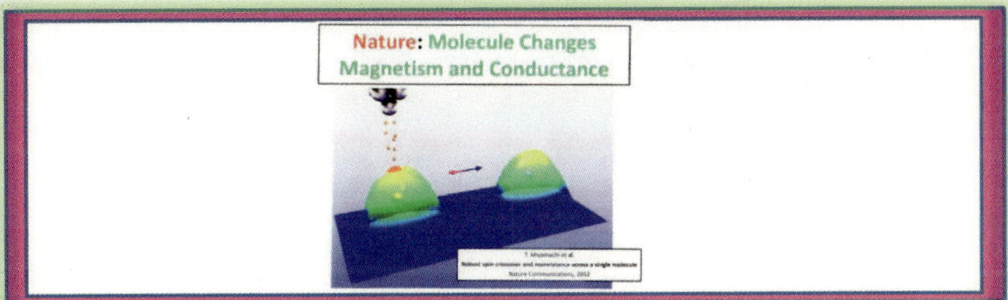

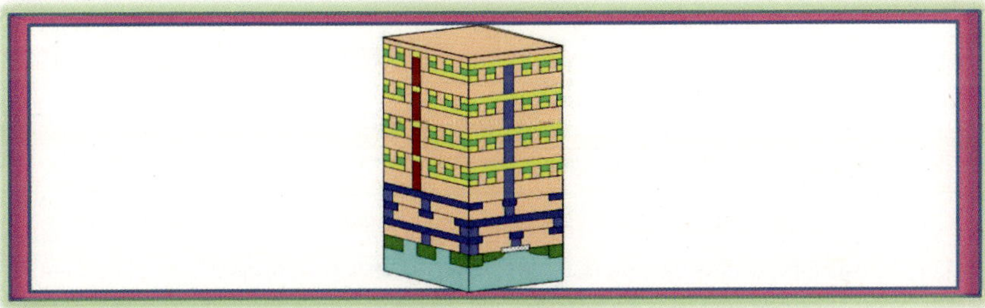

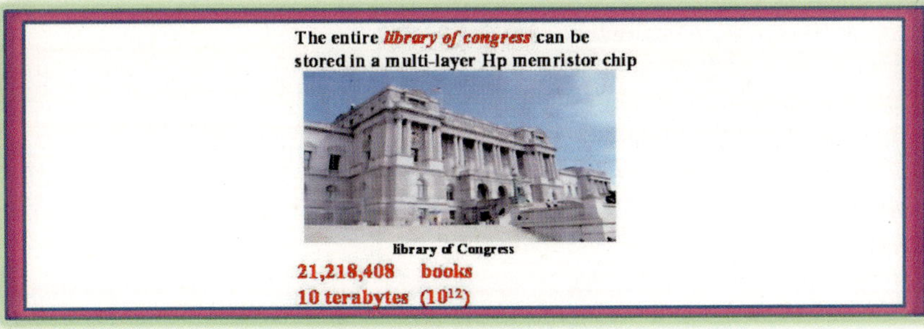

Nature: Molecule Changes
Magnetism and Conductance

The entire library of congress can be
stored in a multi-layer Hp memristor chip
library of Congress
21,218,408 books
10 terabytes (10^12)

Brian Josephson
1973 Nobel Prize for Physics for discovering the Josephson junction device and circuit model.

Modeling Quasi-Particle Pair Interference Current in Josephson Junctions

In the quantum-mechanical analysis of the Josephson junction, a small contribution to the device current is derived by Josephson to be given by

$$i = M(\cos \varphi)\, v$$

where M is a constant which depends on the device parameters. This equation, which is usually negligible, represents a **memristor** defined by

$$q = M \sin \varphi$$

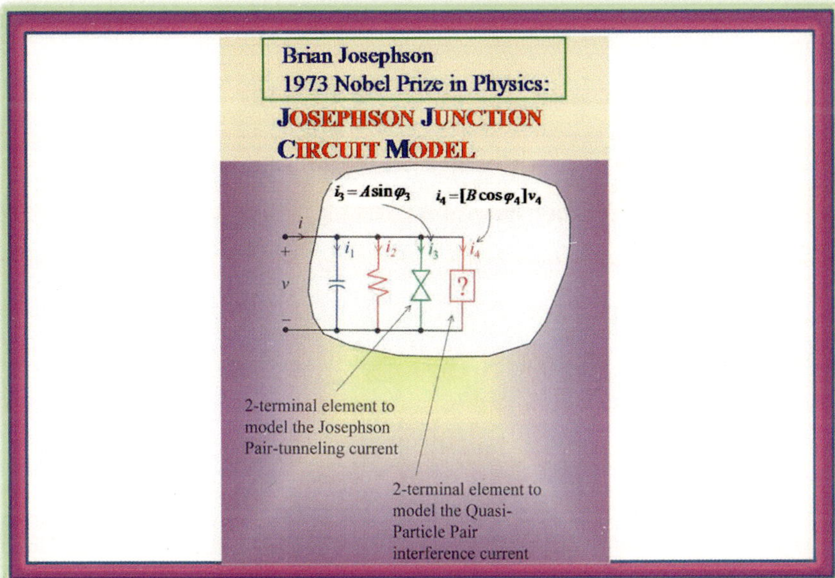
Brian Josephson
1973 Nobel Prize in Physics:
JOSEPHSON JUNCTION
CIRCUIT MODEL
$i_3 = A \sin \varphi_3$
$i_4 = [B \cos \varphi_4] v_4$
2-terminal element to
model the Josephson
Pair-tunneling current
2-terminal element to
model the Quasi-
Particle Pair
interference current

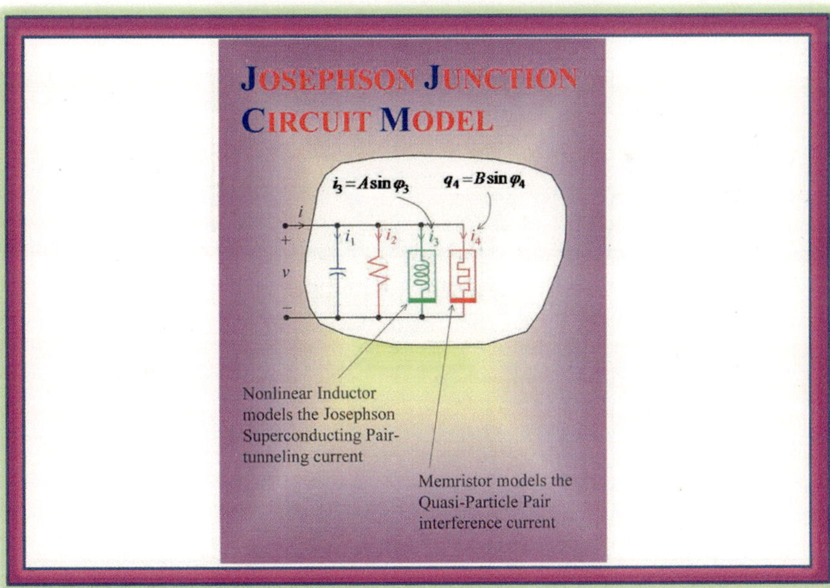
JOSEPHSON JUNCTION
CIRCUIT MODEL
$i_3 = A \sin \varphi_3$
$q_4 = B \sin \varphi_4$
Nonlinear Inductor
models the Josephson
Superconducting Pair-
tunneling current
Memristor models the
Quasi-Particle Pair
interference current

How to Make a Memristor ?

A memristor with *any* φ-vs.-q characteristic can be synthesized via a mutator and a nonlinear resistor.

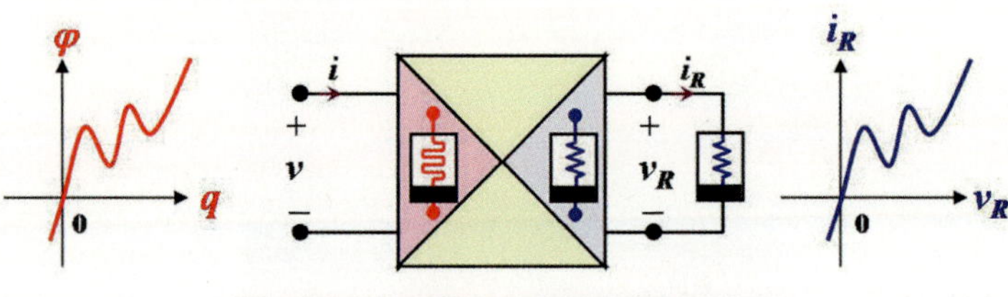

Resistor-to-Memristor Mutator

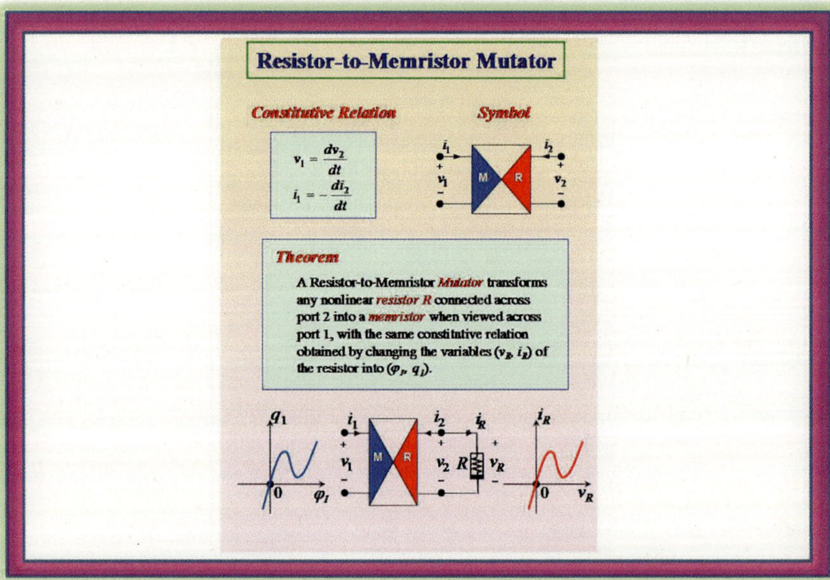

Resistor-to-Memristor Mutator
Constitutive Relation
$v_1 = \frac{dv_2}{dt}$
$i_1 = -\frac{di_2}{dt}$
Symbol
i_1
v_1
M
R
i_2
v_2
Theorem
A Resistor-to-Memristor Mutator transforms any nonlinear resistor R connected across port 2 into a memristor when viewed across port 1, with the same constitutive relation obtained by changing the variables (v_R, i_R) of the resistor into (φ_1, q_1).
q_1
0
φ_1
i_1
v_1
i_2
v_2
R
i_R
v_R
0

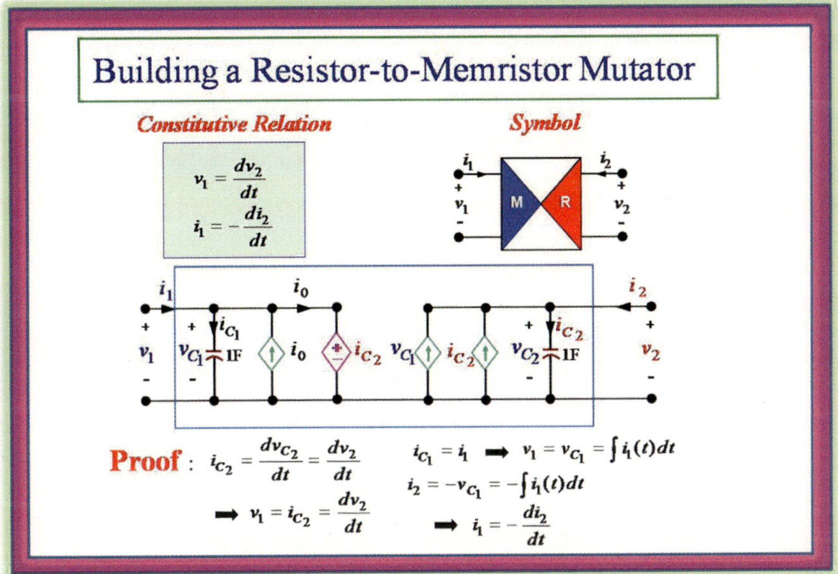
Building a Resistor-to-Memristor Mutator
Constitutive Relation
$v_1 = \frac{dv_2}{dt}$
$i_1 = -\frac{di_2}{dt}$
Symbol
M
R
1F
1F
Proof : $i_{C_2} = \frac{dv_{C_2}}{dt} = \frac{dv_2}{dt}$
$\Rightarrow v_1 = i_{C_2} = \frac{dv_2}{dt}$
$i_{C_1} = i_1 \Rightarrow v_1 = v_{C_1} = \int i_1(t)dt$
$i_2 = -v_{C_1} = -\int i_1(t)dt$
$\Rightarrow i_1 = -\frac{di_2}{dt}$

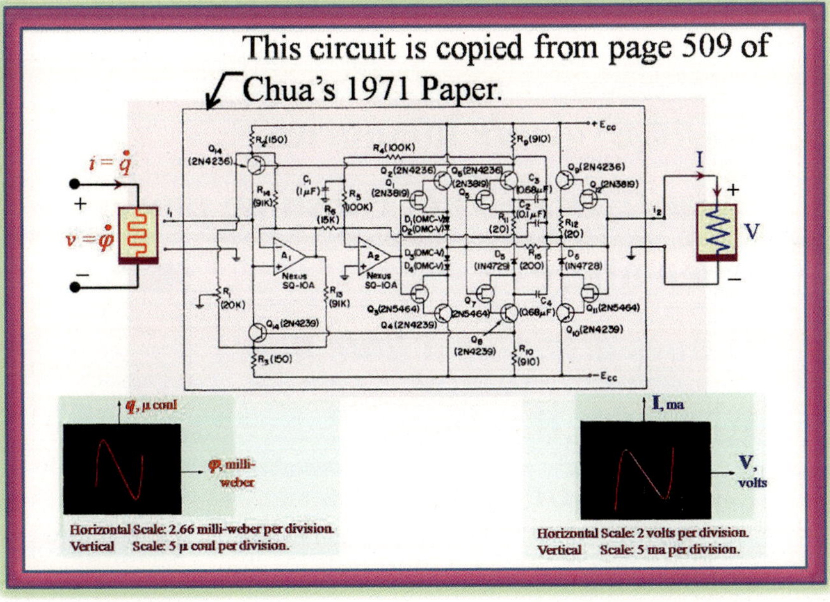
This circuit is copied from page 509 of
Chua's 1971 Paper.
$i = \dot{q}$
$v = \dot{\varphi}$
I
V
q, μ coul
φ, milli-weber
Horizontal Scale: 2.66 milli-weber per division.
Vertical Scale: 5 μ coul per division.
I, ma
V, volts
Horizontal Scale: 2 volts per division.
Vertical Scale: 5 ma per division.

What's Next ?

Memristors are not Lossless

As *non-volatile* memories, *memristors* do not consume power when idle.

It does *dissipate a little heat* whenever it is being "*written*" or "*read*"

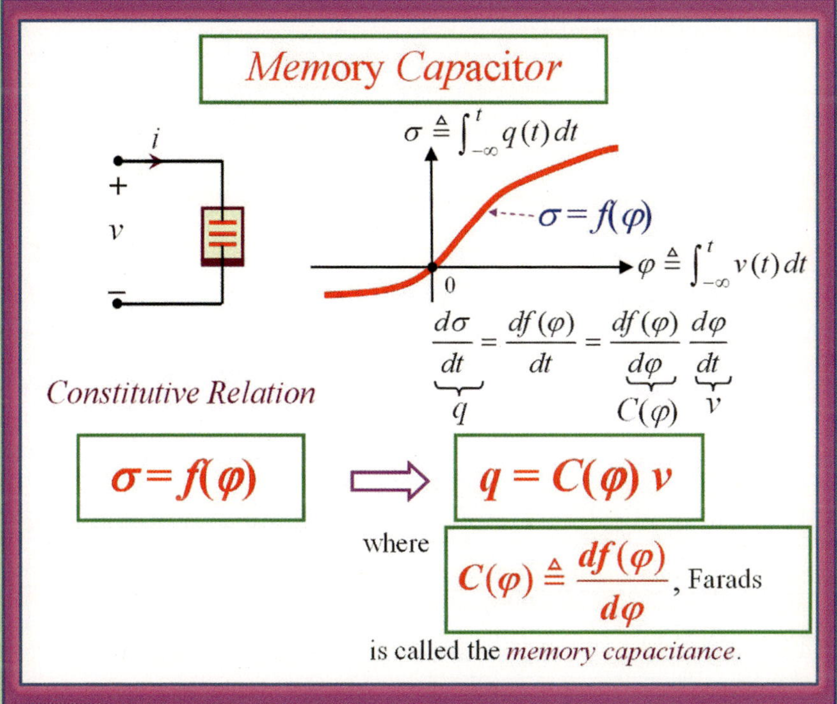

Memory Capacitor
$\sigma \triangleq \int_{-\infty}^{t} q(t)\,dt$
i
+
v
−
$\sigma = f(\varphi)$
$\varphi \triangleq \int_{-\infty}^{t} v(t)\,dt$
0
$\underbrace{\frac{d\sigma}{dt}}_{q} = \frac{df(\varphi)}{dt} = \underbrace{\frac{df(\varphi)}{d\varphi}}_{C(\varphi)} \underbrace{\frac{d\varphi}{dt}}_{v}$
Constitutive Relation
$\sigma = f(\varphi)$
$q = C(\varphi)\,v$
where
$C(\varphi) \triangleq \frac{df(\varphi)}{d\varphi}$, Farads
is called the memory capacitance.

Theorem

The *memory capacitor*

is a *lossless* element

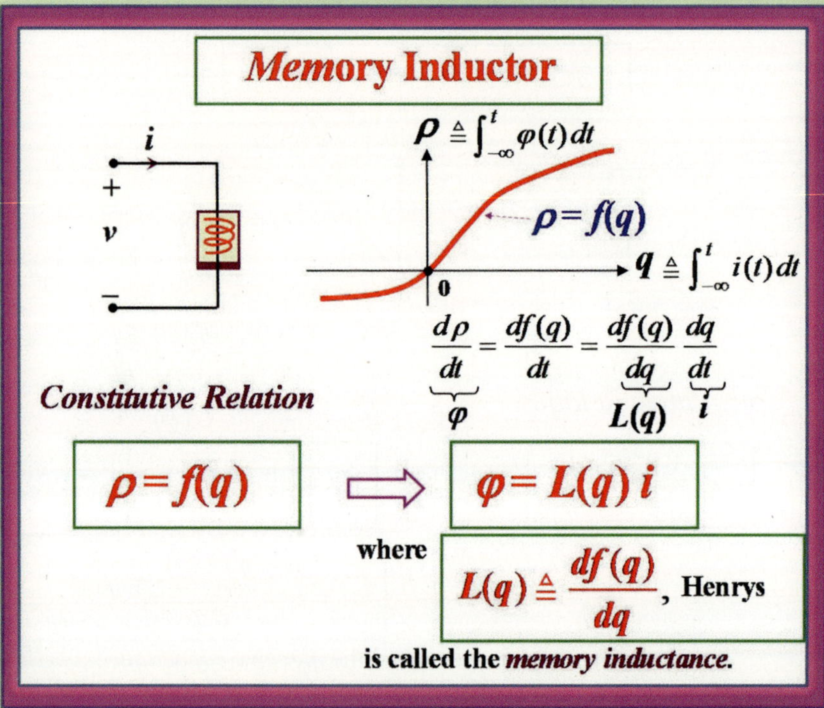

Memory Inductor
i
+
v
−
$\rho \triangleq \int_{-\infty}^{t} \varphi(t)\,dt$
$\rho = f(q)$
$q \triangleq \int_{-\infty}^{t} i(t)\,dt$
0
$\underbrace{\frac{d\rho}{dt}}_{\varphi} = \frac{df(q)}{dt} = \underbrace{\frac{df(q)}{dq}}_{L(q)} \underbrace{\frac{dq}{dt}}_{i}$
Constitutive Relation
$\rho = f(q)$
$\varphi = L(q)\, i$
where
$L(q) \triangleq \frac{df(q)}{dq}$, Henrys
is called the memory inductance.

Theorem

The *memory inductor*

is a *lossless* element

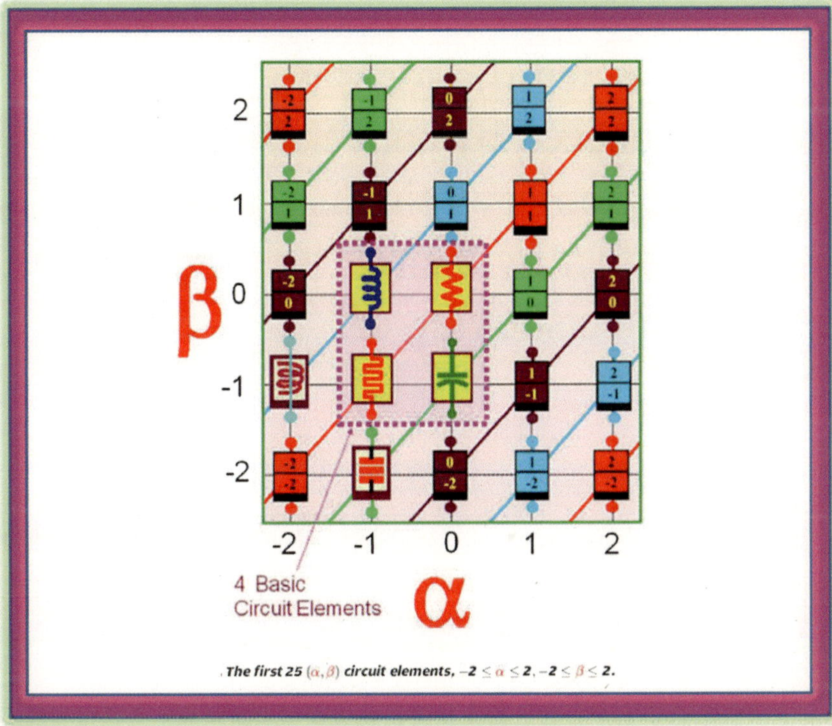

The first 25 (α, β) circuit elements, $-2 \leq \alpha \leq 2, -2 \leq \beta \leq 2$.

Frequently Asked Question

Why did it
take so long
for your ***memristor***
theory to be
validated ?

New Scientific ideas do not succeed by converting contemporary scientists, but rather by their opponents' dying off

Max Planck

Max Planck

New theories have four stages of acceptances:

J. B. S. Haldane

I. this is worthless nonsense;

II. this is interesting, but perverse;

III. this is true, but quite unimportant;

IV. I always said so.

J. B. S. Haldane

Part II
Theory, Modeling, and Simulation

Chapter 3
The Art and Science of Constructing a Memristor Model

R. Stanley Williams and Matthew D. Pickett

3.1 Introduction

The methods utilized here for constructing a useful model of a memristor [6, 7] are adapted from those described by Prof. Leon O. Chua [8, 9] to illustrate the more general problem of device modeling for any nonlinear circuit elements. The basic reason why creating a device model is so difficult is that one is essentially trying to solve an inverse problem in a complex nonlinear system. Experience has shown that the total time required for analyzing a particular system theoretically to capture the essential device physics, collecting a substantial and robust electrical data set, testing various hypotheses and finally iterating to a model that reproduces the measured nonlinear behavior of a memristor to satisfactory accuracy generally requires more than one year of effort for a small research team. This laborious task is rewarded by having a mathematical representation that provides significant intuition into how the device works the way it does, reveals the most important physical processes that determine its behavior, and can be used in computer programs such as SPICE to successfully design and numerically simulate complex integrated circuits that utilize the device.

A nonlinear device model [8, 9] consists of a set of ideal circuit elements appropriately connected together that can replicate the experimentally measured electrical properties of the physical device to a desired accuracy. The ideal elements are defined mathematically and may include nonlinear algebraic, ordinary differential, partial differential and integral equations. These can be thought of as providing the basis set of solutions for the device model and should be chosen from carefully defined relations to be as complete and relevant as possible. For nonlinear devices, Chua has constructed a *Circuit-Element-Array* [8], also called a

R.S. Williams (✉) • M.D. Pickett
Hewlett-Packard Laboratories, Palo Alto, CA, USA
e-mail: stan.williams@hp.com; matthew.pickett@hp.com

R. Tetzlaff (ed.), *Memristors and Memristive Systems*,
DOI 10.1007/978-1-4614-9068-5_3,

doubly periodic table of ideal circuit elements [10], that covers the *mathematically possible* relationships between measurements of current (i) and voltage (v) in a two-terminal device. He has also proven that these elements are independent (no individual element can be emulated by any combination of other two-terminal element types) and any combination of two or more elements of the same type belongs to that type (element closure). A critical realization is that there is no perfect model for any physical device [8]—all models are approximations, and in fact a particular physical device may be described by different models depending on the operating environment. The best guide is parsimony with iteration; find the simplest model that produces satisfactory results, but modify it if new measurements are not predicted with sufficient accuracy.

Chua has described a set of five qualities that a realistic device model should possess [8]. It must be *well-posed*, or in other words it should not yield any nonphysical solutions, such as a voltage or current that tends to infinity in a finite time. It must have *simulation capability*, i.e. a computational solution of the model should yield results that are judged to be close enough to experimentally measured data for the physical device. The model should have *qualitative similarity* to the device, especially with respect to any limiting or asymptotic behaviors that are observed. A truly useful model must have *predictive ability*, in that simulations of operating environments for which no previous experimental data have been collected prove to be accurate when corresponding measurements are performed. The *structural stability* of the model means that its properties do not change qualitatively when small changes are made to the parameters of the model. An additional requirement is that any parameters in the model determined by fitting to experimental data should be intrinsic to the device and not depend on the external measurement circuitry. Finally, there is the issue of judgment—if a model appears to violate a law of physics or involves parameters that diverge or oscillate wildly while the voltage and current of the physical device are well behaved, the model is not useful and a different model needs to be formulated, likely using a different set of ideal circuit elements.

The necessity and importance of analyzing nonlinear circuits and extracting valid device models from them are growing dramatically as feature sizes on integrated circuits rapidly approach single digit nanometer dimensions [10] and neurophysiologists strive to understand the signaling that occurs in biological networks [11, 13, 14, 16]. However, there is a substantial benefit that accrues to a successful confrontation of the difficulties. The complexity of nonlinear systems also means that there is a great deal more information about the circuit in the measurement data than can be obtained from the analysis of a purely linear network. For example, no set of external electrical measurements on a three-terminal black box can distinguish between a wye (Y) and a delta (Δ) network of three linear impedances. However, if such a black box contains one nonlinear element, it is easy to determine a unique model structure if one makes the appropriate measurements [9]. This realization can be tremendously helpful in constructing the wiring diagram of the brain, for example, since there is much that can be inferred from the measured signaling data to complement the structural information about neuron connectivity.

Chua has described two different ideal procedures, each with four basic steps, for finding a device model: the physical principles approach and the black box or measurement approach [8]. The physical approach assumes that one understands the physics and operating mechanisms of the device to a sufficiently high level of confidence that one can derive the device model from known information. The four-step process involves the following [8]: (1) device physics analysis and partitioning, (2) physical equation formulation, (3) equation simplification and solution, and (4) nonlinear network synthesis. The black box approach assumes that nothing is known about the physics of the device and thus one must characterize the device experimentally and find a mathematical model that reasonably reproduces the measurements. In this case, the four steps are as follows [8]: (1) experimental observations, (2) mathematical modeling, (3) model validation, and (4) nonlinear network synthesis. For memristors, it is critical that the experiments include time (or frequency)-dependent measurements over many orders of magnitude—static measurements (or quasi-static measurements in which a current or voltage is varied slowly and perhaps irregularly without a time stamp) will not contain enough information to specify a valid model. The end goal of either path is a predictive model that can be used in a dynamical simulation of a complex circuit that contains one or more instances of the nonlinear device, and thus step four, utilizing the model, is the same for both paths. In practice, the actual route to finding a device model will be some hybrid of the two ideal procedures. In order to ensure that the model is constrained by physics, one needs to include as much information as possible on what is known about the operating mechanisms, especially what the expected functional forms and asymptotic behavior are for the model. However, especially for nanoscale or biological systems, it is unlikely that a sufficient model can be derived from scratch, so some set of measurements will have to be performed to be able to fit parameters in a derived model or even to distinguish among several different possible functional forms in the model. As a classical example, Hodgkin and Huxley appear to have followed a similar heuristic in the development of their famous model for the action potential of the squid axon [19], which was informed by the physical process of transmembrane ion flow but used ad hoc differential equations and state variables to fit their experimentally measured electrical behavior.

After carefully formulating a mathematically rigorous treatment of nonlinear networks [5], Chua realized that the set of basis functions for describing general nonlinear circuit elements was incomplete and first postulated the existence of a memory resistor or memristor [6], although no physical example was known at the time. This formulation was generalized to define a class of nonlinear dynamical circuit elements originally called memristive systems [7], but in his paper "The Fourth Element" [15], he simplified the nomenclature by using the term "memristor" for the more general element and recommended using the phrase "ideal memristor" to describe the originally defined circuit element if necessary. He has recently written or co-written several tutorials to clear up confusion that has arisen to describe how to recognize a memristor from its electrical properties [2, 15]. In particular, the current–voltage (i–v) characteristic of a memristor will display a "pinched hysteresis loop" if energized by a sinusoidal voltage or current source, with the hysteresis loops collapsing with increasing frequency of the excitation (in contrast to capacitive hysteresis, for which any loops will get larger and/or more complex). In another tutorial, "Resistance switching memories are memristors" [12], he again describes how to recognize a memristor via the pinched hysteresis loop and shows that the memristor equations provide the basis for modeling the behavior of a wide variety of devices presently under investigation for nonvolatile memories, even though the devices are based on completely different physical principles.

A memristor is thus defined to be any dynamical electronic circuit element that obeys the following Chua memristor equations [7], independent of the underlying physics in the system that gives rise to the relations:

$$v = R(w,i)\,i \quad \text{and}$$

$$\mathrm{d}w/\mathrm{d}t = \mathrm{F}(w,i)\,,$$

where the first equation is the quasi-static resistance equation, or state-dependent Ohm's law, and the second is the dynamical equation that describes the temporal evolution of the state variable (or variables) w as a function F of w and possibly the current (or alternately voltage if one is considering the conductance of the element). In these equations, i is an independent input function and often a function of time. The resistance R is a function of the physical state w, which imparts memory to the device, and also possibly the current i, which results in a nonlinear Ohm's law.

Even though they are both constructed from transition metal oxide materials, the two example devices to be analyzed in this chapter have very different physical properties and the approaches used to their construct memristor models were quite distinct, so they serve to illustrate the general approach for creating nonlinear device models quite well.

3.2 Locally Active Niobium Dioxide Memristor: Physical Principles Approach

Niobium dioxide, NbO_2, represents a class of Mott insulator materials that enable locally active memristors, which were described by Itoh and Chua [20] and shown to facilitate the construction of an oscillator circuit. Although NbO_2 has been known to exhibit a current-controlled negative differential resistance (NDR) for decades [4, 18], only recently was it demonstrated that this system can be modeled by the Chua memristor equations [23]. At room temperature NbO_2 is an insulator. Sandwiched in a crosspoint device (Fig. 3.1a, insert), this material will warm up as the voltage across it is increased and current starts to flow through it. As the Joule heating increases, feedback [3] between lower resistance and higher current causes a channel through the device to exceed the electronic phase transition temperature ($\sim$800 C) and the NbO_2 in this channel becomes metallic, as illustrated in Fig. 3.1b,

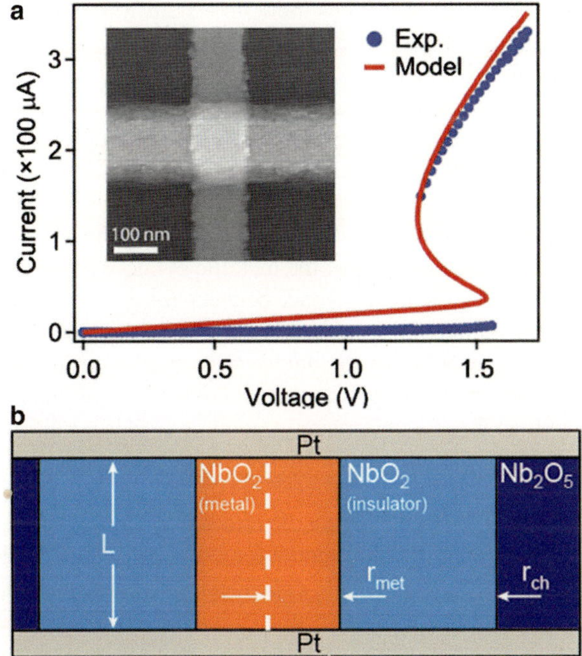

Fig. 3.1 Niobium dioxide Mott memristor. (**a**) Experimentally measured (*blue dots*) and modeled (*red line*) current–voltage curve for a $110 \times 110\,\text{nm}^2$ niobium dioxide memristor with an inset SEM image of a crosspoint device. (**b**) Schematic cross section of the device model during operation showing an NbO_2 channel that consists of two concentric cylindrical electronic phases, the inner core heated above the insulator-to-metal transition temperature and the outer lower temperature Mott insulating material [23]

with roughly a four order of magnitude increase in conductivity. This rapid decrease in resistance gives rise to the NDR—the voltage across the device decreases as the current increases, as shown in Fig. 3.1a, until all the NbO_2 in the device has become metallic and the slope of the *i*–*v* characteristic becomes positive again.

At the same time, the heat capacity and latent heat of transformation of the NbO_2 provides a delay or transient memory in the time required for the device to warm up or cool down. The result is a locally active memristor—a device that has the capacity to inject energy into a circuit when it is biased within the regime of negative *i*–*v* characteristic slope. A simple analytical model for the Chua memristor equations of this system is easily derived [23]. The resistance R is a function of a state variable w that in this case represents the normalized radius (i.e., the radius of the metallic phase r_{met} divided by the radius of the total NbO_2 cylinder r_{ch} in the device, Fig. 3.1b) of the material heated above the electronic transition temperature by Joule heating:

$$v = R_{ch}(w) \cdot i.$$

The dynamical equation describes how the state variable w increases or decreases, depending on if the power dissipation in the device is larger than or smaller than the thermal conductance:

$$\frac{dw}{dt} = g\,(w, i) = \left(\frac{d\Delta H}{dw}\right)^{-1} \cdot \left(R_{ch}(w)i^2 - \Gamma_{th}(w)\Delta T\right).$$

The dynamical equation has three parts: the Joule heating Ri^2 of the material, the heat conduction from the surface of the metallic cylinder to the boundary of the device $\Gamma_{th}(w)\Delta T$, where ΔT is the difference between the electronic phase transition temperature and the ambient temperature of the surroundings, and the derivative of the net enthalpy change ΔH of the device with respect to the radius w of the metallic channel. The agreement between the model and experimentally measured quasi-static data (Fig. 3.1a) is surprisingly good, considering the simplifications and the extremely large temperature gradient assumed. Sometimes, one gets lucky. Most of the parameters in the model were obtained from the literature for bulk properties, with a fitting parameter being the effective radius r_{ch} of the entire NbO_2 channel [23]. However, since this is a model for a dynamical device, it is necessary to compare simulation results to time-dependent data. Using SPICE, both the frequency and the shape of the $v(t)$ traces of a Pearson–Anson oscillator constructed from a Mott memristor and a capacitor were very accurately simulated [23]. This observation was critically important, because there was no inductor in either the experimental circuit or the model; the conclusion is that the Mott memristor has a positive reactance under certain biasing conditions and can be used to ensure that current in a circuit is continuous in time. The SPICE model has also been used to very accurately model the threshold, gain, and spiking behavior of a neuristor [17], an electronic device that emulates the action potential in an axon of a neuron, built with two Mott memristors and two capacitors [24]. The very good agreement between the time-dependent simulations and experimental measurements provides the verification of the memristor model, as shown in Fig. 3.2.

3.3 Nonvolatile Titanium Dioxide Memristor: Measurement Approach

Titanium dioxide (TiO_2) was the first material explicitly identified to exhibit memristor properties [26]. The simple model first used to explain why this device displayed a pinched hysteresis loop was the drift of positively charged oxygen vacancies, which act as dopants in many semiconducting transition metal oxides, in an applied electric field. However, this model left out several important processes involved in the motion of oxygen vacancies, in particular Fick diffusion [27, 28] and thermophoresis [29]. Writing down the full set of equations that describe the coupled motion of electrons and ions as well as the generation of heat and thermal gradients leads to a complex system of coupled differential equations, and a priori it is not possible to predict which terms are important or how to approximate

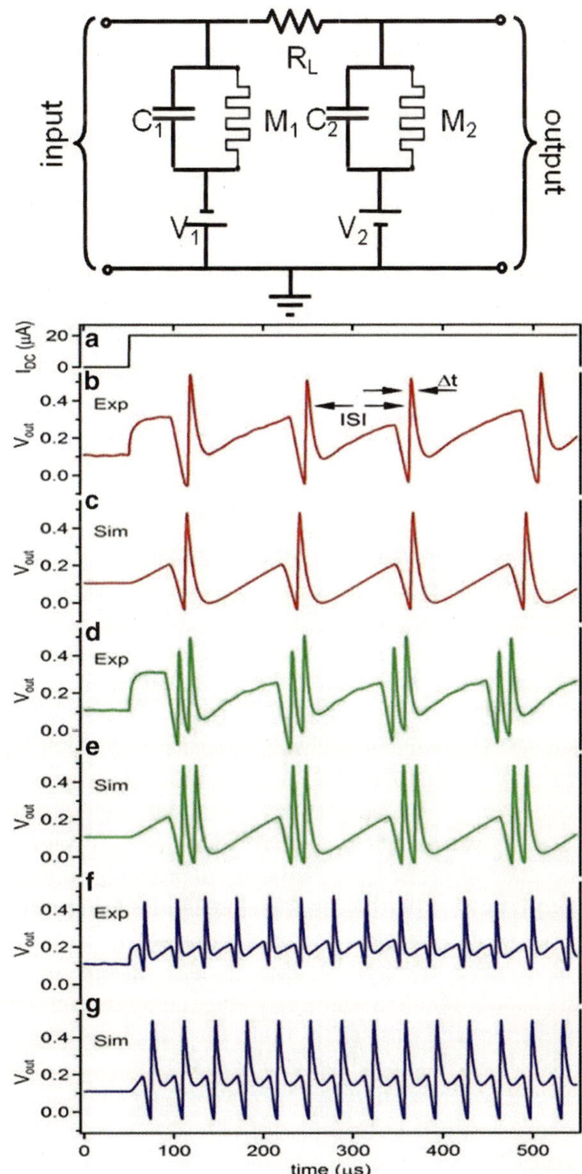

Fig. 3.2 (*Top*) Schematic diagram of the neuristor circuit, which includes two nominally identical Mott memristors M_1 and M_2, two capacitors C_1 and C_2 that in general are different from each other, two bias voltages with the same magnitude but opposite polarity and a load resistor R_L that couples the two oscillators. (*Bottom*) For a current sourced neuristor with $I_{DC} = 20$ μA (*a*), the experimental (*b, d, f*) and simulated (*c, e, g*) voltage outputs vs. time plots are shown. By tuning the channel capacitances C_1 and C_2, the neuristor exhibits (*b, c*) regular spiking ($C_1 = 5.1$ nF, $C_2 = 0.75$ nF), (*d, e*) chattering ($C_1 = 5.1$ nF, $C_2 = 0.5$ nF), and (*f, g*) fast spiking ($C_1 = 1.6$ nF, $C_2 = 0.5$ nF) modes of operation. C_1 controls the inter spike intervals (ISI) and C_2 controls the spike width (Δt)

the equations with a reasonable analytical form. Thus, two sets of experiments are necessary: one to understand the quasi-static current–voltage relation and the other to capture the dynamics of the state variable(s). To determine the quasi-static $R(w,i)$ relation, we should measure the underlying transport physics without interference from dynamical or switching processes, which requires a series of low amplitude current–voltage measurements to avoid switching or otherwise perturbing the device. We need to identify what physical property of the system is represented by the state variable(s) w and understand how to control the state so that we can perform current–voltage measurements over a wide range of w. Since the transport is likely to be nonlinear, the range of voltages and currents over which the measurements are made must be large enough to properly capture and quantify the current–voltage relation, identify w, and produce a self-consistent analytical expression for $R(w,i)$, but low enough to avoid state drift or switching.

Determining a dynamical expression for $F(w,i)$ is even more demanding because of the large number of measurements required. Since the devices exhibit stochastic behavior in their switching characteristics [21], each measurement should be performed several times in order to determine the mean or median behavior of the device. Since the dynamics of a device that operates via ion motion can be affected by a complex combination of drift, diffusion, and thermophoresis, finding an equation that reproduces the data accurately may require testing hundreds of different possible models.

The dynamical measurements to determine $F(w,i)$ require that the resistance of a device be measured as a function of time over many orders of magnitude and for many different applied voltages (or currents), which results in a very large two-dimensional data set (resistance vs. voltage and time). This is done by first setting the device to a known initial state, or value of w. Then, an approximately constant voltage (or current) is applied to the device and both the current and voltage are measured as a function of time to follow the evolution of the state. Once a set of measurements have been collected, the device is reset back to its original state w, and the measurements are performed again with a different value of the voltage (or current) applied to the device. This procedure must be repeated for a wide range of voltage (or current) amplitudes and both positive and negative polarities, in order to obtain a complete picture of both the "ON" and "OFF" switching of the device if it is bipolar. If the device is unipolar, in principle only one set of voltage polarities is needed but it is best to check both polarities to see if there is an asymmetry.

These measurements can be performed in a pulsed mode, in which the pulse width increases exponentially in order to sample many decades of time intervals, and a small amplitude current–voltage measurement is made to determine the device state after each pulse [22]. Alternately, the measurements can be made in real time by applying a step excitation (voltage or current) to the device, and following both the current and voltage as the device state evolves in time [25]. In both of these approaches, it is critical that all of the voltage and current data have an accurate time stamp. It is also necessary to use an effective four-point probe type of analysis so that the temporal evolution of both the current and the voltage can be obtained independently, with as little contribution from series resistance as possible. The result of all these measurements is a two-dimensional array of values of $w(i,t)$. We then either find an analytical function that closely approximates w and differentiate with respect to t or numerically differentiate the array of values to obtain a new two-dimensional array $\mathrm{d}w(i,t)/\mathrm{d}t$. We then invert the function $w(i,t)$ to create the array $\mathrm{d}w(w,i)/\mathrm{d}t$. The art is to find the analytical equation $F(w,i)$ that best matches the derivative of w. In order to have a high quality predictive model that does not break

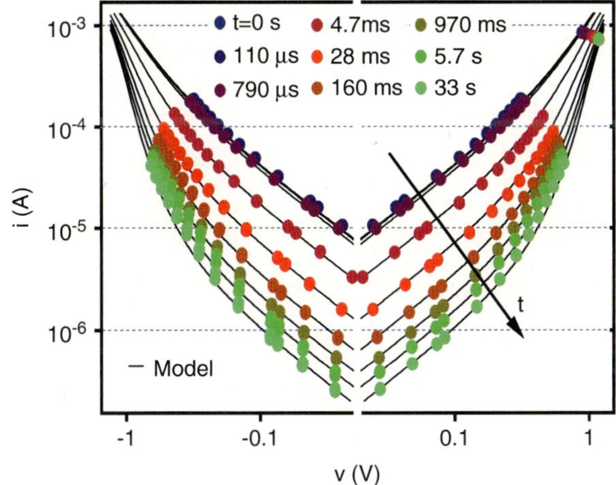

Fig. 3.3 Current–voltage characteristics of a TiO_2 nonvolatile memristor for a selection of times during a 4.5 V OFF-switching state-test. The curves are best fits to a series resistor plus Simmons tunnel barrier. The legend lists the total time the device was under the applied voltage in seconds, after which the *i–v* data were collected. Each fit yielded a value of the state variable *w*, the width of the tunnel barrier after the total time under bias [22]

down in unanticipated regions of device parameters, we should be guided in our choice of functions by the known or at least plausible physics-based state equations. We employed the pulsed state-test to experimentally determine the Chua memristor equations for a TiO_2-based nonvolatile memristor [22]. After each pulse to change the state of the device, we measured its *i–v* characteristic with a low voltage ramp to determine the resistance. We found that the *i–v* curves were best described by the Simmons tunneling equation shown below with the constants defined in [22], and fit many hundreds of *i–v* curves to this equation while only allowing the width of the tunnel barrier to vary. We obtained excellent fits over all the *i–v* curves, with a few examples shown below in Fig. 3.3.

$$i = \frac{j_0 A}{\Delta w^2}\left\{\phi_I \exp\left(-B\phi_I{}^{1/2}\right) - (\phi_I + ev_g)\exp\left(-B(\phi_I + ev_g)^{1/2}\right)\right\}$$

As shown in Fig. 3.4a, we then plotted each value of *w* obtained from the tunneling fits as a function of time for all the different voltages applied to have a graphical representation of *w*(*v*,*t*). These curves were then numerically differentiated and plotted vs. *w* to obtain the desired plots of d*w*(*w*,*v*)/dt, as shown in Fig. 3.4b. Given that the current–voltage characteristic (tunneling) was exponential in nature and the switching speed was also exponential, we explored a set of double exponential functions in order to obtain our model, shown as solid lines in Fig. 3.4b, to the data.

The agreement between the model and the data was quite good, and the equation utilized for Fig. 3.4b and shown below has been used in a SPICE simulation to successfully model the titanium dioxide memristor electronic properties [1].

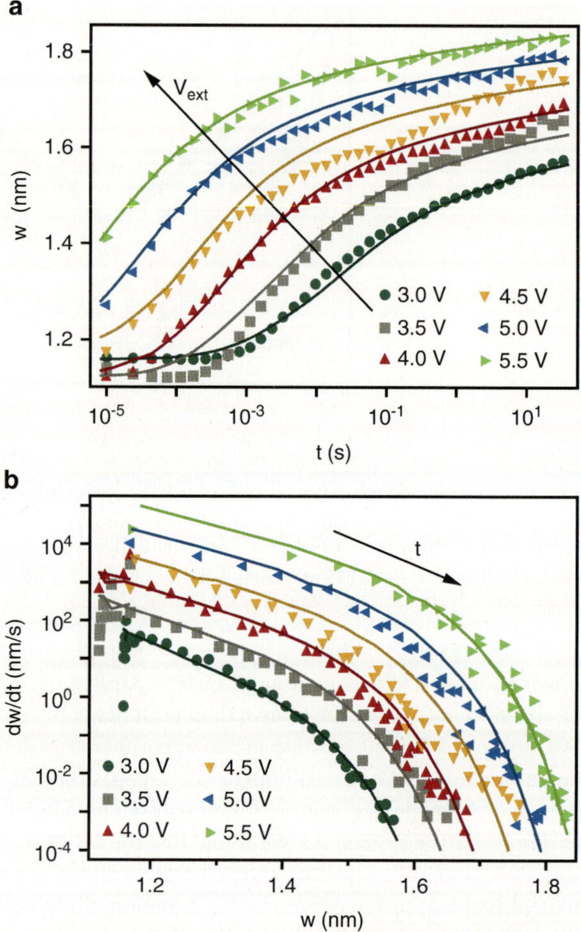

Fig. 3.4 Dynamical behavior of the tunnel barrier width w for the titanium dioxide memristor. The state variable w evolves as a function of time for different applied voltages for a series of (**a**) OFF-switching state tests. The legends indicate the applied external voltage. The lines are the numerical solution to the respective switching differential equations described in the text. (**b**) Shows the numerical derivative of the data in (**a**) plotted as a function of w for the different applied voltages. The lines are calculated from the differential equations using the measured values of w and i at each point in time. The derivative of the state variable is interpreted as the speed of the oxygen vacancy front as the applied voltage pushes it away from the top electrode [22]

The OFF-switching equation is shown below (there is a similar equation with different constants for ON-switching).

$$\dot{w} = f_{OFF} \sinh\left(\frac{i}{i_{OFF}}\right) \exp\left(-\exp\left(\frac{w - a_{OFF}}{w_c} - \frac{|i|}{b}\right) - \frac{w}{w_c}\right)$$

3.4 Conclusions

At this time, constructing a compact model of a memristor that is useful for circuit design and simulation is a complex and arduous task. Perhaps after several examples of successful models have been derived and the methodology for creating them has become clearer, it will be more straightforward. The examples described here show that there is no set procedure that one can follow blindly. There really is both an art and a science to memristor modeling.

References

1. H. Abdalla, M.D. Pickett, SPICE modeling of memristors, in *IEEE International Symposium on Circuits Systems (ISCAS)*, pp. 1832–1835 (2011)
2. S.P. Adhikari, M.P. Sah, H. Kim, L.O. Chua, Three fingerprints of memristor. IEEE Trans. Circuits Syst. I (2013)
3. C.N. Berglund, Thermal filaments in vanadium dioxide. IEEE Trans. Electron Devices **16**, 432 (1969)
4. K.L. Chopra, Current-controlled negative resistance in thin niobium oxide films. Proc. IEEE **51**, 941–942 (1963)
5. L.O. Chua, *Introduction to Nonlinear Network Theory* (McGraw-Hill, New York, 1969)
6. L.O. Chua, Memristor—the missing circuit element. IEEE Trans. Circuits Theory **18**, 507–519 (1971)
7. L.O. Chua, S.M. Kang, Memristive devices and systems. Proc. IEEE **64**, 209–223 (1976)
8. L.O. Chua, Device modeling via basic nonlinear circuit elements. IEEE Trans. Circuits Syst. **CAS-27**, 1014–1044 (1980)
9. L.O. Chua, Nonlinear circuits. IEEE Trans. Circuits Syst. **CAS-31**, 69–87 (1984)
10. L.O. Chua, Nonlinear foundations for nanodevices, Part I: The four-element torus. Proc. IEEE **91**, 1830–1859 (2003)
11. L.O. Chua, Local activity is the origin of complexity. Int. J. Bifurcation Chaos **15**, 3435–3456 (2005)
12. L.O. Chua, Resistance switching memories are memristors. Appl. Phys. A **102**, 765–783 (2011)
13. L.O. Chua, V. Sbitnev, H. Kim, Hodgkin-Huxley axon is made of memristors. Int. J. Bifurcation Chaos **22**, 1230011 (2012)
14. L.O. Chua, V. Sbitnev, H. Kim, Neurons are poised near the edge of chaos. Int. J. Bifurcation Chaos **22**, 1250098 (2012)
15. L.O. Chua, The fourth element. Proc. IEEE **100**, 1920–1927 (2012)
16. L.O. Chua, Memristor, Hodgkin-Huxley, and edge of chaos. Nanotechnology **24**(38), 383001 (2013)
17. H.D. Crane, The neuristor. IRE Trans. Electron. Comput. **EC-9**, 370–371 (1960)
18. D.V. Geppert, A new negative-resistance device. Proc. IEEE **51**, 223 (1963)
19. A.L. Hodgkin, A.F. Huxley, A quantitative description of membrane current and its application to conduction and excitation in nerve. J. Physiol. **117**, 500–544 (1952)
20. M. Itoh, L.O. Chua, Memristor oscillators. Int. J. Bifurcation Chaos **18**, 3183–3206 (2008)
21. G. Medeiros-Ribeiro, F. Perner, R. Carter, H. Abdalla, M.D. Pickett, R.S. Williams, Lognormal switching times for titanium dioxide bipolar memristors: origin and resolution. Nanotechnology **22**, 095702 (2011)
22. M.D. Pickett, D.B. Strukov, J.L. Borghetti, J.J. Yang, G.S. Snider, D.R. Stewart, R.S. Williams, Switching dynamics in titanium dioxide memristive devices. J. Appl. Phys. **106**, 074508 (2009)
23. M.D. Pickett, R.S. Williams, Sub-100 femtoJoule and sub-nanosecond thermally-driven threshold switching in niobium oxide crosspoint nanodevices. Nanotechnology **23**, 215202 (2012)
24. M.D. Pickett, G. Medeiros-Ribeiro, R.S. Williams, A scalable neuristor built with Mott memristors. Nat. Mater. **12**, 114–117 (2013)

25. J.P. Strachan, A.C. Torrezan, G. Medeiros-Ribeiro, R.S. Williams, Measuring the switching dynamics and energy efficiency of tantalum oxide memristors. Nanotechnology **22**, 505402 (2011)
26. D.B. Strukov, G.S. Snider, D.R. Stewart, R.S. Williams, The missing memristor found. Nature **453**, 80–83 (2008)
27. D.B. Strukov, R.S. Williams, Exponential ionic drift: fast switching and low volatility of thin-film memristors. Appl. Phys. A **94**, 515–519 (2009)
28. D.B. Strukov, J.L. Borghetti, R.S. Williams, Coupled ionic and electronic transport model of thin-film semiconductor memristive behavior. Small **5**, 1058–1063 (2009)
29. D.B. Strukov, F. Alibart, R.S. Williams, Thermophoresis/diffusion as a mechanism for unipolar resistive switching in metal-oxide-metal memristors. Appl. Phys. A **107**, 509–518 (2012)

Chapter 4
Fourth Fundamental Circuit Element: SPICE Modeling and Simulation

Dalibor Biolek and Zdenek Biolek

4.1 Introduction

The memristor was originally defined as an electric element which provided the last vacant connection between voltage, current, and their time-domain integrals, i.e. between flux and charge [1]. It is rightly described as the fourth fundamental circuit element next to the resistor, capacitor, and inductor.

It turns out that scientists met with the signs of memristive behavior long before the discovery of the memristor itself. These signs were described and modeled using means among which the memristor was missing. Immediately after the introduction of the concept of memristor in 1971 this hypothetical element was used for modeling certain processes that exhibited the attributes of memory behavior. The memristor was used for the modeling of devices based on varied and mutually unrelated physical principles. An example is the work of Oster [2], which shows that the tapered dashpot or an electrochemical system can be modeled as a memristor. Since the 1970s the memristor has gained a firm position as a standard modeling tool in branches which utilize the methods of network thermodynamics [3].

So it soon turned out that the concept of memristor is also useful for fields of study other than electrical engineering. The researchers involved in the modeling of systems that are composed of subsystems of different physical nature, as is common, for example, in electromechanics, now use the so-called generalized memristor [4]. It establishes a link between two physical quantities, which are in the given physical field of study analogous to the flux and the charge; for example in the

D. Biolek (✉)
Department of Electrical Engineering/Microelectronics, University of Defence/Brno University of Technology, Brno, Czech Republic
e-mail: dalibor.biolek@unob.cz

Z. Biolek
Department of Microelectronics, Brno University of Technology, Brno, Czech Republic
e-mail: zdenek.biolek@gmail.com

R. Tetzlaff (ed.), *Memristors and Memristive Systems*,
DOI 10.1007/978-1-4614-9068-5_4,

case of mechanics, they are the effort and the position. Using the memristor we can therefore model processes in a variety of systems differing in their physical nature, be it mechanical [5], hydraulic [6], neuromorphic [7], or other systems. With this general conception, the memristor can be seen as an element that guarantees a clear correlation between the accumulated effort (momentum, the integral of effort) and the state achieved (displacement, integral of flow).

After a report was published on the successful implementation of memristor in TiO_2 [8] in 2008, the interest in this element skyrocketed. However, the memristor comes to be considered specifically as an element of electrical nature. Moreover, according to the original definition of the memristor, the state from which the instantaneous value of memristance is derived is the charge (or the flux). In the course of time, however, it was discovered that in real systems the memristive behavior can be caused by complicated physical processes the instantaneous state of which can be determined by an entire vector of values of various physical origins [9]. Memory realized by magnetic spin, processes in TiO_2, etc. can be given as examples. Though in terms of resistive port, the voltage and current appear in the memristor equation, the specific physical principle that ensures memory behavior is projected into the equation by the state of x, where x is, generally, the vector of non-electrical quantities. This is a significant generalization of the original definition of the memristor, in which the electric charge was considered the only state of the system. The idea of generalized memristor as an element of general non-electrical nature, whose usefulness for the description of real systems had shown even before the discovery in Palo Alto, thus begins to affect one of the two equations of memristor—the equation of state. In terms of port equation, however, the memristor is still regarded as an electrical element.

Another complex process was set in motion with the discovery that the physical principles, governed by equations of the memristor and still suitable for the implementation of electronic nonvolatile memory in the real world, are very rare. Most of the discovered principles rather fall into the category of memristive systems that differ from the memristor in two practical ways.

The first difference is that the memristance of an element depends not only on the state of the system, but it is also affected by the instantaneous values of voltage or current. This, however, contradicts the original definition of memristor, whose memristance must depend only on the state of the system. The nonlinearity of an element due to voltage or current would mean that the instantaneous state of an element should not be automatically taken for the instantaneous state of memory. In terms of the nonvolatile memory, an element remembers the instantaneous state even when disconnected from the external power source. If the state depended on the value of voltage or current, it would mean that when the power source is disconnected it could change.

The second difference concerns the dynamics of an element. Actually, the rate of change of the state is usually not only dependent on the values of voltage and current but is often also influenced by the instantaneous state of the element. In the case of a true memristor the rate of change does not depend on the state itself. Otherwise, in the aftermath of voltage or current excitation, its own dynamics would cause another change of state, which would mean that the element would not necessarily be applicable as analog nonvolatile memory.

Today, we already know that the HP memristor is actually a complex memristive system which can be, according to all available data, used as nonvolatile memory. A similar situation also occurred with some other discoveries, when it turned out that the element either behaved as a memristor only when the operating point remained in a predefined area or this did not happen at all because of the nonlinearities related to voltage or current. These difficulties have led to efforts by some scientists to review the original definition of memristor such that it does not strictly distinguish memristors from less specific memristive systems [10]. It is a very delicate question that remains to be answered. The idea of the memristor as a fundamental element that logically complements the missing element in the mosaic of the basic elements of electrical engineering is so valuable that it is appropriate to distinguish between the "ideal" memristor in terms of the original definition of 1971 [1] and its generalized versions, which can be summarized as memristive systems [11]. Since the memristor in the sense of [1] is, in principle, an ideal circuit element, this attribute will not be used in this chapter, and the bare term "memristor" will be used for the fourth fundamental element.

The issue of appropriately modeling a memristor became topical immediately after the publication of the discovery in Palo Alto [12–14]. Due to the unavailability of memristors, it was necessary to develop sufficiently accurate models that would allow experimenting with the memristor via simulation programs and studying situations that would occur in real experiments in laboratory conditions. Over time, some researchers have realized that developing more and more sophisticated models of samples from HP laboratories is not enough. The knowledge of the fundamental principles of ideal memristor was lacking and so was the contact with real time experimentation with a component that fully complies with the definition of the memristor. Since the properties of a memristor are determined by the shape of its constitutive relation, there is no such thing as a universal generic memristor. Despite all the diversity of behavioral expressions resulting from an almost arbitrary choice of the constitutive relation, memristors show specific behavioral characteristics, the so-called fingerprints (FPs), which can be effectively examined through computer experiments.

This chapter discusses two stages of studying the memristor: the creation of a model and the simulation of the behavior of an element by way of using the model and the software tools. The initial stage, i.e. modeling of two-terminal components that are related to memristors, must be based on the elementary pillars of L. Chua's "correct modeling" [15]. These are referred to in the introductory part of the chapter. The starting point for the creation of the "correct models" is the defining characteristics of the memristor, especially the port and state equations, constitutive relations, and the "parameter vs. state map" [10], which are so versatile that they determine the behavior of memristors in the general situation. From the defining characteristics of memristor some specific features are apparent and can be identified in their behavior. The definition of these FPs is useful because the "fingerprints" can help to assess whether the model of memristor, no matter whether mathematical, software or hardware model, behaves correctly. In the next section of this chapter are described some possible ways of implementing memristor models into simulation programs such as SPICE, including examples of simulation results.

4.2 Starting Points of a "Correct Modeling" of Memristors

4.2.1 *Memristor Definition by Its Port and State Equations*

The principles of L. Chua's "correct modeling" can be briefly summarized in the statement that a correct model of the system must only depend on the system and not on its environment [15]. These dependencies are usually expressed either by the constitutive relation (CR) or by the parameter vs. state map (PSM) or by port and state equations.

As for memristors controlled by voltage or current (or by flux or charge) or memristors of "non-electrical" nature, whose port quantities are not necessarily voltages and currents, it is possible to generalize the above definition of memristor characteristics (i.e. the starting points of its correct modeling) as follows.

The memristor can be defined by a port equation which denotes the relationship between the excitation variable u and the response y:

$$y(t) = g(x)u(t). \tag{4.1}$$

The symbol $g(x)$ denotes a nonlinear function of the state variable x whose derivative with respect to time is equal to the excitation quantity:

$$\frac{d}{dt}x = u. \tag{4.2}$$

The above port (4.1) and state (4.2) equations may define a wide variety of known memristors of electrical and non-electrical nature, depending on the choice of physical quantities u and y. It is obvious from (4.2) that the state variable x is the time-domain integral of the excitation quantity u (the abbreviation is TIU), which can also be denoted u_I. For further explanation, it will be helpful to consider a fourth variable, the time-domain integral of the response y (TIY), denoted by the symbol y_I.

4.2.2 *Memristor as Electrical Two-Terminal Device*

Table 4.1 shows such a selection of the u and y quantities that Eqs. (4.1) and (4.2) correspond to the well-known classical definitions of "electrical" memristors controlled by voltage (VCMR) and current (CCMR). The state variable is the flux φ for VCMR and the charge q for CCMR.

Table 4.1 Voltage- and current-controlled memristors (VCMR and CCMR) and their circuit and state variables

Memristor	Input (u)	Output (y)	TIU = state ($x = u_I$)	TIY (y_I)	PSM [$g(x)$]	CR [$y_I = F(u_I)$]
VCMR	v	i	φ	q	$G_M(\varphi)$	$q(\varphi)$
CCMR	i	v	q	φ	$R_M(q)$	$\varphi(q)$

4.2.3 *Parameter vs. State Map (PSM), Constitutive Relation (CR), and Pinched Hysteresis Loop (PHL) of the Memristor*

The function of state, $g(x)$, is introduced in the penultimate column of Table 4.1 by the acronym PSM, i.e. parameter vs. state map [10]. This parameter is a memductance GM for VCMR and memristance RM for CCMR. PSM can be considered as the basic memristor characteristic, which is independent of the way the memristor interacts with its environment. This characteristic can be derived from the constitution relation (CR) of the memristor. According to the last column of Table 4.1, CR can be defined as a relationship between TIY and TIU, i.e.

$$y_I = F\left(u_I\right) = F(x). \tag{4.3}$$

Differentiating (4.3) with respect to time yields

$$\frac{d}{dt}y_I(t) = y(t) = \frac{d}{dt}F(x) = \frac{d}{dx}F(x)\frac{dx}{dt} = \frac{d}{dx}F(x)u(t). \tag{4.4}$$

Comparing this result with (4.1), one can verify the statement from [10] that the CR and PSM are equivalent characteristics for the description of memristor, and that the following relationship holds between them:

$$g(x) = \frac{d}{dx}F(x). \tag{4.5}$$

For the needs of modeling such memristors, some of whose parameters, for example the memristance, may also show discontinuous dependence on the state variable, it can be useful to consider $g(x)$ as a piecewise-continuous function with potential points of step discontinuities. Then it results from (4.5) that CR can be represented by a piecewise differentiable function $F(x)$ with possible points of discontinuity of its derivative with respect to x.

The y versus u relationship is a third frequently used memristor characteristic, namely $v(i)$ or $i(v)$ for the current- or voltage-controlled memristor. When driving the memristor by the controlling signal $u(t)$, the typical pinched hysteresis loop (PHL), drawn in the $[y,u]$ coordinates, strongly depends on the type of excitation. That is why, in contrast to CR and PSM, it does not represent a "correct model" of the memristor [10]. Regarding the hysteresis, the $v(i)$ and $i(v)$, and thus also $y(u)$

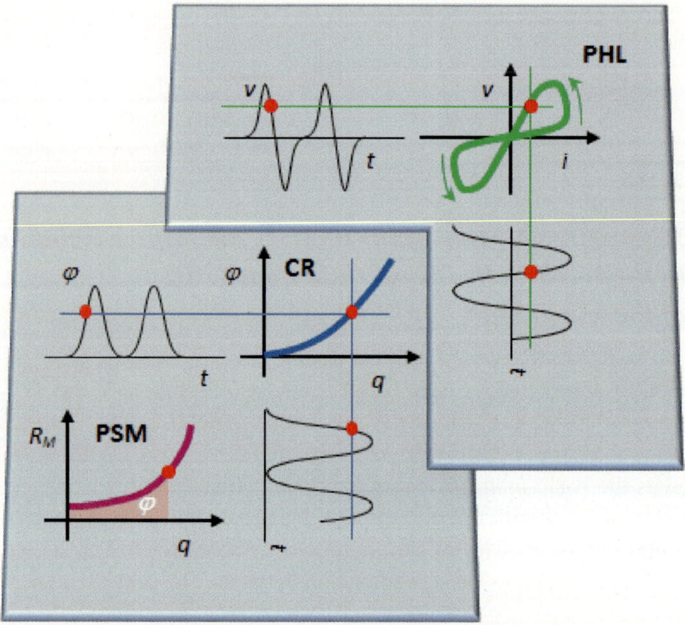

Fig. 4.1 Examples of CR, PSM, and PHL of current-controlled memristor

dependence relations cannot be modeled by single-valued mathematical functions in the classical sense of the word. The essence of the memristor is thus coded in its CR or also PSM. Some FPs described below result from the general rules which are related to these characteristics (Fig. 4.1).

4.2.4 Memristor Generalization for Non-Electric Domains

From a broader perspective, the voltage and current are physical quantities, which are specified in the bond graph theory [2] as effort (E) and flow (F). Table 4.2 summarizes the definitions of E and F variables for physical sub-domains in the port-Hamiltonian framework [4, 5]. It is obvious that VCMR and CCMR from Table 4.1 are only special cases of memristors, which could be abbreviated to ECRM and FCMR (effort-controlled memristor and flow-controlled memristor). For example, a hydraulic "WC memristor" controlled by "Flow Rate" is given in [6]. An interesting analysis is given in [2], where the mechanical tapered dashpot is represented by unambiguous nonlinear dependence of the integral of applied force and the piston displacement, whereas a pinched hysteresis loop appears in the force–velocity coordinates. This indicates the mechanical memristor. The port and state equations (4.1) and (4.2), PSM and CR discussed above then represent universal characteristics of all these generalized memristors across various domains.

Table 4.2 Effort and flow and their time-domain integrals in various domains as port quantities and native state variables of generalized memristors

	Port variables		Native state variables	
Domain	Effort E	Flow F	Momentum = TIE	Displacement = TIF
Electrical	Voltage	Current	Flux	Charge
Mechanical, translation	Force	Linear velocity	Momentum	Position
Mechanical, rotation	Torque	Angular velocity	Angular momentum	Angle
Hydraulic	Pressure	Volumetric flow	Pressure momentum	Volume
Thermodynamic	Temperature	Entropy flow	Temperature momentum	Entropy
Chemical	Chemical potential	Molar flow	Chemical momentum	Moles

4.2.5 State Variables of Memristors

The knowledge resulting from Eq. (4.2), namely that the natural state quantity alias the memristor memory is the time-domain integral of the excitation quantity, is expressed in Table 4.2 in the column "native state variables." Depending on whether the excitation quantity of the memristor is a quantity of the effort or flow type, the state variable is either its time-domain integral (TI), denoted as momentum (TIE), or displacement (TIF).

It should be noted that the values of the above state variables are theoretically unbounded. The total value of the electric charge passed through the electric memristor is a typical example. For an unambiguous dependence of memristance on the amount of this charge, the memristor would need to have an unlimited memory. In an actually existing system, this memory is put into effect by a specific physical principle. For example, the state of the TiO_2 memristor is determined by the position of the boundary between the undoped and doped layers [8]. This boundary can move only within strictly given limits. Although there is a link between this position and the charge flowing through, this dependence is strongly nonlinear near the limit positions, which can be simply described as a "thickening of memory" near the limit states. Because there is a direct connection between the position of the boundary and the resistance of the memristor, the memristance can be directly selected as the state variable of the TiO_2 memristor.

The freedom in the choice of various state variables of the same system, which does not change the port behavior of the system, shows the usefulness of denoting the memristor state variables of the momentum and displacement types, listed in Table 4.2, as *native state variables of the memristor*, and thus distinguishing them from all other potential optional state variables. The choice of the state variable other than the native one may cause that the equations of a particular memristor are not in a unified form (4.1) and (4.2). Therefore, it is useful to know the answer to the question whether the port and the state equations of a concrete system, which do not meet the formal definition equations (4.1) and (4.2), actually describe a memristor. The answer is positive if the transformation from the given into the native state variable leads to a unified form of the memristor equations (4.1) and (4.2). If these equations are in a unified form, then the system can be described by the characteristics of the CR and PSM types, which do not depend on anything other than this system. This is an unmistakable sign of the memristor.

To illustrate this problem, consider the memristor described by Eqs. (4.1) and (4.2), whose memory is implemented by a certain physical principle. The corresponding state variable x', modeling this principle, is limited within the interval (x'_{min}, x'_{max}) and it is in an unambiguous algebraic functional relationship with the native state variable x

$$x = S\left(x', x_0'\right), \tag{4.6}$$

where S is a strictly monotonous function with respect to the variable x'. The symbol $x_0{}'$ denotes the initial value of the state variable x' at the time instant when the exciting quantity u is applied to the memristor or, in other words, at the moment when the native state variable x is x_0 (see Fig. 4.2a). The coordinate $x_0{}'$ of the point of intersection of the curve in Fig. 4.2a with the x' axis thus denotes the initial state of the physical memory of the memristor prior to signal excitation. This state can be set via an appropriate vertical shift of the given curve. Its strict monotonicity ensures an unambiguous reversed dependence of the state x' on the native state variable x (see also Fig. 4.2b)

$$x' = H\left(x, x_0'\right), \tag{4.7}$$

where the function H can be obtained via the inversion of the function S with respect to the argument x'. The function H models a causal dependence of the physically implemented state x' on the native state variable. The initial state $x_0{}'$ at the beginning of the memristor excitation can be modeled via a proper horizontal shift of the curve in Fig. 4.2b.

Note that if the function H reached its limit values x'_{min} or x'_{max} for finite values of the native state variable x as, for example, illustrated in Fig. 4.2c, then the condition of the strict monotonicity would be violated in the sections with zero-valued derivative of the function H, and the change of the native state variable would not cause any change of the state variable x'. Since in really existing systems, in contrast to the state variable x', the native state variable is not stored, such a system loses the attribute of the boundless memory and, strictly speaking, it is not a memristor.

Taking into account equation (4.6), the port equation (4.1) of a memristor will be described by the state-dependent Ohm's law, where the memristance or memductance will depend on a new state variable x':

$$y(t) = g(x)u(t) = g\left(S\left(x', x'_0\right)\right)u(t). \tag{4.8}$$

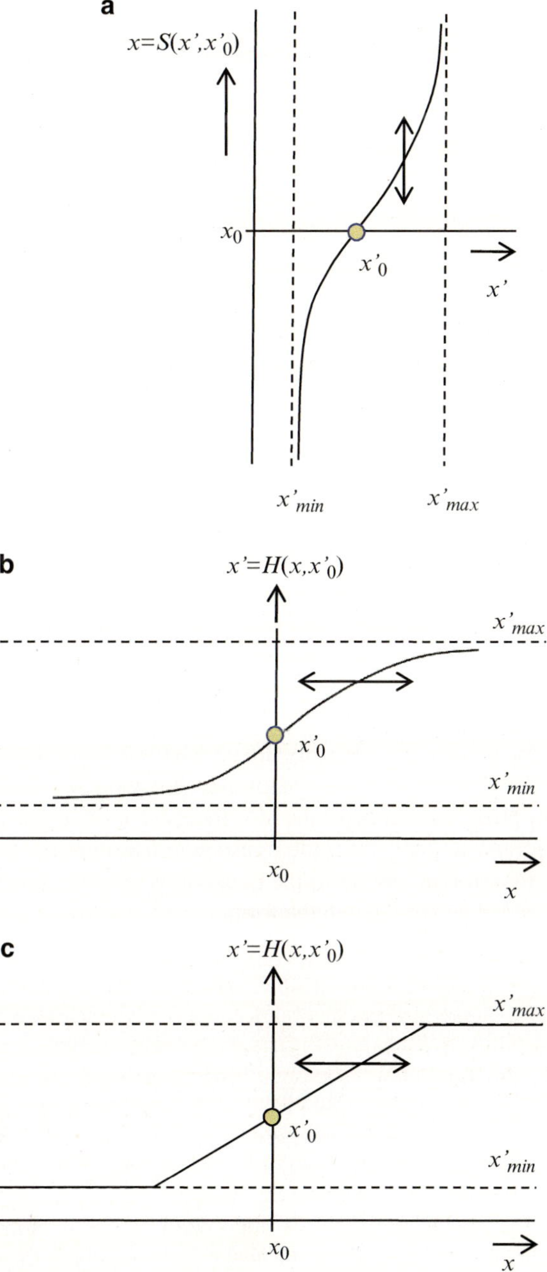

Fig. 4.2 Example of nonlinear relationships between native state variable x of memristor and a state variable x' which models a concrete way of physical implementation of memristor memory: (**a**) dependence (4.6), (**b**) inverse dependence (4.7), (**c**) example of the dependence (4.7) for a system with limited memory

Considering Eq. (4.6), the state equation (4.2) will be transformed as follows:

$$\frac{d}{dt}S\left(x',x'_0\right)=\frac{dS\left(x',x'_0\right)}{dx'}\frac{dx'}{dt}=u, \tag{4.9}$$

thus

$$\frac{d}{dt}x'=f\left(x'\right)u,\ f\left(x'\right)=\frac{1}{\frac{dS(x',x'_0)}{dx'}} \tag{4.10}$$

For a strictly monotonic function S, its derivative in (4.10) is not equal to zero and the function f attains only finite values for $x\in R$. This ensures that without excitation, i.e. for $u=0$, the right-hand side of state equation (4.10) is always zero and the state of the system would therefore not be changing (an attribute of non-volatility). An example of the system, described by a state equation of type (4.10), is the well-known TiO_2 memristor, provided that the state variable x' is the normalized width of doped layer and that nonlinear dopant drift is modeled by simple window functions [8, 13, 14, 16].

It is obvious that, according to Eq. (4.10), the controllability of the state x' will depend on the attributes of the function f. For example, consider that $f(x')=0$ for some state x'. This would happen for an infinite derivative of the transforming function S with respect to x'. Then no excitation u would be able to change this state since the derivative of the state would be zero regardless of the value of the excitation u. The system with limited memory, having the characteristic as in Fig. 4.2c, is a typical example. Probably the best-known example is the problem of fixing the limit states of the TiO_2 memristor, which is modeled, for instance, by the Joglekar window [16], if one accepts the idea that such a limit case can occur at all.

If, on the contrary, we start from the nonstandard-form state equation (4.10) of the system, and look for an answer to the question whether it models a memristor, then this equation can be arranged to the form

$$\int\frac{dx}{f(x)}=\int udt=u_I,\ f(x)\neq 0. \tag{4.11}$$

It is thus obvious that if we select the integral on the left side of Eq. (4.11) as the state variable of the system, and consider the limitation (4.11) for the function f, which follows from the strict monotonicity of the functions S and H, then the derivative of this variable with respect to time will be equal to the excitation variable u. The new state equation will therefore be exactly in the unified form (4.2) and the system is thus a memristor. For the TiO_2 memristor, which is modeled as current-controlled memristor, the integral on the left side of (4.11) is the charge, i.e. the native state variable of this memristor.

It is also apparent from Eq. (4.10) and from Fig. 4.2 that, due to the limitation of the state variable x' within the bounded interval (x'_{min}, x'_{max}), the rate of change of the state variable approaches zero when x' approaches the borders of this interval.

As a result, these limit states are not actually accessible in the course of a finite time. In the case of the "true" memristor, i.e. ideal hypothetical element, the problem of the "state fixation" cannot occur because such a memristor cannot be put into this state via an external signal within a finite time.

4.2.6 General Memristive System Versus Memristor

In terms of utilizing the memristor as a memory, it is important that the state of the memory is determined by the native state variable. However, it is usually not observable directly but only through another currently chosen state variable, which is related to the physical implementation of the memristor. The most easily accessible variable is memristance or memductance. The defining Eqs. (4.1) and (4.2) provide the non-volatility of the memory in two ways: (1) the state derivative with respect to time does not depend on the state itself [see Eq. (4.2)], (2) the memristor parameter g is unambiguously determined by the state and it does not depend on the instantaneous values of the port quantities [see Eq. (4.1)].

In general, these attributes are not fulfilled for general memristive systems that are describable by the generalized port and state equations in the form

$$y = g(\mathbf{x}, u)\ \ u, \tag{4.12}$$

$$\frac{d}{dt}\mathbf{x} = f(\mathbf{x}, u). \tag{4.13}$$

The symbol $\boldsymbol{x}$ represents here an n-dimensional vector of state variables. Since the system parameter g already depends on the excitation quantity, the PSM characteristics cannot be simply defined as a mere function of the state. The constitutive relation, which is closely related to the PSM, cannot be defined either. In addition, the state derivative with respect to time depends on the state as well as on the excitation, and the nature of this dependence can be flexibly modified via the function f. That is why we cannot speak of a memristor in the sense of the definitions (4.1) and (4.2).

In the following text, we focus on memristors in the sense of their conventional definition. If some of the parts are related to more general memristive systems, this will be explicitly specified.

4.3 Fingerprints

The so-called memristor fingerprints (FPs) are mentioned already in the original paper [1]. FPs are useful because they can serve to identify the memristive nature of the system from experimental data. The v–i pinched hysteresis loop (PHL), measured under periodical excitation, is the most widely known memristor FP. These PHLs, observable in the past for systems of different nature [10], can be of miscellaneous shapes. Some of their attributes were analyzed, for example, in [10, 11] but from the point of view of general memristive systems. The loop classification into "crossing type" (CT) and "non-crossing type" (NCT) was introduced in [9]. The latter loops are observable, for example, for thermistors [9]. However, it is shown in [17] that such loops cannot be the FPs of ideal memristors, thus the thermistor is not a memristor in the sense of the original definition in [1] but is a memristive system according to the more general definition in [11].

The paper [18] is an illustration of the dispute whether a given element is or is not a memristor, and whether it is or is not a memristive system. The FPs of these systems are specified there over the frame of the definitions from [1, 11], for example that the PHLs must be observable for any initial conditions [18], etc. In the invited lecture at the third Memristor and Memristive Symposium [19], L. Chua specifies the basic FPs of memristive systems, including one degenerate case of DC behavior of "ideal" memristors.

Therefore it turns out to be useful to analyze the basic FPs of memristive systems, focusing on "ideal" memristors in the first step. This section thus follows the memristor definition in Sect. 4.2, which determines the memristor essence, knowing that all possible manifestations of memristor behavior are governed by its basic characteristics. Some memristor FPs (i.e., consequences of the memristor essence) are pointed out for the memristor DC and AC excitations.

4.3.1 Memristor in DC (Pseudo) Steady State

It follows from Eq. (4.2) that there are no theoretical limitations to the evolution of the native state variable of the ideal memristor. For a constant input signal u, the state variable tends to infinity. Any nonlinear limitation, which is natural for real systems, would violate formula (4.2).

Suppose that at time $t_0 = 0$, when the native state variable is set to the value x_0, the memristor will be driven by a constant signal $u(t) = U$. Then Eqs. (4.1) and (4.2) can be rewritten in the form

$$y = g(x)\ U,\ \ x = x_0 + \ Ut. \tag{4.14}$$

The state variable increases ($U > 0$) or decreases ($U < 0$) ad infinitum or it remains constant ($U = 0$).

From the point of view of the native state of the memristor, the true steady state can only be established for a zero-valued controlling signal.

From the point of view of an external observer who is working only with the u and y quantities, the internal state can be unobservable. If the following limits exist,

$$G_+ = \lim_{x\to\infty} g(x),\ \ G_- = \lim_{x\to-\infty} g(x), \tag{4.15}$$

then the circuit containing such a memristor moves towards a DC pseudo-steady state. For this state, the memristor behaves as a linear resistor with conductance (4.15) (voltage-controlled memristor), or resistance (4.15) (current-controlled memristor). The conductance/resistance will depend on the polarity of the controlling signal.

It should be noted that the integration of the native state variable is permanently in progress during this DC pseudo-steady state. To recover the original state of the memristor, we would need a longer time period than the time interval for which the memristor persisted in the previous state.

It is obvious from the above analysis that the true DC steady state, which is characterized by steady values of not only port but also state variables, is only for $u=0$ and $y=0$. Then the following FPs can be formulated:

FP1: In the DC steady state, when voltage, current, and memristor state variables are steady, the memristor behaves as a nullor, i.e. with zero volts across its terminals and zero current flowing through.

FP2: In the DC pseudo-steady state, when the memristor state variables are not settled, the memristor behaves as a linear resistor, described by limits (4.15).

FP3: For memristors with undefined limits (4.15), the DC pseudo-steady state does not exist.

FP1 has been authored by L. Chua [19]. The pseudo-steady states, which can cause various paradoxical manifestations of ideal hypothetical elements, can be avoided, for example, via a direct excitation of the memristor by periodical signals with zero DC components.

It follows from the analysis in Sect. 4.2.5 that if a relationship between the native state variable x and the state variable x' as defined here exists, then a continuous integration of both state variables is in progress, and thus a true steady state cannot be reached from the point of view of the state variable x' either. However, within the frame of existing real systems, when the memory implementation can be done with some finite precision, the true DC steady state is achieved within a finite time which depends on the precision of fulfilling the condition $f(x')=0$ [see Eq. (4.10)]. Among other things, this can be a reason for the seemingly paradoxical behavior of memristor models during their computer simulation (see Sect. 4.4 for more details): Owing to the finite precision in the representation of the numerical value of the state variable x', its value can be set to its boundary value during the analysis. Then, however, $f(x')=0$, the state derivative is zero, and the system can extricate itself from this "dead regime" only with the help of other potential numerical inaccuracies. The model behavior near such boundary states then has not much in common with the behavior of the memristor which has been originally modeled.

4.3.2 Memristor in Periodical Steady State

Consider the memristor (4.1), (4.2) driven by a periodical signal u with a repeating period T, repeating frequency $F = 1/T$ or angular frequency $\Omega = 2\pi F$. For a zero DC component, the signal has its Fourier series

$$u(t) = \sum_{k=1}^{\infty} \left[A_k \cos(k\Omega t) + B_k \sin(k\Omega t)\right], \tag{4.16}$$

where A_k and B_k are the amplitudes of the cosine and sine terms of the kth harmonics.

Assume that at time $t_0 = 0$, at which the memristor state is $x = x_0$, the controlling signal (4.16) is applied. Then the state variable $x(t)$ for $t \geq 0$ will be as follows:

$$x(t) = x_{DC} + \frac{1}{\Omega}\sum_{k=1}^{\infty}\left[\frac{A_k}{k}\sin(k\Omega t) - \frac{B_k}{k}\cos(k\Omega t)\right], \quad x_{DC} = x_0 + \sum_{k=1}^{\infty}\frac{B_k}{k\Omega}. \tag{4.17}$$

The state variable and also the output variable y, which is derived from the state variable via algebraic equation (4.1), behave periodically without any initial transients. We conclude that

FP4: If a periodical signal $u(t)$ with zero DC component is applied to the memristor, the state and output variables $x(t)$ and $y(t)$ proceed immediately to the periodical steady state without any transient phenomena.

FP4 can be utilized as a quick test whether the memristive system, analyzed by a simulation program, is or is not a memristor according to [1]. For example, the SPICE models of the so-called potassium and sodium ion-channel memristors [19], driven by a sinusoidal signal, exhibit gradual transition to periodical steady state. The PHL then settles into a stable figure in the course of several repeating periods of excitation. Therefore, according to FP4, these are not memristors defined in [1]. On the other hand, when modeling a system satisfying formally the Eqs. (4.1) and (4.2), and the simulation outputs exhibit transient processes, this can indicate computational errors due to possible numerical problems. However, without considering FP4, the simulator outputs can be misinterpreted as correct.

Now consider that the frequency of the signal (4.16) increases ad infinitum. It follows from (4.17) and (4.1) that the state variable converges to a constant value x_0 and that a simple linear relationship $y = g(x_0)u$ exists between the output and input. Then the hysteresis effect in the y–u coordinates disappears. These conclusions, which are well known for simple sinusoidal excitation, can be generalized as follows:

FP5: When driving the memristor by a general periodical signal with zero DC component, and when the frequency increases ad infinitum, the area of the corresponding PHL tends to zero, and the memristor behaves as linear resistor. Its resistance is equal to the initial memristance at the beginning of excitation.

Now consider that the input signal (4.16) contains only sine-type harmonic components, i.e. it is described as an odd function, thus $u(-t) = -u(t)$. Then the state signal $x(t)$ (4.17) is an even function, thus $x(-t) = x(t)$. According to (4.1), the output signal $y(t)$ must be an odd function. Then the following FP6 can be conceived:

FP6: When driving the memristor via a periodical signal described by an odd function of time, then the y–u PHL will be odd-symmetric.

Two less obvious FPs result from FP6: For odd-type excitation, the memristor response always has zero DC component, and the corresponding PHL is always of the "crossing type" [9].

The sine waveform belongs to frequently used odd-type signals for memristor testing. Many papers contain examples of PHLs, obtained under such excitation, which violate the odd symmetry either quite evidently (the loops located in first and third quadrants are asymmetrical) or latently (the loops have incorrect orientations, thus the PHLs are of the non-crossing type [9]).

In the following, consider a general memristive system excited by a sinusoidal signal with a single nonzero Fourier coefficient B_1 in (4.16). Round the origin of the y–u coordinates, the hysteresis loop can be pinched in the following ways: (1) *crossing type* (CT), (2) *non-crossing type* (NCT), (3) *degenerating type* (DT). The first two types are defined in [9]. DT denotes the boundary case when the contour lines of the loop merge into one single-valued curve (the hysteresis has vanished) [10].

The type of pinching can be deduced from the derivatives of the difference signal $y_d(t) = y(t) - y(T/2 - t)$ at time 0. The condition $y_d = 0$ implies the loop passing through the origin. The nonzero first derivative indicates CT. If the first derivative is zero with nonzero second derivative, the PHL is NCT. With the first two derivatives being zero and the third derivative nonzero, a point of inflexion appears, which indicates a CT loop. These conditions can be arranged into the following unclosed scheme:

CT(0): $dy_d(t)/dt \neq 0$
NCT(1): $dy_d(t)/dt = 0, d^2y_d(t)/dt^2 \neq 0$
CT(2): $dy_d(t)/dt = 0, d^2y_d(t)/dt^2 = 0, d^3y_d(t)/dt^3 \neq 0$
NCT(3): $dy_d(t)/dt = 0, d^2y_d(t)/dt^2 = 0, d^3y_d(t)/dt^3 = 0, d^4y_d(t)/dt^4 \neq 0$
...

The numbers inside braces denote the so-called *order of touching*, i.e. the maximum order of the zero derivative of y_d at the origin. Starting from the first-order touching, we talk about the tangential CT or NCT, when the PHL passes through the origin in both directions with the same slope.

As regards the memristor, its output signal $y(t)$ is an odd function of time. Then it can be shown that all even-order derivatives of y_d at $t = 0$ are zero. According to the above scheme, the existence of NCT loops for the memristor is ruled out. The degenerate PHL (DT) appears when all the derivatives are zero. It is obvious that there can be loops with tangential crossing of even order (second or higher).

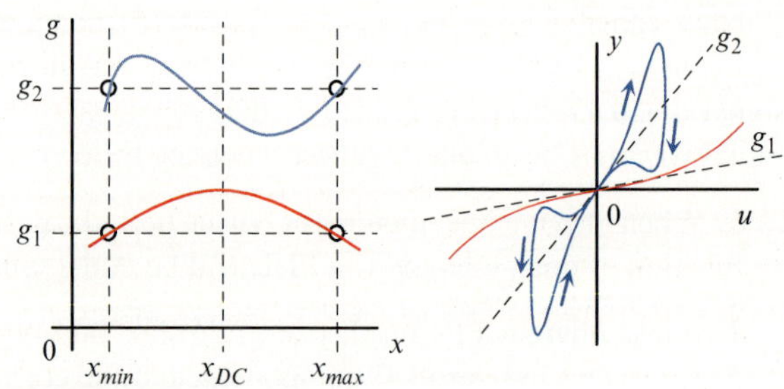

Fig. 4.3 Examples of degenerate loop (g_1) and tangential CT loop (g_2)

The following FPs can be summarized from the above analysis:

FP7: The *y–u* PHL of a memristor driven by a sinusoidal signal can be either CT with even-order crossing or DT.

FP8: The necessary and also sufficient condition for the existence of tangential CT loop of a memristor driven by a sinusoidal signal is as follows: the boundary points of the range of the state variable $x \in <x_{min}, x_{max}>$ on the curve $g(x)$ must lie on a horizontal line (see Fig. 4.3, case "g_2").

FP9: When driving the memristor by a sinusoidal signal, then the necessary and also sufficient condition for the existence of the tangential CT loop with 2*n*-order of touching, where *n* is a positive integer, is

$$\frac{d^i g\left(x_{\min}\right)}{dx^i} = (-1)^i \frac{d^i g\left(x_{\max}\right)}{dx^i} \tag{4.18}$$

for all $i = 0,1,2,\ldots,n{-}1$.

FP10: The necessary and also sufficient condition for the existence of DT loop of a memristor driven by a sinusoidal signal is as follows: the curve $g(x)$ must be symmetrical with regard to the vertical line which demarcates the center of the range of the state variable x (see Fig. 4.3, case "g_1").

FP8 means that at any time instant when the PHL passes through the *y–u* origin, the g parameter of the memristor, i.e. memristance or memductance, is of the same value. The corresponding slopes of the contour curves of the PHL are thus equal in both directions.

FP9 can be understood as follows: the 2*n*-order touching at the origin means that the derivatives d^iy/du^i, $i = 0,1,2,3,\ldots,2n$ must be identical for each passage of the *y–u* characteristic through the origin, and they only start differing with the $(2n{+}1)$ order. Consider the state defined according to (4.2) as a time-domain integral of the input signal. The Faà di Bruno's Formula holds for the derivative of the $(2n{+}1)$ order [20]:

$$\frac{d^{2n+1}y}{du^{2n+1}} = \frac{d^{2n}g\left(x(u)\right)}{du^{2n}} = \sum a_k \frac{d^k g}{dx^k} \tag{4.19}$$

where

$$a_k = \frac{(2n)!}{k_1!k_2!\cdot\cdot k_{2n}!}\prod_{i=1}^{2n}\left(\frac{1}{i!}\frac{d^i x}{du^i}\right)^{k_i} \tag{4.20}$$

and the sum is applied over all nonnegative integer roots $k_1, \ldots, k_{2n}$ of the Diophantine equation $k_1 + 2k_2 + \ldots + 2nk_{2n} = 2n$, with k being the sum $k = k_1 + k_2 + \ldots + k_{2n}$. It can be proved that, with regard to the zero-valued odd derivatives of the function $x = x(u)$ at the origin, this series contains only n nonzero terms which correspond to all roots k_i for $k = 1,2,\ldots, n$. Then Eq. (4.19) can be simplified to the form

$$\left.\frac{d^{2n+1}y}{du^{2n+1}}\right|_{t=0} - \left.\frac{d^{2n+1}y}{du^{2n+1}}\right|_{t=\frac{T}{2}} = \sum_{k=1}^{n} a_k\left(\frac{d^k g\left(x_{\min}\right)}{dx^k} - (-1)^k\frac{d^k g\left(x_{\max}\right)}{dx^k}\right) \tag{4.21}$$

where the a_k coefficient is evaluated for the state x_{min}. Supposing that the odd-order derivative (4.21) is nonzero and all lower-order derivatives are zero, then the statement from FP9 according to (4.18) must hold.

FP10 is a limiting case of FP9 for $n \to \infty$. It can be derived based on the consideration that the degenerate scenario holds for the equality $y(t) = y(T/2 - t)$. For $u(t) = B_1 \sin(\Omega t)$ and considering FP9, Eqs. (4.1) and (4.2) lead to the equality

$$g\left(x_{DC} - \frac{B_1}{\Omega}\cos\left(\Omega t\right)\right) = g\left(x_{DC} + \frac{B_1}{\Omega}\cos\left(\Omega t\right)\right) \tag{4.22}$$

Since Eq. (4.22) holds for any arbitrary time t, the function g must be symmetrical with regard to the vertical line $x = x_{DC}$.

FP8 to FP10 state that loops with tangential crossing and also degenerate loops cannot occur in memristors with monotonic functions $g(x)$, e.g. in the well-known TiO_2 memristor.

4.4 Memristor Models and Their Software Implementation

The method for modeling a memristor as described below is based on a simple conception which can be expressed by the sentence:

> Memristor models must accurately reflect all FPs of the "ideal" memristor.

This underlines the fact that what is being modeled is the fourth fundamental element and not a generalized memristive or other system. Such models are essential for the analysis of the behavior of the fundamental, i.e. ideal circuit elements, where any deflection from the theoretical model leads to an erroneous behavior of the object under study. Possible modeling of the real properties of an application circuit

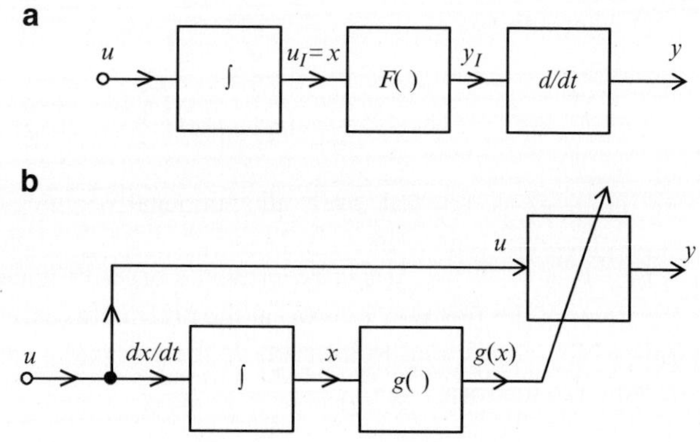

Fig. 4.4 Models of ideal memristors, based on (**a**) constitutive relation $y_I = F(u_I)$, (**b**) PSM $g(x)$ or port equation (4.1) and state equation (4.2)

is done by adding extra blocks of models that reflect these influences. Memristor models in the above sense of the word should therefore be behavioral models, which are based on port and state equations or on the PSM or CR characteristics.

Another possible approach is to create a model of a specific physically existing system, which we believe acts as a memristor. In this case, we usually proceed from the state equation for the state variable which is given by a physical principle, such as the coordinate of the boundary between doped and undoped layers in the TiO_2 memristor. Then it is appropriate to verify if it is actually a memristor (i.e., to check whether there is a clear connection between this variable and the native state variable).

The next step is to implement these models in a suitable simulation program, e.g. one from the group of SPICE-family programs.

Models based directly on the memristor PSM or CR can of course be used only for the "ideal" memristors, for which these characteristics are defined. Models based on equations are universal for all memristive systems. In the following text, we will demonstrate that models based on the knowledge of CR are not recommendable for practical simulations, and that the other two memristor models, based either on circuit equations or PSM, have the same structure. Further on, we will point out a potential numerical problem which can be associated with the computer simulation of systems based on state equations of the type of (4.10), and the way how to overcome it.

4.4.1 Block Diagrams of Models

The model shown in Fig. 4.4a, based on the knowledge of CR, proceeds from the fact that CR determines the integral quantity y_I from the integral quantity u_I through a nonlinear function F according to Eq. (4.3). This calculation is therefore modeled by the block $F(\,)$. It is therefore necessary to integrate the exciting signal u prior to

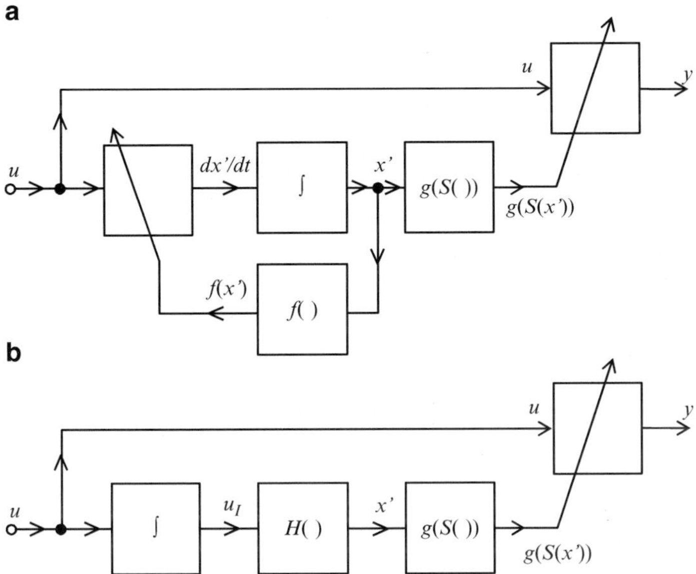

Fig. 4.5 Memristor model according to (**a**) Eqs. (4.8) and (4.10), (**b**) Eqs. (4.8), (4.2), and (4.7)

its further processing in the nonlinear block F. On the contrary, we must differentiate the output of the block F with respect to time to get the response y. With regard to the well-known problems associated with numerical differentiation, it is a weak point of this modeling scheme. Therefore, it is preferable to use the model in Fig. 4.4b, which is based on the fact that the excitation quantity is also the time derivative of the state variable. By integrating it, we get a state x which is converted in the nonlinear block g to the memristor parameter $g(x)$. If we multiply this parameter by the excitation u, we will get the response y.

It is obvious that by using the function $g(x)$, which is the derivative of $F(x)$ with respect to x, we can avoid the problematic differentiating circuit. Furthermore, it is clear that the model in Fig. 4.4b can be constructed directly from the knowledge of PSM regardless of the physical nature of the memristor, because x is directly the native state variable. At the same time, however, the model structure is directly related to port equation (4.1) and state equation (4.2), and the model can also be derived from the physical model of the system.

Figure 4.5a represents a model generalization for the case when the memristor is modeled by a modified state equation (4.10) with the use of another state variable x', which is derived from the native state variable x by a nonlinear transformation (4.7). The derivative of the state with respect to time is now given by the product of the excitation and the nonlinear state function $f(x')$. The state-dependent Ohm's law (4.1) is now represented by the composed nonlinear function $g(S(\,))$, see Eq. (4.8). A number of published TiO_2 memristor SPICE models, e.g. [13], are based on the block diagram in Fig. 4.5a.

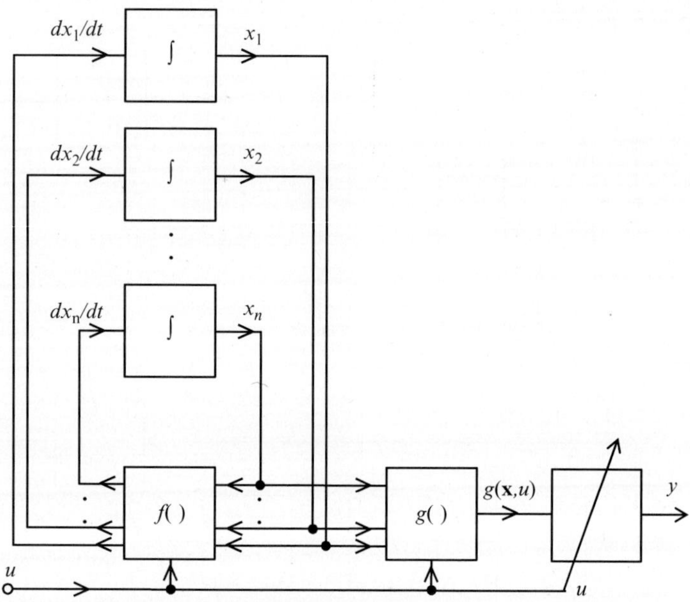

Fig. 4.6 Model of general memristive system based on Eqs. (4.12) and (4.13)

A different approach is shown in Fig. 4.5b. Use is made here of the fact that the state x' can be derived from the native state variable using a transforming function H according to (4.7). By integrating the excitation we get the signal u_I, which is the native state variable, from which the state x' can be obtained via a nonlinear block H. From the state, the parameter g is calculated and by multiplying it by the excitation quantity we get the response y. In Sect. 4.4.2 we demonstrate the fundamental advantage of such modeling over the generally used model in Fig. 4.5a. Let us be aware though that with the inclusion of an integrator in the model in Fig. 4.5a we have incorporated a native state variable into the model, which represents an unbounded memory. Then the model can act as a memristor, even if the transformed state variable x' represented a limited memory (see, e.g., Fig. 4.2c). The state equation of the system in Fig. 4.5b is a classical state equation of memristor for the native variable, and the port equation is at the same time a classical port equation of memristor (4.2), where $g(x) = g(S(H(x)))$.

It is obvious that the standard method of dynamic system modeling with the aid of integrators and nonlinear blocks is also usable for modeling general n-th order memristive systems [11]. An example of the direct programming of Eqs. (4.12) and (4.13) is shown in Fig. 4.6. We will not be concerned with other variants of the model which would be based on alternatively selected state variables.

4.4.2 Implementation of Models in PSPICE

Examples of the implementation of memristor models in the PSpice simulation program are described in this section. With slight modifications, given by syntactic differences, it is possible to implement these models also in other programs such as LTSpice, WinSpice, and HSpice.

These programs deal with systems of equations where the unknown and the input quantities have their physical dimensions of voltages and currents. When modeling memristors, whose exciting, output, and state quantities may be of a different physical nature, it is therefore necessary to use the following approaches:

1. The method of representing the exciting and response quantities: As for a memristor of electrical nature, we assemble the model such that it can be fully used for the modeling of more complex applications. In other words, we model the memristor as a floating two-terminal device. For the voltage-controlled memristor (VCMR), the exciting quantity will be the differential voltage across its terminals and the response quantity will be the current. In the case of the current-controlled memristor (CCMR), it will be the other way round. When considering memristors with non-electrical nature of their port quantities, it is necessary to consider how to define the excitation and the response. It is useful to do so in accordance with Table 4.2, i.e. to represent the quantity of the effort type by electrical voltage and the quantity of the flow type by electric current. However, if we do not intend to combine different partial models into larger units and if the object of the simulation is only a single memristive system, it is also possible to express the excitation and the response arbitrarily, e.g. by a pair of voltages.
2. The method of representing the state and other internal quantities: In principle, it is irrelevant whether the internal quantity is represented by voltage or current. In the SPICE-family programs, the former is usually preferred. Since internal quantities are determined based on formulae, controlled sources participate in their representation.
3. Most SPICE-family programs do not allow a direct modeling of resistors with variable resistances. Therefore, when modeling memristors of electrical nature, it is necessary to use an indirect way of modeling a memristive port by a controlled source. Consider, for example, the model in Fig. 4.4b, where, for the voltage-controlled memristor, the exciting quantity u would be a voltage and the response quantity y would be a current. Then we connect a controlled current source to the memristive port. Its current will be calculated according to the formula $g(x)u$. For the current-controlled memristor, we connect a controlled voltage source to the memristive port, and its voltage will be determined by the formula $g(x)i$, where i is the port current.
4. The usual methods of modeling an integrator in the SPICE-type programs are as follows. Consider that it is necessary to model the integration block with input u and output x.

The first method consists in utilizing a controlled source of voltage or current, depending on whether we are going to represent the variable x in the form of voltage or current. A formula for calculating the time-domain integral of u is assigned to this source. In the PSPICE language, for example, the formula would be SDT(V(in1, in2)). The SDT function calculates the time-domain integral of the memristor port voltage between the pins in1 and in2. We can add a term to this formula that defines the initial state of the integrator at the starting point of the transient analysis.

The second method is based on the knowledge that if a 1-Farad capacitor is charged by the current $i(t)$, then the capacitor voltage is equal to the time-domain integral of this current. So irrespective of its physical nature, the quantity we intend to integrate can be used to control the current of that source. The integral of this quantity, no matter what its physical unit is, will be numerically equal to the voltage across the charged capacitor in volts. The initial state of the integrator can be set by the IC attribute of the capacitor or by the .IC command.

Note that memristive systems are being simulated especially in the time domain, and their models should be designed such that they operate reliably during the Transient analysis.

The model implementation in PSpice will be demonstrated on several specific examples. A complete list of the source code of particular PSpice subcircuits is given for each example. In addition, PSpice input files for each simulation are included, with references to the library *memristor.lib*, summarizing the above subcircuits.

4.4.2.1 Memristor Switching Memory

Figure 4.7 illustrates an example of a piecewise-linear constitutive relation $q = q(\varphi)$ and two-state PSM $G_M = G_M(\varphi)$ of a memristor for the RRAM binary memory [10]. The memristor parameter is a memductance G_M, which is dependent on the flux φ and on its threshold value Φ according to the relation

$$G_M = \begin{matrix} G_{M1} & \text{if abs}(\varphi) < \Phi \\ G_{M2} & \text{otherwise} \end{matrix} \tag{4.23}$$

Since the native state variable is used for memristor definition, we can use the model in Fig. 4.4b. It is obvious that the exciting variable u is now the terminal voltage v of the memristor, the output variable y is the memristor current, and the state variable x is the flux φ, i.e. the time-domain integral of voltage.

One of the possible ways of specifying the block diagram for the purposes of SPICE modeling is shown in Fig. 4.8a.

The floating memristor is modeled at the + and – terminals using the controlled current source G. Its current is calculated as the product of the memristor voltage V(+, –) and the variable V(GM), which is the voltage at the node GM. The controlled voltage source EGM is connected to this node. Its voltage is calculated using formula (4.23). This way the memductance, which depends on flux, is modeled. The flux is represented in the model as a voltage of the node *phi*, thus

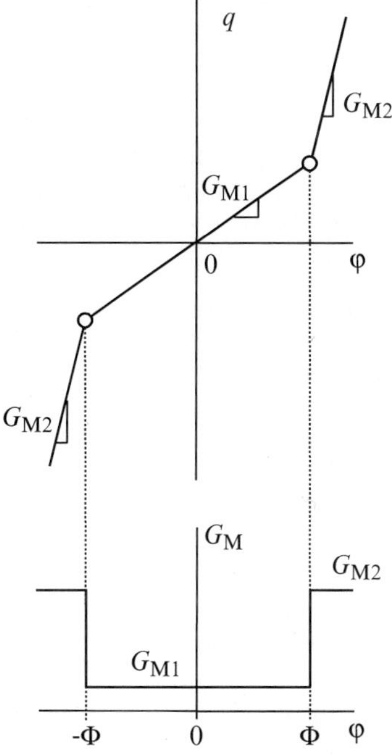

Fig. 4.7 Model of voltage- (or flux-) controlled memristor for binary memristive memories

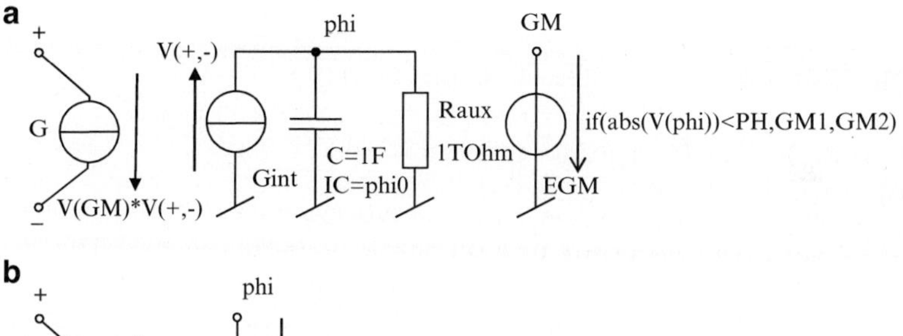

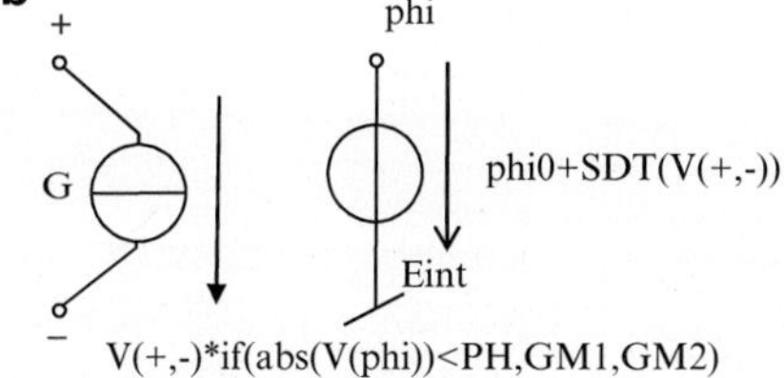

Fig. 4.8 Specification of block diagram from Fig. 4.4b for modeling memristor from Fig. 4.7 in PSpice, (**a**) complete, (**b**) economical version

V(phi). This voltage is obtained by integrating the terminal voltage of the memristor with respect to time. This integration is accomplished via the controlled current source Gint, which charges the capacitor C. The initial state of the flux can be adjusted via selecting the variable phi0. The auxiliary resistor Raux is necessary for providing the DC path between the node phi and the ground.

Figure 4.8b demonstrates a possible simplification of the model. The integrator is now implemented by a controlled voltage source Eint, whose voltage is calculated using the internal function SDT. Individual calculation of the voltage V(GM), expressing the memductance, is now omitted, and it is applied directly to the calculation of the current of the controlled source G. Note that, compared to the model in Fig. 4.8a, the users would not have any direct data about the memductance after finishing the simulation, but they may obtain them indirectly from the voltage and current.

Based on the model in Fig. 4.8a, it is possible to compile a SPICE subcircuit with the source code as below:

```
SUBCKT 1:
.subckt memristor_PWL + – params: GM1=20u GM2=100u
+ PH=0.2 phi0=0
G + – value={V(GM)*V(+,-)}
Gint 0 phi value={V(+,-)}
C phi 0 1 IC={phi0}
Raux phi 0 1 T
EGM GM 0 value={if(abs(V(phi))<PH,GM1,GM2)}
.ends memristor_PWL
```

The source code of the model from Fig. 4.8b is simpler:

```
SUBCKT 2:
.subckt memristor_PWL_simple + – params:
+ GM1=20u GM2=100u PH=0.2 phi0=0
G + – value={V(+,-)*if(abs(V(phi))<PH,GM1,GM2)}
Eint phi 0 value={phi0+SDT(V(+,-))}
.ends memristor_PWL_simple
```

If we excite such a memristor by a sinusoidal 2 V/1 Hz voltage source, then the complete input file for the simulation may be as follows:

```
CIR 1:
PWL memristor, sin input
.param Vmax 2 f 1
Vmem mem 0 SIN 0 {Vmax} {f}
Xmem mem 0 memristor_PWL params: GM1=20u GM2=100u
+ PH=0.2 phi0=0
.lib memristor.lib
.tran 0 5 0 0.5m skipbp
.probe
.end
```

The following PWL memristor is defined on the fourth line: It is connected between the nodes mem and 0, with the conductances GM1 = 20 μs, GM2 = 100 μs, and with the parameter PH = 200 mV s (the threshold flux Φ from Fig. 4.7). The initial flux is zero (phi0 = 0). According to PSM in Fig. 4.7, the corresponding

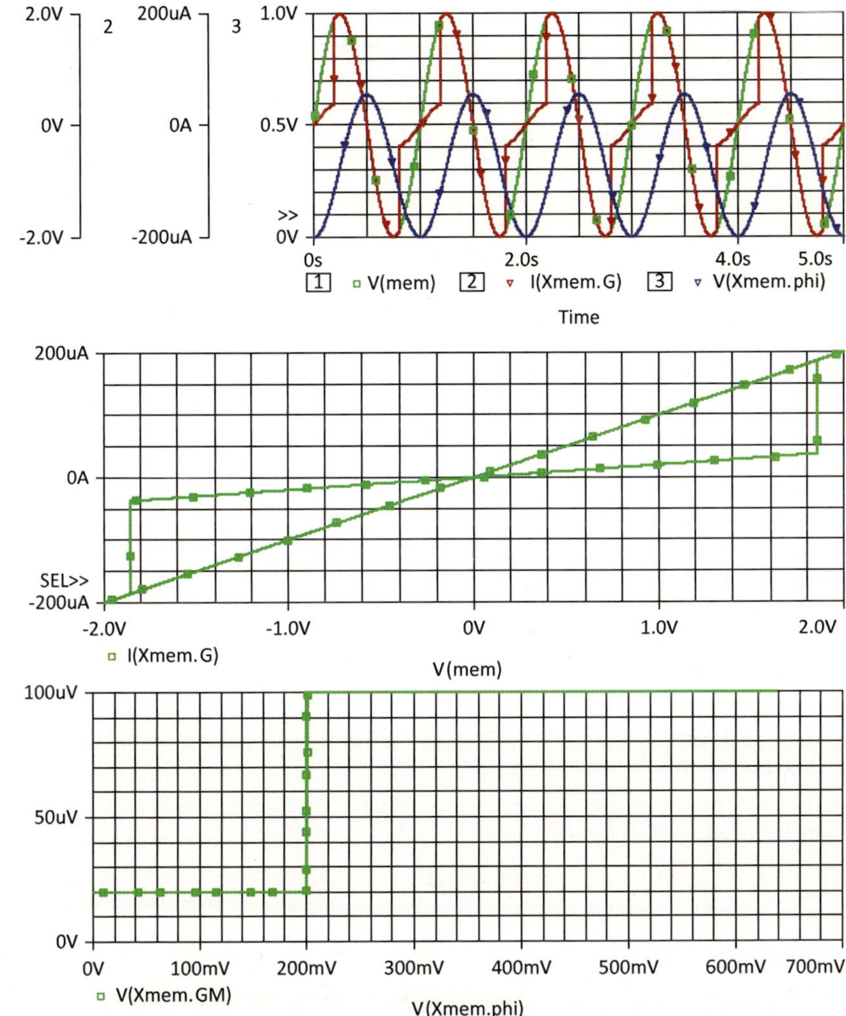

Fig. 4.9 Results of transient analysis of PWL memristor with parameters GM1 = 20 μs, GM2 = 100 μs, Φ = 200 mV s, excited by 2 V/1 Hz sinusoidal voltage. Quantities V(mem), I(Xmem.G), V(Xmem.phi), and V(Xmem.GM) represent memristor voltage, current, flux, and conductance. At time 0, memristor is in state phi0 = 0

conductance is GM1. Since the terminal voltage of the memristor has a sinusoidal character, its time-domain integral will always be nonnegative, and the flux φ will move the operating points at the characteristics in Fig. 4.7 only within the region $\varphi \geq 0$.

The simulation results are shown in Fig. 4.9. Note that the quantities V(mem), I(Xmem.GM), V(Xmem.phi), and V(Xmem.GM) represent the voltage, current, flux, and conductance of the memristor. The flux waveform V(Xmem.phi) in the top plot confirms that the flux is nonnegative and oscillates between zero and ca 636 mV s. The bottom part of the figure, demonstrating the conductance versus flux relationship, indicates a step change in the memristor conductance for the flux threshold value PH = 0.2 mV s. The middle plot shows the corresponding PHL.

If the operating point is forced to move in the GM vs. flux characteristics also in the area of negative flux, and thus all three piecewise-linear segments of CR are used, then the effect which is known from the behavior of the so-called complementary resistive switch will be produced [21]. It can be accomplished by modifying the initial state of the memristor, e.g. via selecting phi0 = – 0.3 Vs on the fourth line of the input file. The corresponding results are given in Fig. 4.10.

Also note that the memristor passes immediately into the periodical steady state, which is in accordance with FP4. By gradually increasing the excitation frequency *f*, it is possible to verify the validity of FP5: Since the flux swing gradually diminishes, then, after exceeding some upper limit frequency, the working point can no longer be switched between different conduction states. The memristor will then behave as a linear resistor, thus without any hysteresis in the *i*–*v* characteristic.

4.4.2.2 The TiO_2 Memristor

The well-known simple physical model of a memristor [8] in Fig. 4.11 consists of a thin TiO_2 double-layer (thickness D is about 10 nm) between a pair of platinum electrodes. One of the TiO_2 layers is doped with oxygen vacancies and therefore it behaves as a semiconductor. The second, undoped layer has insulating properties. As a result of complex processes in the material, the width w of the doped layer varies in dependence on the amount of electric charge that passes through the memristor. When the current passes in a given direction, the boundary between the two layers is moving in the same direction. The resulting resistance R_{mem} of the memristor is given by the sum of resistances of doped and undoped layers, which can be described as

$$R_{mem}\left(x'\right) = R_{on}x' + R_{off}\left(1 - x'\right) = R_{off} - \Delta R x', \;\; \Delta R = R_{off} - R_{on} \tag{4.24}$$

where

$$x' = \frac{w}{D} \in (0,1) \tag{4.25}$$

is the width of the doped layer, scaled to the total width D, and R_{off} and R_{on} are the memristor resistance limit values for $w = 0$ and $w = D$. The ratio of both resistances is usually given between 10^2 and 10^3.

Ohm's law holds between the memristor voltage and current:

$$v(t) = R_{mem}\left(x'\right) \cdot i(t). \tag{4.26}$$

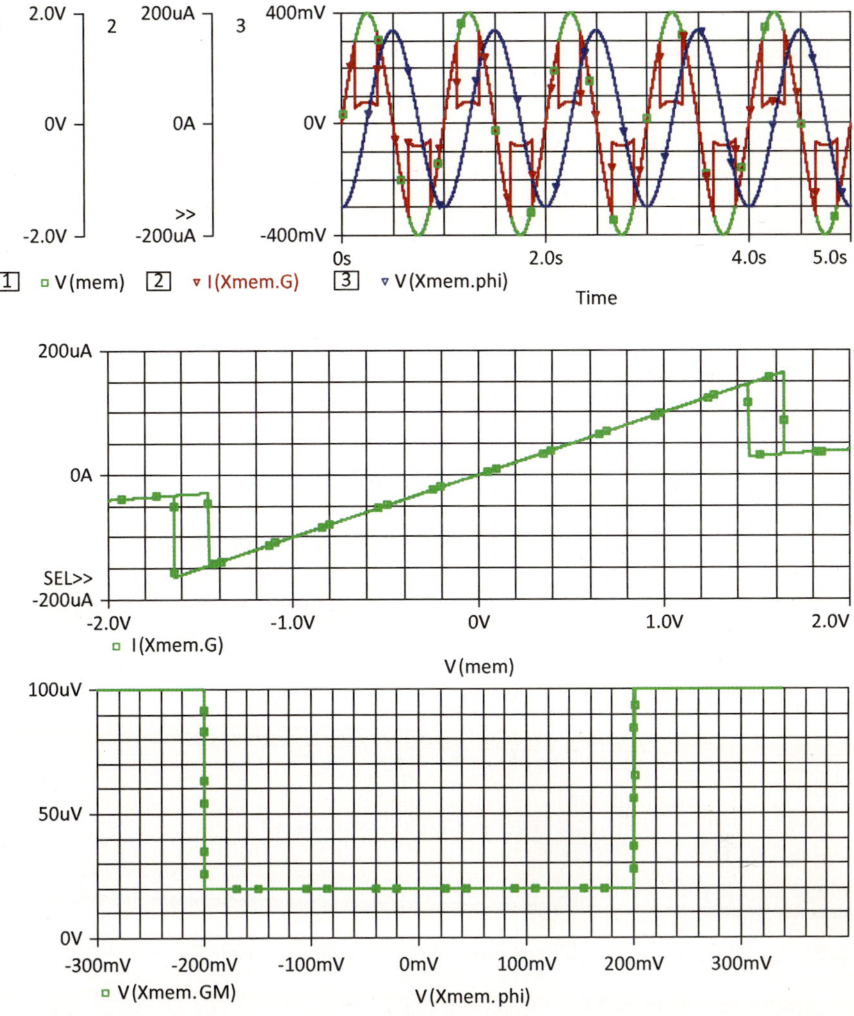

Fig. 4.10 Results of transient analysis of PWL memristor with parameters GM1 = 20 μs, GM2 = 100 μs, Φ = 200 mV s, excited by 2 V/1 Hz sinusoidal voltage. Quantities V(mem), I(Xmem.G), V(Xmem.phi), and V(Xmem.GM) represent memristor voltage, current, flux and conductance. At time 0, memristor is in state phi0 = –0.3 V s

The speed of the boundary between the two layers can be simply described by the state equation [16]

$$\frac{dx'}{dt} = ki(t)f_W\left(x'\right), \quad k = \frac{\mu_v R_{on}}{D^2} \tag{4.27}$$

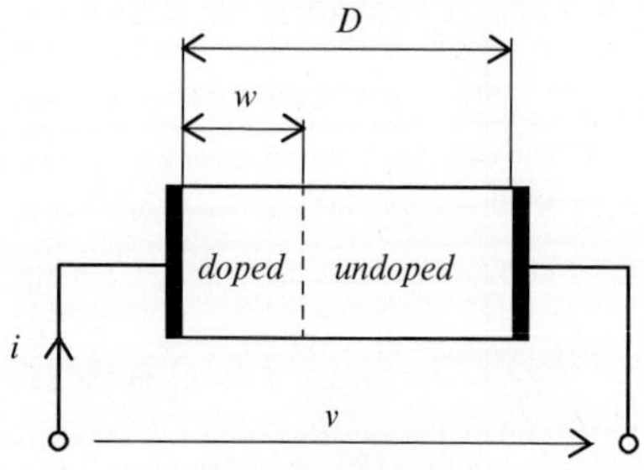

Fig. 4.11 Simple model of TiO_2 memristor according to [8]

where μ_v is the so-called average mobility of dopants with values of about 10^{-14} m^2 s^{-1} V^{-1}. In nano-devices, even small voltages may cause enormously high strengths of electric field to occur that subsequently induce significant nonlinearities in ion transport [8]. These nonlinearities have a specific impact at the edges of a thin film when the speed of the boundary gradually decreases to zero when approaching both edges of the layers. This phenomenon, known as nonlinear drift, can be modeled in a simplified way by the so-called window function $f_W(x')$ on the right side of differential equation (4.27).

From the existing window functions, let us mention the rectangular (f_{WR}), Joglekar (f_{WJ}) [16], and Biolek (f_{WB}) [13] functions, see Fig. 4.12:

$$f_{WR}\left(x'\right) = \text{stp}\left(x'\right) - \text{stp}\left(x' - 1\right), \tag{4.28}$$

$$f_{WJ}\left(x'\right) = 1 - \left(2x' - 1\right)^{2p}, \tag{4.29}$$

$$f_{WB}\left(x', i\right) = 1 - \left(x' - \text{stp}\left(-i\right)\right)^{2p}, \tag{4.30}$$

where p is a positive integer and i is the current flowing through the memristor. The current direction is considered positive if this current increases the width of the doped layer, thus $x' \to 1$. The *stp* symbol denotes the step function

$$\text{stp}(i) = \begin{cases} 1 & \text{for } i \geq 0 \\ 0 & \text{for } i < 0 \end{cases} \tag{4.31}$$

With p growing towards infinity, the shape of window functions (4.29) and (4.30) is changing towards the form of a rectangular window. For $p \to \infty$, the Biolek window is passing to the Corinto window [22]. When comparing the Joglekar and the Biolek windows, it is useful to mention that the curve of the Joglekar window starts from the origin of the coordinates with a slope of $4p$, whereas the slope of the curve of the Biolek window, pointing to the origin, is only $2p$. The Biolek window should therefore be modeled with twice the parameter p for the Joglekar window in order to maintain the same slopes.

A serious problem with the rectangular and Joglekar windows consists in the fact that if the system got into one of its limit states, i.e. $x' = 0$ or $x' = 1$, then, according to (4.27), this would mean a zero derivative condition and thus the impossibility

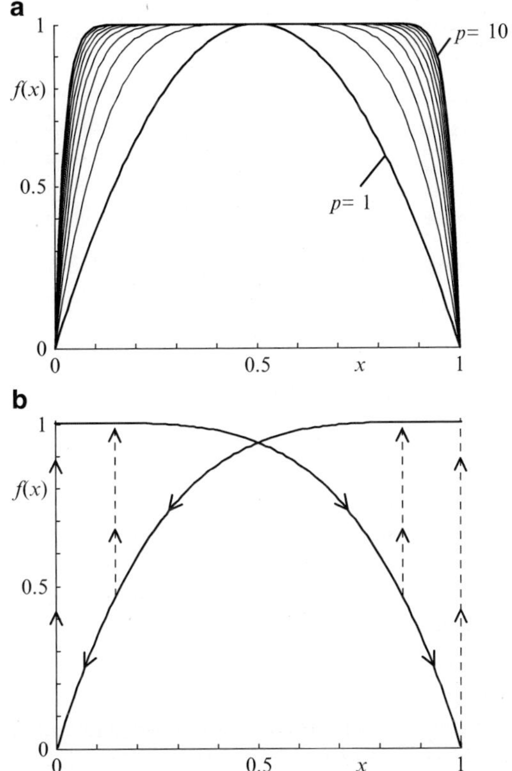

Fig. 4.12 (**a**) Joglekar window from $p = 1$–10, (**b**) Biolek window for $p = 2$. For $p \to \infty$, both windows are tending to rectangular window

to change this status by any external excitation (see the analysis at the end of Sect. 4.3.1). Some authors circumvent this problem by considering nonzero values of the window function at limit states [23]. However, then another problem appears: If the system reaches a limit state that cannot be got over due to physical reasons, the state cannot change anymore and its derivative with respect to time must therefore be zero. This can be changed only by changing the direction of the exciting quantity. The Biolek window is based on this fact. This window function depends on the state as well as on the direction of the current in the circuit.

Note for completeness that other windows for modeling the nonlinear dopant drift were also published, for example Strukov [8], Prodromakis [24], and TEAM, or Kvatinski [25] windows.

When analyzing Eqs. (4.24)–(4.27), we conclude that they formally correspond to memristor equations (4.8) and (4.10) in the nonstandard form. Comparing the state equations (4.27) and (4.10), and taking into account the window functions (4.28)–(4.30), we come to the following conclusions:

When modeling the TiO_2 memristor by the rectangular or the Joglekar window, the state equation corresponds to the equation of the current-controlled memristor, i.e. $u = i$, in the nonstandard form (4.10): the state derivative is equal to the product of the controlling quantity (current) and the nonlinear function of the state $kf_W(x')$.

An analysis of the rectangular window shows that the state variable is directly proportional to the native state variable. More specifically, the state variables are proportional within an admissible range (4.25), and the state variable (4.25) shows limitations at both boundaries. This is the case from Fig. 4.2c. The result is that the system can smoothly pass into states with zero-derivatives of these states within a finite time, but it cannot extricate itself from these states. That is why it is not a memristor in the true sense of the word. It acts as a memristor only if the state x' moves within the limits (4.25). The model is therefore problematic, for instance, when modeling the hard-switching effects.

For the finite values of the parameter p, the Joglekar window leads to a strictly monotonic transformation between the native state variable x and the state variable x' as shown in Fig. 4.2b, and the corresponding model is thus a model of a memristor.

It is obvious from the above analysis that the respective window functions are just another way of expressing the nonlinear relationship H or S between the native and the "physical" state variables, and that the nonlinear functions H (4.7) could be used instead of the window functions for modeling the nonlinear drift in the TiO_2 memristor. This approach will be utilized below.

When modeling the TiO_2 memristor with the Biolek or the Corinto window, it is not a memristor model any more but the model of a more general first-order memristive system according to Eqs. (4.10) and (4.11): The state derivative is a nonlinear function of the state and also the controlling quantity $kif_{WB}(x',i)$. This also means that the actual TiO_2 memristor, for which the above fixation of the state is not observable (which can be well modeled using these windows), cannot, generally speaking, be considered the fourth fundamental circuit element, and therefore some of the FPs from Sect. 4.3 do not apply to it.

Over time, these simple models were elaborated in order to correlate more closely with experimental data. However, the characterization was done only for a few specifically manufactured memristive systems excited particularly by sinusoidal and triangular signals. It is therefore not obvious to what extent the models are valid under general conditions [26]. In other words, the given models probably still do not show the attributes of the correct modeling by Prof. Chua. A typical representative of such modeling is a refined model of the memristive port from [27]

$$i = x^n c_1 \sinh(d_1 v) + c_2 \left(e^{d_2 v} - 1\right) \tag{4.32}$$

where x is the normalized width of the doped layer according to (4.25), understood here as a state variable, while the symbols n, c_1, d_1, c_2, and d_2 denote fitting constants.

The state equation is considered in the form

$$\frac{dx}{dt} = a f_W(x) v^q \tag{4.33}$$

where a is a fitting constant, f_w is the window function, and q is a positive odd number.

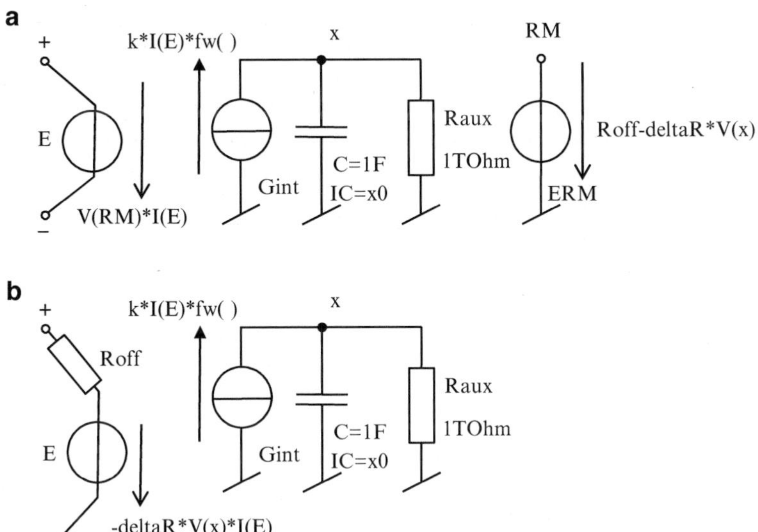

Fig. 4.13 Specification of circuit equations (4.24)–(4.27) for modeling TiO_2 memristor in PSpice, (**a**) complete, (**b**) economical version

We can conclude from the analysis of Eqs. (4.32) and (4.33) that the memristive port is described by the port equation (4.12) for the voltage-controlled generalized first-order memristive system, where the memductance depends on the state x as well as on the voltage v:

$$g(x,v) = \begin{matrix} x^n c_1 d_1 + c_2 d_2 & ..v = 0 \\ x^n c_1 \frac{\sinh(d_1 v)}{v} + c_2 \frac{e^{d_2 v} - 1}{v} & ..v \neq 0 \end{matrix} \tag{4.34}$$

The state equation (4.33) corresponds to the state equation of the first-order memristive system (4.13). That is why it is not a memristor. The mathematical model shows that it is a nonvolatile memory because the state derivative is zero for $v = 0$, and the given state then determines the memductance according to formula (4.34) for $v = 0$. It results from (4.33), however, that if we use some of the classical window functions of the type of (4.28) or (4.29), the model will suffer from the “fixed state” problem. Instead of this, it is useful to use a function of the type of (4.30) which will depend not only on the state but also on the excitation quantity, the voltage in this case. An example is described in [26].

In the following, the procedures of modeling the basic memristor equations (4.24)–(4.29) and (4.31) are given. The SPICE modeling of the TiO_2 memristive systems in terms of Eqs. (4.32)–(4.34) and others is described in detail in [26, 28].

The model of the TiO_2 memristor, given by Eqs. (4.24)–(4.27), has been redrawn to the diagram in Fig. 4.13a with the aim of implementing it subsequently in the SPICE program.

In the model in Fig. 4.13a, the memristance is calculated from formula (4.24) as the terminal voltage of the auxiliary source ERM. The state variable, on which the memristance is dependent, is represented by the voltage of the node x. In accordance with the state equation (4.27), this voltage is computed by integrating the expression kif_W, which is the value of the current of the source Gint, charging the capacitor C. The memristor resistance between the + and – terminals is modeled by the controlled source E. Its voltage is calculated as the product of the memristance and the current flowing through the memristive port.

Figure 4.13b shows one of the economical alternatives, when the source ERM for computing the memristance is omitted. When modeling the memristive port, use is made of the fact that the memristance is equal to the difference between the fixed resistance R_{off} and the expression ΔRx. It can be modeled by a serial combination of the resistor R_{off} and the reversed voltage source with the voltage equal to the product of ΔRx and the current flowing through the memristor. The advantage of such a modeling is that the memristive port is not modeled by an ideal source and thus there is no risk of a potential conflict resulting from an improper connection of ideal sources when integrating the memristor in an application circuit.

The code of the SPICE subcircuit (SUBCKT 3) of the model in Fig. 4.13b is as follows.

```
SUBCKT 3:
.subckt memristor_TiO2_1 + - params:
+ Ron=100 Roff=100k Rini=10k p=10
.param uv 10F D 10n
+ k {uv*Ron/D**2} deltaR {Roff-Ron}
+ x0 {(Roff-Rini)/deltaR}
.func fwJ(x) {1-(2*x-1)**(2*p)}; Joglekar window
.func fwB(x,i) {1-(x-stp(-i))**(2*p)}; Biolek window
Roff + aux {Roff}
E aux - value={-deltaR*V(x)*I(E)}
* for using the Biolek window instead Joglekar window,
* replace fwJ(v(x)) by fwB(v(x),I(E)) on line below
Gint 0 x value={k*I(E)*fwJ(v(x))}
C x 0 1 IC={x0}
Raux x 0 1T
.ends memristor_TiO2_1
```

The input file CIR2 listed below models the memristor excitation from the source of 0.2 V/1 Hz sinusoidal voltage. A simple Joglekar window for $p = 1$ is used here for modeling the nonlinear dopant drift. The initial memristance $R_{ini} = 9$ kΩ is set close to the upper limit $R_{max} = 10$ kΩ.

The auxiliary controlled voltage sources Eq and Ephi are used for computing the time-domain integrals of memristor current and voltage, i.e. for computing the charge and the flux. They can be used for a convenient visualization of these variables in a PROBE postprocessor.

```
CIR 2:
TiO2 memristor, sin input
.param Vmax 0.25 f 1
VmemJ memJ 0 SIN 0 {Vmax} {f}
XmemJ memJ 0 memristor_TiO2_1 params:
+ Ron=1k Roff=10k Rini=9k p=1
Eq q 0 value={sdt(-i(VmemJ))}
Ephi phi 0 value={sdt(V(memJ))}
.lib memristor.lib
.tran 0 5 0 0.5m skipbp
.probe
.end
```

The simulation results are shown in Fig. 4.14. The simulation confirms several memristor FPs. The flux-charge CR is an unambiguous curve (the top plot). The same applies to the memristance vs. charge map (second picture from the top). The PHL in current–voltage characteristics (third plot from the top) forms a simple closed pattern, indicating that the circuit has immediately passed into the periodical steady state (see the fingerprint FP4). The PHL exhibits the odd symmetry (see FP6) and is of the crossing type (see FP7). The limit points of the PSM curve do not lie on a horizontal straight line, so that the appropriate PHL cannot be of the crossing type with tangential touching (see FP8) and cannot be degenerated (see FP10).

If we repeat the simulation for an increasing frequency of the excitation signal, we can verify the validity of FP5 (the area within the loop gradually disappears).

The simulation may be repeated for the Biolek window (when modifying the line defining Gint in the SUBCKT3 as indicated in the corresponding note) under the same conditions except for $p=2$. The results are given in Fig. 4.15. It is obvious that several memristor FPs have now been violated. Both CR and PSM are not unambiguous functions of a native state variable. PHL does not show the odd symmetry, and, also, a gradual, not immediate transition to the periodical steady state is evident. On the other hand, this window indicates its potential for modeling real memristive systems that naturally violate the memristor FPs (for example, they generate asymmetric PHLs).

The results of another simulation with the Joglekar window are shown in Fig. 4.16. The simulation parameters are identical to those for the first simulation in Fig. 4.14, only the initial value of the memristance R_{ini} was reduced to 3 kΩ. It results in a shift of the initial position of the boundary between the doped and undoped layers towards the "right-side" edge of the TiO_2 memristor in Fig. 4.11, and thus in increasing the initial value of the state variable x'. The results in Fig. 4.16 are surprising at first sight: a number of memristor FPs are violated. The waveform of the state variable x′, V(XmemJ.x), manifests a slow transition to the periodical steady state. It violates FP4. This subsequently leads to ambiguities in other characteristics. Nevertheless, the model corresponds to a memristor model.

This contradiction can be explained by numerical errors that occur during the simulation, especially during the first repeating period of the excitation signal, when

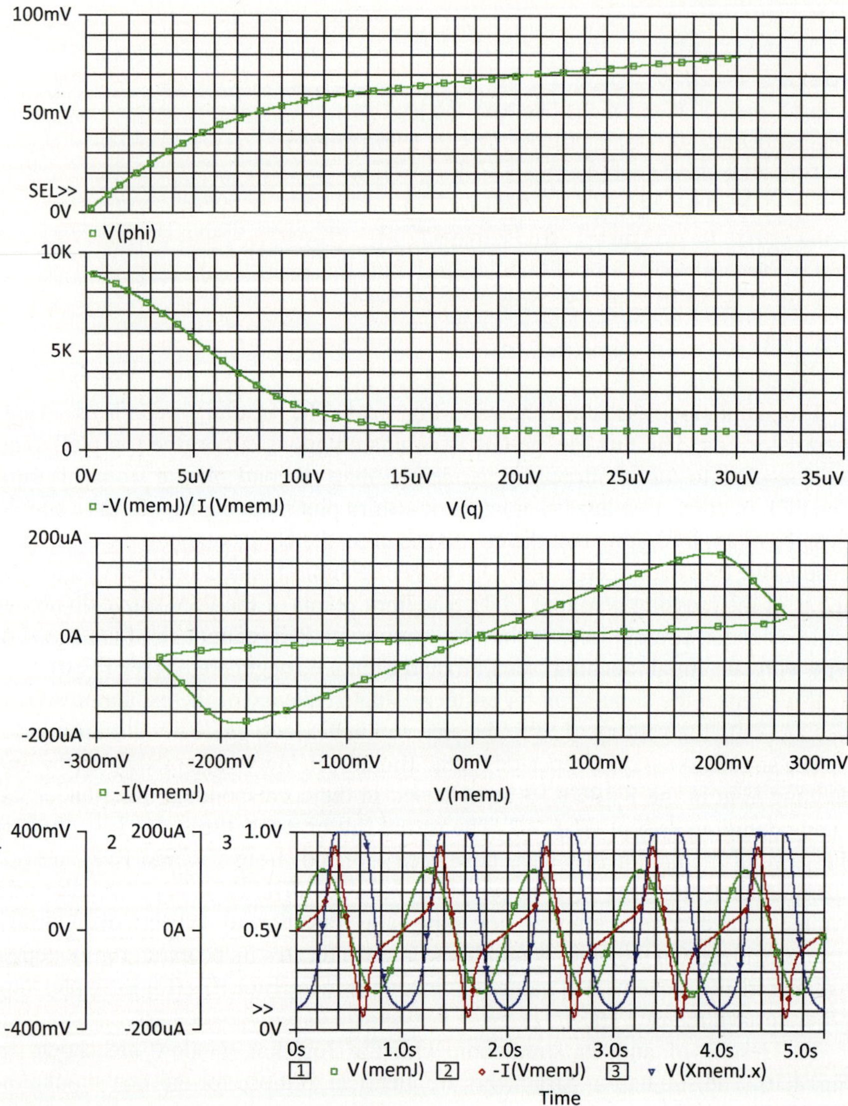

Fig. 4.14 Results of analysis of TiO_2 memristor according to data from CIR2. Joglekar window with $p = 1$ is used to model nonlinear drift. First two plots from top are flux-charge CR and memristance vs. charge map. Below them is the v–i PHL. Bottom plot shows waveforms of voltage, current, and state variable x'

the differences between the calculated state variable x' and its theoretical maximum 1 are so small that they cannot be expressed, within the given accuracy of the number representation, other than by zero. However, the state derivative is then zero and the already mentioned "fixed state condition" occurs. The system can extricate itself

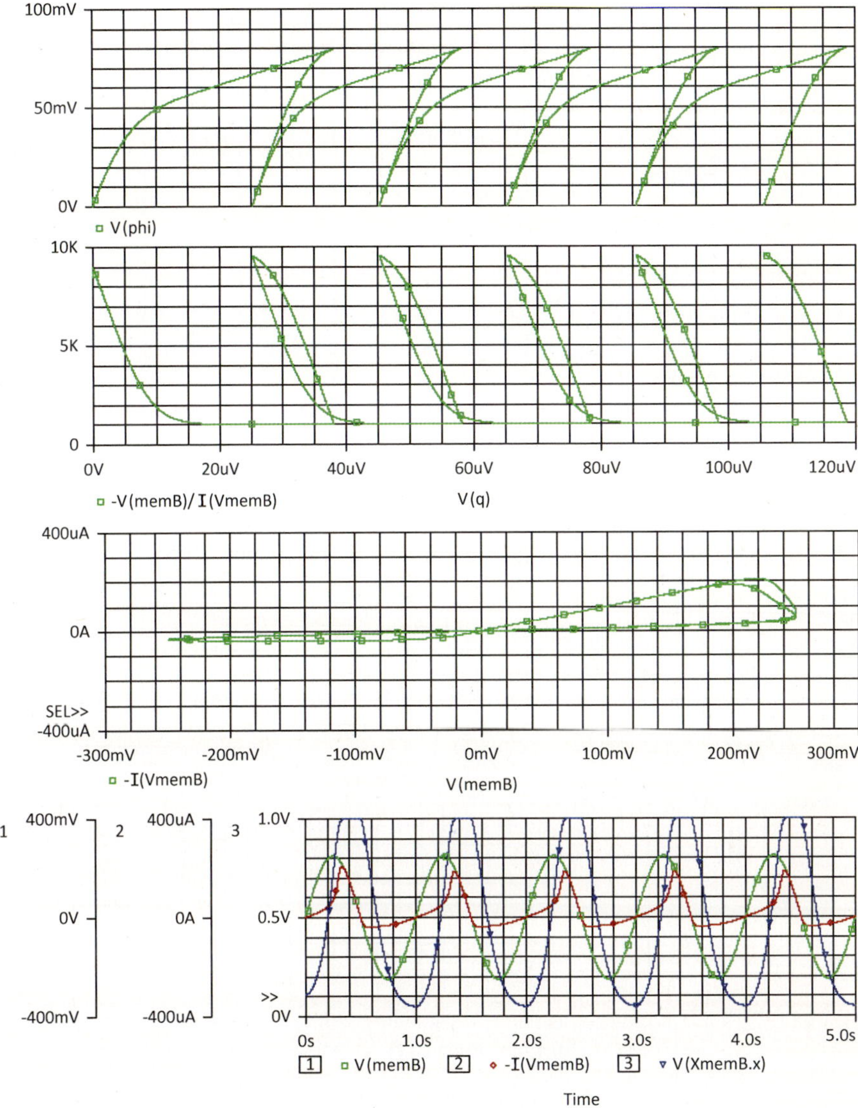

Fig. 4.15 Results of analysis of TiO_2 memristor. Simulation parameters are similar to those in Fig. 4.14. Biolek window with $p = 2$ is used to model nonlinear drift

from such a state only by the influence of similar numerical errors. If somebody does not know FP4, they could easily consider the results in Fig. 4.16 correct, drawing a mistaken conclusion that the system being modeled is not a memristor.

Unfortunately, this incorrect behavior of the simulation program cannot be avoided, even if the accuracy of calculations is increased by tweaking the error

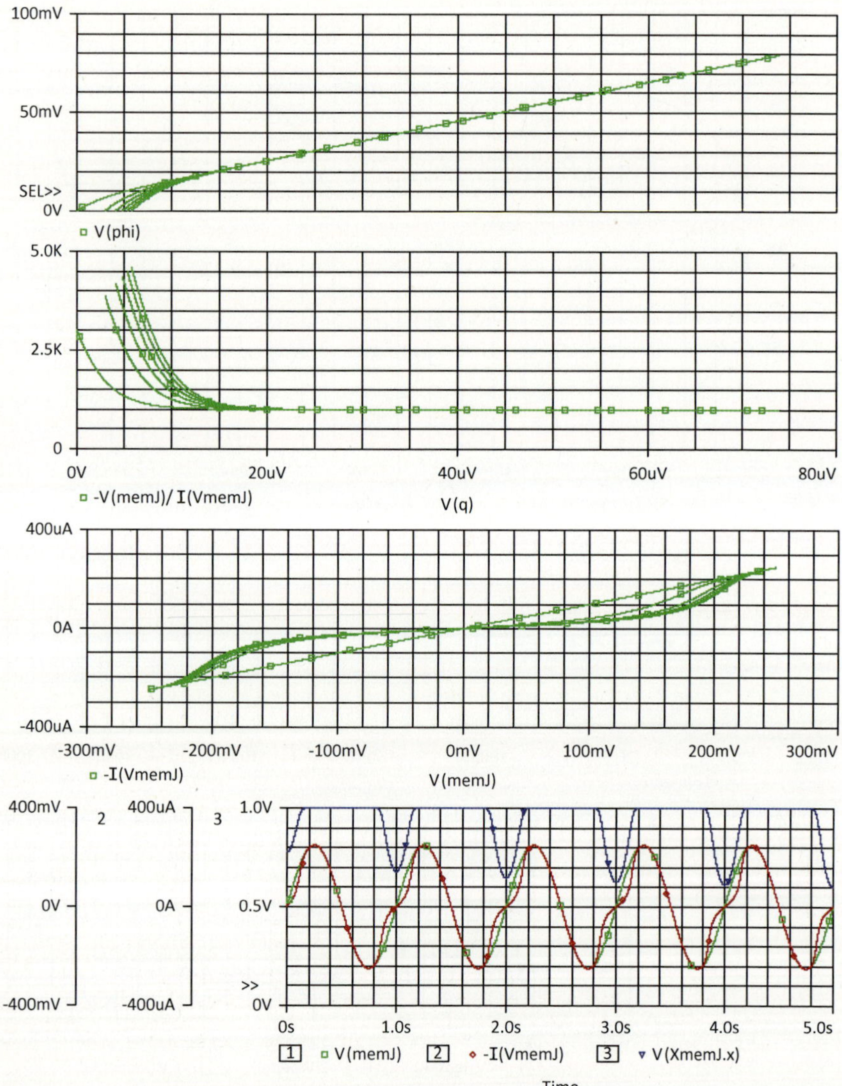

Fig. 4.16 Results of analysis of TiO_2 memristor. Simulation parameters are similar to those in Fig. 4.14 with one exception: initial value of memristance R_{ini} is reduced from 9 to 3 kΩ. Results are burdened with unacceptable numerical errors

criteria such as RELTOL, ABSTOL, and VNTOL, or the step ceiling is reduced. Therefore, it is necessary to look for other, nontraditional ways. One of them is to change the philosophy of memristor modeling by changing from the diagram (a) to the diagram (b) in Fig. 4.5. The source of the numerical errors lies in the effort to solve numerically the differential state equation (4.27). The model in Fig. 4.5b does not use this equation at all. Instead, a primitive differential equation (4.2) is being modeled for the native state variable x, and it is solved by a simple integration. The state variable x' and the memristance derived from x' are determined from the native state variable by simple arithmetic calculations.

To implement the model in Fig. 4.5b, it is necessary to determine the transforming function H between the state variables x and x'. The procedure will be demonstrated for the Joglekar window with $p = 1$, which was used in previous models. Comparing Eqs. (4.10) and (4.27), we find that the transforming function S for the TiO_2 memristor must comply with the rule

$$\frac{dS(x',x'_0)}{dx'} = \frac{1}{kf_{WJ}(x')} = \frac{1}{k}\frac{1}{1-(2x'-1)^2} \tag{4.35}$$

or

$$S(x',x'_0) = \frac{1}{k}\int_{x'_0}^{x'} \frac{1}{1-(2x'-1)^2} = \frac{1}{4k}\ln\left(\frac{x'}{x'_0}\frac{1-x'_0}{1-x'}\right). \tag{4.36}$$

After inverting the S function with respect to the variable x' we get a formula for the transforming function H:

$$H(x,x'_0) = \frac{1}{1+\left(\frac{1}{x'_0}-1\right)e^{-4kx}} \tag{4.37}$$

which can be modified to the form

$$H(x,x_0) = \frac{1}{1+e^{-4k(x-x_0)}},\quad x_0 = \frac{1}{4k}\ln\left(\frac{1}{x'_0}-1\right). \tag{4.38}$$

It is evident that the graph of the transforming function H between the native state variable and the variable x' (see Fig. 4.2b) for the TiO_2 memristor and for the Joglekar window with $p = 1$ corresponds to the well-known sigmoid (logistic) function of the type $1/(1+e^{-x})$, shifted along the x axis in dependence on the initial value of the state x'_0.

The SPICE subcircuit (SUBCKT4) shows the method of modeling the TiO_2 memristor on the basis of the method in Fig. 4.5b, utilizing the sigmoid function (4.38), which is defined on the fourth line. The memristive port is modeled the same as in the previous SUBCKT3 (fifth and sixth lines). The purpose of the source Eq is to calculate the native state variable, i.e. the charge, and thus the time-domain integral of the current flowing through the memristive port. The computed charge is then expressed as the voltage of the node q. The source Ex is used to convert the native state variable to the boundary position via the sigmoid function H.

```
SUBCKT 4:
.subckt memristor_TiO2_H + - params: Ron=100 Roff=100k
+ Rini=10k
.param uv 10F D 10n
+ k {uv*Ron/D**2} deltaR {Roff-Ron}
+ x0 {(Roff-Rini)/deltaR}
.func H(x,x0) {1/(1+(1/x0-1)*exp(-4*k*x))}
Roff + aux {Roff}
E aux - value={-deltaR*V(x)*I(E)}
Eq q 0 value={sdt(I(E))}
Ex x 0 value={H(v(q),x0)}
.ends memristor_TiO2_H
```

The simulation from Fig. 4.16 was repeated after replacing the classical model of the memristor from SUBCKT3 with the SUBCKT4 model. The results are shown in Fig. 4.17. It is obvious that the numerical errors from the previous simulation have been removed.

A more detailed analysis of the effects of the numerical errors discussed leads to the knowledge that these errors can completely corrupt the simulated outputs. Examples are given in Fig. 4.18, which were generated on the basis of the input file CIR3:

```
CIR 3:
TiO2 memristors, sin input
.param Vmax 1 f 1 Ron 1k Roff 10k Rini 9k p 1
VmemJ memJ 0 SIN 0 {Vmax} {f}
XmemJ memJ 0 memristor_TiO2_1 params:
+ Ron={Ron} Roff={Roff} Rini={Rini} p={p}
VmemH memH 0 SIN 0 {Vmax} {f}
XmemH memH 0 memristor_pokus params:
+ Ron={Ron} Roff={Roff} Rini={Rini}
.lib memristor.lib
.tran 0 5 0 0.5m skipbp
.probe
.end
```

The amplitude of the excitation voltage is now increased to 1 V. Two memristors (XmemJ with the standard model and XmemH with the sigmoid model) are simulated simultaneously. Even though the models are mathematically equivalent, the simulation results are absolutely different. The standard model leads to a rapid transition of the state x' to its upper limit, which causes the above numerical errors. These errors then burden all further calculations. The system is moving to a steady state which can be evaluated as a "hard-switching effect." After comparing it with the upper pair of plots, it is clear that the memristor is in a different mode. The curves of the evolution of the position of the boundary between doped and undoped layers overlap only at the beginning of the transient analysis until the numerical problem occurs.

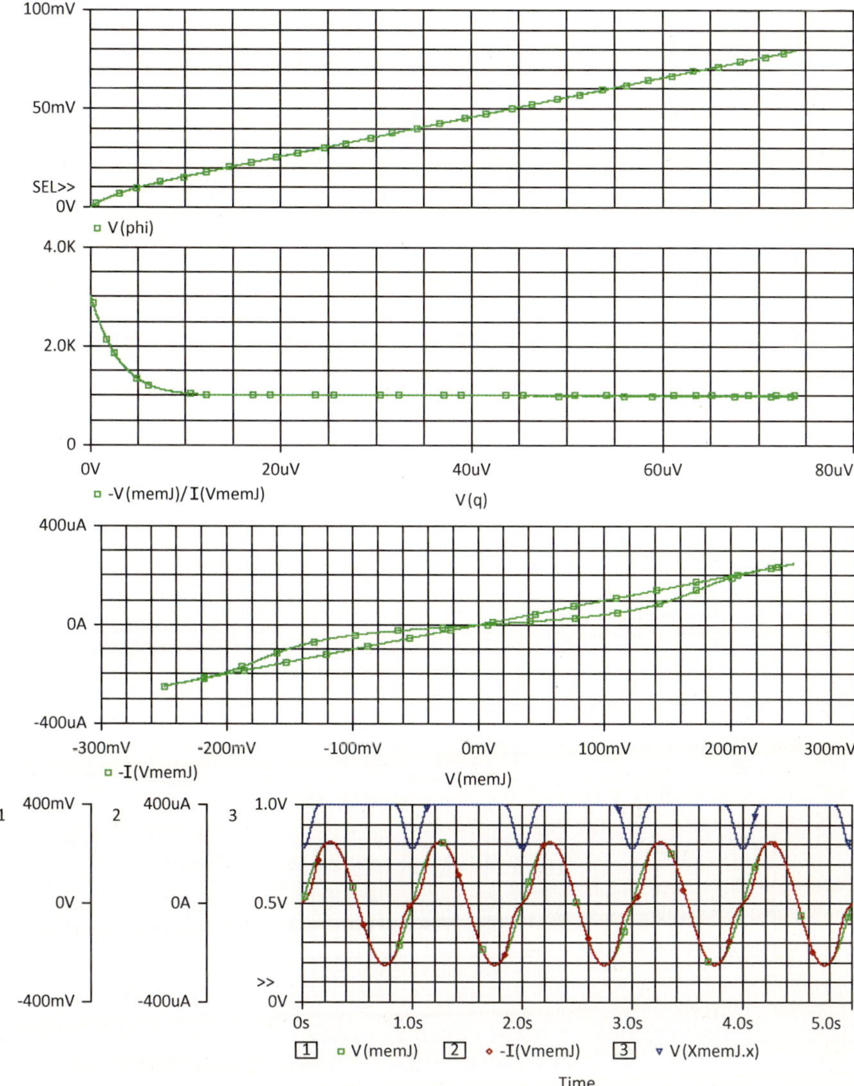

Fig. 4.17 Results of analysis of TiO_2 memristor. Simulation parameters are similar to those in Fig. 4.16, but SUBCKT4 is used for memristor modeling. Outputs are not affected by any numerical errors as in Fig. 4.16

As can be seen, modeling a memristor by means of function that transforms the state variables can be preferable to modeling it on the basis of the window functions. Tweaking the window functions by their parameters (see the parameter p of the Joglekar window) is accompanied by modifying the shapes of the transforming

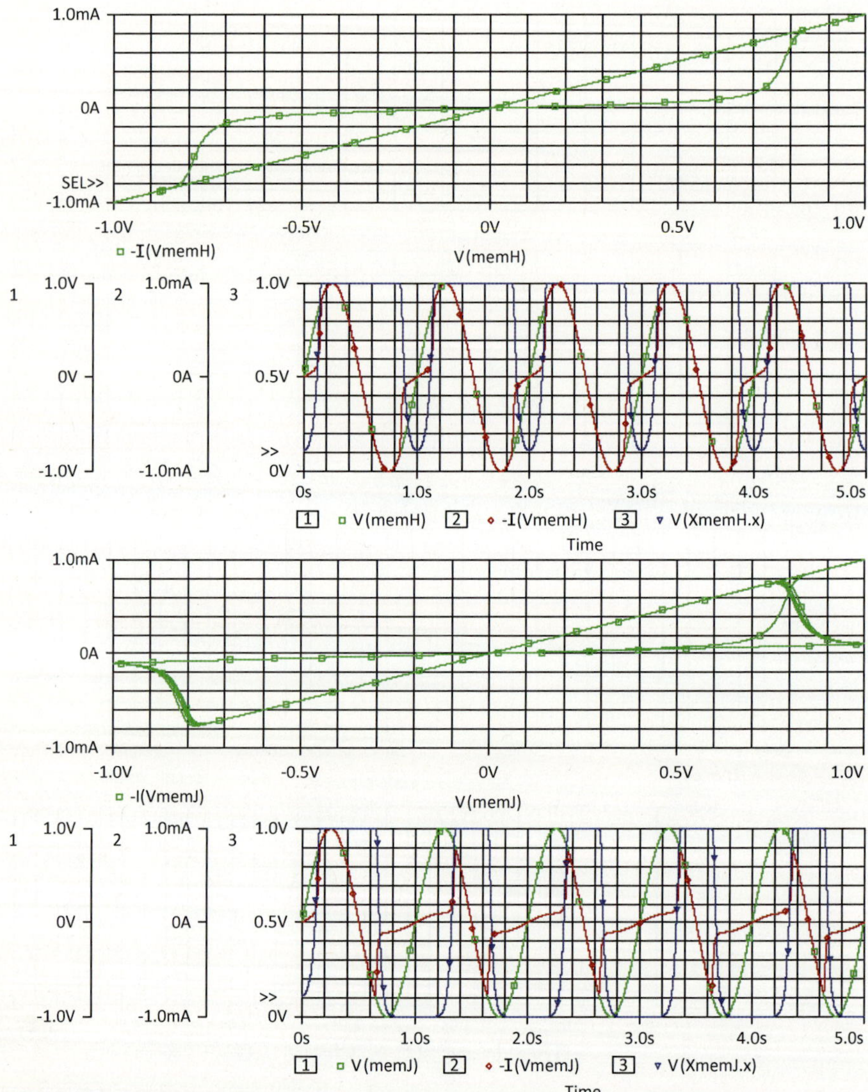

Fig. 4.18 Results of analysis of TiO_2 memristor with two different but mathematically equivalent models. Simulation parameters are defined in CIR3. First two plots show behavior of model utilizing functions transforming states. Next two plots reveal unacceptable errors generated by standard memristor model using window functions

functions *S* and *H*, which prompts the idea of searching for such transformation functions (not window functions) that would lead to the required behavior of the memristor model.

4.4.2.3 Memristors with Polynomial CR

Consider a memristor with CR (4.3) expressed in the polynomial form

$$y_I = \sum_{k=1}^{\infty} g_k x^k \tag{4.39}$$

where g_k, $k = 1, 2, \ldots$ are real numbers. The series does not contain an absolute term. Then the CR graph will pass through the origin of the coordinates.

According to (4.5), the memristor PSM is

$$y_I = \sum_{k=1}^{\infty} g_k x^k. \tag{4.40}$$

Listed below is an example of the SPICE subcircuit (SUBCKT5) and input file (CIR4) for the simulation of the memristor with CR (4.39), considering a fifth-order polynomial. By selecting various values of the coefficients g_k, we can choose various shapes of the constitutive relations and the corresponding waveforms and *i–v* PHLs. Figure 4.19 shows examples of some simulation outputs generated from the data in CIR4. The PHL has the crossing points also outside the origin of the coordinates.

```
SUBCKT 5:
*Model of flux controlled memristor with polynomial CR
.subckt memristor_poly 1 2 params:
+ g1=1.5m g2=0.2 g3=-3 g4=0 g5=0 phi0=0
Ephi phi 0 value={phi0+SDT(v(1,2))}
Gm in 0 value={(g1+2*g2*v(phi)+
+ 3*g3*v(phi)**2+4*g4*v(phi)**3+5*g5*v(phi)**4)*v(1,2)}
.ends memristor_poly
CIR 4:
polynomial memristor, sin input
.param Vmax 1 f 10
Vmem mem 0 SIN 0 {Vmax} {f}
Xmem mem 0 memristor_poly params:
+ g1=10m g2=0.7 g3=10 g4=-240 g5=-10k phi0=0
EQ q 0 value={sdt(-i(Vmem))}
Ephi phi 0 value={sdt(V(mem))}
.lib memristor.lib
.tran 0 0.2 0 0.2m skipbp
.probe
.end
```

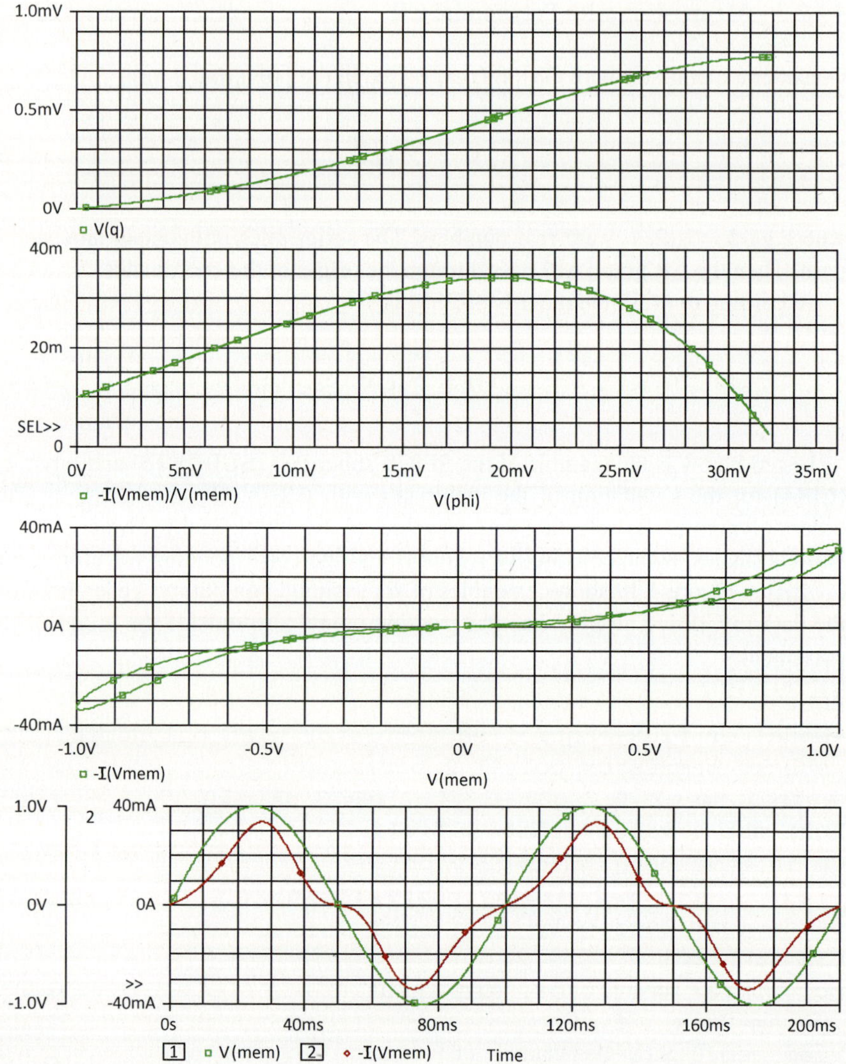

Fig. 4.19 Results of analysis of memristor with polynomial CR according to data from CIR4. From *top* to *bottom*: charge-flux CR, memductance-flux PSM, current–voltage PHL, and voltage and current waveforms

Figure 4.20 shows the simulation results based on the task defined in CIR5.

```
CIR 5:
polynomial memristors, sin input
.param Vmax 1 f 10
Vmem1 mem1 0 SIN 0 {Vmax} {f}
Xmem1 mem1 0 memristor_poly params:
```

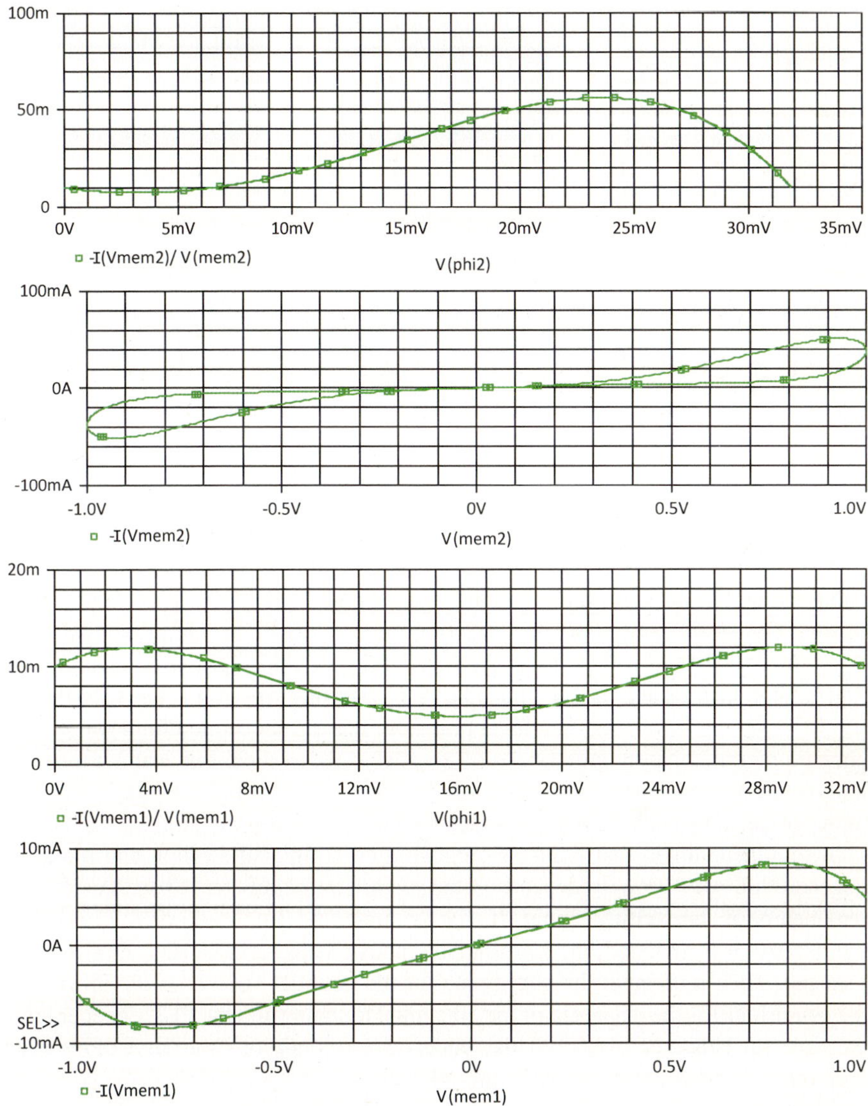

Fig. 4.20 Results of analysis of two memristors with polynomial CRs according to data from CIR5. First two plots show the memductance vs. flux PSM and current vs. voltage PHL of Xmem2 memristor with PHL having tangential touching at origin. Next two figures show similar characteristics of memristor Xmem1 with degenerated PHL

```
+ g1=10m g2=0.7 g3=-100 g4=4022 g5=-50536
EQ1 q1 0 value={sdt(-i(Vmem1))}
Ephi1 phi1 0 value={sdt(V(mem1))}
Vmem2 mem2 0 SIN 0 {Vmax} {f}
```

Fig. 4.21 Hydraulic memristor. Propeller is conducted within threaded rod. It is driven by liquid flow which moves it between two extreme positions $w = 0$ and $w = w_{max}$ (see also Fig. 4.22)

```
Xmem2 mem2 0 memristor_poly params:
+ g1=10m g2=-0.72 g3=70 g4=1000 g5=-57655
EQ2 q2 0 value={sdt(-i(Vmem2))}
Ephi2 phi2 0 value={sdt(V(mem2))}
.lib memristor.lib
.tran 0 0.2 0 0.2m skipbp
.probe
.end
```

Two memristors with different coefficients are simulated simultaneously, each being excited from a source of identical harmonic signal. The boundary points of the PSM of the memristor Xmem2 (the upper plot) lie on the horizontal line. According to FP8, the PHL must then pass through the origin with a tangential touching of its arms and the loop must be of the crossing type (CT). The second plot from the top confirms this. The third plot from the top shows that the PSM of the Xmem1 memristor is symmetrical with respect to the vertical line leading through the middle of the swing of the native state variable. According to FP10, it is a sign of degenerated hysteresis loop. This fact is confirmed in the last plot.

It should be noted that both the tangential crossing at the origin and the loop degeneration may strongly depend on the memristor parameters as well as on the parameters of its exciting signal. If, for example, the amplitude or frequency of the excitation is changed, the swing of the native state variable (i.e., flux) is also changed. Then it changes the coordinates of the boundary points of the PSM, and the conditions for the existence of the tangential touching and the PHL degeneration are violated. This example shows that, under certain conditions, the curve of the loop area versus the frequency can pass through zero points.

4.4.2.4 Hydraulic Memristor

Consider a hydraulic system according to Fig. 4.21. A part of the liquid flowing through a tube of a conic profile rotates a propeller that moves a plug, led through a bolt thread in one or the other direction, following the direction of the liquid stream. The active cross-area changes with the plug movement, and this modifies the resistance that the tube puts up to the flowing liquid. If the liquid stops flowing, the movement also stops and the hydraulic device remembers its state (and also its hydraulic resistance) until the liquid continues flowing and moving the plug again.

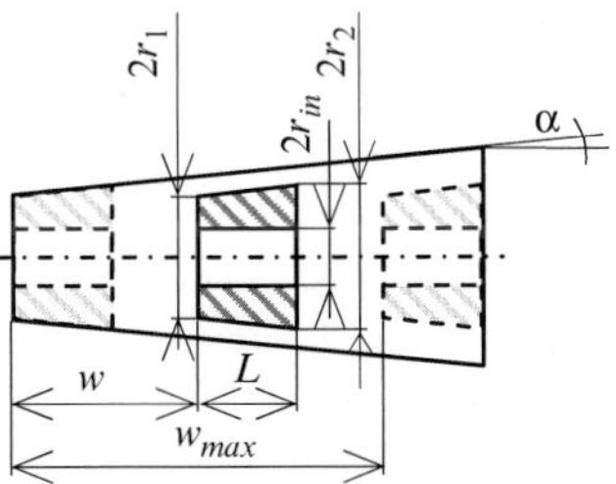

Fig. 4.22 Sizes of hydraulic memristor components, and dimensioning of plug position w; w can change within interval $[0, w_{max}]$. Propeller is rotated by liquid flowing through cylindrical opening of radius r_{in}. Inner walls of tube and surface of outer part of plug have identical tapering $\alpha = \text{atn}[(r_2 - r_1)/L]$

The propeller remembers the total volume of the liquid that passed through its inner opening via its immediate position w (see Fig. 4.22), and, based on this, it continuously controls the hydraulic resistance R of the entire system. The maximum resistance of the tube R_{off} is for the state $w = 0$ according to the left part of Fig. 4.21, when it is closed by the plug and the flow passes just through the propeller opening. In the state $w = w_{max}$, shown on the right of Fig. 4.21, the liquid flows not only via the propeller but also around the plug, and the resistance of the system has the minimum value R_{on}. We will demonstrate that if due to the liquid flow the plug moves alternately to either side such that it still remains inside the tube (see the middle part of Fig. 4.21), the system behaves as an ideal hydraulic memristor, i.e. hydraulic resistor continuously changing value of its resistance in dependence on the amount of liquid q that passed through.

Assuming laminar liquid flow, an analogy of Ohm's law for hydraulic systems can be described as follows:

$$p = Ri \tag{4.41}$$

where p [Pa] is the difference between pressures at the ends of the tube, i [m^3/s] is the rate of the liquid flow through the tube, and R is the hydraulic resistance of the system, which consists of the dominant resistance R_{in} of the central opening with the radius r, and the resistance R_{out} between the tube walls and the plug.

The following formulae can be derived for both resistances:

$$R_{in} = \frac{8\mu L}{\pi r_{in}^4}, \quad R_{out} = \frac{6\mu}{\pi w^3 \tan^4 \alpha} \ln\left(\frac{r_2}{r_1}\right) \tag{4.42}$$

where μ is the liquid viscosity.

Formula (4.42) for R_{in} is the well-known Hagen–Poiseulle relation [29]. The formula for R_{out} holds if all along the tube the space between the plug and the walls of the tube is negligible compared to the diameter of the plug.

Note that the resistors R_{in} and R_{out} act in parallel in the face of the liquid flow. Then the relationship between the current i_{in}, rotating the propeller, and the total current i is as follows:

$$i_{in} = \frac{i}{1 + \frac{R_{in}}{R_{out}}} \tag{4.43}$$

The propeller driven by the current i_{in} moves the plug in accordance with the rule

$$\frac{dw}{dt} = ki_{in} \tag{4.44}$$

where k is the proportionality constant given by the lead of the screw thread. The state equation

$$\frac{dw}{dt} = f(w)i \tag{4.45}$$

follows from Eqs. (4.42)–(4.44). It is formally identical to Eq. (4.10), where

$$f(w) = \frac{k}{1 + Aw^3}, \quad A = \frac{4L}{3}\left(\frac{\tan\alpha}{r_{in}}\right)^4 \ln^{-1}\left(\frac{r_2}{r_1}\right). \tag{4.46}$$

The function $q = S(w, w_0)$ from relation (4.6) between the native state variable, i.e. the flow $q = \int i dt$, and the plug position w, can be found by integrating according to (4.11)

$$q = S(w, w_0) = \int_{w_0}^{w} \frac{1 + Aw^3}{k} dw = \frac{1}{k}\left[w - w_0 + \frac{A}{4}\left(w^4 - w_0^4\right)\right] \tag{4.47}$$

where w_0 is the initial position of the plug.

The analytical expression of the inverse function $w = H(q, w_0)$ is rather difficult in this case. It is technically inapplicable for constructing the model in Fig. 4.5b. A method for solving this inversion directly within the SPICE environment will be evident from the following example of a SPICE macro model of hydraulic memristor.

```
SUBCKT 6:
*Model of hydraulic memristor
.subckt hydro_memristor in+ in- params:
+ Vinit=0 viscosity=1m L=50m ri=3m r2=6m r1=5m
+ w0=0k=2000
.param Rin {8*viscosity*L/(pi*ri**4)} tang {(r2-r1)/L}
+ ln21 {log(r2/r1)} A {4/3*(tang/ri)**4*L/ln21}
.func State(w,w0) {(w-w0+A/4*(w**4-w0**4))/k}
EQ Volume 0 value={Vinit+SDT(i(Vaux))}
```

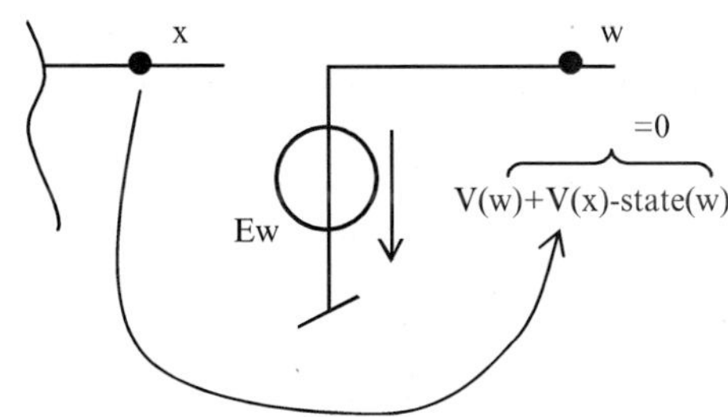

Fig. 4.23 Controlled voltage source searches for value obtained by inverting the function *state*( )

```
Ememductance Memductance 0 value=
+ {1/Rin+pi*tang**4*v(w)**3/(6*viscosity*ln21)}
Gmem in+ aux value={V(in+,in-)*(V(Memductance))}
Vaux aux in- 0
Ew w 0 value={V(Volume)-State(v(w),w0)+V(w)}
.ends hydro_memristor
```

The model parameters are chosen with a view to the requirement of laminar liquid flow, i.e. the model works with low pressures and with low flow velocities. The resistance between the plug and both ends of the tube is neglected. The conductance of the tube is continuously calculated using a controlled voltage source Ememductance, according to the current geometry of the memristor. The function *S*( ) is denoted s*tate* and it is defined in a compact form as a SPICE function. The inversion function *H*( ) is calculated automatically by a recursively specified controlled source Ew. The general algorithm for calculating the inverse function is shown in Fig. 4.23. A zero-voltage Vaux dummy source is used for sensing the current flowing through the memristive port. The volume flow is calculated from this current by integration using an EQ source.

The sought variable w is generated by the voltage source Ew, which generates the voltage as a solution of the recurrent formula $w = w + x - S(w)$. The simulator is therefore continually forced to find such a value w for which the equality $S(w) = x$ holds.

The following input file (CIR6) describes the hydraulic memristor connected by a pipe of a total length of 5 m and inner radius of 5 mm to the source of pressure with a harmonic waveform, having an amplitude of 10 kPa and a frequency of 1 Hz.

```
CIR6:
hydraulic memristor, sin input
*
Vin in 0 sin 0 10k 1Hz
Rs in pipe {8*viscosity*Ls/pi/rs**4}
.param Ls 5 rs 5m viscosity 1m
Xhydropipe 0 hydro_memristor
.lib hydro_memristor.lib
.tran 0.04 2 0 2m UIC
.probe
.end
```

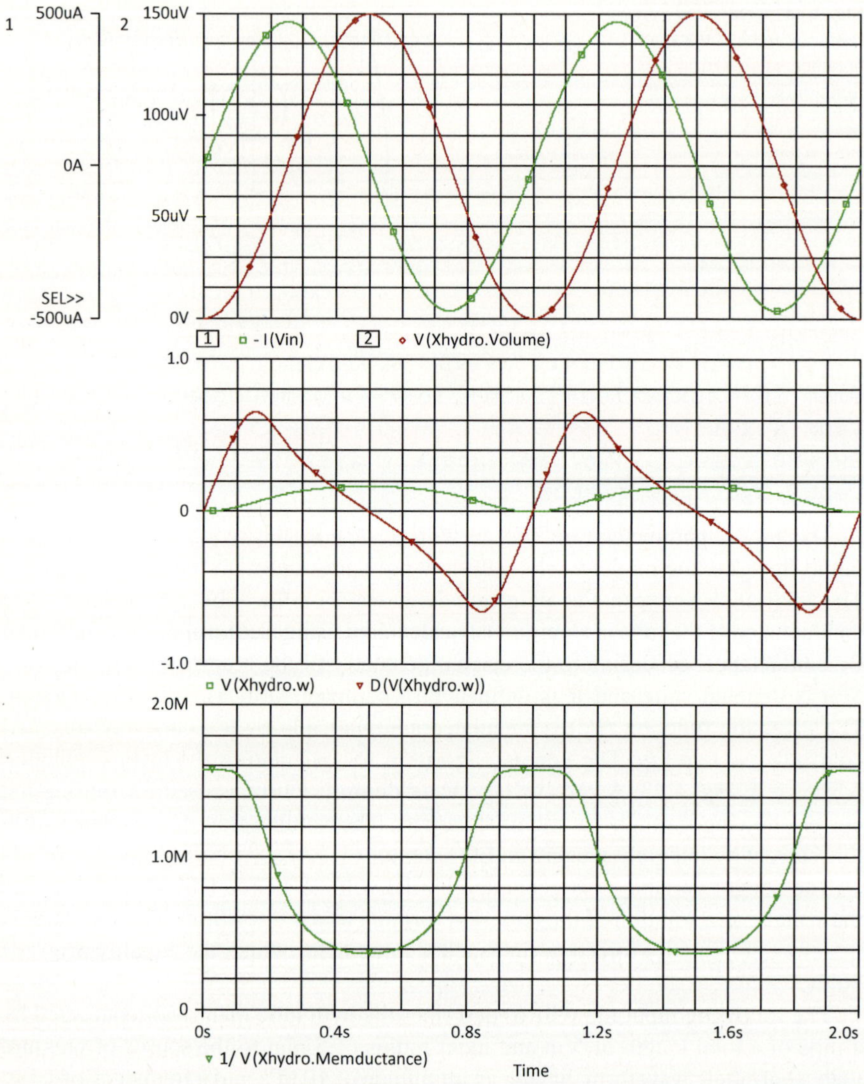

Fig. 4.24 Behavior of model of hydraulic memristor. From *top* to *bottom*: time dependence of liquid flow rate [m^3/s] through memristor and amount of liquid passing through [m^3] (*top plot*), position of plug [m] and rate of plug movement [m/s] (*middle plot*), and hydraulic resistance (*bottom plot*)

The simulation results are shown in Fig. 4.24. The initial state is $w = 0$, i.e. the plug is in an extreme position when the hydraulic resistance is maximum. The memristor achieves the minimum resistance at each odd half-period, when the deflection of the plug is maximum, approximately 165 mm. The rate of the liquid flow reaches a maximum value of 474 mL/s. The maximum speed of the plug movement is 0.65 m/s. For each half-period, 150 mL of liquid will flow through the memristor.

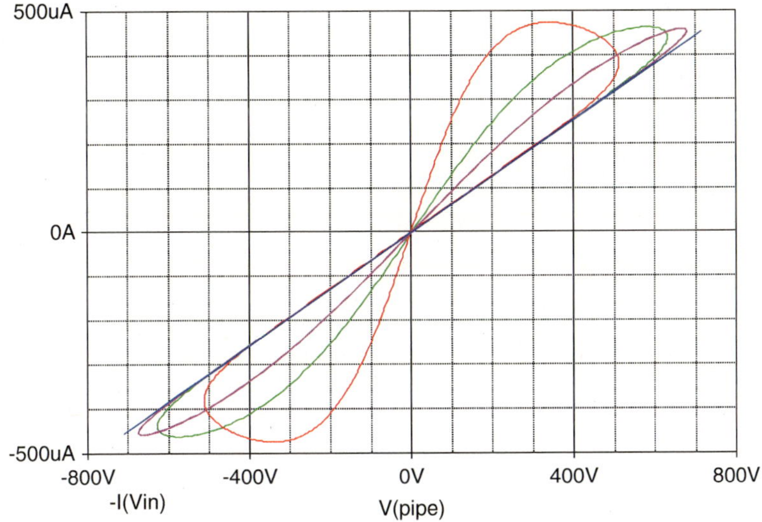

Fig. 4.25 Demonstration of FP5 for hydraulic memristor. Loops are drawn for frequencies in ascending order: 1 Hz (*red curve*), 2 Hz (*green*), 3 Hz (*violet*), 20 Hz (*blue*)

The results of a simulation checking the validity of FP5 are in Fig. 4.25. The loop area decreases with increasing frequency of harmonic excitation. In the limit case, the loop degenerates into a line with a slope corresponding to the value of the conductance of the tube with a plug in the position w_0. For the parameters specified in the netlist, the conductance is 636.2 nm^3 Pa^{-1} s^{-1}.

4.4.2.5 Sodium and Potassium Memristive Systems

The Hodgkin–Huxley (H–H) mathematical model from [30], describing how action potentials in neurons are propagated along the axon, is considered one of the great achievements of modern biophysics. The axon model is made up from a line of identical H–H cells, which are coupled by identical passive resistors [31]. Two components from the H–H cell were identified by Hodgkin and Huxley as time-varying resistors. However, L. Chua and S. Kang pointed out in the paper [11] the one-port denoted as R_K is a potassium channel first-order memristive system, and the other one-port R_{Na} is identified as a sodium channel second-order memristive system. The papers [31, 32] contain an excellent analysis of H–H cells from the new perspective of the memristive Hodgkin–Huxley axon model, which clarifies many hitherto unresolved anomalous phenomena, connected to the original incorrect concept of time-varying resistors inside the H–H cell.

The potassium channel is described as follows [31]:

$$i_K = G_K(n)v_K, \tag{4.48}$$

$$\frac{dn}{dt} = f(n, v_K), \tag{4.49}$$

where

$$G_K(n) = \overline{g}_K n^4, \; f(n, v_K) = \alpha_n(v_K)(1-n) - \beta_n(v_k)n, \tag{4.50}$$

$$\alpha_n(v_K) = 0.01\frac{v_K + E_K + 10}{e^{\frac{v_K+E_K+10}{10}} - 1}, \; \beta_n(v_K) = 0.125\; e^{\frac{v_K+V_K}{80}}. \tag{4.51}$$

In (4.51), the coefficients α_n and β_n are given in reciprocal msec and v_K and E_K are in mV. As given in [31], $E_K = 12$ mV. For the purpose of SPICE simulation, it is advantageous to modify (4.51) such that α_n and β_n will be computed in reciprocal sec, with v_K and E_K being in volts:

$$\alpha_n(v_K) = 100\frac{100(v_K + E_K) + 1}{e^{[100(v_K+E_K)+1]} - 1}, \; \beta_n(v_K) = 125\; e^{12.5(v_K+E_K)} \tag{4.52}$$

The dimensionless state variable n in (4.48), (4.49), and (4.50) can vary between 0 and 1. The constant $\overline{g}_K$ in (4.50) is approximately 36 mS/cm^2 [31].

Comparing (4.48), (4.49) with (4.12), (4.13), we can conclude that the potassium channel is a first-order voltage-controlled memristive system.

The PSpice subcircuit of the potassium ion-channel memristive system, based on Eqs. (4.48)–(4.52), together with self-explanatory notes, is given below.

```
SUBCKT 7:
.subckt Potassium_Ion_Channel_Mems in1 in2
+ params: Ek 12m ninit 0
*defining parameters and functions, see (4.48)-(4.50),
*(4.52)
.param gk 36m
.func GKn(n) {gk*n**4}
.func alphan(v)={100*(100*(v+Ek)+1)/
+ (exp(100*(v+Ek)+1)-1)}
.func betan(v)={125*exp(12.5*(v+Ek))}
.func fn(n,v)={alphan(v)*(1-n)-betan(v)*n}
* defining memristive port via Eq. (4.48)
G in1 in2 value={v(in1,in2)*GKn(v(n))}
*computing derivative of state variable according to
*(4.49)
Ed nd 0 value={fn(v(n),v(in1,in2))}
*computing state variable via integrating its
*derivative
E n 0 value={ninit+SDT(v(nd))}
.ends Potassium_Ion_Channel_Mems
```

Via the variable *ninit* we can define the initial state of the variable n. It can be useful, for example, for setting the proper value of the initial condition, which corresponds to the steady state, skipping long transients. The system state is computed via a controlled source denoted E which provides the time-domain integration of the derivative of the state variable, stored in the variable $v(nd)$.

The sodium channel is modeled as follows [31]:

$$i_{Na} = G_{Na}(m,h)\, v_{Na}, \tag{4.53}$$

$$\frac{dm}{dt} = f_1(m, v_{Na}), \quad \frac{dh}{dt} = f_2(h, v_{Na}), \tag{4.54}$$

where

$$\begin{aligned} G_{Na}(m,h) &= \overline{g}_{Na} m^3 h, \\ f_1(m, v_{Na}) &= \alpha_m(v_{Na})(1-m) - \beta_m(v_{Na})\, m, \\ f_2(h, v_{Na}) &= \alpha_h(v_{Na})(1-h) - \beta_h(v_{Na})\, h, \end{aligned} \tag{4.55}$$

$$\alpha_m(v_{Na}) = 0.1 \frac{v_{Na} - E_{Na} + 25}{e^{\frac{v_{Na}-E_{Na}+25}{10}} - 1}, \quad \beta_h(v_{Na}) = \frac{1}{e^{\frac{v_{Na}-E_{Na}+30}{10}} + 1} \tag{4.56}$$

$$\alpha_h(v_{Na}) = 0.07\ e^{\frac{v_{Na}-E_{Na}}{20}}, \quad \beta_m(v_{Na}) = 4\ e^{\frac{v_{Na}-E_{Na}}{18}}$$

In (4.56), the alpha and beta coefficients are given in reciprocal msec and v_{Na} and E_{Na} are in mV. In [31], $E_{Na} = 115$ mV. The modified Eq. 4.56 for α_m, β_m, α_h, and β_h, computed in reciprocal sec for v_{Na} and E_{Na} in volts are as follows:

$$\alpha_m(v_{Na}) = 2500 \frac{40(v_{Na} - E_{Na}) + 1}{e^{[100(v_{Na}-E_{Na})+2.5]} - 1}, \quad \beta_h(v_{Na}) = \frac{1000}{e^{[100(v_{Na}-E_{Na})+3]} + 1}, \tag{4.57}$$

$$\alpha_h(v_{Na}) = 70\ e^{50(v_{Na}-E_{Na})}, \quad \beta_m(v_{Na}) = 4000\ e^{\frac{v_{Na}-E_{Na}}{0.018}}.$$

The dimensionless state variables m and h in (4.53), (4.54), and (4.55) can vary between 0 and 1. The constant $\overline{g}_{Na}$ in (4.55) is approximately 120 mS/cm^2 [32].

Comparing (4.53), (4.54) with (4.12), (4.13), we can conclude that the sodium channel is a second-order voltage-controlled memristive system.

The PSpice subcircuit of the sodium ion-channel memristive system, based on Eqs. (4.53)–(4.57), is as follows:

```
SUBCKT 8:
.subckt Sodium_Ion_Channel_Mems in1 in2
+ params: Ena 115m minit 0 hinit 0
*defining parameters and functions, see (4.53)-(4.55),
*(4.57)
```

```
.param gna 120m
.func Gn(m,h) {gna*m**3*h}
.func alpham(v)={2500*(40*(v-Ena)+1)/
+ (exp(100*(v-Ena)+2.5)-1)}
.func betam(v)={4000*exp((v-Ena)/0.018)}
.func alphah(v)={70*exp(50*(v-Ena))}
.func betah(v)={1000/(exp(100*(v-Ena)+3)+1)}
.func f1(m,v)={alpham(v)*(1-m)-betam(v)*m}
.func f2(h,v)={alphah(v)*(1-h)-betah(v)*h}
*defining memristive port via Eq. (4.53)
G in1 in2 value={v(in1,in2)*Gn(v(m),v(h))}
*computing derivatives of state variables according to
*(4.54)
Emd md 0 value={f1(v(m),v(in1,in2))}
Ehd hd 0 value={f2(v(h),v(in1,in2))}
*computing state variables via integrating their
*derivatives
Em m 0 value={minit+SDT(v(md))}
Eh h 0 value={hinit+SDT(v(hd))}
.ends Sodium_Ion_Channel_Mems
```

Since it is the second-order memristive system, two first-order state differential equations must be modeled and solved according to the general block diagram in Fig. 4.5.

Figure 4.26 demonstrates the transient analysis of potassium ion-channel memristive device which is excited by sinusoidal voltage waveform under the conditions specified in the circuit file CIR7:

```
CIR 7:
Potassium Ion Channel Memristive System
Vin in 0 sin 0 50m 1k
Xmem in 0 Potassium_Ion_Channel_Mems
.lib memristor.lib
.tran 0 30m 0 30u skipbp
.probe
.end
```

Note that for a rather high repeating frequency of the exciting signal (1 kHz, see Fig. 4.26a), the steady state of the potassium mem-system is established after a long transient. This is apparent from the curve $v(n)$, which represents the evolution of the state variable. The corresponding trajectories in the voltage–current coordinates (see the upper part of the figure) converge to steady-state PHLs. It is an evident violation of FP4 about the immediate establishing the periodical steady state for memristors, indicating that the analyzed device is more general memristive system.

For lower frequencies (100 Hz, see Fig. 4.26b), when the repeating periods are comparable to or lower than the system time constants, the steady state is reached within a few periods.

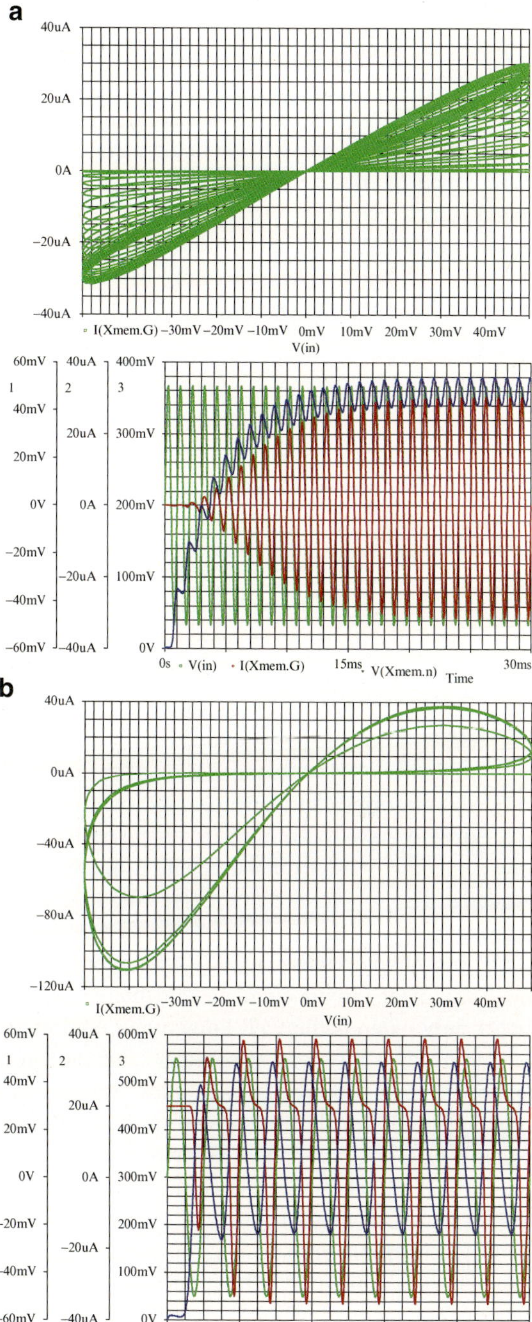

Fig. 4.26 PSpice simulation of potassium ion-channel memristive system. Transient analyses from zero initial conditions, for sinusoidal 50 mV input voltage with frequency (**a**) 1 kHz, (**b**) 100 Hz

Similar observations can be made for the sodium ion-channel memristive system. The simulation results in Fig. 4.27 have been obtained from the data in the circuit file CIR8.

```
CIR 8:
Sodium Ion Channel Memristive System
Vin in 0 sin 0 50m 1k
Xmem in 0 Sodium_Ion_Channel_Mems
.lib memristor.lib
.tran 0 10m 0 10u skipbp
.probe
.end
```

In accordance with the simulations described in [31], sinusoidal voltage sources were used with various repeating frequencies, and the corresponding PHLs were analyzed, including the phenomena of the hysteresis disappearing with growing frequency. Figure 4.28 shows the results of the transient analysis under the conditions specified in the figure caption. Note that both devices exhibit the well-known FPs of the memory elements: *v–i* PHLs, which shrink to straight lines if the frequency of exciting signal is growing to infinity.

4.5 Conclusions

A simplified but today frequent view of the memristor as an element manufactured in HP laboratories can cause a range of misunderstandings, possibly resulting in two extreme views on the memristor. According to the first one, the memristor is only a specific two-terminal element, utilizing the memory effect in the TiO_2 nano-structure. The second view perceives the memristor as an arbitrary element which exhibits the *v–i* PHLs.

In this chapter the reader is reminded that the term memristor denotes an ideal nonlinear circuit element with unambiguous flux-charge characteristic, and that it only extends the well-known and commonly utilized set of ideal elements, containing also resistors, capacitors, and inductors. Even though these elements cannot be manufactured in their ideal representations, they are indispensable as modeling tools for describing and understanding the essence of processes within existing systems. It is shown below that the memristor concept can also be useful in technical branches of non-electrical nature, particularly in mechanics, hydraulics, and thermodynamics. In terms of the original definition [1] and the generalizing views [2–5], the memristor is an element which provides a one-to-one connection between the full histories of the effort and flow. That way it becomes the very first fundamental element which correctly models [15] nonvolatile memory processes in systems of both animate and inanimate nature. If this basic building block is omitted when making up the model of a concrete system, it must be substituted via some nonstandard framework. The model is then unnecessarily complicated and need not

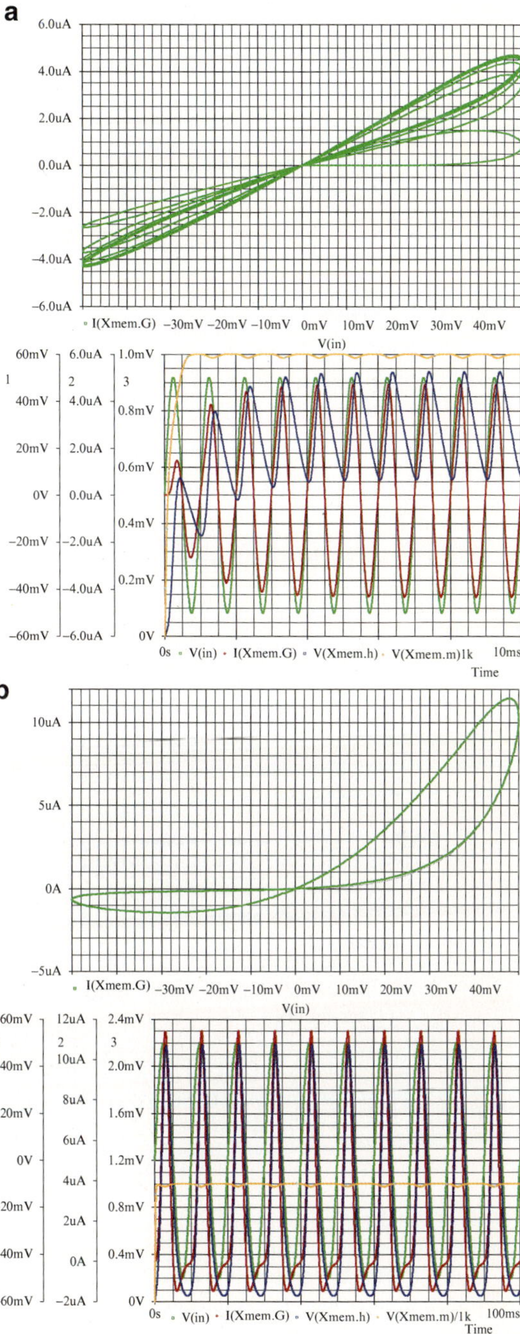

Fig. 4.27 PSpice simulation of sodium ion-channel memristive system. Transient analyses from zero initial conditions, for sinusoidal 50 mV input voltage with frequency (**a**) 1 kHz, (**b**) 100 Hz

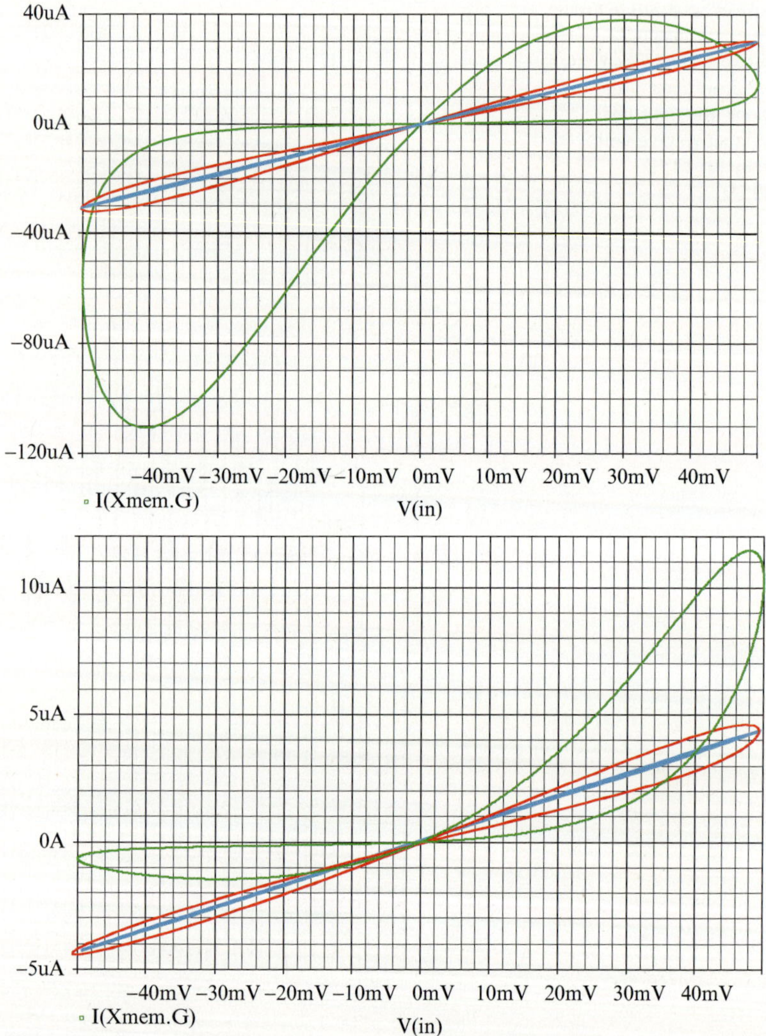

Fig. 4.28 Pinched hysteresis loops of (**a**) potassium, (**b**) ion-channel memristive system. Steady-state transient analyses for sinusoidal 50 mV input voltage with frequency of 100 Hz (*green lines*), 1 kHz (*red lines*), 10 kHz (*blue lines*)

be sufficiently credible. As one of the most illustrative example, let us mention the famous axon model from 1952 [30]. At that time, the memristor concept had not been introduced yet and thus it could not be used in this model.

It is shown in this chapter that a physical system can be modeled as a memristor if the system behavior can be described by quantities of the effort and flow types, which conform to Eqs. (4.1) and (4.2) defining the memristor. This consideration allowed introducing the so-called memristor native state variable, which can be the time-domain integral of effort or flow. The native state variable is naturally boundless, that way expressing the memristor's unbounded memory. We study here its relation to another state variable, which represents a concrete method of implementing the memory in an existing system. Connections to the window functions for modeling nonlinear dopant drift in the TiO_2 memristor are also analyzed. The results are then utilized to find a method for increasing the precision of computer simulations. This is the case, for example, of avoiding fatal numerical errors when simulating hard-switching effects in the TiO_2 memristor. Such errors can occur when utilizing the conventional methods of modeling.

An analysis of selected ten fingerprints (FPs) of the memristor precedes the Sect. 4.4 about the modeling and computer simulation of memristors. These FPs are subsequently used as useful tools for verifying the correctness of the simulation outputs as well as the correctness of the mathematical models. For example, the application of a simple FP4 reveals the hidden effects of the above numerical errors generated by the simulation program. Some FPs clarify, for the very first time, the conditions of the occurrence of certain abnormalities in PHLs. This is particularly related to FP7–10, which deal with the types of the loop crossing at the origin and with the degenerate loops.

Section 4.4, which is concerned with memristor models and their implementation in SPICE-like simulation programs, starts from the general block diagrams of memristors, and these diagrams result from the general memristor characteristics, in particular the PSM (parameter vs. state map), and from the port and state equations. The methodology of generating the models, their software implementation, and their utilization in the simulation program are demonstrated for memristors of various types: switching memory memristor, TiO_2 memristor, memristor with polynomial CRs (Constitutive Relations), hydraulic memristor, and sodium a potassium memristive systems.

Acknowledgments This work was partially supported by the Czech Science Foundation under grant No P102/10/1614, and by the project for development of K217 Dept., UD Brno.

References

1. L.O. Chua, Memristor—the missing circuit element. IEEE Trans. Circuit Theory **18**(5), 507–519 (1971)
2. G.F. Oster, D.M. Auslander, The memristor: a new bond graph element. J. Dyn. Syst. Meas. Control **94**(3), 249–252 (1972)
3. D.C. Mikulecky, Network thermodynamics and complexity: a transition to relational systems theory. Comput. Chem. **25**(4), 369–391 (2001)
4. D. Jeltsema, A.J. van der Schaft, Memristive port-Hamiltonian systems. Math. Comput. Model. Dyn. Syst. **16**(2), 75–93 (2010)
5. D. Jeltsema, A. Dòria-Cerezo, Port-Hamiltonian formulation of systems with memory. Proc. IEEE **100**(6), 1928–1937 (2012)
6. Z. Biolek, D. Biolek, V. Biolková, Analytical solution of circuits employing voltage- and current-excited memristors. IEEE Trans. Circuits Syst. Regul. Pap. **59**(11), 2619–2628 (2012)

7. Y.V. Pershin, S. Fontaine, M. Di Ventra, Memristive model of amoeba's learning. Phys. Rev. E. **80**, 021926/1–021926/6 (2009)
8. D.B. Strukov, G.S. Snider, D.R. Stewart, R.S. Williams, The missing memristor found. Nature **453**, 80–83 (2008)
9. Y.V. Pershin, M. Di Ventra, Memory effects in complex materials and nanoscale systems. Adv. Phys. **60**, 145–227 (2011)
10. L.O. Chua, Resistance switching memories are memristors. Appl. Phys. A. **102**, 765–783 (2011)
11. L.O. Chua, S.M. Kang, Memristive devices and systems. Proc. IEEE **64**(2), 209–223 (1976)
12. S. Benderli, T.A. Wey, On SPICE macromodeling of TiO_2 memristors. Electron. Lett. **45**(7), 377–379 (2009)
13. Z. Biolek, D. Biolek, V. Biolková, Spice model of memristor with nonlinear dopant drift. Radio Eng. **18**(2), 210–214 (2009)
14. A. Rák, G. Cserey, Macromodeling of the memristor in SPICE. IEEE Trans. Comput. Aided Des. Integr. Circuits Syst. **29**(4), 632–636 (2010)
15. L.O. Chua, Nonlinear circuit foundations for nanodevices, Part I: The four-element torus. Proc. IEEE **91**(11), 1830–1859 (2003)
16. Y.N. Joglekar, S.J. Wolf, The elusive memristor: properties of basic electrical circuits. Eur. J. Phys. **30**(4), 661–675 (2009)
17. D. Biolek, Z. Biolek, V. Biolkova, Pinched hysteresis loops of ideal memristors, memcapacitors and meminductors must be 'selfcrossing'. Electron. Lett. **47**(25), 1385–1387 (2011)
18. H. Kim, M. P. Sah, S. P. Adhikari, Pinched hysteresis loops is the fingerprint of memristive devices. arXiv:1202.2437v2 (2012)
19. L.O. Chua, in Hodgkin-Huxley, memristor and the edge of chaos, in *Invited Lecture at the 3rd Memristor and Memristive Symposium*, Turin, Italy, 2012
20. H.N. Huang, S.A.M. Marcantognini, N.J. Young, Chain rules for higher derivatives. Math. Intell. **28**(2), 1–12 (2006)
21. E. Linn, R. Rosezin, C. Kügeler, R. Waser, Complementary resistive switches for passive nanocrossbar memories. Nat. Mater. **2010**(9), 403–406 (2010)
22. F. Corinto, A. Ascoli, A boundary condition-based approach to the modeling of memristor nano-structures. IEEE Trans. Circuits Syst. Regul. Pap. **59**(11), 2713–2726 (2012)
23. S. Shin, K. Kim, S.M. Kang, Compact models for memristors based on charge-flux constitutive relationships. IEEE Trans. Comput. Aided Des. Integr. Circuits Syst. **29**(4), 590–598 (2010)
24. T. Prodromakis, B.P. Peh, C. Papavassiliou, C. Toumazou, A versatile memristor model with non-linear dopant kinetics. IEEE Trans. Electron Devices **58**(99), 1–7 (2011)
25. S. Kvatinsky, E.G. Friedman, A. Kolodny, U.C. Weiser, TEAM: ThrEshold adaptive memristor model. IEEE Trans. Circuits Syst. Regul. Pap. **60**(1), 211–221 (2013)
26. R. Kozma, R.E. Pino, G.E. Pazienza (eds.), *Advances in Neuromorphic Memristor Science and Applications* (Springer, New York, 2012)
27. E. Lehtonen, M. Laiho, CNN using memristors for neighborhood connections, in *12th International Workshop on Cellular Nanoscale Networks and Their Applications (CNNA)*, Berkeley, CA, 2010
28. K. Eshraghian, O. Kavehei, K.R. Cho, J.M. Chappell, A. Iqbal, S.F. Al-Sarawi, D. Abbott, Memristive device fundamentals and modeling: applications to circuits and systems simulation. Proc. IEEE **100**(6), 1991–2007 (2012)
29. N.D. Manring, *Hydraulic Control Systems* (Wiley, USA, 2005), p. 464
30. A.L. Hodgkin, A.F. Huxley, A quantitative description of membrane current and its application to conduction and excitation in nerve. J. Physiol. **117**, 500–544 (1952)
31. L. Chua, V. Sbitnev, H. Kim, Hodgkin-Huxley axon is made of memristors. Int. J. Bifurcation Chaos **22**(3), 1230011-1–1230011-48 (2012)
32. L. Chua, V. Sbitnev, H. Kim, Neurons are poised near the edge of chaos. Int. J. Bifurcation Chaos **22**(4), 250098-1–250098-49 (2012)

Chapter 5
Application of the Volterra Series Paradigm to Memristive Systems

Alon Ascoli, Torsten Schmidt, Ronald Tetzlaff, and Fernando Corinto

5.1 Introduction

The modeling [1] and investigation [2] of the nonlinear dynamics of memristive systems [3, 4] is one of the hottest topics of current research. The nonlinearity of these unique systems calls for the need to employ techniques from nonlinear system theory [5] to investigate and model their behavior and the dynamics occurring in memristor circuits. One of the most well-known theories for modeling dynamical systems is the Volterra series paradigm [6]. Let us consider a dynamical system with input $x(t)$ and output $y(t)$. Its block diagram is shown in Fig. 5.1.

The Volterra series representation of the output $y(t)$ to the system is given by

$$y(t) = \sum_{n=1}^{\infty} y_n(t), \tag{5.1}$$

whose nth-order component $y_n(t)$ is described by

$$y_n(t) = \int_{-\infty}^{+\infty} \dots \int_{-\infty}^{+\infty} h_n(\tau_1, \dots \tau_n) x(t-\tau_1) \dots x(t-\tau_n) d\tau_1 \dots d\tau_n. \tag{5.2}$$

A. Ascoli (✉) • R. Tetzlaff
Technische Universität Dresden, Mommsenstraße 12, 01062 Dresden, Germany
e-mail: alon.ascoli@tu-dresden; ronald.tetzlaff@tu-dresden

T. Schmidt
Hochschule Ansbach, Residenzstraße 8, 91522 Ansbach, Germany
e-mail: torsten.schmidt@hs-ansbach.de

F. Corinto
Politecnico di Torino, Corso Duca degli Abruzzi 24, 10129 Torino, Italy
e-mail: fernando.corinto@polito.it

R. Tetzlaff (ed.), *Memristors and Memristive Systems*,
DOI 10.1007/978-1-4614-9068-5_5,

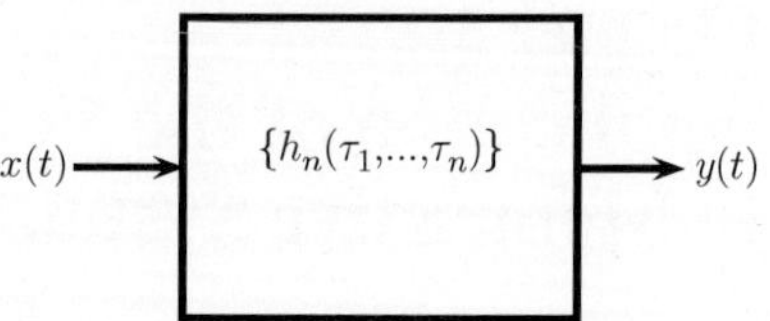

Fig. 5.1 Block diagram of a Volterra system

Here $h_n(\tau_1, \ldots \tau_n)$ is a real valued function of $\tau_1, \ldots, \tau_n$ [7] called *n*th-order Volterra kernel in the time domain. In Fig. 5.1 the whole set of Volterra kernels for $n \in [1, \infty]$ is denoted as $\{h_n(\tau_1, \ldots \tau_n)\}$.

The *n*th-order Volterra kernel in the *s*-domain is defined as [8]

$$H_n(s_1, \ldots, s_n) = \int_{-\infty}^{+\infty} \ldots \int_{-\infty}^{+\infty} h_n(\tau_1, \ldots \tau_n) e^{-(s_1\tau_1 + \cdots + s_n\tau_n)} d\tau_1 \ldots d\tau_n. \quad (5.3)$$

Letting $s_k = j\omega_k$ $(k \in \{1, \ldots, n\})$ (5.3) may be recast as

$$H_n(j\omega_1, \ldots, j\omega_n) = \int_{-\infty}^{+\infty} \ldots \int_{-\infty}^{+\infty} h_n(\tau_1, \ldots \tau_n) e^{-(j\omega_1\tau_1 + \cdots + j\omega_n\tau_n)} d\tau_1 \ldots d\tau_n, \quad (5.4)$$

denoting the *n*th-order Volterra kernel in the frequency domain.

Sect. 5.2 presents a brief historical overview of the Volterra series theory including some of the most recent works and gives some insight on its application for the modeling and investigation of memristive systems. Sects. 5.3 and 5.4, respectively, introduce a systematic approach for modeling the dynamics of a class of single- and two-memristor circuits. Finally conclusions are drawn in Sect. 5.5.

5.2 Application of the Volterra Series Paradigm to Memristive Systems

The determination of solutions of differential equations by applying Volterra series representations has been addressed by many authors. An approach based on a power series expansion of the solution of a differential equation is addressed in the pioneering work of Barrett [9] by separating the linear and nonlinear parts of the considered differential equation. Further contributions have been given by Waddington [10], Krener [11], Lesiak [12] and Gilbert [13]. An overviewing presentation of applied methods, i.e. the Carleman linearization approach, the variational equation method, and the growing exponential approach is outlined in [14]. There, it has been shown that the so-called linear analytic state equations can be represented by a Volterra series expansion. The question whether nonlinear systems and their nonlinear operators can be approximated by Volterra operators has been treated in different contributions of Sandberg, summarized in [15], and giving the important result that, for kernel functions satisfying

$$\int_{[0,\infty)^k} |h_k(\tau_1, \ldots, \tau_k)| d\tau_1 \ldots d\tau_k < \infty,$$

the expansion

$$y(t) = \sum_{k=1}^{\infty} \int_{-\infty}^{t} \cdots \int_{-\infty}^{t} h_k(\tau_1, \dots, \tau_k) x(t - \tau_1) \dots x(t - \tau_k) d\tau_1 \dots d\tau_k, \qquad -\infty < t < \infty$$

converges uniformly for Lebesgue measurable bounded inputs $x(t)$ such that $|x(t)| \leq \delta$ for $-\infty < t < \infty$. In the fundamental work of Boyd [16] it was shown that a system may admit a Volterra series representation provided it possesses the fading memory property. It follows that for a nonlinear circuit with fading memory a Volterra series expansion of the impedance operator may be derived. In yet another work of Boyd [17] it was outlined that it is possible to model a nonlinear circuit by means of the Volterra series theory provided the linearized circuit is exponentially stable and any electrical component within the circuit is defined by analytic constitutive relations.

Essentially, in cases where the dynamics of a system can be modeled by means of the Volterra series, it is important to determine the radius of the convergence region of the series expansion. This problem has been tackled by different authors [18–21]. By assuming that each term of a series expansion is bounded, Barrett [22] has presented a method allowing time domain convergence analysis. Boyd [23] addressed this problem in the frequency domain as well. Recently, Hélie [24] studied single-input finite-dimension analytic systems described by

$$\begin{aligned} \dot{\mathbf{x}} &= \mathbf{A}\mathbf{x} + \mathbf{B}u + \mathbf{P}(\mathbf{x}) + \mathbf{Q}(\mathbf{x}, u), \text{ and} \\ \mathbf{y} &= \mathbf{g}(\mathbf{x}, u), \end{aligned} \tag{5.5}$$

where the state, the input, and the output are, respectively, defined as $\mathbf{x}(t) \in \mathbb{R}^N$, $u(t) \in \mathbb{R}$, and $\mathbf{y}(t) \in \mathbb{R}^M$, the state initial condition is set to $\mathbf{x}(0) = \mathbf{0}$, while $\mathbf{A} \in \mathbb{R}^{N \times N}$ and $\mathbf{0} \neq \mathbf{B} \in \mathbb{R}^{N \times 1}$. Further $\mathbf{g}(\mathbf{x}, u)$ is analytic at $(\mathbf{0}, 0)$, while $\mathbf{P}(\mathbf{x}) \in \mathbb{R}^N$ and $\mathbf{Q}(\mathbf{x}, u) \in \mathbb{R}^N$ are analytic functions given by

$$\begin{aligned} \mathbf{P}(\mathbf{x}) &= \sum_{j=2}^{J} \mathbf{P}_j(\underbrace{\mathbf{x}, \dots, \mathbf{x}}_{j}), \text{ and} \\ \mathbf{Q}(\mathbf{x}, u) &= \sum_{k=2}^{K} \mathbf{Q}_k(\underbrace{\mathbf{x}, \dots, \mathbf{x}}_{k-1}, u), \end{aligned}$$

with $\mathbf{P}_j \in \mathfrak{ML}_j(\mathbb{X}, \mathbb{X})$[1] and $\mathbf{Q}_k \in \mathfrak{ML}_{k-1,1}(\mathbb{X}, \mathbb{R}, \mathbb{X})$. More details are reported in [24]. There it is shown that the Volterra kernel functions of an input-to-state system can be derived by applying a recursive construction method. Theorems allowing

[1] $\mathfrak{ML}(\mathbb{E}_1, \dots, \mathbb{E}_L, \mathbb{F})$ denotes the vector space of multilinear functions. Further the following notation is introduced: $\mathfrak{ML}_{j_1,\dots,j_L}(\mathbb{E}_1, \dots, \mathbb{E}_L, \mathbb{F}) = \mathfrak{ML}(\underbrace{\mathbb{E}_1, \dots, \mathbb{E}_1}_{j_1}, \dots \underbrace{\mathbb{E}_L, \dots, \mathbb{E}_L}_{j_L}, \mathbb{F})$.

the determination of bounds of the radius of the convergence region of the series expansion are presented for several typical norms ($L^\infty([0,T])$, $L^\infty(\mathbb{R}_+)$ as well as exponentially weighted norms). Moreover it is also indicated that an input-to-output series expansion may be derived from the input-to-state series representation of dynamical system (5.5). Comparing now (5.5) to the equations which Chua [25] introduced as a way to unfold the memristor's unique dynamical properties, i.e.

$$\frac{dx}{dt} = a_1 x + a_2 x^2 + \ldots + a_m x^m + b_1 i + b_2 i^2 + \ldots + b_n i^n + \sum_{j,k=1}^{p,r} c_{j,k} x^j i^k, \text{ and}$$
$$v = R(x) i, \tag{5.6}$$

(where $R(x)$ is the state-dependent resistance or memristance, while v and i, respectively, denote memristor voltage and current) it follows that several memristor equations can be represented by single-input finite-dimension analytic equations by letting $n = r = 1$ in (5.6). By considering now one of the examples given in [25] for illustration, one can obtain the original Hewlett Packard [26] memristor equation from (5.5) for the special choice $\mathbf{A} = 0$, $\mathbf{B} = \mu_v \frac{R_{on}}{D}$, $\mathbf{P}(\mathbf{x}) = 0$, $\mathbf{Q}(\mathbf{x}, u) = 0$, $\mathbf{x} = x$ (where the one-dimensional state x denotes the normalized length l of the conductive layer of the memristor nano-structure), and taking the memristor current as input $u = i$, the memristor voltage as output $y = v$, and finally letting $g(\mathbf{x}, u) = R(x)i$. Using (5.6), the original Hewlett Packard memristor equation is derived by posing $a_i = 0$ for all $i \in \{1, \ldots, m\}$, $b_1 = \mu_v \frac{R_{on}}{D}$, $b_i = 0$ for all $i \in \{2, \ldots, n\}$, $c_{j,k} = 0$ for all $j \in \{1, \ldots, p\}$ and for all $k \in \{1, \ldots, r\}$ and assuming, once again, $x = l$.

Another approach has been presented by Boyd [16] and shows that, letting ε be a positive constant, any time-invariant fading memory operator $N\{\cdot\}$ can be represented by a finite Volterra series operator $\hat{N}\{\cdot\}$ such that

$$||N\{u\} - \hat{N}\{u\}|| \le \varepsilon \tag{5.7}$$

for any input $u \in K = \{u \in C(\mathbb{R}) | \, ||u|| \le M_1, \; |u(\tilde{t}) - u(t)| \le M_2(\tilde{t} - t), \; t \le \tilde{t}\}$, where M_1 and M_2 are constants. Here $\hat{N}\{\cdot\}$ is the input–output operator of an exponentially stable, finite-dimension linear system expressed by

$$\dot{\mathbf{z}}(t) = \mathbf{A}\mathbf{z}(t) + \mathbf{B} \cdot u(t) \tag{5.8}$$

with output described by

$$y(t) = P(\mathbf{z}(t)), \tag{5.9}$$

where state $\mathbf{z}(t) \in \mathbb{R}^N$, input $u(t) \in \mathbb{R}$, and output $y(t) \in \mathbb{R}$, while $\mathbf{A} \in \mathbb{R}^{NXN}$, $\mathbf{B} \in \mathbb{R}^{NX1}$, and $P(\mathbf{z}) \in \mathbb{R}$ is a polynomial function. It is obvious that (5.8)–(5.9) have the same form as mathematical models of certain memristive systems (in particular, $P(\mathbf{z})$ may model a memory resistance or conductance in case the input $u(t)$ is in current or voltage form, respectively). It is thus possible to represent such

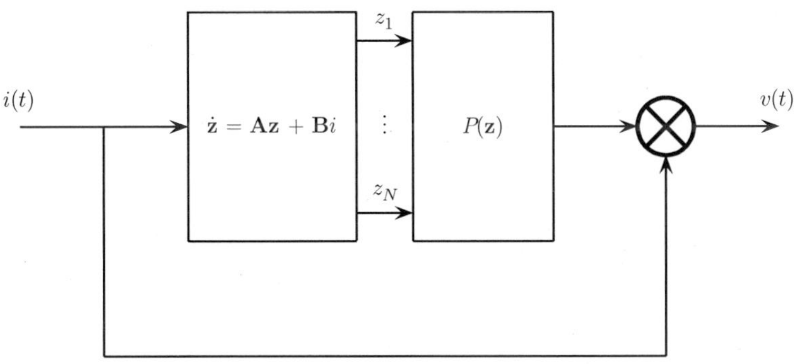

Fig. 5.2 Representation of a current-controlled memristive system as a cascade connection of an exponentially stable linear system with a static nonlinear polynomial system

memristive systems by means of a finite Volterra series expansion. Let us give some more detail. By assuming for the moment a current-controlled memristive system, i.e. $u(t) = i(t)$, then (5.8)–(5.9) may be equivalently represented by the structure shown in Fig. 5.2.

Assuming furthermore that the memristor current $i(t)$ is the output of a linear time-invariant system, i.e. a first-order Volterra system with kernel $h_1(\tau) = \delta(\tau)$, it directly follows that the memristor voltage $v(t)$ can be regarded as the output of a Volterra system. Thus, memristive systems having the structure given in Fig. 5.2 possess a Volterra series representation which can be further used to model fading memory operators $N\{\cdot\}$ pertaining to memristive systems of different complexity. In particular, if a circuit can be modeled as the connection of a number of memristive elements, each of which admits a Volterra series expansion, then the kernel functions of Volterra system referring to the interconnected structure describing the overall circuit can be derived in a straightforward manner as outlined in [14] and [18].

As a final note, it is interesting to note that Corinto and Ascoli [27] have recently discovered memristive behavior in a purely passive electronic system made up of the cascade connection of a static nonlinear two-port and an exponentially stable linear dynamic one-port. Such a system may be represented by a structure very similar to the one in Fig. 5.2.

5.3 Investigation of the Nonlinear Dynamics of a Class of Single-Memristor Circuits

In this section we shall present a systematic methodology to derive the Volterra series representation for a class of current-driven single-memristor circuits of the type shown in Fig. 5.3, where the two-port, an arbitrary linear and dynamic circuit, is embedded between the excitation in current form and a charge-controlled memristor load m. Note that a similar approach may be followed to analyze flux-driven

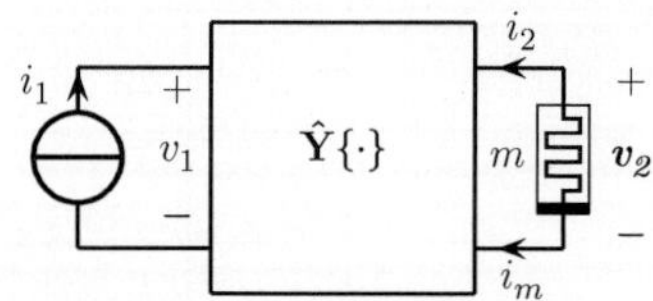

Fig. 5.3 General topology of the single-memristor circuits

single-memristor circuits. Denoting the memristor current with symbol i_m and its time integral, i.e. memristor charge, with symbol q_m, let the charge-controlled memristor m in Fig. 5.3 be characterized by the following memristance function

$$M(q_m(t)) = M\left(\int_{-\infty}^{t} i_m(t')dt'\right) = \sum_{k=1}^{\infty} m_k\,(q_m(t))^k. \tag{5.10}$$

Note that any linear resistance m_0 within the memristance of memristor m, extracted from the time series of $M(q_m)$ in (5.10), will be taken into account by inserting it in series with the memristor, but will be considered as part of the two-port. In this case the actual memristor voltage is the sum of the voltage drops across $M(q_m)$ and m_0. The proposed methodology may be split into three steps, which shall be treated in Sects. 5.3.1–5.3.3:

1. Determination of the dynamic equation;
2. Determination of the Volterra kernel equations using the harmonic probe technique [8];
3. Computation of the system output from the kernel equations.

In Sect. 5.3.4 we shall present a sample circuit from the class of circuits in Fig. 5.3 together with simulation results confirming the accuracy of the Volterra series representation in modeling the circuit dynamics.

5.3.1 Determination of the Dynamic Equation

Expressing the two-port input and output currents—i_1 and i_2, respectively—in terms of the two-port input and output voltages—v_1 and v_2—the system may be described by the following set of equations

$$i_1(t) = \hat{Y}_{11}\{v_1(t)\} + \hat{Y}_{12}\{v_2(t)\}, \text{ and} \tag{5.11}$$

$$i_2(t) = \hat{Y}_{21}\{v_1(t)\} + \hat{Y}_{22}\{v_2(t)\}, \tag{5.12}$$

where $\hat{Y}_{\alpha,\beta}\{\cdot\}$ $(\alpha,\beta \in 1,2)$ are integro-differential operators and $\hat{\mathbf{Y}}\{\cdot\}$ is the matrix composed of them, i.e.

$$\hat{\mathbf{Y}}\{\cdot\} = \begin{bmatrix} \hat{Y}_{11}\{\cdot\} & \hat{Y}_{12}\{\cdot\} \\ \hat{Y}_{21}\{\cdot\} & \hat{Y}_{22}\{\cdot\} \end{bmatrix}. \tag{5.13}$$

Note that the two-port of Fig. 5.3 may alternatively be described in terms of the impedance matrix, defined in Sect. 5.4, where this alternative approach shall be followed for the investigation of the class of two-memristor circuits. Applying $\hat{Y}_{21}\{\cdot\}$ to (5.11) and $\hat{Y}_{11}\{\cdot\}$ to (5.12), subtracting the resulting equations and assuming that $\hat{Y}_{21}\{\hat{Y}_{11}\{\cdot\}\} = \hat{Y}_{11}\{\hat{Y}_{21}\{\cdot\}\}$ in (5.13) leads to

$$\hat{Y}_{21}\{i_1(t)\} - \hat{Y}_{11}\{i_2(t)\} = -\det(\hat{\mathbf{Y}})\{v_2(t)\}, \tag{5.14}$$

where $\det(\hat{\mathbf{Y}})\{\cdot\} = \hat{Y}_{11}\{\hat{Y}_{22}\{\cdot\}\} - \hat{Y}_{21}\{\hat{Y}_{12}\{\cdot\}\}$. Expressing $v_2(t)$ in terms of $i_2(t)$ and noting that $i_2(t) = -i_m(t)$ gives

$$\begin{aligned} v_2(t) &= M(q_m(t))i_m(t) = -M(-q_2(t))i_2(t) \\ &= -M\left(-\int_{-\infty}^{t} i_2(t')dt'\right) i_2(t) = -i_2(t)\sum_{k=1}^{\infty} m_k \left(-\int_{-\infty}^{t} i_2(t')dt'\right)^k . \end{aligned} \tag{5.15}$$

Using this into (5.14) leads to the dynamic equation of the circuit

$$\hat{Y}_{21}\{i_1(t)\} - \hat{Y}_{11}\{i_2(t)\} = \det(\hat{\mathbf{Y}})\left\{i_2(t)M\left(-\int_{-\infty}^{t} i_2(t')dt'\right)\right\}. \tag{5.16}$$

Furthermore, by the notation introduced in the previous section, the circuit of Fig. 5.3 may be viewed as a Volterra system, as shown in Fig. 5.1, where $x(t) = i_1(t)$ and $y(t) = i_2(t)$. As a result, this equation may be recast as

$$\hat{Y}_{21}\{x(t)\} - \hat{Y}_{11}\left\{\sum_{n=1}^{\infty} y_n(t)\right\} = \det(\hat{\mathbf{Y}})\left\{\sum_{n=1}^{\infty} y_n(t)M\left(-\int_{-\infty}^{t}\sum_{n=1}^{\infty} y_n(t)(t')dt'\right)\right\}. \tag{5.17}$$

Inserting (5.10) into (5.17) results into the following dynamic equation for the general circuit of Fig. 5.3

$$\hat{Y}_{21}\{x(t)\} - \hat{Y}_{11}\{y(t)\} = \det(\hat{\mathbf{Y}})\left\{y(t)\sum_{k=1}^{\infty} m_k \left(-\int_{-\infty}^{t} y(t')dt'\right)^k\right\}, \tag{5.18}$$

where $y(t)$ is expressed by (5.1).

5.3.2 Determination of the Volterra Kernels

Let us describe how dynamic equation (5.18) may be used to determine Volterra kernels. One of the most powerful techniques to carry out this task is the harmonic probe technique [8]. Let us explain how one may derive an equation involving the kernel of order m in the frequency domain, i.e. $H_m(j\omega_1,\ldots,j\omega_m)$, for an m-tone complex exponential input of the form[2]

$$x(t) \triangleq x^{(m)}(t) = c\sum_{n=1}^{m} e^{j\omega_n t}, \tag{5.19}$$

where c is a constant introduced to facilitate the final step of the technique (see end of this section). Firstly the computation of each component of order n ($n \in [1,m]$)—let us use symbol $y_n^{(m)}(t)$ to denote it—of the system response $y(t) \triangleq y^{(m)}(t)$ to the m-tone complex exponential input $x^{(m)}(t)$ has to be carried out by means of (5.2). For each value of m the expression for $y_n^{(m)}(t)$ depends on the Volterra kernel of order n in the frequency domain, i.e. $H_n(j\omega_1,\ldots,j\omega_n)$, according to the following general formula, valid as $t \to \infty$ [8]

$$y_n^{(m)}(t) = c^n \sum_{i_1=1}^{m} \cdots \sum_{i_n=1}^{m} H_n(j\omega_{i_1},\ldots,j\omega_{i_n}) e^{(j\omega_{i_1}+\ldots+j\omega_{i_n})t}. \tag{5.20}$$

Next dynamic equation (5.18) is rewritten so as to involve only terms with factor c^m. This operation, facilitated by the previous introduction of constant c within the expression of the m-tone input, consists of unfolding left- and right-hand sides of (5.18) as follows. Let us first consider the left-hand side of dynamic equation (5.18). Since the expression for the input—see (5.19)—has a c factor for any value of m, $x(t)$ shall be set to $x^{(m)}(t)$ for $m = 1$ and to 0 for any other value of m. Further, within the series expressing $y(t)$ only the mth-order component $y_m^{(m)}(t)$ shall be considered. The right-hand side of dynamic equation (5.18) is set to 0 for $m = 1$—since for this choice of m there exists no term with factor c—while for $m > 1$ it consists of a series of integrals with factor c^m—where the number of terms in the series increases with m (for $m = 1$ only one term is present). After this procedure is completed, some algebraic calculations are carried out to combine all terms with common harmonic content expressed by factor $e^{j(\omega_1+\ldots+\omega_m)t}$ on each side of the resulting equation. Equating these two groups of terms allows the determination of $H_m(j\omega_1,\ldots,j\omega_m)$.

5.3.2.1 First-Order Volterra Kernel

Let us first determine an equation for the kernel of order 1 (thus here $m = 1$). Reformulating (5.18) so that only terms with factor c appear in it, in accordance with Sect. 5.3.2, yields

$$\hat{Y}_{21}\{x^{(1)}(t)\} - \hat{Y}_{11}\{y_1^{(1)}(t)\} = 0. \tag{5.21}$$

[2]Note that for $m > 1$ the technique requires the preliminary derivation of all kernels of order $1,\ldots,m-1$.

In this equation we set $x^{(1)}(t) = ce^{j\omega t}$ and express $y_1^{(1)}(t)$ as follows from (5.20) with $m = 1$ and $n = 1$, i.e. as

$$y_1^{(1)}(t) = ce^{j\omega t} H_1(j\omega). \tag{5.22}$$

By using properties of integro-differential operators applied to periodic signals, from (5.21) we may finally derive the Volterra kernel of order 1 in the frequency domain for the class of circuits of Fig. 5.3. This is found to be described by

$$H_1(j\omega) = \frac{\hat{Y}_{21}(j\omega)}{\hat{Y}_{11}(j\omega)}, \tag{5.23}$$

where $Y_{\alpha\beta}(j\omega)$ $(\alpha, \beta \in \{1,2\})$ are the elements of the admittance matrix $\hat{\mathbf{Y}}(j\omega)$, expressed—as it is standard in circuit theory—in the frequency domain and arranged as

$$\hat{\mathbf{Y}}(j\omega) = \begin{bmatrix} \hat{Y}_{11}(j\omega) & \hat{Y}_{12}(j\omega) \\ \hat{Y}_{21}(j\omega) & \hat{Y}_{22}(j\omega) \end{bmatrix}. \tag{5.24}$$

The determinant of $\hat{\mathbf{Y}}(j\omega)$ is defined as

$$\det(\hat{\mathbf{Y}}(j\omega)) = \hat{Y}_{11}(j\omega)\hat{Y}_{22}(j\omega) - \hat{Y}_{12}(j\omega)\hat{Y}_{21}(j\omega).$$

5.3.2.2 Second-Order Volterra Kernel

Let us determine an equation for the kernel of order 2 (here $m = 2$ and thus $x^{(2)}(t) = c(e^{j\omega_1 t} + e^{j\omega_2 t})$). Using the methodology described in Sect. 5.3.2, extracting terms with factor c^2 from (5.18), we get

$$-\hat{Y}_{11}\{y_2^{(2)}(t)\} = \det(\hat{\mathbf{Y}}) \left\{ y_1^{(2)}(t) m_1 \left(-\int_{-\infty}^{t} y_1^{(2)}(t')dt' \right) \right\}. \tag{5.25}$$

In this equation we express $y_1^{(2)}(t)$ and $y_2^{(2)}(t)$ as follows from (5.20) when $m = 2$ and n is, respectively, set to 1 and 2, i.e. as

$$y_1^{(2)} = c(e^{j\omega_1 t} H_1(j\omega_1) + e^{j\omega_2 t} H_1(j\omega_2)), \tag{5.26}$$

and

$$y_2^{(2)} = c^2 \Big(H_2(j\omega_1, j\omega_1)e^{2j\omega_1 t} + H_2(j\omega_2, j\omega_2)e^{2j\omega_2 t} + (H_2(j\omega_1, j\omega_2) + H_2(j\omega_2, j\omega_1))\, e^{(j\omega_1 + j\omega_2)t} \Big). \tag{5.27}$$

After some algebraic manipulation, under the assumption of a symmetric second-order kernel, (5.25) yields the following closed-form expression for the second-order Volterra kernel

$$H_2(j\omega_1, j\omega_2) = \frac{m_1}{j\omega_1} \frac{\det(\hat{\mathbf{Y}}(j(\omega_1+\omega_2)))}{\hat{Y}_{11}(j(\omega_1+\omega_2))} H_1(j\omega_1)H_1(j\omega_2). \tag{5.28}$$

In the general case of non-symmetric kernels, mathematical operations on (5.25) do not lead to a closed-form solution for the second-order kernel. The resulting equation is reported in Appendix 5.6.1. Note that, under no assumption on the symmetry properties of the kernels, a closed-form expression for the nth-order kernel in the frequency domain, i.e. $H_n(j\omega_1, \ldots, j\omega_n)$, may be given only for $n = 1$. However, the proposed Volterra-series modeling approach may be easily applied to this general case, as explained in Appendix 5.6.

5.3.2.3 Third-Order Kernel

Finally, let us derive an equation for the kernel of order 3 (here m is set to 3 and thus $x^{(3)}(t) = c(e^{j\omega_1 t} + e^{j\omega_2 t} + e^{j\omega_3 t})$). Unfolding (5.18) so as to contain only terms with factor c^3 gives

$$-\hat{Y}_{11}\{y_3^{(3)}(t)\} = \det(\hat{\mathbf{Y}}) \left\{ y_2^{(3)}(t) m_1 \left(-\int_{-\infty}^{t} y_1^{(3)}(t')dt' \right) + y_1^{(3)}(t) m_1 \left(-\int_{-\infty}^{t} y_2^{(3)}(t')dt' \right) + y_1^{(3)}(t) m_2 \left(-\int_{-\infty}^{t} y_1^{(3)}(t')dt' \right)^2 \right\}. \tag{5.29}$$

Using (5.20) with $m = 3$ and n, respectively, set to 1, 2, and 3 to express $y_n^{(3)}(t)$ $(n = 1,2,3)$ in (5.29), evaluating the integrals in this equation, carrying out some algebraic manipulation and then equating left- and right-hand side of the resulting equation with respect to terms with harmonic content $e^{j(\omega_1+\omega_2+\omega_3)t}$ (all other harmonics are of no interest), under the assumption of a symmetric third-order kernel, we derive the following closed-form expression for the third-order Volterra kernel in the frequency domain

$$H_3(j(\omega_1, \omega_2, \omega_3)) = \frac{\det(\hat{\mathbf{Y}}(j(\omega_1+\omega_2+\omega_3)))}{\hat{Y}_{11}(j(\omega_1+\omega_2+\omega_3))} \left\{ m_1 \frac{H_1(j\omega_3)H_2(j\omega_1, j\omega_2)}{j\omega_3} + m_1 \frac{H_1(j\omega_1)H_2(j\omega_2, j\omega_3)}{j(\omega_2+\omega_3)} - m_2 \frac{H_1(j\omega_1)H_1(j\omega_2)H_1(j\omega_3)}{j\omega_2 j\omega_3} \right\}. \tag{5.30}$$

In the non-symmetric case mathematical operations on (5.29) do not lead to a closed-form solution for the third-order kernel. The resulting equation is reported in Appendix 5.6.2.

5.3.3 *Computation of the System Output from the Knowledge of the Kernels*

Inserting a specific waveform to the input of a Volterra system of the type shown in Fig. 5.1 its response $y(t)$ may be computed using (5.1). For example, for a one-tone sine wave input—let $x(t) = x_0 \sin(\omega t) = \frac{1}{2j}\sum_{i=-1}^{+1} x_i sgn(i) e^{j\omega_i t}$ where $x_i = x_0$ for $i = -1, 1$ and $\omega_{-1} = -\omega_1$—it follows that the nth-order component of $y(t)$ may be expressed as [28]

$$y_n(t) = \frac{1}{(2j)^n} \sum_{i_1=-1,\ i_1\neq 0}^{+1} \cdots \sum_{i_n=-1,\ i_n\neq 0}^{+1} x_{i_1} sgn(i_1) \ldots x_{i_n} sgn(i_n) \\ H_n(j\omega_{i_1}, \ldots j\omega_{i_n}) e^{j(\omega_{i_1}+\ldots+\omega_{i_n})t}. \tag{5.31}$$

Defining $\omega_1 \triangleq \omega$, denoting with $\Re\{\cdot\}$ and $\Im\{\cdot\}$ the operations extracting real and imaginary parts, this equation yields the following expressions for the first three components of $y(t)$

$$y_1(t) = x_0 \Im\{H_1(j\omega)e^{j\omega t}\}, \tag{5.32}$$

$$y_2(t) = -\frac{x_0^2}{2}\left(-\Re\{H_2(j\omega, -j\omega)\} + \Re\{H_2(j\omega, j\omega)e^{2j\omega t}\}\right), \tag{5.33}$$

and

$$y_3(t) = -\frac{x_0^3}{4}\Big(\Im\{H_3(j\omega, j\omega, j\omega)e^{3j\omega t}\} + \\ \Im\{(H_3(-j\omega, j\omega, j\omega) + H_3(j\omega, j\omega, -j\omega) + H_3(j\omega, -j\omega, j\omega))\, e^{j\omega t}\}\Big). \tag{5.34}$$

Finally, retaining the first $l \geq 1$ components of the output response $y(t)$ of the Volterra system to the sine input $x(t) = x_0 \sin(\omega t)$ into (5.1) yields the lth-order Volterra series approximation of the actual output current of a two-port in the class of Fig. 5.3. In the next section we choose an element from this class and show how well the output of the Volterra system reconstructs the time waveform of current $i_2(t)$, derived through numerical integration of the nonlinear ordinary differential equations describing the one-memristor circuit under study.

Our results show that the level of accuracy of the reconstruction improves as a larger number of components is retained in the Volterra series of $y(t)$, i.e. as l is increased. This is even more important if the number of nonzero coefficients m_k in the memristance function $M(q_m)$ is increased, since the nonlinearity of the system gets stronger, and, consequently, the output current $i_2(t)$ contains a larger number of frequency components.

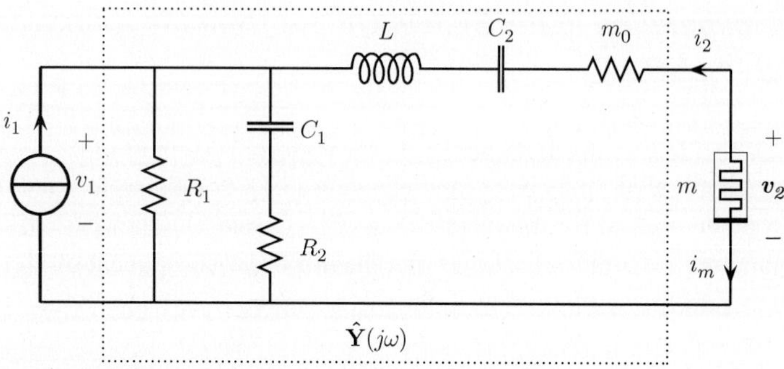

Fig. 5.4 Single-memristor oscillator

5.3.4 Example

Let us consider a circuit in the class of Fig. 5.3, i.e. the single-memristor oscillator depicted in Fig. 5.4.

The two-port within the dotted rectangular box is characterized by an admittance matrix described by (5.24), where, defining

$$G_1 = R_1^{-1},$$

$$Y_1(j\omega) = \frac{j\omega C_1}{1 + j\omega C_1 R_2}, \text{ and}$$

$$Y_2(j\omega) = \frac{j\omega C_2}{1 + m_0 j\omega C_2 + LC_2(j\omega)^2},$$

we have

$$\hat{Y}_{11}(j\omega) = G_1 + Y_1(j\omega) + Y_2(j\omega), \text{ and}$$
$$\hat{Y}_{12}(j\omega) = \hat{Y}_{21}(j\omega) = -\hat{Y}_{22}(j\omega) = -Y_2(j\omega).$$

In order to show the accuracy of the Volterra series-based model of the circuit of Fig. 5.4, we shall compare the dynamical behavior predicted by the Volterra series-based approach with that resulting from numerical integration of the dynamical equations under the same setting for input waveform and circuit component values. The dynamical equations of the non-autonomous dynamical system of Fig. 5.4 may be expressed as

$$\dot{y}_1 = \frac{1}{C_1}\left(-\frac{1}{R_1+R_2}y_1 - \frac{R_1}{R_1+R_2}y_3 + \frac{R_1}{R_1+R_2}i_1(t)\right),$$

$$\dot{y}_2 = \frac{1}{C_2}y_3,$$

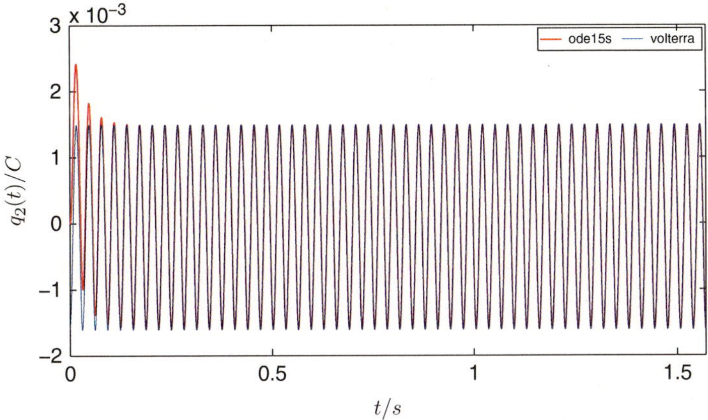

Fig. 5.5 Time waveform of $q_2(t)$ from numerical integration of (5.35) (*in red*) and from Volterra-series based estimate (*in blue*)

$$\dot{y}_3 = \frac{1}{L}\left(\frac{R_1}{R_1+R_2}y_1 - \left(\frac{R_1R_2}{R_1+R_2} + m_0 + M(y(4))\right)y_3 - y_2 + \frac{R_1R_2}{R_1+R_2}i_1(t)\right), \text{ and}$$

$$\dot{y}_4 = y_3, \tag{5.35}$$

where the states are defined as $y_1 = v_{C1}$, $y_2 = v_{C2}$, $y_3 = i_L$ and $y_4 = q_L$.

Let us apply a current input $i_1(t) = i_{10}\sin\omega t = x_0 \sin\omega t$ (where we let $i_{10} = x_0$) to the circuit of Fig. 5.4. The amplitude and angular frequency of the excitation are, respectively, set to $x_0 = 1A$ and $\omega = 200\,rads^{-1}$. Denoting the time period of the input sine wave as $T = \frac{2\cdot\pi}{\omega}$, the time step is chosen as $\frac{T}{1000}$, the initial time is taken as $0s$, and the simulation time is taken as $50\cdot T$. The values of the electrical components are selected as follows: $R_1 = 10k\Omega$, $C_1 = 2\mu F$, $C_2 = 1\mu F$, $R_2 = 100\Omega$, and $L = 5mH$. Regarding the choice for the coefficients in the memristance expression given in (5.10), only m_0 and m_1 are set to non-null values. In particular we let $m_0 = 2k\Omega$ and $m_1 = 0.83\cdot 10^6\,\Omega C^{-1}$. This choice refers to a simple model for Williams' memristor [26]. In the numerical simulation to follow the ordinary differential equation solver $ode15s$, available in the Matlab software package, was employed to solve (5.35) with initial conditions $[y_1(0), y_2(0), y_3(0), y_4(0)]' = [0,0,0,0]'$. Regarding the proposed Volterra series-based methodology, the system output is approximated by expanding the Volterra series in (5.1) up to $n = 4$. Figure 5.5 shows the time waveform of q_2 according to the numerical simulation of (5.35) (red curve) and to the Volterra series-based estimate (blue curve). It is evident that the Volterra series-based estimate accurately tracks the dynamics of the numerical result.

Figure 5.6 shows the agreement between the memristor voltage-current characteristic from numerical simulation of (5.35) (red curve) and from Volterra series estimation (blue curve). Note that the memristor voltage includes the drop across linear resistor m_0, as it is clear from Fig. 5.4.

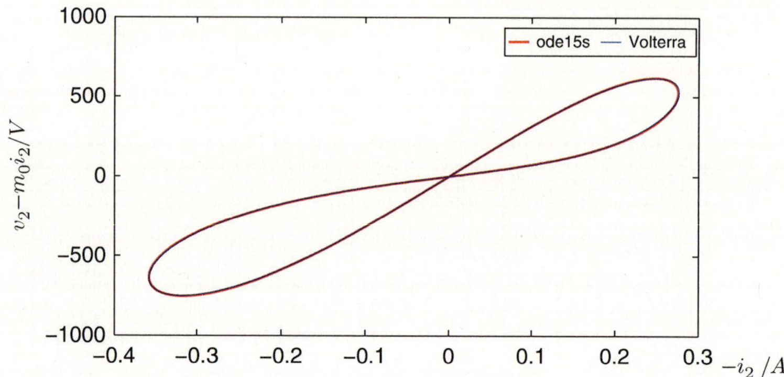

Fig. 5.6 Typical memristor voltage-current bow-tie: matching between numerical solution (*in red*) and Volterra series-based estimate (*in blue*)

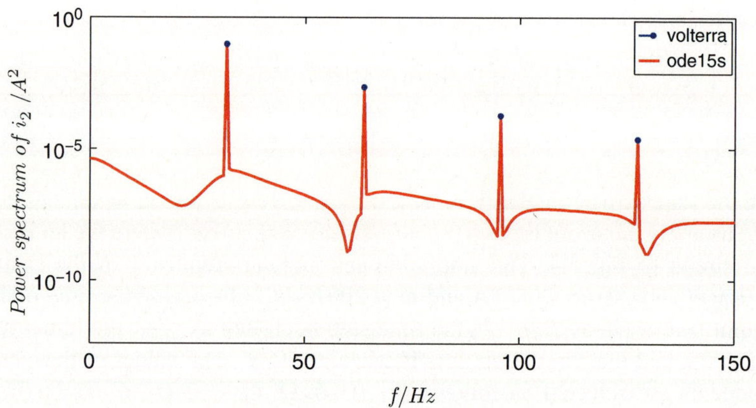

Fig. 5.7 Power spectrum of $i_2(t)$ from numerical simulation (*in red*) and first four coefficients of the series of cosine functions in which the output $y(t)$ of the Volterra system may be unfolded (*blue dots*). The y-axis is in logarithmic scale

The red curve in Fig. 5.7 shows the power spectrum of the two-port output current of the circuit of Fig. 5.4, computed by means of the fast Fourier transform of the time series of i_2 recorded from the above numerical simulation of (5.35). The blue dots denote the coefficients—let us call them as Y_n—of the series of cosine functions of the form $\sum_{n=0}^{m} Y_n \cos(n\omega t + \varphi_n)$, in which we may recast the Volterra series-based predicted output $y(t) = \sum_{n=1}^{m=4} y_n(t)$. The agreement between the blue dots and the peaks of the red curve further demonstrates the goodness of the proposed approach.

5.4 Investigation of the Nonlinear Dynamics of a Class of Two-Memristor Circuits

In this section we extend the methodology proposed in Sect. 5.2 to the class of two-memristor circuits shown in Fig. 5.8, where $i_p(t)$ and $v_p(t)$ denote current and voltage of port $p = \{1,2,3\}$ of a three-port network described by means of the impedance matrix $\hat{\mathbf{Z}}\{\cdot\}$ (alternatively the admittance matrix $\hat{\mathbf{Y}}\{\cdot\}$ may be used), composed of integro-differential operators $\hat{Z}_{\gamma\delta}\{\cdot\}$ $(\gamma, \delta \in 1,2,3)$ according to

$$\hat{\mathbf{Z}}\{\cdot\} = \begin{bmatrix} \hat{Z}_{11}\{\cdot\} & \hat{Z}_{12}\{\cdot\} & \hat{Z}_{13}\{\cdot\} \\ \hat{Z}_{21}\{\cdot\} & \hat{Z}_{22}\{\cdot\} & \hat{Z}_{23}\{\cdot\} \\ \hat{Z}_{31}\{\cdot\} & \hat{Z}_{32}\{\cdot\} & \hat{Z}_{33}\{\cdot\} \end{bmatrix}. \tag{5.36}$$

Port 1 of the three-port is closed on an input source providing current $i_1(t)$, while ports 2 and 3 are, respectively, loaded with charge-controlled memristors m_2 and m_3. Note that a similar approach may be followed to analyze flux-driven two-memristor circuits. Let the memristance function of memristor m_r $(r \in \{2,3\})$ be expressed by

$$M(q_{m_r}) = M\left(\int_{-\infty}^{t} i_{m_r}(t')dt'\right) = \sum_{k=1}^{\infty} m_{rk}\, q_{m_r}^{k}, \tag{5.37}$$

where q_{m_r} and $i_{m_r} = -i_r$, respectively, denote charge and current flowing through memristor m_r at port $r \in \{2,3\}$, while symbol m_{rk} indicates the kth coefficient in the series of powers of q_{m_r} in which the memristance function $M(q_{m_r})$ is unfolded in (5.37).

In the class of dynamical systems of Fig. 5.8 current i_r at port r may be seen as the output of a Volterra system associated with memristor m_r $(r = \{2,3\})$ and excited with input $i_1(t)$. In the remaining part of this paper the Volterra system associated with memristor m_r shall be referred to as Volterra system $r-1$ $(r \in \{2,3\})$. The nth-order kernels of Volterra system $r-1$ in the time and frequency domain are, respectively, denoted as $h_{rn}(\tau_1, \ldots, \tau_n)$ and $H_{rn}(j\omega_1, \ldots, j\omega_n)$.

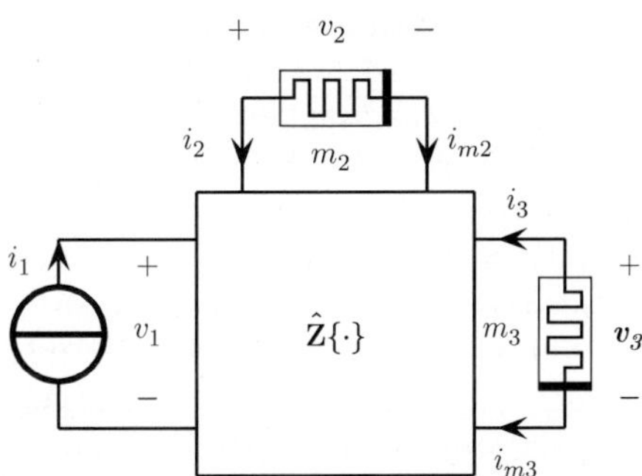

Fig. 5.8 General topology of the class of two-memristor circuits

5.4.1 Determination of the Dynamic Equations

Expressing the port voltages in terms of the port currents, the three-port in Fig. 5.8 is described according to

$$v_1(t) = \hat{Z}_{11}\{i_1(t)\} + \hat{Z}_{12}\{i_2(t)\} + \hat{Z}_{13}\{i_3(t)\}, \tag{5.38}$$

$$v_2(t) = \hat{Z}_{21}\{i_1(t)\} + \hat{Z}_{22}\{i_2(t)\} + \hat{Z}_{23}\{i_3(t)\}, \text{ and} \tag{5.39}$$

$$v_3(t) = \hat{Z}_{31}\{i_1(t)\} + \hat{Z}_{32}\{i_2(t)\} + \hat{Z}_{33}\{i_3(t)\}. \tag{5.40}$$

Equation (5.38) for port 1 closed on the excitation needs not be considered any further in the analysis to follow. On the other hand, for port r, $r = \{2,3\}$, we express v_r as $v_r = M(q_{m_r})i_{m_r}$, where $M(q_{m_r})$ is described by (5.37), and, as mentioned in Sect. 5.4, we introduce Volterra system $r-1$ with input $x(t) = i_1(t)$ and output $y_r(t) = i_r(t)$, expanded in Volterra series as

$$y_r(t) = \sum_{n=1}^{\infty} y_{rn}(t), \tag{5.41}$$

Equations (5.39) and (5.40) may be recast as

$$-\sum_{k=1}^{\infty} m_{2k}\left(-\int_{-\infty}^{t} y_2(t')dt'\right)^k y_2(t) = \hat{Z}_{21}\{x(t)\} + \hat{Z}_{22}\{y_2(t)\} + \hat{Z}_{23}\{y_3(t)\}, \text{ and} \tag{5.42}$$

$$-\sum_{k=1}^{\infty} m_{3k}\left(-\int_{-\infty}^{t} y_3(t')dt'\right)^k y_3(t) = \hat{Z}_{31}\{x(t)\} + \hat{Z}_{32}\{y_2(t)\} + \hat{Z}_{33}\{y_3(t)\}. \tag{5.43}$$

These are the dynamic equations of the class of circuits of Fig. 5.8.

5.4.2 Determination of the Volterra Kernels

In this section we shall apply the harmonic probe technique reviewed in Sect. 5.3 to the dynamic equations (5.42)–(5.43). The nth-order component $y_{rn}^{(m)}(t)$ ($n \in [1,m]$) of the response $y_r(t) \triangleq y_r^{(m)}(t)$ of the Volterra system $r-1$ ($r \in 2,3$) to an m-tone complex exponential input of the form given in (5.19), computed after tailoring equation (5.20) to the Volterra system $r-1$ (i.e., after replacing in this equation $y(t)$ and $H_n(j\omega_1,\ldots,j\omega_n)$ with $y_r(t)$ and $H_{rn}(j\omega_1,\ldots,j\omega_n)$, respectively), depends upon the nth-order kernel in the frequency domain of the Volterra system $r-1$.

5.4.2.1 First-Order Kernels

Let us first derive the first-order kernel equations for the two Volterra systems identified in Fig. 5.8.

Here $m = 1$ and thus a one-tone input is applied to Volterra system $r-1$ ($r \in \{2,3\}$). Using (5.41) into (5.42)–(5.43) and retaining only terms with factor c, the resulting equations have null left-hand sides and assume the following form

$$0 = \hat{Z}_{21}\{x^{(1)}(t)\} + \hat{Z}_{22}\{y_{21}^{(1)}(t)\} + \hat{Z}_{23}\{y_{31}^{(1)}(t)\}, \text{ and} \tag{5.44}$$

$$0 = \hat{Z}_{31}\{x^{(1)}(t)\} + \hat{Z}_{32}\{y_{21}^{(1)}(t)\} + \hat{Z}_{33}\{y_{31}^{(1)}(t)\}, \tag{5.45}$$

where $y_{r1}^{(1)}(t)$ denotes the first component ($n = 1$) of the output response of Volterra system $r-1$ ($r \in \{2,3\}$) to the one-tone input ($m = 1$). In (5.44)–(5.45) $x^{(1)}(t) = ce^{j\omega t}$, while the expression for $y_{r1}^{(1)}(t)$ is computed after tailoring equation (5.20) to the Volterra system $r-1$ ($r \in \{2,3\}$) and setting $m = 1$ and $n = 1$. It follows that (5.44)–(5.45) lead to

$$0 = \hat{Z}_{21}(j\omega) + \hat{Z}_{22}(j\omega)H_{21}(j\omega) + \hat{Z}_{23}(j\omega)H_{31}(j\omega), \text{ and} \tag{5.46}$$

$$0 = \hat{Z}_{31}(j\omega) + \hat{Z}_{32}(j\omega)H_{21}(j\omega) + \hat{Z}_{33}(j\omega)H_{31}(j\omega), \tag{5.47}$$

where $\hat{Z}_{\gamma,\delta}(j\omega)$ ($\gamma, \delta \in 1,2,3$) are the elements of the impedance matrix $\hat{\mathbf{Z}}(j\omega)$, arranged as

$$\hat{\mathbf{Z}}(j\omega) = \begin{bmatrix} \hat{Z}_{11}(j\omega) & \hat{Z}_{12}(j\omega) & \hat{Z}_{13}(j\omega) \\ \hat{Z}_{21}(j\omega) & \hat{Z}_{22}(j\omega) & \hat{Z}_{23}(j\omega) \\ \hat{Z}_{31}(j\omega) & \hat{Z}_{32}(j\omega) & \hat{Z}_{33}(j\omega) \end{bmatrix}. \tag{5.48}$$

Solving (5.46)–(5.47) for the first-order kernel $H_{r1}(j\omega)$ of Volterra system $r-1$ ($r \in \{2,3\}$) gives

$$H_{21}(j\omega) = \frac{\hat{Z}_{21}(j\omega)\hat{Z}_{33}(j\omega) - \hat{Z}_{23}(j\omega)\hat{Z}_{31}(j\omega)}{\hat{Z}_{32}(j\omega)\hat{Z}_{23}(j\omega) - \hat{Z}_{22}(j\omega)\hat{Z}_{33}(j\omega)}, \text{ and} \tag{5.49}$$

$$H_{31}(j\omega) = \frac{\hat{Z}_{31}(j\omega)\hat{Z}_{22}(j\omega) - \hat{Z}_{32}(j\omega)\hat{Z}_{21}(j\omega)}{\hat{Z}_{32}(j\omega)\hat{Z}_{23}(j\omega) - \hat{Z}_{22}(j\omega)\hat{Z}_{33}(j\omega)}. \tag{5.50}$$

5.4.2.2 Second-Order Kernels

This section determines the equation for the second-order kernel $H_{r2}(j\omega_1, j\omega_2)$ of Volterra system $r-1$ ($r \in \{2,3\}$). Here $m = 2$, therefore a two-tone input $x^{(2)}(t) = c\left(e^{j\omega_1 t} + e^{j\omega_2 t}\right)$ is applied to each of the two Volterra systems. Using (5.41) into (5.42)–(5.43) and keeping only c^2-terms in the resulting equations yields

$$-m_{21}\left(-\int_{-\infty}^{t} y_{21}(t')dt'\right) y_{21}(t) = \hat{Z}_{22}\{y_{22}^{(2)}(t)\} + \hat{Z}_{23}\{y_{32}^{(2)}(t)\}, \text{ and} \tag{5.51}$$

$$-m_{31}\left(-\int_{-\infty}^{t} y_{31}(t')dt'\right) y_{31}(t) = \hat{Z}_{32}\{y_{22}^{(2)}(t)\} + \hat{Z}_{33}\{y_{32}^{(2)}(t)\}, \tag{5.52}$$

where $y_{rn}^{(2)}(t)$ denotes the nth component ($n = \{1,2\}$) of the output response of Volterra system $r-1$ ($r \in \{2,3\}$) to the two-tone input ($m = 2$). Let us express $y_{rn}^{(2)}(t)$ in (5.51)–(5.52) as follows after tailoring equation (5.20) to Volterra system $r-1$ ($r \in \{2,3\}$) and setting $m = 2$ and $n \in \{1,2\}$. Calculating the integrals in (5.51)–(5.52), equating side-by-side terms with harmonic content expressed by factor $e^{j(\omega_1+\omega_2)t}$, and then assuming that Volterra system $r-1$ ($r \in \{2,3\}$) admits a symmetric second-order kernel, some further algebraic manipulation on the resulting equations allows the derivation of the following closed-form expression for $H_{r2}(j\omega_1, j\omega_2)$

$$H_{22}(j\omega_1, j\omega_2) = \frac{1}{j\omega_2}\left(\frac{\hat{Z}_{33}(j(\omega_1+\omega_2))m_{21}H_{21}(j\omega_1)H_{21}(j\omega_2)}{\eta j(\omega_1+\omega_2)} - \frac{\hat{Z}_{23}(j(\omega_1+\omega_2))m_{31}H_{31}(j\omega_1)H_{31}(j\omega_2)}{\eta j(\omega_1+\omega_2)}\right), \tag{5.53}$$

and

$$H_{32}(j\omega_1, j\omega_2) = -\frac{1}{j\omega_2}\left(\frac{\hat{Z}_{32}(j(\omega_1+\omega_2))m_{21}H_{21}(j\omega_1)H_{21}(j\omega_2)}{\eta j(\omega_1+\omega_2)} - \frac{\hat{Z}_{22}(j(\omega_1+\omega_2))m_{31}H_{31}(j\omega_1)H_{31}(j\omega_2)}{\eta j(\omega_1+\omega_2)}\right), \tag{5.54}$$

where $\eta(\cdot) \triangleq \hat{Z}_{33}()\hat{Z}_{22}() - \hat{Z}_{23}()\hat{Z}_{32}()$. Appendix 5.6.3 reports the equation for the second-order kernel of Volterra system $r-1$ ($r \in \{2,3\}$) in the non-symmetric case.

5.4.2.3 Third-Order Kernels

Here we derive the equation for the third-order kernel $H_{r3}(j\omega_1, j\omega_2, j\omega_3)$ of Volterra system $r-1$ ($r \in \{2,3\}$). Letting $m = 3$, a three-tone input $x^{(3)}(t) = c\left(e^{j\omega_1 t} + e^{j\omega_2 t} + e^{j\omega_3 t}\right)$ is applied to each of the two Volterra systems.

Using (5.41) into (5.42)–(5.43), and then retaining only c^3-terms, the resulting equations reduce to

$$-m_{21}\left(-\int_{-\infty}^{t} y_{21}(t')dt'\right) y_{22}(t) - m_{21}\left(-\int_{-\infty}^{t} y_{22}(t')dt'\right) y_{21}(t) - m_{22}\left(-\int_{-\infty}^{t} y_{21}(t')dt'\right)^2 y_{21}(t) = \hat{Z}_{22}\{y_{23}(t)\} + \hat{Z}_{23}\{y_{33}(t)\}, \tag{5.55}$$

and

$$-m_{31}\left(-\int_{-\infty}^{t} y_{31}(t')dt'\right) y_{32}(t) - m_{31}\left(-\int_{-\infty}^{t} y_{32}(t')dt'\right) y_{31}(t) - m_{32}$$
$$\left(-\int_{-\infty}^{t} y_{31}(t')dt'\right)^2 y_{31}(t) = \hat{Z}_{32}\{y_{23}(t)\} + \hat{Z}_{33}\{y_{33}(t)\}, \tag{5.56}$$

where $y_{rn}^{(3)}(t)$ denotes the nth component ($n = \{1,2,3\}$) of the output response of Volterra system $r-1$ ($r \in \{2,3\}$) to the three-tone input ($m = 3$). Let us express $y_{rn}^{(3)}(t)$ in (5.55)–(5.56) as follows from tailoring (5.20) to Volterra system $r-1$ ($r \in \{2,3\}$) and setting $m = 3$ and $n \in \{1,2,3\}$. Calculating the integrals in equations (5.55)–(5.56), equating side-by-side terms with harmonic content expressed by factor $e^{j(\omega_1+\omega_2+\omega_3)t}$, and then assuming that Volterra system $r-1$ ($r \in \{2,3\}$) admits a symmetric third-order kernel, some further algebraic manipulation on the resulting equations allows the derivation of the following closed-form expression for $H_{r3}(j\omega_1, j\omega_2, j\omega_3)$

$$\begin{aligned} H_{23}(j\omega_1, j\omega_2, j\omega_3) = {} & \frac{1}{\eta(j(\omega_1+\omega_2+\omega_3))}\Bigg\{\hat{Z}_{33}(j(\omega_1+\omega_2+\omega_3)) \\ & \Bigg[m_{21}H_{22}(j\omega_2, j\omega_3)H_{21}(j\omega_1)\left(\frac{1}{j\omega_1}+\frac{1}{j(\omega_2+\omega_3)}\right)+\frac{m_{22}}{3} \\ & H_{21}(j\omega_1)H_{21}(j\omega_2)H_{21}(j\omega_3)\left(\frac{1}{\omega_1\omega_2}+\frac{1}{\omega_2\omega_3}+\frac{1}{\omega_1\omega_3}\right)\Bigg] \\ & -\hat{Z}_{23}(j(\omega_1+\omega_2+\omega_3))\Bigg[m_{31}\left(\frac{1}{j\omega_1}+\frac{1}{j(\omega_2+\omega_3)}\right) \\ & H_{32}(j\omega_2, j\omega_3)H_{31}(j\omega_1) \\ & +\frac{m_{32}H_{31}(j\omega_1)H_{31}(j\omega_2)H_{31}(j\omega_3)}{(\omega_1\omega_3)}\Bigg]\Bigg\}, \end{aligned} \tag{5.57}$$

and

$$\begin{aligned} H_{33}(j\omega_1, j\omega_2, j\omega_3) = {} & \frac{1}{-\eta(j(\omega_1+\omega_2+\omega_3))}\Bigg\{\hat{Z}_{32}(j(\omega_1+\omega_2+\omega_3)) \\ & \Bigg[m_{21}H_{22}(j\omega_2, j\omega_3)H_{21}(j\omega_1)\left(\frac{1}{j\omega_1}+\frac{1}{j(\omega_2+\omega_3)}\right) \\ & +\frac{m_{22}H_{21}(j\omega_1)H_{21}(j\omega_2)H_{21}(j\omega_3)}{(\omega_1\omega_3)}\Bigg] \\ & -\hat{Z}_{22}(j(\omega_1+\omega_2+\omega_3))\Bigg[m_{31}\left(\frac{1}{j\omega_1}+\frac{1}{j(\omega_2+\omega_3)}\right) \end{aligned}$$

$$H_{32}(j\omega_2, j\omega_3)H_{31}(j\omega_1) + \frac{m_{32}}{3}H_{31}(j\omega_1)H_{31}(j\omega_2)H_{31}(j\omega_3)$$
$$\left(\frac{1}{\omega_1\omega_2} + \frac{1}{\omega_2\omega_3} + \frac{1}{\omega_1\omega_3}\right)\Bigg]\Bigg\}. \quad (5.58)$$

Appendix 5.6.4 reports the equation for the third-order kernel of Volterra system $r-1$ ($r \in \{2,3\}$) in the non-symmetric case.

5.4.3 Computation of the System Output from the Kernels

The nth-order response y_{rn} of Volterra system $r-1$ ($r \in \{2,3\}$ to a sine wave excitation of the form $x(t) = x_0 \sin(\omega t) = \frac{1}{2j}\sum_{i=-1}^{+1} x_i sgn(i) e^{j\omega_i t}$, where $x_i = x_0$ for $i = -1, 1$ and $\omega_{-1} = -\omega_1$, may be computed after tailoring equation (5.31) to Volterra system $r-1$ ($r \in \{2,3\}$).

Inserting the first $l \geq 1$ components of the output response $y_r(t)$ of Volterra system $r-1$ to sine-wave input $x(t) = x_0 \sin(\omega t)$ into (5.41) yields the lth-order Volterra series approximation of the actual current $i_r(t)$ at port r of the three-port of Fig. 5.8 resulting from the application of current $i_1(t) = i_0 \sin(\omega t)$ ($i_0 = x_0$) at port 1. In the next section, we select one particular element from the class of Fig. 5.8 and then show the agreement between the output current $i_r(t)$ at port r ($r \in 2,3$) derived by means of numerical integration of the nonlinear ordinary differential equations describing the chosen two-memristor circuit and its Volterra series-based approximation $y_r(t)$. The same comments regarding the level of accuracy of the Volterra series-based approach as in the case of single-memristor circuits (see Sect. 5.3.3) apply for the two-memristor circuits.

5.4.4 Example

Figure 5.9 shows an oscillator within the class of two-memristor circuits of Fig. 5.8.

The three-port within the dotted rectangular box is characterized by an impedance matrix expressed by (5.48), where

$$\hat{Z}_{11}(j\omega) = \hat{Z}_{13}(j\omega) = \hat{Z}_{31}(j\omega) = R,$$
$$\hat{Z}_{12}(j\omega) = \hat{Z}_{21}(j\omega) = 0,$$
$$\hat{Z}_{22}(j\omega) = j\omega L + m_0,$$
$$\hat{Z}_{23}(j\omega) = \hat{Z}_{32}(j\omega) = -j\omega L, \text{ and}$$
$$\hat{Z}_{33}(j\omega) = R + m_0 + j\omega L + \frac{1}{j\omega C}.$$

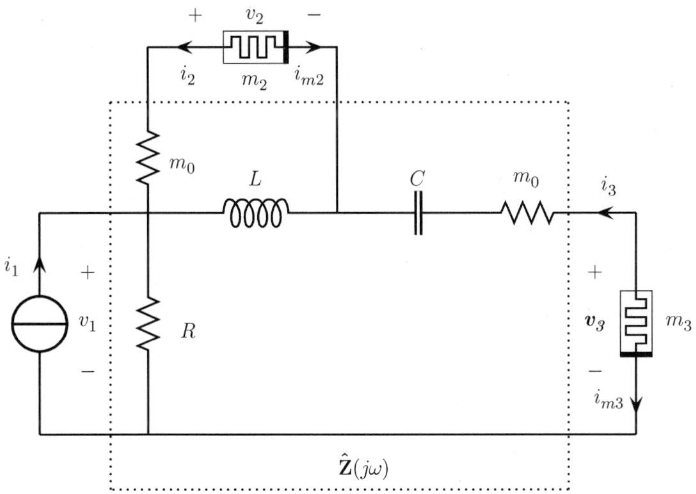

Fig. 5.9 Two-memristor oscillator

The state equations of the circuit of Fig. 5.9 may be written as

$$\dot{y}_1 = \frac{1}{C}\frac{Ri_1(t) - y_1 + (m_0 + M(-y_3))\,y_2}{R + 2m_0 + M(-y_3) + M(-y_4)},$$

$$\dot{y}_2 = \frac{m_0 + M(-y_3)}{L}\,\frac{Ri_1(t) - y_1 - (R + m_0 + M(-y_4))y_2}{R + 2m_0 + M(-y_3) + M(-y_4)},$$

$$\dot{y}_3 = -\frac{Ri_1(t) - y_1 - (R + m_0 + M(-y_4))\,y_2}{R + 2m_0 + M(-y_3) + M(-y_4)}, \text{ and}$$

$$\dot{y}_4 = -\frac{Ri_1(t) - y_1 + (m_0 + M(-y_3))\,y_2}{R + 2m_0 + M(-y_3) + M(-y_4)}, \tag{5.59}$$

where the states are defined as $y_1 = v_C$, $y_2 = i_L$, $y_3 = q_2 = \int_{-\infty}^{t} i_2(t')dt'$ and $y_4 = q_3 = \int_{-\infty}^{t} i_3(t')dt'$. Let us apply a sine-wave current $i_1(t) = i_0 \sin(\omega t) = x_0 \sin(\omega t)$ at port 1 and compare the results of the numerical simulation of (5.59) with the Volterra series-based solutions. The amplitude and angular frequency of the excitation are, respectively, set to $x_0 = 0.5A$ and $\omega = 200\,rads^{-1}$. Denoting the time period of the input sine wave as $T = \frac{2\cdot\pi}{\omega}$, the time step is chosen as $\frac{T}{1000}$, the initial time is taken as $0\,s$, and the simulation time is set to $50 \cdot T$. The values of the circuit components are taken as $R = 10\,k\Omega$, $L = 50\,mH$, and $C = 2\,\mu F$. The non-null coefficients in the memristance function $M(q_{m_r})$ of memristor m_r for each $r \in \{2,3\}$ are $m_0 = 2\,k\Omega$ and $m_1 = 0.83\ 10^6 \Omega C^{-1}$. Equations (5.59) are numerically integrated by means of the $ode15s$ solver with initial condition $[y_1(0), y_2(0), y_3(0), y_4(0)]' = [0,0,0,0]'$. Regarding the proposed Volterra series-based methodology, the system output is approximated by expanding the Volterra series in (5.41) up to $n = 3$.

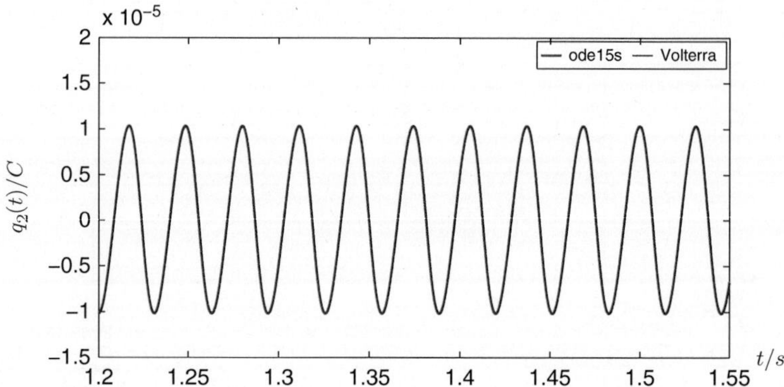

Fig. 5.10 Time waveform of $q_2(t)$ from its Volterra-series representation (*blue curve*) and from the numerical simulation of (5.59) (*red curve*) under the circuit parameter and input setting specified in the text

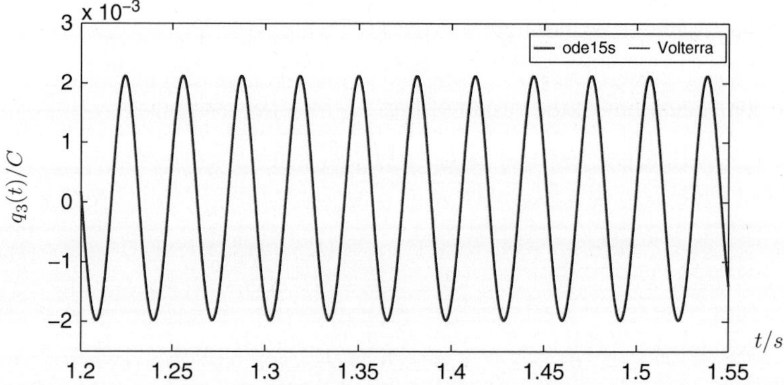

Fig. 5.11 Time behavior of $q_3(t)$ resulting from the numerical integration of (5.59) (*in red*) and from the Volterra series-based analysis (*in blue*)

Figures 5.10 and 5.11, respectively, show the time waveforms of $q_2(t)$ and of $q_3(t)$ as observed in numerical simulation (red curves) and as predicted by the proposed Volterra series technique.

The current-voltage characteristics of memristors m_2 and m_3 are, respectively, plotted in Figs. 5.12 and 5.13, where, once again, the numerical and Volterra series-based solutions are, respectively, shown in red and blue. In each of such figures the memristor voltage includes the drop across the linear resistor m_0. Note that in the simulation scenario under observation memristor m_2 is behaving as a linear resistor. On the other hand memristor m_3 exhibits in the current-voltage plane the typical bow-tie expected of this type of two-terminal element under sinusoidal excitation [3].

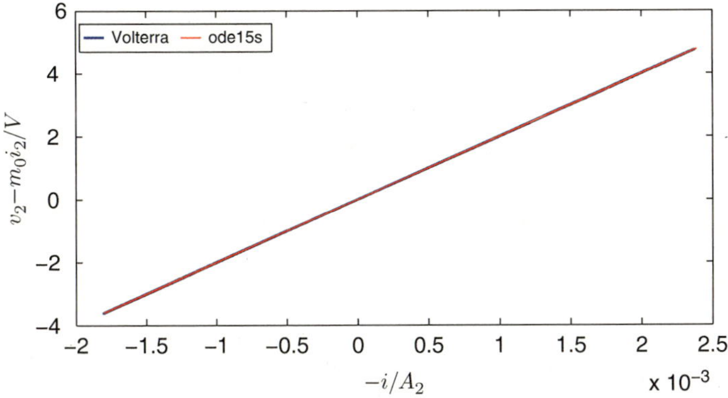

Fig. 5.12 Voltage versus current for memristor m_2. Here the *blue and red curves*, respectively, refer to the Volterra-based representation of the dynamical system under study and to the mathematical model expressed by (5.59)

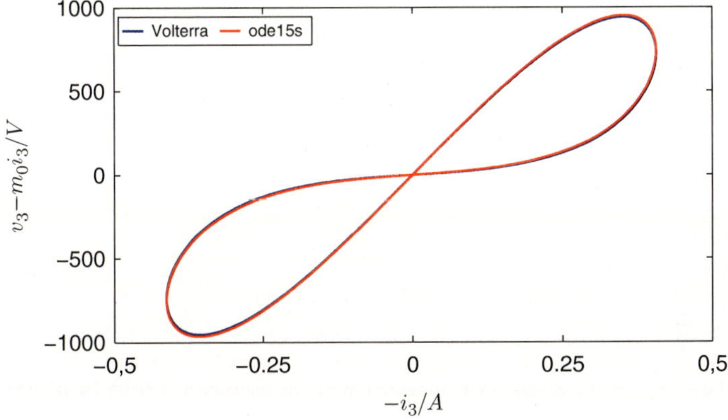

Fig. 5.13 Numerical result (*in red*) and Volterra-series representation (*in blue*) of the current-voltage characteristic of memristor m_3

Finally, the power spectra of currents $i_2(t)$ and $i_3(t)$ are, respectively, shown in Figs. 5.14 and 5.15 for both the ode15s solution to (5.59) (red curve) and the Volterra-series estimate (blue dots).

5.5 Conclusions

A full understanding of the nonlinear dynamics of memristors is one of the most important steps toward a large-scale use of this electrical component in IC design. Due to its unique dynamical behavior novel techniques need to be developed for

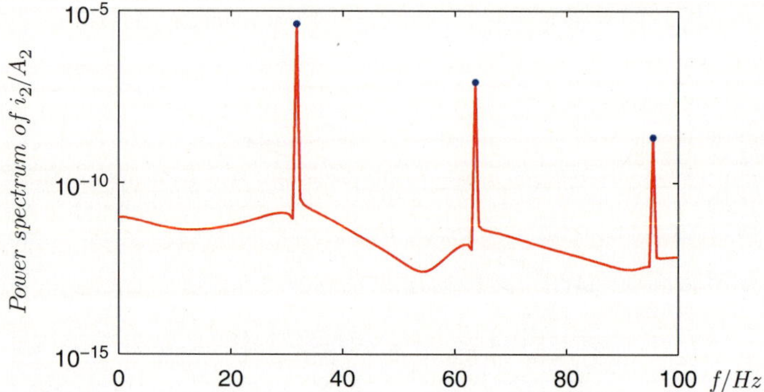

Fig. 5.14 Power spectrum of $i_2(t)$. The legend clarifies the color coding

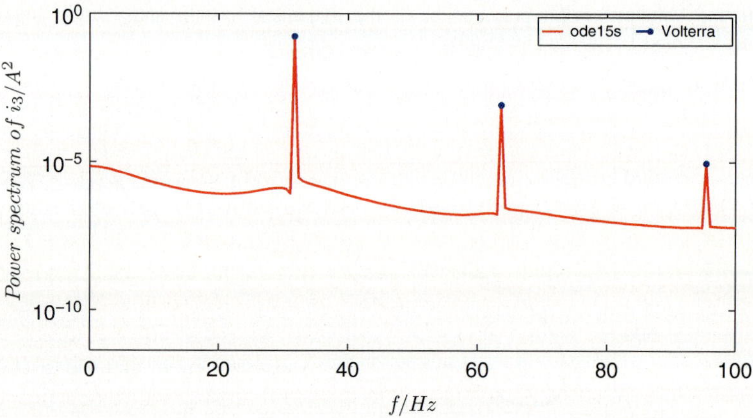

Fig. 5.15 Power spectrum of $i_3(t)$. The same color coding as in Figs. 5.10–5.14 applies here

the investigation and modeling of memristors and circuits based upon them. In this paper we propose an approach based on the Volterra series technique to track the dynamics of classes of single- and two-memristor circuits. The approach is first formulated from a theoretical point of view and then validated through examples. The proposed analytical tool may be easily generalized to investigate the nonlinear dynamics of a class of multi-memristor circuits.

5.6 Appendix

This section reports the equations for the Volterra kernels of orders 2 and 3 for the class of single- and two-memristor circuits, respectively, analyzed in Sects. 5.3 and 5.4, for the general case of non-symmetric kernels. Note that, since, under no assumption on the symmetry properties of the kernels, closed-form expressions for the kernels of a Volterra system may not be given except for the first order case, in order to derive the output response of this system to a given sine-wave input $x(t)$, the following methodology needs to be used. First an expression for the Volterra series-based representation of the nth component $y_n(t)$ of the system output $y(t)$ to the given input is determined by means of (5.31). In this expression the nth-order Volterra kernel in the frequency domain, i.e. $H_n(j\omega_1,\ldots,j\omega_n)$, occurs a number of times and in each occurrence it is sampled at a specific n-tuple of frequencies $(\omega_1,\ldots,\omega_n)$ (see (5.33)–(5.34) for the second- and third-order cases). Using these specific n-tuple of frequencies into the kernel equation of order n, together with results of similar analysis previously carried out for kernel equations of lower orders (i.e. from order 1 up to order $n-1$) allows the final derivation of the expressions of the above-mentioned samples of the nth-order Volterra kernel within the expression for $y_n(t)$.

5.6.1 Equation for the Second-Order Kernel in the Non-Symmetric Case for Single-Memristor Circuits

From Sect. 5.3.2.2 using (5.25), the equation for the second-order Volterra non-symmetric kernel in the frequency domain is found to be

$$H_2(j\omega_1,j\omega_2)+H_2(j\omega_2,j\omega_1)=\frac{\det(\hat{\mathbf{Y}}(j(\omega_1+\omega_2)))}{\hat{Y}_{11}(j(\omega_1+\omega_2))}m_1H_1(j\omega_1)$$
$$H_1(j\omega_2)\frac{j(\omega_1+\omega_2)}{j\omega_1 j\omega_2}. \tag{5.60}$$

5.6.2 Equation for the Third-Order Kernel in the Non-Symmetric Case for Single-Memristor Circuits

From Sect. 5.3.2.3 using (5.29), the equation for the third-order Volterra non-symmetric kernel in the frequency domain is found to be

$$H_3(j\omega_1,j\omega_2,j\omega_3)+H_3(j\omega_1,j\omega_3,j\omega_2)+H_3(j\omega_3,j\omega_2,j\omega_1)+H_3(j\omega_2,j\omega_1,j\omega_3)$$
$$+H_3(j\omega_2,j\omega_3,j\omega_1)+H_3(j\omega_3,j\omega_1,j\omega_2)=-\frac{\det(\hat{\mathbf{Y}}(j(\omega_1+\omega_2+\omega_3)))}{\hat{Y}_{11}(j(\omega_1+\omega_2+\omega_3))}\Bigg\{-m_1$$
$$\Bigg[(H_2(j\omega_2,j\omega_3)+H_2(j\omega_3,j\omega_2))\frac{H_1(j\omega_1)}{j\omega_1}+(H_2(j\omega_1,j\omega_3)+H_2(j\omega_3,j\omega_1))$$

$$\frac{H_1(j\omega_2)}{j\omega_2} + (H_2(j\omega_1, j\omega_2) + H_2(j\omega_2, j\omega_1)) \frac{H_1(j\omega_3)}{j\omega_3} + (H_2(j\omega_2, j\omega_3) +$$

$$H_2(j\omega_3, j\omega_2)) \frac{H_1(j\omega_1)}{j(\omega_2 + \omega_3)} + (H_2(j\omega_1, j\omega_3) + H_2(j\omega_3, j\omega_1)) \frac{H_1(j\omega_2)}{j(\omega_1 + \omega_3)}$$

$$+ \frac{H_1(j\omega_3)}{j(\omega_1 + \omega_2)} (H_2(j\omega_1, j\omega_2) + H_2(j\omega_2, j\omega_1)) \Bigg] + 2m_2 H_1(j\omega_1) H_1(j\omega_2) H_1(j\omega_3)$$

$$\left(\frac{1}{j\omega_2 j\omega_3} + \frac{1}{j\omega_1 j\omega_3} + \frac{1}{j\omega_1 j\omega_2} \right) \Bigg\}. \tag{5.61}$$

5.6.3 Equation for the Second-Order Kernel in the Non-Symmetric Case for Two-Memristor Circuits

From Sect. 5.4.2.2 using (5.51)–(5.52), the equation for the second-order non-symmetric kernel $H_{r2}(j\omega_1, j\omega_2)$ in the frequency domain for Volterra system $r-1$ ($r \in \{2,3\}$) is found to be

$$H_{22}(j\omega_1, j\omega_2) + H_{22}(j\omega_2, j\omega_1) = \frac{j(\omega_1 + \omega_2)}{j\omega_1 j\omega_2}$$
$$\left(\frac{\hat{Z}_{33}(j(\omega_1 + \omega_2)) m_{21} H_{21}(j\omega_1) H_{21}(j\omega_2)}{\hat{Z}_{33}(j(\omega_1 + \omega_2)) \hat{Z}_{22}(j(\omega_1 + \omega_2)) - \hat{Z}_{23}(j(\omega_1 + \omega_2)) \hat{Z}_{32}(j(\omega_1 + \omega_2))} \right.$$
$$\left. - \frac{\hat{Z}_{23}(j(\omega_1 + \omega_2)) m_{31} H_{31}(j\omega_1) H_{31}(j\omega_2)}{\hat{Z}_{33}(j(\omega_1 + \omega_2)) \hat{Z}_{22}(j(\omega_1 + \omega_2)) - \hat{Z}_{23}(j(\omega_1 + \omega_2)) \hat{Z}_{32}(j(\omega_1 + \omega_2))} \right), \tag{5.62}$$

and

$$H_{32}(j\omega_1, j\omega_2) + H_{32}(j\omega_2, j\omega_1) = \frac{j(\omega_1 + \omega_2)}{j\omega_1 j\omega_2}$$
$$\left(\frac{\hat{Z}_{32}(j(\omega_1 + \omega_2)) m_{21} H_{21}(j\omega_1) H_{21}(j\omega_2)}{\hat{Z}_{32}(j(\omega_1 + \omega_2)) \hat{Z}_{23}(j(\omega_1 + \omega_2)) - \hat{Z}_{22}(j(\omega_1 + \omega_2)) \hat{Z}_{33}(j(\omega_1 + \omega_2))} \right.$$
$$\left. - \frac{\hat{Z}_{22}(j(\omega_1 + \omega_2)) m_{31} H_{31}(j\omega_1) H_{31}(j\omega_2)}{\hat{Z}_{32}(j(\omega_1 + \omega_2)) \hat{Z}_{23}(j(\omega_1 + \omega_2)) - \hat{Z}_{22}(j(\omega_1 + \omega_2)) \hat{Z}_{33}(j(\omega_1 + \omega_2))} \right). \tag{5.63}$$

5.6.4 Equation for the Third-Order Kernel in the Non-Symmetric Case for Two-Memristor Circuits

From Sect. 5.4.2.3 using (5.55)–(5.56), the equation for the third-order non-symmetric kernel $H_{r3}(j\omega_1, j\omega_2, j\omega_3)$ in the frequency domain for Volterra system $r-1$ ($r \in \{2,3\}$) is found to be

$$
\begin{aligned}
&H_{23}(j\omega_1,j\omega_2,j\omega_3)+H_{23}(j\omega_1,j\omega_3,j\omega_2)+H_{23}(j\omega_2,j\omega_1,j\omega_3)\\
&+H_{23}(j\omega_2,j\omega_3,j\omega_1)+H_{23}(j\omega_3,j\omega_1,j\omega_2)+H_{23}(j\omega_3,j\omega_2,j\omega_1)\\
&=\frac{1}{-F_1(j\bar{\omega})}\Bigg\{\hat{Z}_{33}(j\bar{\omega})\Bigg[m_{21}\Bigg(F_2(j\omega_2,j\omega_3)\left(\frac{H_{21}(j\omega_1)}{j\omega_1}+\frac{H_{21}(j\omega_1)}{j\omega_2+j\omega_3}\right)\\
&+F_2(j\omega_1,j\omega_3)\left(\frac{H_{21}(j\omega_2)}{j\omega_2}+\frac{H_{21}(j\omega_2)}{j\omega_1+j\omega_3}\right)+\left(\frac{H_{21}(j\omega_3)}{j\omega_3}+\frac{H_{21}(s_3)}{j\omega_1+j\omega_1}\right)
\end{aligned}
\tag{5.64}
$$

and

$$
\begin{aligned}
&H_{33}(j\omega_1,j\omega_2,j\omega_3)+H_{33}(j\omega_1,j\omega_3,j\omega_2)+H_{33}(j\omega_2,j\omega_1,j\omega_3)\\
&+H_{33}(j\omega_2,j\omega_3,j\omega_1)+H_{33}(j\omega_3,j\omega_1,j\omega_2)+H_{33}(j\omega_3,j\omega_2,j\omega_1)\\
&=\frac{1}{F_1(j\bar{\omega})}\Bigg\{\hat{Z}_{32}(j\bar{\omega})\Bigg[m_{21}\Bigg(F_2(j\omega_2,j\omega_3)\left(\frac{H_{21}(j\omega_1)}{j\omega_1}+\frac{H_{21}(j\omega_1)}{j\omega_2+j\omega_3}\right)\\
&+F_2(j\omega_1,j\omega_3)\left(\frac{H_{21}(j\omega_2)}{j\omega_2}+\frac{H_{21}(j\omega_2)}{j\omega_1+j\omega_3}\right)+\left(\frac{H_{21}(j\omega_3)}{j\omega_3}+\frac{H_{21}(j\omega_3)}{j\omega_1+j\omega_2}\right)\\
&F_2(j\omega_1,j\omega_2)\Bigg)-2m_{22}\left(\frac{1}{j\omega_1 j\omega_2}+\frac{1}{j\omega_2 j\omega_3}+\frac{1}{j\omega_1 j\omega_3}\right)H_{21}(j\omega_1)H_{21}(j\omega_2)\\
&H_{21}(j\omega_3)\Bigg]-\hat{Z}_{22}(j\bar{\omega})\Bigg[m_{31}\Bigg(F_3(j\omega_2,j\omega_3)\left(\frac{H_{31}(j\omega_1)}{j\omega_1}+\frac{H_{31}(j\omega_1)}{j\omega_2+j\omega_3}\right)\\
&+F_3(j\omega_1,j\omega_3)\left(\frac{H_{31}(j\omega_2)}{j\omega_2}+\frac{H_{31}(j\omega_2)}{j\omega_1+j\omega_3}\right)+\left(\frac{H_{31}(j\omega_3)}{j\omega_3}+\frac{H_{31}(j\omega_3)}{j\omega_1+j\omega_2}\right)\\
&F_3(j\omega_1,j\omega_2)\Bigg)-2m_{32}\left(\frac{1}{j\omega_1 j\omega_2}+\frac{1}{j\omega_2 j\omega_3}+\frac{1}{j\omega_1 j\omega_3}\right)H_{31}(j\omega_1)H_{31}(j\omega_2)\\
&H_{31}(j\omega_3)\Bigg]\Bigg\},
\end{aligned}
\tag{5.65}
$$

$$F_2(j\omega_1, j\omega_2) - 2m_{22}\left(\frac{1}{j\omega_1 j\omega_2} + \frac{1}{j\omega_2 j\omega_3} + \frac{1}{j\omega_1 j\omega_3}\right) H_{21}(j\omega_1) H_{21}(j\omega_2)$$

$$H_{21}(j\omega_3)\Big] - \hat{Z}_{23}(j\bar{\omega})\Big[m_{31}\left(F_3(j\omega_2, j\omega_3)\left(\frac{H_{31}(j\omega_1)}{j\omega_1} + \frac{H_{31}(j\omega_1)}{s_2+s_3}\right)\right.$$

$$+F_3(j\omega_1, j\omega_3)\left(\frac{H_{31}(j\omega_2)}{j\omega_2} + \frac{H_{31}(j\omega_2)}{j\omega_1+s_3}\right) + \left(\frac{H_{31}(s_3)}{s_3} + \frac{H_{31}(s_3)}{s_1+s_2}\right)$$

$$F_3(j\omega_1, j\omega_2) - 2m_{32}\left(\frac{1}{j\omega_1 j\omega_2} + \frac{1}{j\omega_2 j\omega_3} + \frac{1}{j\omega_1 j\omega_3}\right) H_{31}(j\omega_1) H_{31}(j\omega_2)$$

$$H_{31}(j\omega_3)\Big]\Big\}, \tag{5.66}$$

where we defined

$$\begin{aligned} F_1(j\omega) &\triangleq \hat{Z}_{32}(j\omega)\hat{Z}_{23}(j\omega) - \hat{Z}_{22}(j\omega)\hat{Z}_{33}(j\omega), \\ F_2(j\omega_1, j\omega_2) &\triangleq H_{22}(j\omega_1, j\omega_2) + H_{22}(j\omega_2, j\omega_1), \\ F_3(j\omega_1, j\omega_2) &\triangleq H_{32}(j\omega_1, j\omega_2) + H_{32}(j\omega_2, j\omega_1), \quad \text{and} \\ \bar{\omega} &\triangleq \omega_1 + \omega_2 + \omega_3. \end{aligned} \tag{5.67}$$

References

1. F. Corinto, A. Ascoli, A boundary condition-based approach to the modeling of memristor nanostructures. IEEE Trans. Circuits Syst. I **59**(11), 2713–2726 (2012)
2. F. Corinto, A. Ascoli, M. Gilli, Nonlinear dynamics of memristor oscillators. IEEE Trans. Circuits Syst. *I* **58**(6), 1323–1336 (2011)
3. L.O. Chua, Memristor: the missing circuit element. IEEE Trans. Circuit Theory **18**(5), 507–519 (1971)
4. L.O. Chua, S.M. Kang, Memristive devices and systems. Proc. IEEE **64**(2), 209–223 (1976)
5. H.K. Khalil, *Nonlinear Systems*, 3rd edn. (Prentice Hall, Englewood Cliffs, 2002)
6. M. Schetzen, *The Volterra and Wiener Theories of Nonlinear Systems* (Krieger Publishing Company, Malabar, FL, 2006)
7. L.O. Chua, C.Y. Ng, Frequency domain analysis of nonlinear systems. IEE J. Electron. Circuits Syst. **3**(4), 165–185 (1979)
8. S. Boyd, Y.S. Tang, L.O. Chua, Measuring Volterra kernels. IEEE Trans. Circuits Syst. **CAS-30**(8), 571–577 (1983)
9. J.F. Barrett, The use of functionals in the analysis of non-linear physical systems. J. Electron. Control **15**(6), 567–615 (1963)
10. J. Waddington, F. Fallside, Analysis of non-linear differential equations by the Volterra series. Int. J. Control **3**(1), 1–15 (1966)
11. A.J. Krener, Linearization and bilinearization of control systems. Proceedings of Allerton Conference on Circuit and System Theory, pp. 834–843, University of Illinois, Urbana Champaign, Illinois, 1974
12. C. Lesiak, A. Krener, The existence und uniqueness of Volterra series for nonlinear systems. IEEE Trans. Automat. Control **AC-23**, 1090–1095 (1978)

13. E.G. Gilbert, Functional expansions for the response nonlinear differential systems. IEEE Trans. Automat. Control **AC-22**(6), 909–921 (1977)
14. W.J. Rugh, *Nonlinear System Theory* (The Johns Hopkins University Press, Baltimore, MD, 1981)
15. I.W. Sandberg, A perspective on system theory. IEEE Trans. Circuits Syst., **CAS-31**(1), 88–103 (1984)
16. S.P. Boyd, L.O. Chua, Fading memory and the problem of approximating nonlinear operators with Volterra series. IEEE Trans. Circuits Syst. **CAS-32**(11), 1150–1161 (1985)
17. S.P. Boyd, L.O. Chua, Volterra series for nonlinear circuits. Proceedings of International Symposium on Circuits and Systems, p. 369, Kyoto, 1985
18. S.P. Boyd, L.O. Chua, C.A. Desoer, Analytical foundations of Volterra series. IMA J. Math. Control Info. **1**(3), 243–282 (1984)
19. R.W. Brockett, Volterra series and geometric control theory. Automatica, **12**, 167–176 (1976)
20. I.W. Sandberg, Bounds for discrete-time Volterra series representations. IEEE Trans. Circuits Syst. I **46**(1), 135–139 (1999)
21. T. Siu, M. Schetzen, Convergence of Volterra series representation and BIBO stability of bilinear systems. Int. J. Syst. Sci. **22**(12), 2679–2684 (1991)
22. J.F. Barrett, The use of Volterra series to find region of stability of a non-linear differential equation. Int. J. Control **1**(3), 209–216 (1965)
23. S.P. Boyd, Volterra Series: Engineering Fundamentals. PhD Dissertation Thesis, 1985
24. T. Hélie, B. Laroche, Computation of convergence bounds for volterra series of linear-analytic single-input systems. IEEE Trans. Automat. Control **56**(9), 2062–2072 (2011)
25. L.O. Chua, Resistance switching memories are memristors. Appl. Phys. A **102**, 765–783 (2011)
26. D.B. Strukov, G.S. Snider, D.R. Stewart, R.S. Williams, The missing memristor found. Nature **453**, 80–83 (2008)
27. F. Corinto, A. Ascoli, Memristive diode bridge with LCR filter. Electron. Lett. **48**(14), 824–825 (2012)
28. Z.-Q. Lang, S.A. Billings, Evaluation of output frequency responses of nonlinear systems under multiple inputs. IEEE Trans. Circuits Syst. II **47**(1), 28–38 (2000)

Part III
Memristive Devices and Applications

Chapter 6
Memristive Devices: Switching Effects, Modeling, and Applications

Yuchao Yang, Ting Chang, and Wei Lu

6.1 Introduction

The rapid, exponential growth of modern electronics has brought about profound changes to our daily lives. However, maintaining the growth trend now faces significant challenges at both the fundamental and practical levels [1]. Possible solutions include *More Moore*—developing new, alternative device structures, and materials while maintaining the same basic computer architecture, and *More Than Moore*—enabling alternative computing architectures and hybrid integration to achieve increased system functionality without trying to push the devices beyond limits. In particular, an increasing number of computing tasks today are related to handling large amounts of data, e.g. image processing as an example. Conventional von Neumann digital computers, with separate memory and processer units, become less and less efficient when large amount of data have to be moved around and processed quickly. Alternative approaches such as bio-inspired neuromorphic circuits, with distributed computing and localized storage in networks, become attractive options [2–6].

The reemergence of neuromorphic systems is fueled by two factors. First, more understanding has been obtained on both biological neural networks and manmade networks through experimental and modeling studies [7–10]. Second, the emergence of new classes of nanodevices, particularly two-terminal resistive switching devices (memristive devices) [11–17], makes it possible to build functional neuromorphic hardware that will not only serve to test the various neural network models but also can directly lead to new, effective, high-performance computing hardware.

Y. Yang • T. Chang • W. Lu (✉)
Department of Electrical Engineering and Computer Science,
University of Michigan, Ann Arbor, MI 48109, USA
e-mail: yuchaoy@umich.edu; satich@umich.edu; wluee@umich.edu

R. Tetzlaff (ed.), *Memristors and Memristive Systems*,
DOI 10.1007/978-1-4614-9068-5_6,

In this chapter, we will discuss recent progress in the development of neuromorphic hardware based on nanoscale memristive devices. In particular, we will focus on a few representative device systems in terms of resistive switching characteristics, switching mechanisms, and theoretical modeling and show how two key properties, local adaptive learning and large connectivity, can be obtained in memristive devices that in turn make them well suited for neuromorphic applications.

6.2 Resistive Switching in Memristive Devices

Resistive switching (RS) phenomena have been reported as early as the 1960s [18] and today such devices are extensively studied as resistive random-access memory (RRAM) for future nonvolatile memory [14–17, 19–22]. These devices generally are simple in structure (typically two-terminal) and nanoscale in dimensions (scaling <10 nm has been demonstrated [23]), while at the same time offering excellent performance in terms of switching speed [24] and write/erase cycling [25]. Tremendous work has been performed to understand the various types of switching mechanisms that are responsible for RS phenomena in different material systems [14–17, 19–21, 26–30], which can be broadly categorized into valence change, electrochemical metallization, phase change, thermo-chemical, ferroelectric, and nanomechanical effects [15], as shown in Fig. 6.1. Here we focus on devices that show "analog" memristive behavior, i.e. devices that are particularly suitable for neuromorphic applications. Experimentally, devices that fall in the first

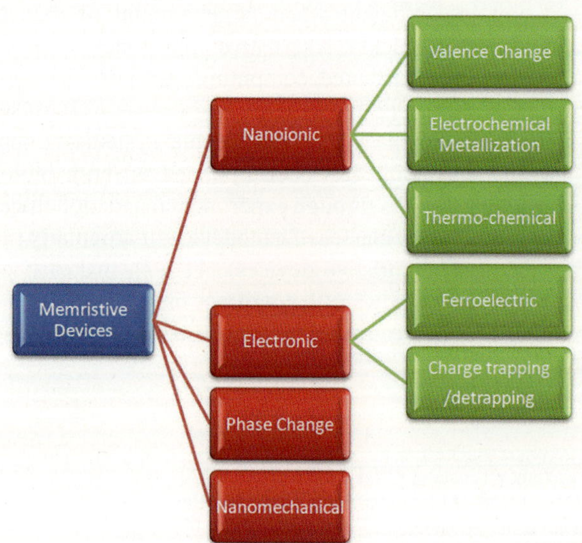

Fig. 6.1 Classification of memristive devices according to the working mechanism

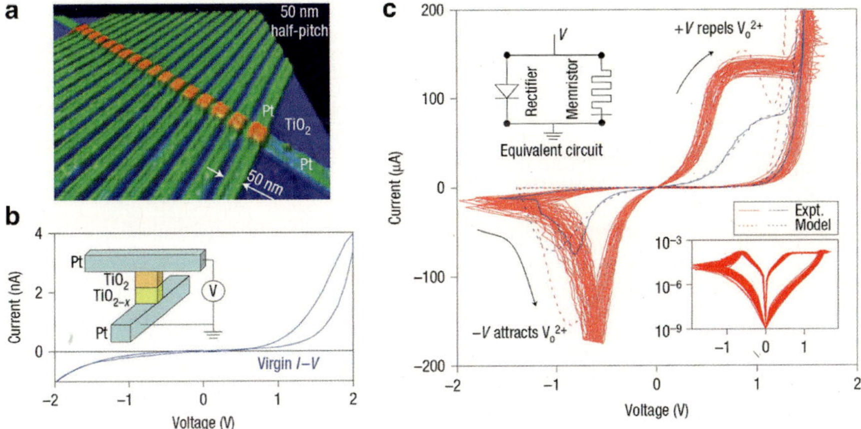

Fig. 6.2 Bipolar resistive switching in TiO_2-based memristors. (**a**) An atomic force microscope image of 1 × 17 crosspoint devices with 50 nm half-pitch. Pt nanowires were fabricated by nanoimprint lithography and sandwich a 50-nm-thick TiO_2 insulating film. (**b**) Initial *I–V* curve of the device in virgin state, showing a rectifying characteristic. *Inset*: the device structure. (**c**) Experimental (*solid*) and modeled (*dotted*) switching *I–V* curves. *Lower inset*: the switching *I–V* curves in log-scale. *Upper inset*: the equivalent circuit model consisting of a rectifier in parallel with a memristor used for modeling. Reprinted with permission from [21]. Copyright (2008) Nature Publishing Group

three categories (namely, valence change, electrochemical metallization, and phase change effects) have been extensively studied for this purpose and below we present results from a few representative studies.

6.2.1 Memristive Devices Based on Valence Change

The connection between resistance switching and memristive effects (definition given later) was made by HP labs in 2008, where the behaviors of nanoscale TiO_2 crosspoint devices were shown to fit the descriptions of memristive devices (memristors for short throughout this Chapter) [13]. Specifically, the devices exhibited pinched hysteresis loops (so no energy is stored) and frequency and history-dependent programming (see Fig. 6.2), in agreement with the predictions of memristors. Physically, RS of TiO_2 is a result of local stoichiometric change caused by the migration of oxygen vacancies (V_Os) and can be assigned into the valence change category. As V_Os act as donors in TiO_2 the accumulation/depletion of V_Os can cause an increase/decrease of the local conductance which in turn results in the modulation of the overall device conductance. Specifically, in the study of Yang et al. [21], the drift of oxygen vacancies towards the Pt/TiO_2 Schottky interface, driven by the external electric field, results in the accumulation of V_Os and the creation of conducting channels that shunt the barrier for electron injection. When applying electric fields with a reverse polarity, the oxygen vacancies are driven away from the Schottky barrier, hence dissolving the conducting channels and switching the device back to off state. Later it was demonstrated that the aggregation of V_Os in TiO_2 in the conductive region may in fact lead to the formation of a new oxygen-deficient Magnéli phase (Ti_4O_7) that is metallic, as directly revealed by TEM studies [26, 30].

Since the devices operate by the redistribution of V_Os, the creation of a definitive V_O distribution profile will help obtain reliable device operations. Conventionally this is achieved in a high-voltage "electroforming process" to create the oxygen vacancies and define the V_O distribution. The forming process involves electro-reduction of oxygen ions, which produces oxygen gas and leaves oxygen vacancies in the oxide films [31]. On the other hand, high-voltage forming is not desirable in practical device applications, is hard to control, and reduces device yield and reliability. Based on the improved understanding of the electroforming process and the resistive switching mechanism, the forming process can be successfully eliminated either by reducing the thickness of the oxide film to keep just the switching interface or by intentionally introducing an oxygen-deficient layer next to the switching layer in the device fabrication process as a reservoir of oxygen vacancies [31].

Improved understanding of the switching mechanism has been obtained through a series of material analyses that offered useful information on film composition, microcrystalline structure, and switching locations [29, 30] as well as modeling that attempted to match experimental results with known physical effects, for example, the drift and diffusion current equations [27, 32], which will be discussed later. It also needs to be noted that although the conducting channels may be formed locally, the RS process, as determined by external stimuli, can be affected by the global device design such as the electrode materials, layer stacks, and structures around the switching regions [33, 34]. For example, the reactions between TiO_2 and the metal electrodes (particularly with the thermally diffused Ti adhesion layer) can create additional oxygen vacancies in TiO_2 and largely modulate the electronic barrier at the interface, which in turn affect the resistive switching behavior. The localized diffused adhesion material may also serve as seeds for switching channel formation, leading to improved switching reliability [34].

Similar memristive behaviors have also been observed in other oxide materials. Here we choose WO_3 as an example. In modern CMOS processing, W is widely used as a contact material so WO_3 can be easily incorporated into standard manufacturing processes [35]. Additionally, WO_3 has been a long-studied material with well-known characteristics and preparation methods [36]. Similar to TiO_2, WO_3 is also a transition metal oxide with many sub-stoichiometric states making them n-type semiconductors. This material system thus offers many attractive properties for device applications and has been used for both digital [37] and analog [36, 38] types of memory studies.

Typical analog-type resistive switching in WO_x-based memristive devices is shown in Fig. 6.3 in a study performed by Chang et al. [36]. The device is composed

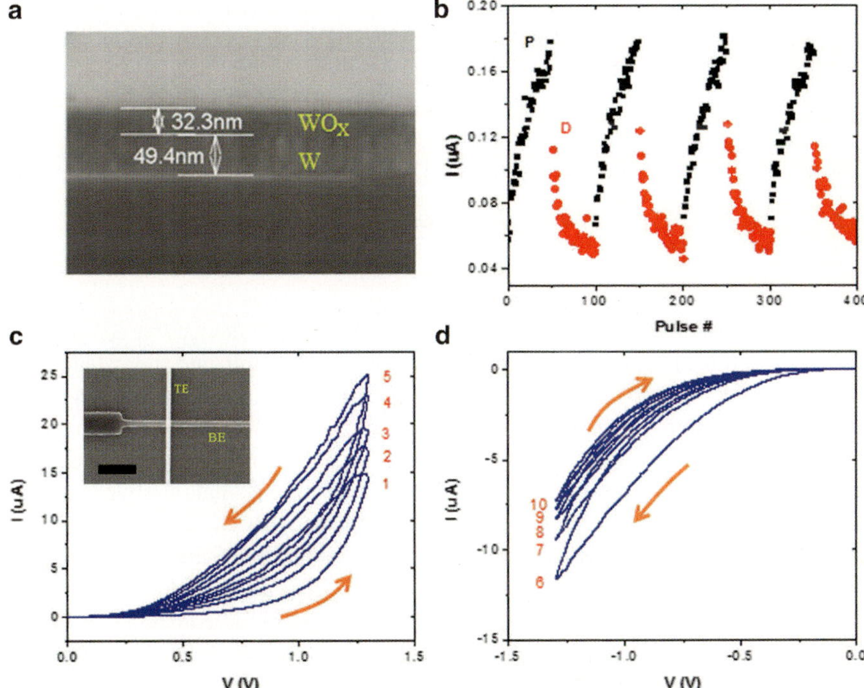

Fig. 6.3 Memristive switching in WO_x-based devices. (**a**) Cross-sectional SEM image of the oxidized W film, showing a WO_x film formed on top of W. (**b**) Pulse response of the WO_x device showing conductance potentiation and depression. (**c–d**) Analog resistive switching of the WO_x device in the positive (negative) voltage regions. *Inset* to (**c**): top-view SEM image of a crosspoint WO_x device. Scale bar: 2 μm

of a Pd top electrode (TE) and a W bottom electrode (BE) sandwiching a WO_x film, as shown by the top-view scanning electron microscope (SEM) image in Fig. 6.3c. A rapid thermal annealing (RTA) step was adopted to partly oxidize the W bottom electrodes, therefore directly forming a WO_x layer on top of the W bottom electrodes, as shown in Fig. 6.3a. This oxidation process naturally introduced a continuous concentration gradient of oxygen vacancies [39], unlike the cases of abrupt V_O concentration profiles formed in other bilayer structures. By tuning the V_O distribution with external biases, analog-type resistive switching can be reliably obtained (see Fig. 6.3b–d) [36].

The quasi-continuous tuning of the device resistance with memory effects is the key enabling factor of memristor-based neuromorphic circuits. Briefly, the conductance modulation in memristors can be thought as being analogous to the plastic synaptic weight changes in biological synapses, so memristors can be used to emulate biological functions. This behavior is more clearly demonstrated in pulse measurements, as shown in Fig. 6.3b, where positive and negative pulses (i.e., spike inputs) cause the memristor conductance to increase or decrease by incremental amounts, corresponding to the potentiation (P) and depression (D) effects, respectively. Further characterizations and modeling confirmed the oxygen vacancy motion origin of the resistive switching effects in these devices [36], which will be discussed in more detail in the next section.

6.2.2 Memristive Devices Based on Electrochemical Metallization

Besides oxygen vacancy (anion)-driven devices, memristive devices based on metal ion (cation) redistribution have also been reported. They fall into the category of electrochemical metallization cells. The Ag_2S atomic switch reported in 2005 was an ingenious design to control RS within a 1 nm spacing [40]. The electrochemical (redox) effect between the Ag_2S electrode and the Ag deposits determines whether a conductive bridge is formed or disrupted, thus determining what resistive (conductive) state the device is in. Later, it was found that not only the conductive bridge connection can be controlled, but the size (width/diameter) of the bridge can also be modulated to produce multilevel (analog) switching effects. Hence, the Ag_2S devices also potentially have the ability to emulate learning and memory functions of synapses [41, 42]. Specifically, electrical transport in the atomic switches can be divided into two regimes: tunneling regime and contact regime (Fig. 6.4). When the metal bridge has not connected the counter electrode, the electrical transport of the device is dominated by tunneling through the gap between the front of the metal bridge and the counter electrode. After the metal bridge forms contact with the counter electrode, further application of electric signals will result in lateral growth of the bridge, so the conductance is determined by the width of the bridge [41]. As a consequence, the initial contact between the metal bridge and the counter electrode will cause an abrupt transition of conduction from tunneling to contact, hence leading to a digital-type switching behavior, while further growth of the metal bridge in the contact area will only expand the size of the bridge and lead to analog switching. This was indeed observed experimentally, as shown in Fig. 6.4a. The device was initially in the tunneling regime, at about −0.2 V the contact was achieved, accompanied by a sharp increase in current. Afterwards the current was increased gradually in subsequent sweeps, corresponding to the expansion of the bridge size. The digital and analog changes in resistance in the two regimes were also clearly observed by recording the resistance values at the end of each sweep (Fig. 6.4b, c).

One of the first demonstrations of memristor-based synaptic functions was achieved by Jo et al. in Ag/a-Si-based memristive devices. Similar to the Ag_2S device, the Ag/a-Si devices rely on the redistribution of Ag ions within the a-Si matrix. In digital-type devices, Ag conducting filaments can be formed/dissolved in the a-Si matrix and high-performance metrics have been demonstrated for nonvolatile memory applications, including high device yield of 99 %, scaling

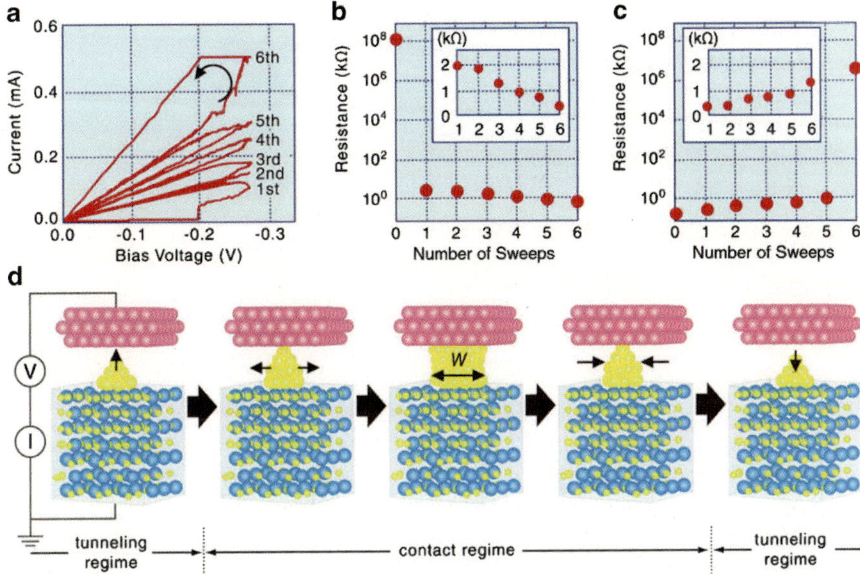

Fig. 6.4 Analog resistive switching in Ag_2S-based atomic switches. (**a**) Experimental results showing an initial abrupt change followed by gradual increase in current. (**b**) Resistance values at the end of each negative sweep. The *inset* shows the resistance change in the contact regime in linear scale. (**c**) Resistance values at the end of each positive sweep, during which the atomic bridge was disconnected. The *inset* shows the resistance change in the contact regime in linear scale. (**d**) Operation model of the atomic switch. Application of a negative bias voltage to the Pt electrode widens the metal atomic bridge, while a positive bias results in a thinner one. Reprinted with permission from [41]. Copyright (2010) Wiley-VCH Verlag GmbH & KGaA, Weinheim

potential of <50 nm, fast programming speed of 5 ns, high endurance of 10^8, long retention of ~7 years, and multilevel storage capability [20, 43, 44]. To make the device suitable for neuromorphic applications, that is, to achieve analog switching, co-sputtering was employed to controllably incorporate Ag into the a-Si film to achieve a more gradual Ag concentration gradient as schematically shown in Fig. 6.5a [45]. Incremental analog switching, as demonstrated by both pulse measurements and DC sweeps was reliably achieved in these devices. Figure 6.5b shows the incremental adjustment of the conductance of the memristor device by a series of potentiating (3 V, 500 μs) and depressing (−2.6 V, 500 μs) pulses, while Fig. 6.5c, d show the evolution of the pinched hysteresis loops under DC voltage sweeps, where consecutive positive (negative) sweeps lead to gradual conductance potentiation (depression). The analog switching can be understood by electric field-driven migration of Ag ions and resultant movement of the conduction front between the Ag-rich and Ag-poor regions in the active layer. Indeed, simulation results based on this simple model satisfactorily fitted the experimental data in Fig. 6.5c. Through careful material engineering, robust RS behaviors were obtained and the devices were still functional after 1.5×10^8 P/D pulses.

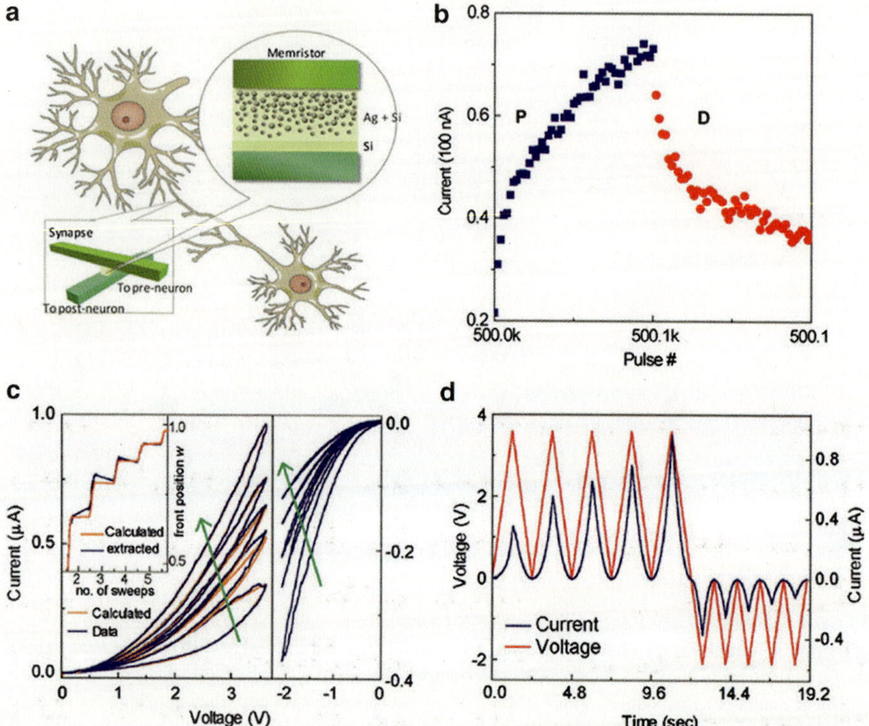

Fig. 6.5 Analog resistive switching in Ag/a-Si-based memristive devices. (**a**) Schematic of using a memristor as a synapse between two neurons. (**b**) Pulse responses of Ag/a-Si-based memristive devices. (**c**) Measured (*blue*) and calculated (*orange*) *I*–*V* characteristics of the Ag/a-Si-based memristor. *Inset*: calculated (*orange*) and extracted (*blue*) values of the normalized conduction front position during positive DC sweeps. (**d**) Current and voltage data as functions of time. Reprinted with permission from [45]. Copyright (2010) American Chemical Society

6.2.3 Memristive Devices Based on Phase-Change Materials

Another group of candidates for synaptic emulation are based on phase change memory (PCM) [46]. By applying electric pulses to generate enough heat and induce local phase transitions, phase change materials can exhibit resistance switching behaviors between amorphous (high resistivity) and crystalline (low resistivity) states and this change can be quite fast and stable. GST ($Ge_2Sb_2Te_5$) is the most commonly used material for PCM. Here, phase change is achieved by heating and quenching of GST through Joule heating. Although mostly studied as a digital memory, when programming and erasing conditions are carefully designed, the GST devices can exhibit analog-type resistance changes if the amount of material being melted and re-crystallized can be controlled [47].

Shown in Fig. 6.6 is a demonstration of analog resistance changes in GST-based devices in a study performed by Kuzum et al. [47]. The analog switching

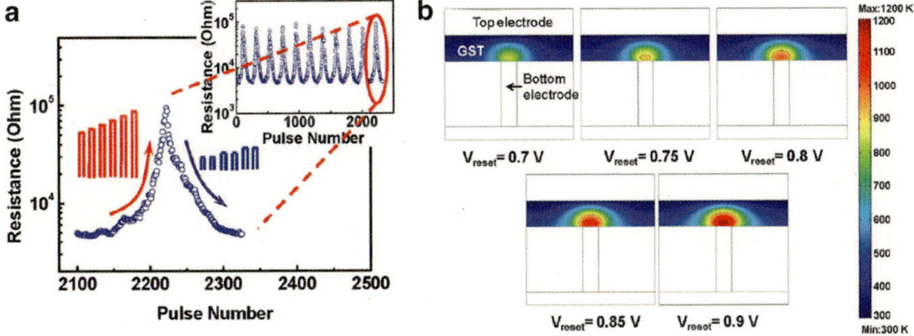

Fig. 6.6 Analog resistive switching in GST-based devices. (**a**) Gradual modulation of device resistance by voltage pulses. (**b**) Finite element simulations of the temperature across the GST region for reset voltages ranging from 0.7 to 0.9 V. The regions with $T > 900$ K are mapped using a solid black line and would turn into amorphous phase. Reprinted with permission from [47]. Copyright (2012) American Chemical Society

was achieved by the application of voltage pulses with incrementally increasing amplitudes, i.e., reset was performed by voltage pulses with increasing amplitude in the range of 2–4 V with 20 mV steps, while set process was achieved by using repeated staircase pulses (20 continuous pulses with amplitudes of 0.5, 0.6, 0.7, 0.8, and 0.9 V). The corresponding incremental resistance changes are shown in Fig. 6.6a for both the set and reset processes [47]. Figure 6.6b shows the temperature distribution in the cell based on finite element simulations, illustrating that the voltage pulses incrementally expand the region with temperature ($T > 900$ K) above the melting point of GST. These high temperature regions will be amorphized after the pulses are removed and account for the gradual increase in resistance during reset [47].

6.3 Modeling of Memristive Devices

6.3.1 General Modeling

Memristor as a device concept was first introduced by the pioneering work of Leon Chua [11] and further formalized by Chua and Kang [12]. Briefly, a device can be called a "memristive system" if it satisfies a set of equations below, with properly chosen state variables:

$$v = R(w,i)\,i \tag{6.1}$$

$$\frac{\mathrm{d}w}{\mathrm{d}t} = f(w,i) \tag{6.2}$$

Here the first equation is the normal I–V equation that relates voltage with current through a resistive term. However, for memristive systems, the resistance R (termed memristance) depends not only on the instantaneous inputs v and i but also depends on one or a set of internal state variable(s) w. The state variable w is in turn governed in the second equation, which specifies how the state variable changes according to the current state and the instantaneous inputs. Since here only the dynamics of w (i.e., its rate) is determined, the full value of w can only be obtained from a *time integral*, i.e. Eq. (6.2) implies that the state of the memristor is history dependent, as observed from the experiments on various materials. This is the key difference between a memristor and other two-terminal devices.

The memristor model is based on abstract mathematical equations (6.1) and (6.2), without necessarily specifying the physics behind them. By mapping these equations to the actual physical processes during device operation such as the ionic diffusion/drift processes during conduction channel formation, realistic device operations can be modeled within the memristor framework. Here again the key is to identify the internal state variable(s) and the corresponding dynamic equation (6.2). The advantage of describing the device operations in the memristor framework, vs. other phenomenological models is that not only does this approach provide an analytical description that can be readily ported into circuit simulators, but also it helps one to identify the driving factors behind the switching effects so more accurate descriptions, particularly those governing the device dynamics, can be obtained.

6.3.2 State Variable as the Conduction Channel Length

The first approach to use the memristor model to explain resistive switching was performed by Strukov et al. at HP Labs when trying to characterize the switching behaviors of TiO_2 devices [13]. Recall that RS in TiO_2 is due to the movement of oxygen vacancies inside the TiO_2 film. To connect the memristor equations with physical processes, Strukov assumed that the memristor is composed of two resistors in series, one is undoped with high resistance and the other is doped by V_O, thus having low resistance [13], as schematically shown in Fig. 6.7a, b. The total thickness of the film D is separated into the doped and undoped regions, and the total resistance is the sum of the two regions. The length of the doped region is taken as the state variable (w), which can be changed by moving the boundary between the two regions under external field due to the drift of V_Os. By assuming Ohmic electronic conduction in both regions and a linear ionic drift with an average ion mobility μ_v in a uniform field, the two memristor equations can be written as:

$$v(t) = \left(R_{ON}\frac{w(t)}{D} + R_{OFF}\left(1 - \frac{w(t)}{D}\right)\right) i(t) \tag{6.3}$$

$$\frac{dw(t)}{dt} = \mu_V \frac{R_{ON}}{D} i(t) \tag{6.4}$$

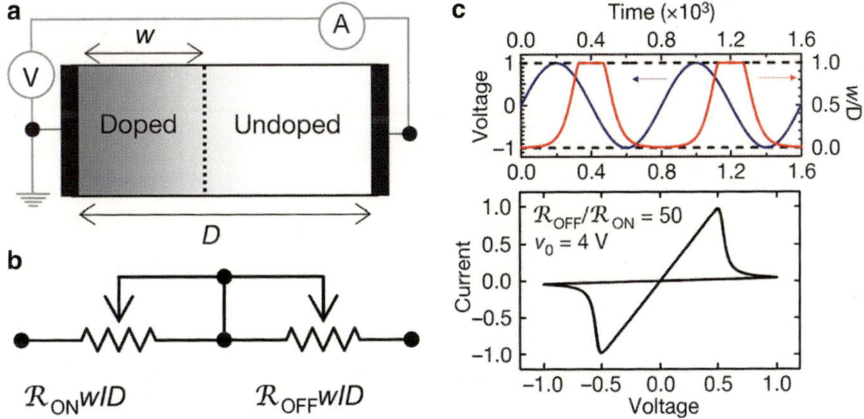

Fig. 6.7 Memristor model for the TiO_2 devices consisting of two resistors in-series. Here the state variable corresponds to the length of the conducting (doped) region. (**a**) Schematic of the model. (**b**) Equivalent circuit of (**a**). (**c**) Simulation results incorporating nonlinear ionic drift. *Upper*: the voltage stimulus (*blue*) and the corresponding change in normalized state variable w/D (*red*), as functions of time. *Lower*: simulated I–V characteristics. Reprinted with permission from [13]. Copyright (2008) Nature Publishing Group

Although both assumptions—linear ionic drift and a uniform conduction front—seem oversimplified the model can still successfully predict the pinched hysteresis and the history-dependent resistance changes. The model was soon refined to include nonlinear effects at high fields such as exponential ionic drift which are important for these devices even under normal operations [32]. In one approach, the nonlinear ionic drift was taken into account by multiplying the right side of Eq. (6.4) with a window function $w(D-w)/D^2$ [13]. The corresponding simulation results are shown in Fig. 6.7c, where binary resistive switching can be apparently observed. It should be noted that the memristor equations used here correspond to the "current-controlled" devices which are easy to implement mathematically but are not as practical as the "voltage-controlled" devices since currents are typically much harder to control than voltages in practice, especially in large networks.

6.3.3 State Variable as the Conduction Channel Area (Width)

Instead of modeling the state variable as the conduction channel length, which typically leads to nonuniform conductance changes (i.e., the conductance scales as 1/length), an alternative is to model the state variable as the conduction channel area (or equivalently, width). This approach seems more natural for many devices since the conduction channels typically form locally and in parallel, instead of creating a uniform front. Additional programming either increases the area (width) of the conduction channel or increases the number of conduction channels. Both effects

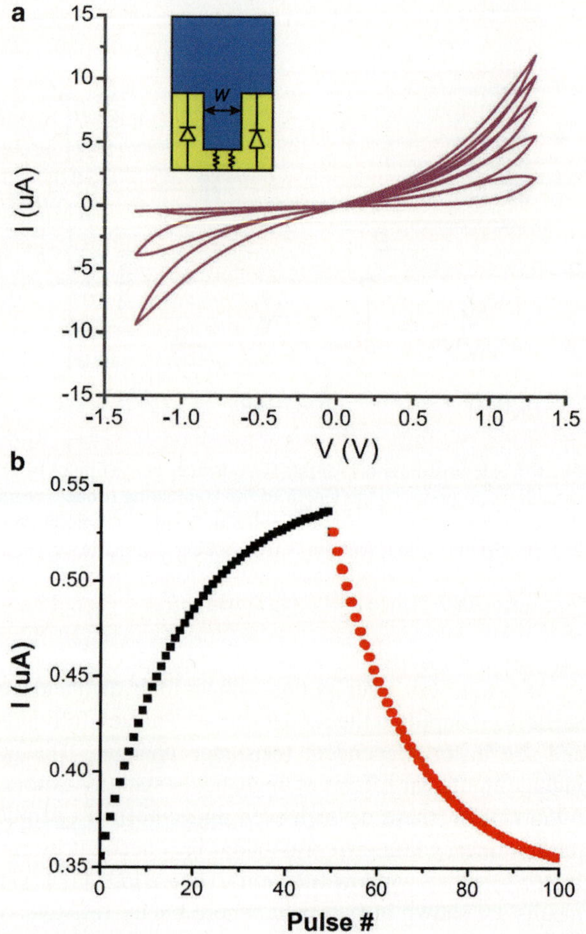

Fig. 6.8 (**a**) Simulated *I–V* characteristics when considering the state variable as the conduction channel area. *Inset*: schematic of the model. Reprinted with permission from [36]. Copyright (2011) Springer. (**b**) Simulated pulse responses using the same set of parameters. Potentiation pulse: 1.2 V, 5 μs; depression pulse: −1.2 V, 5 μs; read pulse: 0.3 V, 3 ms

are equivalent mathematically and lead to an increase of the effective conduction channel area. Using the conduction channel area as the state variable also leads to more uniform conductance changes, in better agreement with experimental observations.

The first attempt to explain resistance switching using the conduction channel area as the state variable was carried out by Chang et al. [36] to explain the resistance switching in WO_3. Additionally, a model incorporating an exponential ionic drift equation was used to model oxygen vacancy movement at high fields [32]. In general, the overall device conductance can be calculated as the sum of the conducting regions in parallel with the Schottky barrier formed in the resistive regions, weighted by the state variable (w, which is the conducting region area normalized over the total device area), as shown in the inset of Fig. 6.8a:

$$i = (1 - w)\alpha[1 - \exp(-\beta v)] + w\gamma \sinh(\delta v) \tag{6.5}$$

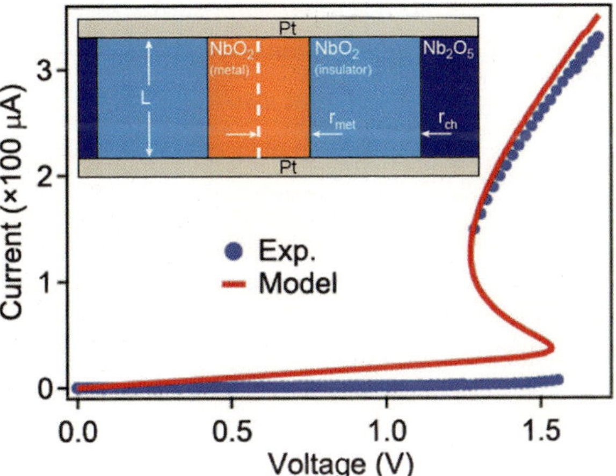

Fig. 6.9 Threshold switching and modeling of NbO_2-based devices. *Dots* and *line* represent experimental and simulated *I–V* characteristics, respectively. The *inset* shows schematic of the device model showing a channel that consists of two cylindrical phases during operation. Reprinted with permission from [48]. Copyright (2012) IOP

Here the first and second terms represent the contributions from Schottky emission in the resistive regions and tunneling-type conduction in the conducting channels, respectively. α, β, γ, and δ are all positive-valued fitting parameters determined by the material properties. The rate equation is determined by the expansion rate of the conduction region area which is related to the exponential ionic drift:

$$\frac{\mathrm{d}w}{\mathrm{d}t} = \lambda \sinh(\eta v) - \frac{w}{\tau} \tag{6.6}$$

The second term w/τ is introduced here in the dynamic equation to account for the lateral diffusion of ions that constitute the conduction channels (i.e., V_Os here). The spontaneous diffusion causes the weakening of the conduction channels and leads to limited retention, an effect that will be discussed in more detail later on.

Figure 6.8 shows the simulated DC *I–V* characteristics and pulse responses based on Eqs. (6.5) and (6.6), where analog bipolar resistive switching can be reliably predicted using a single set of parameters. The simulation results also show decay in retention, evidenced by the overlap between consecutive *I–V* curves during positive voltage sweeps, agreeing well with the experimental data [36].

The use of conduction channel area as the state variable to explain RS behavior was recently applied to other oxide systems. For example, NbO_2 is another material that exhibits interesting RS behaviors [48]. Unlike TiO_2 or WO_3-based devices, switching in NbO_2 is attributed to the thermally driven insulator-to-metal phase transition. Instead of showing memory switching (i.e., the device memorizes the new state after removal of the programming voltage), NbO_2 displays threshold switching characteristics (i.e., the device shows abrupt resistance changes but does not memorize the new state after removal of the programming voltage and leads to negative differential resistance (NDR)), as shown in Fig. 6.9 in a study performed by Pickett and Williams at HP labs. To model the switching behavior, the device was assumed to have a cylindrical core-shell geometry with a conducting core formed through the metal-insulator transition, and the state variable (u) was chosen as the normalized radius of the conducting core $u = r_{met}/r_{ch}$ (inset of Fig. 6.9). The corresponding equations are provided below.

$$v = f(u,i) = R_{ch}(u)i \tag{6.7}$$

$$\frac{du}{dt} = g(u,i) = \left(\frac{d\Delta H}{du}\right)^{-1} \left(R_{ch}(u)i^2 - \Gamma_{th}(u)\Delta T\right) \tag{6.8}$$

where R_{ch} is the total device resistance, ΔH is the total enthalpy change in the channel, Γ_{th} is the thermal conductance of the insulating shell, and T is the absolute temperature. This model successfully captured the threshold switching characteristics of NbO_2 devices, as displayed by the excellent agreement between simulation and experimental results shown in Fig. 6.9. Furthermore, circuit simulations based on the memristor model successfully predicted the behavior of these devices in active circuits such as the "neuristor" behaviors that can effectively emulate the lossless transmission in axons [49, 50].

6.4 Artificial Neuromorphic Applications Utilizing Memristive Devices

The plasticity in conductance and the large connectivity that can be offered by memristors make them well suited for physical implementation of synaptic functions in neuromorphic circuits. Biologically, the weight of a synapse is jointly determined by the firing patterns of both the pre-synaptic and post-synaptic neurons connecting to it. The synapse could either be potentiated or depressed, i.e. having their connections being strengthened or weakened, depending on the neuron spike patterns, the transmission/diffusion of ions (e.g., Ca^{2+}), the activation of receptors, and many other factors [51]. Several fundamental learning rules, including rate-dependent synaptic plasticity, timing-dependent synaptic plasticity, and cooperativity [52], have been discovered and believed to be critical for the efficient operation of biological systems. While these learning rules may vary according to the specific type and location of a synapse, some commonly accepted and well-established rules are discussed here. These important local learning rules have also been successfully demonstrated in memristors.

6.4.1 Spike-Timing-Dependent Plasticity

Spike-Timing-Dependent Plasticity (STDP) is an important synaptic learning rule which states that the synaptic weight is modulated according to the relative timing

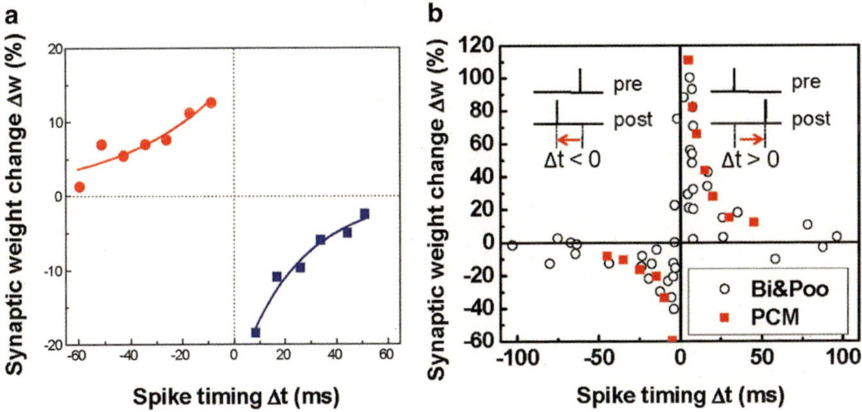

Fig. 6.10 (**a**) Implementation of STDP with Ag/a-Si-based memristors. Reprinted with permission from [45]. Copyright (2010) American Chemical Society. (**b**) Implementation of STDP with GST-based PCM cells. Reprinted with permission from [47]. Copyright (2012) American Chemical Society

of the pre- and post-synaptic neuron firing [53, 54]. It was first postulated in 1949 by D. Hebb who described the effect as "*neurons that fire together, wire together*" [55]. In general, if the spike (action potential) from the pre-synaptic neuron arrives at the synaptic cleft before that of the post-synaptic neuron, potentiation will be induced. Otherwise depression will be induced. How effectively the potentiation and depression take place in turn depends on how far apart the pre- and post-synaptic spikes arrive.

Implementation of STDP with memristors requires the careful design of neuronal input signals at both the pre- and post-terminals. The key is to find a means to translate the relative spike timing into a specific voltage (current) waveform that controls the flux or charge through the memristors, which directly controls the memristor conductance change. Triangular waves, complementary square waves, pulses with exponentially decreasing tails, or other asymmetric waveforms are common examples of the proposed neuronal inputs which could either be realized through test programs or hardware circuits [6, 45, 47, 56–59]. In all of these attempts, the relative timing between the signals from the post-synaptic neuron and that of the pre-synaptic neuron determines how the signals overlap at the memristor junction, which in turn tunes the memristor conductance accordingly.

The first work of this kind was performed by Jo et al. using a hybrid CMOS/memristor circuit, as shown in Fig. 6.10a [45]. Timing information is encoded using the time division multiplexing (TDM) scheme which allocates events to occur at their prescribed time slots, while keeping the spike amplitude constant [60]. This approach, with a fixed spike amplitude but modulated effective width, is more effective for digital neuron circuits. STDP curves with controlled time constants and incremental changes can be obtained, and by choosing the right device parameters results in good agreement with recorded data from biological synapses can be obtained (Fig. 6.10a). Subsequent work by Kuzum et al. implemented using GST-based phase change cells used a dedicated pulsing scheme that precisely controls the crystallization states of the GST devices, as shown in Fig. 6.10b [47]. By adjusting the pulsing waveforms to cause different overlapped signals in amplitude seen by the memristor, a few different forms of STDP curves were observed, similar to observations in neurobiological experiments. Several other similar studies have also been reported to obtain STDP behaviors in different material systems [56, 59].

6.4.2 Rate-Dependent Plasticity

In addition to STDP, rate-dependent plasticity is also a widely observed synaptic learning rule across different kinds of synapses [52]. It is not surprising—both the pre- and post-neurons fire at seemingly randomly time instants and one can either group the signals arriving at the synapse by pairs from the two neurons, or by spike trains from individual neurons. In the second picture, the spike rate should have a significant effect on the synaptic plasticity. It is natural to expect that the more frequently a synapse is being stimulated, the stronger it becomes (and remains strong). For example, by stimulating biological synapses at varying rates, post-tetanic potentiation (PTP) and paired-pulse facilitation (PPF) effects were studied systematically with their according decay time constants identified [61, 62].

The key to the implementation of the rate-dependent learning is identifying an internal decaying element. Without decay the device will only respond to the total number of stimulations, insensitive to the frequency (rate). However, with a decaying mechanism, the synaptic adaptation is a result of the competing effects between the internal decay and the external stimulation and how effective the learning is then critically depends on the stimulation rate (with respect to the decay rate).

The first report on rate-dependency of memristive devices was by Alibart et al. [63] The memristive devices were based on organic nanoparticle transistors in which the synaptic plasticity was achieved by trapping and detrapping of charged carriers which in turn affect the transistor drain current. The nanoparticles were alternatively charged during the pulse period and discharged (decayed) during the intervals between pulses. The total trapped charges then depend on not only the number of pulses but also the frequency. As shown in Fig. 6.11a, in the high stimulating rate case, such as 2, 5, and 10 Hz, the number of holes that are trapped in the nanoparticles exceeds that of detrapped ones. As a result, the number of holes present in the channel is increased, depressing the output current. Adversely, the output current is increased in the low stimulating rate case such as 0.5 and 1 Hz. This charge trapping/detrapping process is argued to be analogous to the rate-dependent synaptic plasticity caused by consumption and recovery of the finite chemical neurotransmitters in biological synapses [64] and can potentially be used to implement rate-dependent learning rules.

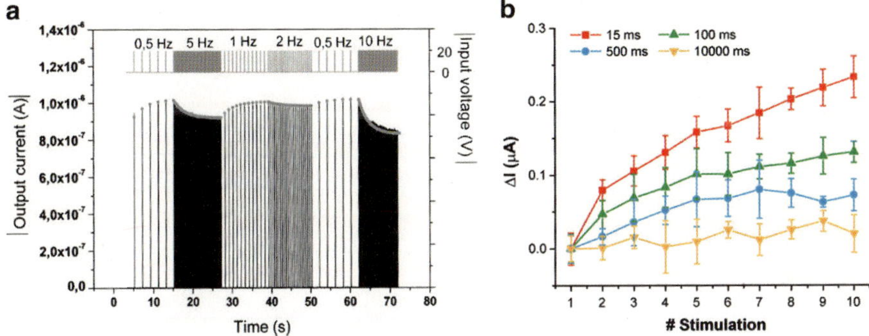

Fig. 6.11 (**a**) Rate-dependent plasticity in organic nanoparticle transistors. Reprinted with permission from [63]. Copyright (2010) Wiley-VCH Verlag GmbH & KGaA, Weinheim. (**b**) Rate-dependent plasticity in tungsten oxide-based memristors. The different symbols correspond to different interval between pulses. Reprinted with permission from [38]. Copyright (2011) American Chemical Society

The nanoparticle transistor devices are, however, three-terminal devices which make scaling more difficult. A more systematic study on two-terminal memristor devices were performed by Chang et al. on WO_3 memristors [38]. As shown in Fig. 6.11b, a high stimulation rate (smaller interval between pulses) leads to more significant conductance enhancement while the device barely responds to stimulations at low rate, even though both the strength and number of stimulations are identical in each case. Here the decay term is the spontaneous diffusion of the oxygen vacancies forming the conduction channels, discussed in Eq. (6.6). PTP and PPF effects with behaviors similar to those in biological systems have also be obtained in these oxide memristor devices [38]. The internal decay of the conduction channels in turn suggests that the devices possess short memory retention (analogous to short-term memory in biological systems). Although short retention is obviously not desirable for nonvolatile data storage, the decay and short-term memory can be desirable properties for neuromorphic circuit implementations, as will be discussed next.

6.4.3 *Short-Term and Long-Term Plasticity*

While long-term plasticity (LTP) is obviously needed for storing the processed information, it is believed that short-term plasticity (STP) helps the system process information by releasing resources that are no longer needed [51]. Biologically, it is still debatable whether these two kinds of plasticity take place at the same location and how one may trigger the other. However, it seems clear that there exist short- and long-memory regimes, separated by their retention times. Additionally, short-term memory can be converted to long-term memory after sufficient training.

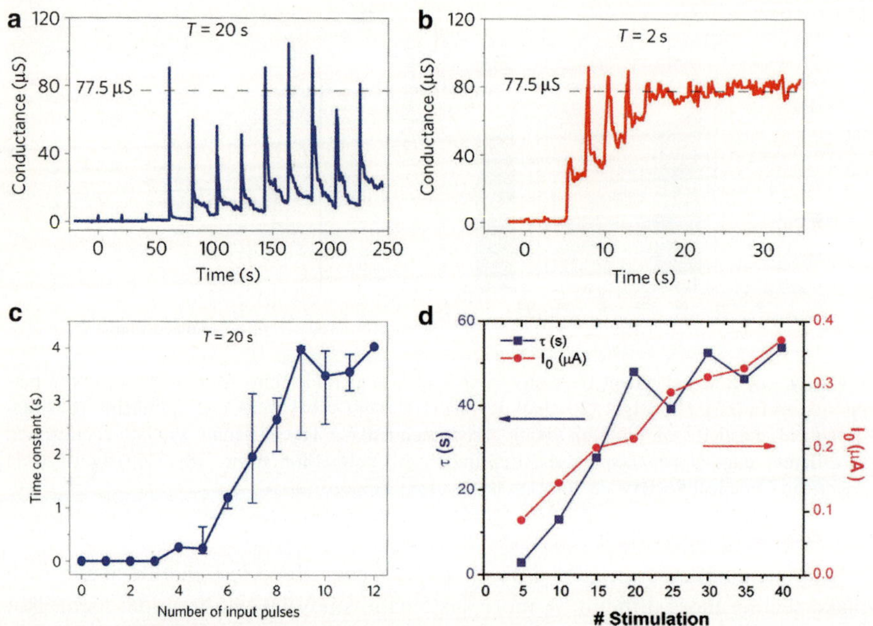

Fig. 6.12 (**a**, **b**) Decay and stabilization of the conductance of Ag_2S-based devices when the input pulses were applied with intervals of 20 s (**a**) and 2 s (**b**). (**c**) Decay time constant as a function of the number of input pulses for Ag_2S-based devices. Reprinted with permission from [42]. Copyright (2011) Nature Publishing Group. (**d**) Time constant and synaptic weight as a function of the number of input pulses for WO_x-based devices. Reprinted with permission from [38]. Copyright (2011) American Chemical Society

In computation, this corresponds to a very practical mechanism to allocate limited resources (e.g., the number of physical synapses) for the most efficient use since the more crucial the information, the longer the memory is kept.

Two independent studies were performed that showed the existence of STP and the transition of STP to LTP in memristive devices. These effects were observed first by Ohno et al. in Ag_2S devices [42], shown in Fig. 6.12a. One can see that the device conductance always decays back to its initial value when low-repetition rate pulses are applied, suggesting STP. However, when the repetition rate is increased to a certain value, a stable high conductance state can be achieved (Fig. 6.12b), indicating a transition from STP to LTP and corresponding to the formation of a stable conducting bridge in the Ag_2S device. It was found that the decay time of the STP state (i.e., how fast STP loses its information) also depends on the training conditions, with longer time constants obtained when the number of input pulses is increased, as displayed in Fig. 6.12c.

In another study reported at roughly the same time, Chang et al. observed STP and STP to LTP transition in WO_3 devices, as shown in Fig. 6.12d. Briefly, the conduction channel area in the memristors is increased through repeated stimulation, which leads to not only an increased conductance (weight) but also longer retention time [38]. This effect corresponds well to the memory enhancement effects observed in biological systems which lead to the transition from STP to LTP [65]. Additionally, both the stimulation rate and the total number of stimulations are found to strongly affect the memory enhancement. With sufficient stimulations, the retention time can be increased by several factors, indicating a transition from STP to LTP can be achieved in the memristors [38].

6.4.4 Associative Learning

Memristors are perhaps the ideal candidate to implement local, synaptic learning rules for the implementation of efficient neuromorphic circuits. However, assembling a large number of individual working devices does not necessarily make a system functional. How functions evolve from biological networks is of course still an ongoing research in the field of neuroscience and research on memristor-based neuromorphic hardware in this field is still at the very early stage.

Associative learning is a form of Hebbian learning experimentally demonstrated by the training of Pavlov's dog [66]. By pairing conditioned stimulus (e.g., sound of a bell) with unconditioned stimulus (e.g., sight of food), the dog learns to associate both events and responds (e.g., salivates) to both stimuli. Associative learning is especially important as it is believed to be behind how brains correlate individual events and how neural networks perform certain tasks very effectively. First proposed by Pershin et al., synaptic emulators and specially designed circuitry were developed to demonstrate associative learning [67]. Figure 6.13 shows the results from a study by Pershin et al., where before learning the "salivation" neuron only responds to the "sight of food" neuron input, i.e. only synapse S1 is on. By simultaneously applying stimulations to both the "sight of food" and "sound" neurons in the learning phase, synapse S2 between the "sound" neuron and the "salvation" neuron is turned on. As a result, stimulus from the "sound" neuron alone is able to excite the "salivation" neuron, therefore establishing an association between the conditioned and unconditioned stimuli. Zeigler et al. later demonstrated associative learning with $Ge_{0.3}Se_{0.7}$ memristors [68]. Along with their neuron-mimicking circuit, both associative and non-associative learning could be implemented.

6.4.5 Emergent Behaviors

Emergent behaviors arise when individual elements interact with each other in a random, complex network, achieving collective behaviors that are not expected from the simple sum of individual elements. Since biological neural networks are highly complex systems with numerous interconnected elements, the evolution of

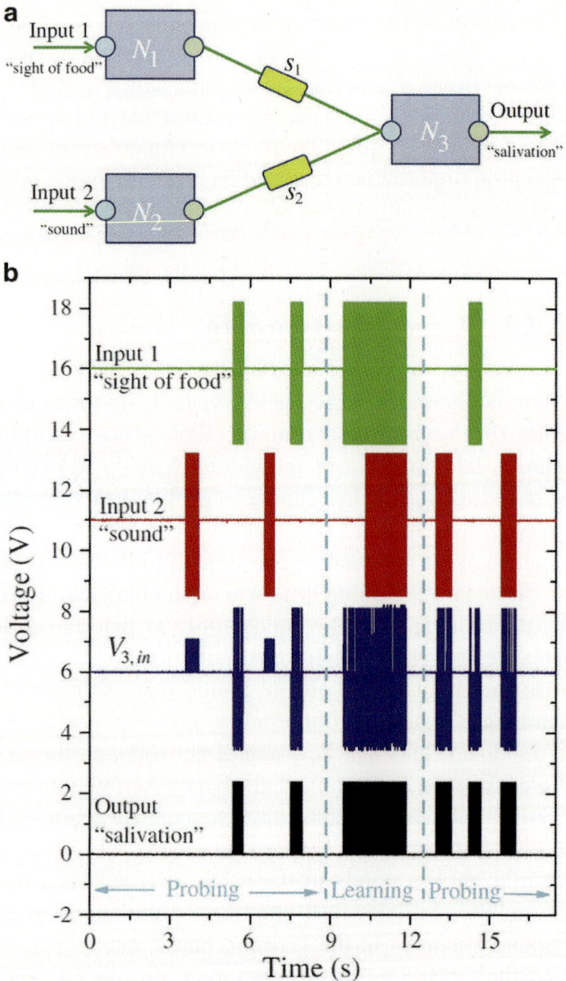

Fig. 6.13 Associative learning implemented with electronic emulators. (**a**) Schematic of the setup. (**b**) Development of associative memory simulated by the setup in (**a**). Reprinted with permission from [67]. Copyright (2010) Elsevier

emergent behaviors is believed to be key to the development of network functions. A primitive attempt to achieve emergent behaviors in memristor networks was performed by Stieg et al. when studying self-assembled Ag nanowire networks that are interconnected randomly [69]. A wide range of discrete, metastable conductance states were observed that were explained by the dynamic reorganization of the interconnected atomic switch network. Such collective behavior is believed to be similar to that of complex neural networks and if can be understood and controlled, can hold potential for efficient memory, information transmission, and adaptability.

Emergence is one-step further in neuromorphic development because it only appears at the network level, never at the synaptic or neuronal level. However, the unpredictability of emergent behaviors also implies that there might not be a "correct" answer; instead, the behaviors of the network evolve with the characteristics of

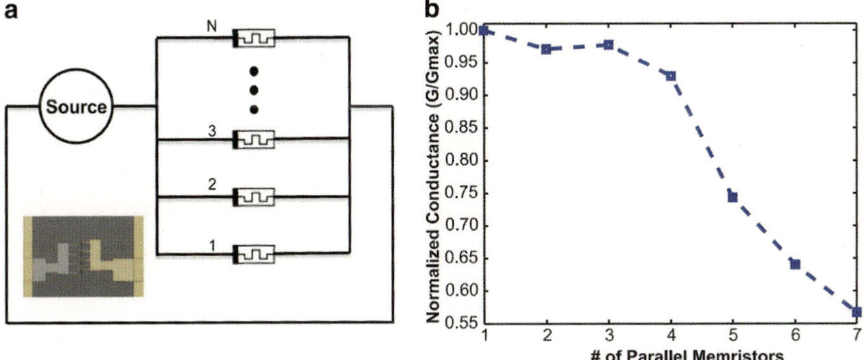

Fig. 6.14 (**a**) Schematic of the memristor circuit used to demonstrate interference and memory capacity effects. The circuit is composed of parallel-memristors driven by a current-limited source. *Inset*: optical microscope image of the parallel memristor array. (**b**) Normalized memristor conductance after training as a function of the array size. Reprinted with permission from [70]. Copyright (2013) American Institute of Physics

individual elements, dynamic environments, and the physical network connectivity (which can be enormous and random). As a result, useful information may only be obtained from statistics, and through a large number of controlled experiments.

6.4.6 Limited Capacity Effect

Another reason that unexpected results may arise when individual devices are connected in a network is that the different devices now share and compete for the same resources. Depending on the specific physical configurations, inputs and dynamics of the network and the devices, diverse results can thus be obtained when individual elements, small networks, or large networks are formed.

The competition effect was recently demonstrated by Hermiz et al. in a study focused on how resources are distributed among competing elements in a network, and how different results can be obtained depending on the network size (Fig. 6.14) [70]. Here a circuit of parallel-memristors connected to a constant current source was studied both through simulation and experiments using WO_3 devices. Specifically, how well the memristors are trained was found to depend on the network size, due to the competition between training (which is amplified by the interference effect) and internal state decay in each memristor. The more memristors sharing the limited current source, the less firmly each memristor is trained, as shown in Fig. 6.14b. Essentially, this study demonstrated the limited memory capacity effect found in psychology studies—the ability to recall information stored in short-term memory falls off significantly beyond a certain list size. By tuning the circuit parameters, different critical list sizes can also be obtained (e.g., a critical list size of 4 matching that found in psychology studies is shown in Fig. 6.14b).

6.5 Large-Scale Memristor Crossbar Network Hardware

Hardware implementation of more complex synaptic functions demands larger-scale memristor networks with inherently high connectivity. Several studies have been carried out recently to construct memristor arrays based on the success on single devices, mostly focusing on implementing basic memory [19, 71] and logic [72] functions so far. The simple two-terminal structure of memristive devices allows them to be integrated into crossbar networks, composed of two sets of parallel nanowire electrodes crossing each other with a memristive device formed at each crosspoint. This configuration potentially provides both the high density and high connectivity that are required to emulate neuromorphic systems. For example, a human cortex has about 10^{10} synapses/cm^2 in density and can be achieved in crossbar arrays with 100 nm pitch.

Figure 6.15a shows an SEM image of a 32×32 memristor crossbar array fabricated on SiO_2/Si substrates by Jo et al. [19]. Efforts were soon extended to

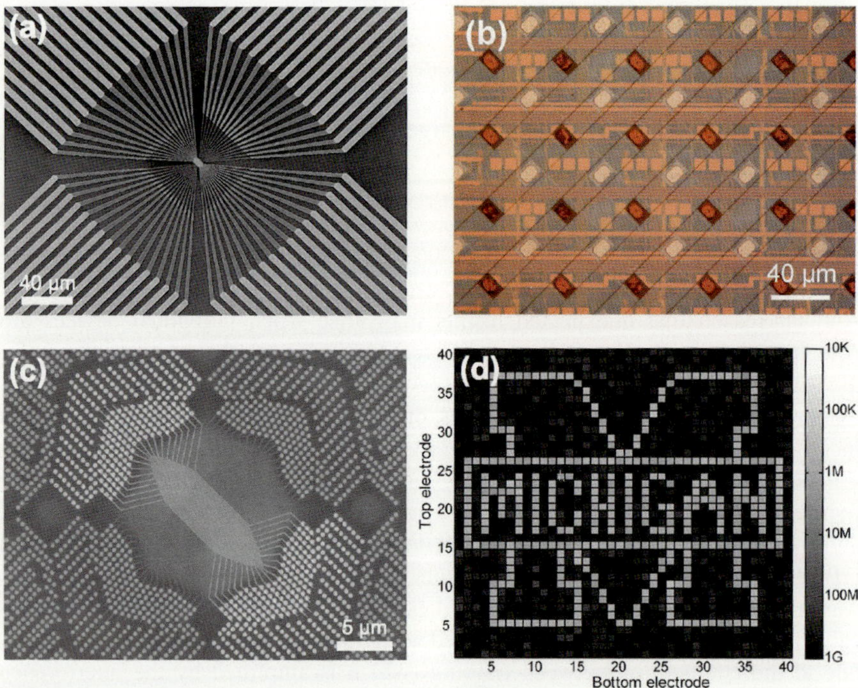

Fig. 6.15 (**a**) SEM image of a 32×32 Ag/a-Si/p-Si crossbar array on a SiO_2/Si substrate. Reprinted with permission from [19]. Copyright (2009) American Chemical Society. (**b**) Optical micrograph of a Pt/TiO_2/Pt crossbar array on CMOS chip. Reprinted with permission from [72]. Copyright (2009) American Chemical Society. (**c**) SEM image of a 40×40 Ag/a-Si/SiGe crossbar array on CMOS chip. (**d**) A bitmap image by storing and retrieving data in the 40×40 crossbar array in (**c**). Reprinted with permission from [71]. Copyright (2012) American Chemical Society

building similar crossbar arrays on top of CMOS chips in order to combine the functionalities of CMOS circuits with the properties of memristive devices. Such hybrid memristor/CMOS crossbar arrays have been demonstrated with both cation and anion migration-based devices, e.g. using Ag/a-Si/SiGe [71] and Pt/TiO_2/Pt [72] devices, respectively. Figure 6.15b shows the optical graph of a Pt/TiO_2/Pt array fabricated on CMOS substrate via nanoimprint lithography, following an approach based on CMOL [73, 74]. The hybrid memristor/CMOS system demonstrated field programmable gate array-like functionalities [72]. Figure 6.15c shows a 40×40 Ag/a-Si/SiGe crossbar array was fabricated on a CMOS chip using local interconnects by Kim et al. [71], showing reliable memory operations (Fig. 6.15c, d). These hybrid memristor/CMOS arrays can potentially accommodate large crossbar networks and can thus provide a good platform for studying complex neuromorphic functions in these networks. However, we note that studies so far are only limited to relatively simple memory and binary logic demonstration, while the development of functional memristor-based neuromorphic networks requires much more complex neuron designs and better understanding and control of the memristor dynamics, as well as better understanding of how functions and emergent behaviors evolve in large networks.

6.6 Summary and Outlook

In summary, tremendous progress has been made in the last a few years on the development of memristive devices and the employment of such devices in neuromorphic systems. Device operation mechanism, performance optimization, and modeling have been extensively studied. A diverse range of local learning rules have been demonstrated and prototype memristor crossbar arrays with very large connectivity and hybrid memristor array/CMOS neuron systems have been demonstrated. However, despite the rapid progress, the field is still only at its infancy. For example, studies of the network dynamics and emergent behaviors have just started, although preliminary results obtained from the few experimental and simulation studies to date are already quite exciting and indicate more breakthroughs to come.

Looking into the future, we have every reason to believe this field will continue to enjoy exponential growth. It is likely that in a few years memristor-based neuromorphic hardware will be available to anyone interested in them, like carbon nanotube or graphene is today to device researchers. This will further fuel the development of modeling, new algorithms, and architectures to most efficiently utilize this new class of hardware, which will in turn speed up device research to take advantage of the new algorithms. It is intriguing to imagine a future where smart neuromorphic chips significantly improve our quality of life. However, this Hercules task can only be achieved through close collaborations among material scientists, device physicist, electrical and computer engineers, and computer scientists.

References

1. International Technology Roadmap for Semiconductors (ITRS), 2011 Edition. http://www.itrs.net/Links/2011ITRS/Home2011.htm
2. C. Mead, Neuromorphic electronic systems. Proc. IEEE **78**, 1629–1636 (1990)
3. C.-S. Poon, K. Zhou, Neuromorphic silicon neurons and large-scale neural networks: challenges and opportunities. Front. Neurosci. **5**, 108–108 (2011)
4. D.S. Modha, R. Ananthanarayanan, S.K. Esser, A. Ndirango, A.J. Sherbondy, R. Singh, Cognitive computing. Commun. ACM **54**, 62–71 (2011)
5. G. Indiveri, B. Linares-Barranco, T.J. Hamilton, A. van Schaik, R. Etienne-Cummings, T. Delbruck, S.-C. Liu, P. Dudek, P. Hafliger, S. Renaud, J. Schemmel, G. Cauwenberghs, J. Arthur, K. Hynna, F. Folowosele, S. Saighi, T. Serrano-Gotarredona, J. Wijekoon, Y. Wang, K. Boahen, Neuromorphic silicon neuron circuits. Front. Neurosci. **5**, 73–73 (2011)
6. Y.V. Pershin, M. Di Ventra, Neuromorphic, digital, and quantum computation with memory circuit elements. Proc. IEEE **100**, 2071–2080 (2012)
7. G. Rachmuth, H.Z. Shouval, M.F. Bear, C.-S. Poon, A biophysically-based neuromorphic model of spike rate- and timing-dependent plasticity. Proc. Natl. Acad. Sci. USA **108**, E1266–E1274 (2011)
8. X. Guardiola, A. Diaz-Guilera, M. Llas, C.J. Perez, Synchronization, diversity, and topology of networks of integrate and fire oscillators. Phys. Rev. E **62**, 5565–5570 (2000)
9. Y. Moreno, A.F. Pacheco, Synchronization of Kuramoto oscillators in scale-free networks. Europhys. Lett. **68**, 603–609 (2004)
10. H.Z. Shouval, M.F. Bear, L.N. Cooper, A unified model of NMDA receptor-dependent bidirectional synaptic plasticity. Proc. Natl. Acad. Sci. USA **99**, 10831–10836 (2002)
11. L.O. Chua, Memristor – the missing circuit element. IEEE Trans. Circuit Theory **18**, 507–519 (1971)
12. L.O. Chua, S.M. Kang, Memristive devices and systems. Proc. IEEE **64**, 209–223 (1976)
13. D.B. Strukov, G.S. Snider, D.R. Stewart, R.S. Williams, The missing memristor found. Nature **453**, 80–83 (2008)
14. R. Waser, M. Aono, Nanoionics-based resistive switching memories. Nat. Mater. **6**, 833–840 (2007)
15. R. Waser, R. Dittmann, G. Staikov, K. Szot, Redox-based resistive switching memories - nanoionic mechanisms, prospects, and challenges. Adv. Mater. **21**, 2632–2663 (2009)
16. J.J. Yang, D.B. Strukov, D.R. Stewart, Memristive devices for computing. Nat. Nanotechnol. **8**, 13–24 (2013)
17. T. Chang, Y. Yang, W. Lu, Building neuromorphic circuits with memristive devices. IEEE Circuits Syst. Mag. **13**, 56–73 (2013)
18. J.G. Simmons, R.R. Verderbe, New conduction and reversible memory phenomena in thin insulating films. Proc. R. Soc. Lond. A Mater. **301**, 77–102 (1967)
19. S.H. Jo, K.H. Kim, W. Lu, High-density crossbar arrays based on a Si memristive system. Nano Lett. **9**, 870–874 (2009)
20. S.H. Jo, K.H. Kim, W. Lu, Programmable resistance switching in nanoscale two-terminal devices. Nano Lett. **9**, 496–500 (2009)
21. J.J. Yang, M.D. Pickett, X.M. Li, D.A.A. Ohlberg, D.R. Stewart, R.S. Williams, Memristive switching mechanism for metal/oxide/metal nanodevices. Nat. Nanotechnol. **3**, 429–433 (2008)
22. L. Chua, Resistance switching memories are memristors. Appl. Phys. A Mater. Sci. Process. **102**, 765–783 (2011)
23. B. Govoreanu, G.S. Kar, Y.Y. Chen, V. Paraschiv, S. Kubicek, A. Fantini, I.P. Radu, L. Goux, S. Clima, R. Degraeve, N. Jossart, O. Richard, T. Vandeweyer, K. Seo, P. Hendrickx, G. Pourtois, H. Bender, L. Altimime, D.J. Wouters, J.A. Kittl, M. Jurczak, 10×10 nm^2 Hf/HfO$_x$ crossbar resistive RAM with excellent performance, reliability and low-energy operation. IEDM 729–732 (2011)

24. A.C. Torrezan, J.P. Strachan, G. Medeiros-Ribeiro, R.S. Williams, Sub-nanosecond switching of a tantalum oxide memristor. Nanotechnology **22**, 485203 (2011)
25. M.-J. Lee, C.B. Lee, D. Lee, S.R. Lee, M. Chang, J.H. Hur, Y.-B. Kim, C.-J. Kim, D.H. Seo, S. Seo, U.I. Chung, I.-K. Yoo, K. Kim, A fast, high-endurance and scalable non-volatile memory device made from asymmetric Ta_2O_{5-x}/TaO_{2-x} bilayer structures. Nat. Mater. **10**, 625–630 (2011)
26. D.H. Kwon, K.M. Kim, J.H. Jang, J.M. Jeon, M.H. Lee, G.H. Kim, X.S. Li, G.S. Park, B. Lee, S. Han, M. Kim, C.S. Hwang, Atomic structure of conducting nanofilaments in TiO_2 resistive switching memory. Nat. Nanotechnol. **5**, 148–153 (2010)
27. D.B. Strukov, J.L. Borghetti, R.S. Williams, Coupled ionic and electronic transport model of thin-film semiconductor memristive behavior. Small **5**, 1058–1063 (2009)
28. Y. Yang, P. Gao, S. Gaba, T. Chang, X. Pan, W. Lu, Observation of conducting filament growth in nanoscale resistive memories. Nat. Commun. **3**, 732 (2012)
29. F. Miao, J.P. Strachan, J.J. Yang, M.-X. Zhang, I. Goldfarb, A.C. Torrezan, P. Eschbach, R.D. Kelley, G. Medeiros-Ribeiro, R.S. Williams, Anatomy of a nanoscale conduction channel reveals the mechanism of a high-performance memristor. Adv. Mater. **23**, 5633–5640 (2011)
30. J.P. Strachan, M.D. Pickett, J.J. Yang, S. Aloni, A.L.D. Kilcoyne, G. Medeiros-Ribeiro, R.S. Williams, Direct identification of the conducting channels in a functioning memristive device. Adv. Mater. **22**, 3573–3577 (2010)
31. J.J. Yang, F. Miao, M.D. Pickett, D.A.A. Ohlberg, D.R. Stewart, C.N. Lau, R.S. Williams, The mechanism of electroforming of metal oxide memristive switches. Nanotechnology **21**, 215201 (2010)
32. D.B. Strukov, R.S. Williams, Exponential ionic drift: fast switching and low volatility of thin-film memristors. Appl. Phys. A Mater. Sci. Process. **94**, 515–519 (2009)
33. J.J. Yang, J.P. Strachan, F. Miao, M.-X. Zhang, M.D. Pickett, W. Yi, D.A.A. Ohlberg, G. Medeiros-Ribeiro, R.S. Williams, Metal/TiO_2 interfaces for memristive switches. Appl. Phys. A Mater. Sci. Process. **102**, 785–789 (2011)
34. J.J. Yang, J.P. Strachan, Q.F. Xia, D.A.A. Ohlberg, P.J. Kuekes, R.D. Kelley, W.F. Stickle, D.R. Stewart, G. Medeiros-Ribeiro, R.S. Williams, Diffusion of adhesion layer metals controls nanoscale memristive switching. Adv. Mater. **22**, 4034–4038 (2010)
35. W.C. Chien, Y.C. Chen, E.K. Lai, F.M. Lee, Y.Y. Lin, A.T.H. Chuang, K.P. Chang, Y.D. Yao, T.H. Chou, H.M. Lin, M.H. Lee, Y.H. Shih, K.Y. Hsieh, C.-Y. Lu, A study of the switching mechanism and electrode material of fully CMOS compatible tungsten oxide ReRAM. Appl. Phys. A Mater. Sci. Process. **102**, 901–907 (2011)
36. T. Chang, S.-H. Jo, K.-H. Kim, P. Sheridan, S. Gaba, W. Lu, Synaptic behaviors and modeling of a metal oxide memristive device. Appl. Phys. A Mater. Sci. Process. **102**, 857–863 (2011)
37. C. Ho, C.-L. Hsu, C.-C. Chen, J.-T. Liu, C.-S. Wu, C.-C. Huang, C. Hu, F.-L. Yang, 9 nm half-pitch functional resistive memory cell with <1 μA programming current using thermally oxidized sub-stoichiometric WO_x film. IEDM 436–439 (2010)
38. T. Chang, S.-H. Jo, W. Lu, Short-term memory to long-term memory transition in a nanoscale memristor. ACS Nano **5**, 7669–7676 (2011)
39. Y. Yang, P. Sheridan, W. Lu, Complementary resistive switching in tantalum oxide-based resistive memory devices. Appl. Phys. Lett. **100**, 203112 (2012)
40. K. Terabe, T. Hasegawa, T. Nakayama, M. Aono, Quantized conductance atomic switch. Nature **433**, 47–50 (2005)
41. T. Hasegawa, T. Ohno, K. Terabe, T. Tsuruoka, T. Nakayama, J.K. Gimzewski, M. Aono, Learning abilities achieved by a single solid-state atomic switch. Adv. Mater. **22**, 1831–1834 (2010)
42. T. Ohno, T. Hasegawa, T. Tsuruoka, K. Terabe, J.K. Gimzewski, M. Aono, Short-term plasticity and long-term potentiation mimicked in single inorganic synapses. Nat. Mater. **10**, 591–595 (2011)
43. S.H. Jo, W. Lu, CMOS compatible nanoscale nonvolatile resistance switching memory. Nano Lett. **8**, 392–397 (2008)

44. K.H. Kim, S.H. Jo, S. Gaba, W. Lu, Nanoscale resistive memory with intrinsic diode characteristics and long endurance. Appl. Phys. Lett. **96**, 053106 (2010)
45. S.H. Jo, T. Chang, I. Ebong, B.B. Bhadviya, P. Mazumder, W. Lu, Nanoscale memristor device as synapse in neuromorphic systems. Nano Lett. **10**, 1297–1301 (2010)
46. M. Wuttig, N. Yamada, Phase-change materials for rewriteable data storage. Nat. Mater. **6**, 824–832 (2007)
47. D. Kuzum, R.G.D. Jeyasingh, B. Lee, H.S.P. Wong, Nanoelectronic programmable synapses based on phase change materials for brain-inspired computing. Nano Lett. **12**, 2179–2186 (2012)
48. M.D. Pickett, R.S. Williams, Sub-100 fJ and sub-nanosecond thermally driven threshold switching in niobium oxide crosspoint nanodevices. Nanotechnology **23**, 215202 (2012)
49. M.D. Pickett, G. Medeiros-Ribeiro, R.S. Williams, A scalable neuristor built with Mott memristors. Nat. Mater. **12**, 114–117 (2013)
50. W. Lu, Memristors: going active. Nat. Mater. **12**, 93–94 (2013)
51. W.M. Cowan, T.C. Südhof, C.F. Stevens (eds.) *Synapses* (Johns Hopkins University Press, Baltimore, 2001)
52. P.J. Sjostrom, G.G. Turrigiano, S.B. Nelson, Rate, timing, and cooperativity jointly determine cortical synaptic plasticity. Neuron **32**, 1149–1164 (2001)
53. G.Q. Bi, M.M. Poo, Synaptic modifications in cultured hippocampal neurons: dependence on spike timing, synaptic strength, and postsynaptic cell type. J. Neurosci. **18**, 10464–10472 (1998)
54. R.C. Froemke, Y. Dan, Spike-timing-dependent synaptic modification induced by natural spike trains. Nature **416**, 433–438 (2002)
55. D.O. Hebb, *The Organization of Behavior: A Neuropsychological Theory* (Wiley, New York, 1949)
56. K. Seo, I. Kim, S. Jung, M. Jo, S. Park, J. Park, J. Shin, K.P. Biju, J. Kong, K. Lee, B. Lee, H. Hwang, Analog memory and spike-timing-dependent plasticity characteristics of a nanoscale titanium oxide bilayer resistive switching device. Nanotechnology **22**, 254023 (2011)
57. P. Krzysteczko, J. Muenchenberger, M. Schaefers, G. Reiss, A. Thomas, The memristive magnetic tunnel junction as a nanoscopic synapse-neuron system. Adv. Mater. **24**, 762–766 (2012)
58. Z.Q. Wang, H.Y. Xu, X.H. Li, H. Yu, Y.C. Liu, X.J. Zhu, Synaptic learning and memory functions achieved using oxygen ion migration/diffusion in an amorphous InGaZnO memristor. Adv. Funct. Mater. **22**, 2759–2765 (2012)
59. S. Yu, Y. Wu, R. Jeyasingh, D. Kuzum, H.S.P. Wong, An electronic synapse device based on metal oxide resistive switching memory for neuromorphic computation. IEEE Trans. Electron Devices **58**, 2729–2737 (2011)
60. G.S. Snider, Spike-timing-dependent learning in memristive nanodevices, in *2008 IEEE/ACM International Symposium on Nanoscale Architectures*, Anaheim (2008), pp. 85–92
61. K.L. Magleby, Effect of repetitive stimulation on facilitation of transmitter release at frog neuromuscular junction. J. Physiol. Lond. **234**, 327–352 (1973)
62. P.P. Atluri, W.G. Regehr, Determinants of the time course of facilitation at the granule cell to Purkinje cell synapse. J. Neurosci. **16**, 5661–5671 (1996)
63. F. Alibart, S. Pleutin, D. Guerin, C. Novembre, S. Lenfant, K. Lmimouni, C. Gamrat, D. Vuillaume, An organic nanoparticle transistor behaving as a biological spiking synapse. Adv. Funct. Mater. **20**, 330–337 (2010)
64. M. Tsodyks, K. Pawelzik, H. Markram, Neural networks with dynamic synapses. Neural Comput. **10**, 821–835 (1998)
65. R.M. Shiffrin, R.C. Atkinson, Storage and retrieval processes in long-term memory. Psychol. Rev. **76**, 179–193 (1969)
66. I. Pavlov, *Conditioned Reflexes: An Investigation of the Physiological Activity of the Cerebral Cortex* (translated by G. V. Anrep) (Oxford University Press, London, 1927)

67. Y.V. Pershin, M. Di Ventra, Experimental demonstration of associative memory with memristive neural networks. Neural Netw. **23**, 881–886 (2010)
68. M. Ziegler, R. Soni, T. Patelczyk, M. Ignatov, T. Bartsch, P. Meuffels, H. Kohlstedt, An electronic version of Pavlov's dog. Adv. Funct. Mater. **22**, 2744–2749 (2012)
69. A.Z. Stieg, A.V. Avizienis, H.O. Sillin, C. Martin-Olmos, M. Aono, J.K. Gimzewski, Emergent criticality in complex Turing B-type atomic switch networks. Adv. Mater. **24**, 286–293 (2012)
70. J. Hermiz, T. Chang, C. Du, W. Lu, Interference and memory capacity effects in memristive systems. Appl. Phys. Lett. **102**, 083106 (2013)
71. K.-H. Kim, S. Gaba, D. Wheeler, J.M. Cruz-Albrecht, T. Hussain, N. Srinivasa, W. Lu, A functional hybrid memristor crossbar-array/CMOS system for data storage and neuromorphic applications. Nano Lett. **12**, 389–395 (2012)
72. Q.F. Xia, W. Robinett, M.W. Cumbie, N. Banerjee, T.J. Cardinali, J.J. Yang, W. Wu, X.M. Li, W.M. Tong, D.B. Strukov, G.S. Snider, G. Medeiros-Ribeiro, R.S. Williams, Memristor-CMOS hybrid integrated circuits for reconfigurable logic. Nano Lett. **9**, 3640–3645 (2009)
73. D.B. Strukov, K.K. Likharev, CMOL FPGA: a reconfigurable architecture for hybrid digital circuits with two-terminal nanodevices. Nanotechnology **16**, 888–900 (2005)
74. G.S. Snider, R.S. Williams, Nano/CMOS architectures using a field-programmable nanowire interconnect. Nanotechnology **18**, 035204 (2007)

Chapter 7
Redox-Based Memristive Devices

Vikas Rana and Rainer Waser

Over the past few decades, MOSFET-based nonvolatile memories have played a significant role in the growth of the portable electronic market. However, aggressive device scaling trends are about to reach their limits. In the quest for the next generation nonvolatile memory device, several mechanisms such as redox-based, phase-change, magnetic-junction, and ferroelectrics have recently been extensively investigated. A highly promising candidate that is expected to succeed the flash memory device is the redox-based resistive random access memory (ReRAM). The fundamental requirements of a nonvolatile memory are nondestructive write/read operations at a speed comparable to current logic devices, infinite retention, low energy consumption, and integration capability with the current CMOS process. In this chapter, we will describe the current understanding of the physical mechanism of redox-based resistive switching and address several technological aspects of metal-oxide ReRAMs.

7.1 Metal-Oxide ReRAM

Silicon-based nonvolatile memory technology, i.e. flash memory [1], has been extensively used, for example, in mobile storage, digital audio players, digital cameras, video games, scientific instrumentation, industrial robotics, medical electronics, and so on. In the quest for faster speed and lower cost, this technology

V. Rana (✉)
Peter Grünberg Institut -7, Forschungszentrum Jülich GmbH, Jülich 52425, Germany
e-mail: v.rana@fz-Juelich.de

R. Waser
Peter Grünberg Institut -7, Forschungszentrum Jülich GmbH, Jülich 52425, Germany

Faculty of Electrical Engineering and Information Technology, RWTH Aachen University, Aachen 52062, Germany & JARA-FIT, Germany
e-mail: waser@IWE.RWTH-Aachen.de

R. Tetzlaff (ed.), *Memristors and Memristive Systems*,
DOI 10.1007/978-1-4614-9068-5_7,

suffers from low endurance and high voltage in writing operations (16–20 V) and is approaching its fundamental scaling limits due to the increasing difficulty of retaining electrons in the shrinking dimensions [2]. These concerns demand an alternative low-cost and low-power memory device. The expected characteristics of an ideal nonvolatile memory comprise write and (preferably nondestructive) read operations at speeds comparable to those of logic devices, low energy consumption, infinite retention, and infinite number of read and write cycles. Various technologies such as magnetic random access memory (MRAM) [3], ferroelectric random access memory (FRAM) [4], spin-transfer torque random access memory (STT-RAM) [5], and redox-based resistive random access memory (ReRAM) [6] are competing to become a mainstream memory technology. In order to dominate the nonvolatile memory market, a future technology must meet the requirements of high performance, robustness, integration capabilities, and low-cost. The high performance of a memory system is defined in terms of high speed, low power consumption, and high reliability.

A promising candidate for future nonvolatile memory is the ReRAM, based on oxide materials which show a significant change in resistance upon application of a voltage bias above a critical threshold and switch back to the original state. The first scientific report on the resistive switching was published in the early 1960s. However, research activities decreased in the 1980s due to the fact that the interpretation of the microscopic mechanism involved in the resistive switching was inherently limited by contemporary analytical tools. The activities were again revived by Asamitsu et al. [7], Kozicki et al. [8] and Beck et al. [9]. So far, the resistive switching phenomena have been observed in a variety of material systems including TiO_2 [10], HfO_2 [11], NiO [12], Al_2O_3 [13], Nb_2O_5 [14], $SrTiO_3$ [15], $Pr_{0.7}Ca_{0.3}MnO_3$ [16], CuO_2 [17], Ag_2S [18], and AgGeSe [19].

The ReRAM structure is simply composed of a resistance-changeable material, especially transition metal oxides sandwiched between two-terminal metal electrodes. Although the switching can be achieved by a current or voltage pulse applied to the electrodes, the switching behavior is highly dependent on the oxide materials as well as on the type of metal electrodes. Due to the simplest atomic structure and conventional CMOS processing compatibility, the binary oxides are the obvious choice for resistive switching applications and could become the dominant nonvolatile memory technology. The ReRAM technology, which combines the features of the high-speed performance of present SRAMs with the non-volatility of the flash memory, can be realized at low cost and low power consumption. Recent research activities show that the metal-oxide ReRAMs can be programmed with a very low write current at an ultra-fast speed in the order of sub-nanoseconds [20]. Figure 7.1 shows a comparison of the ReRAM cell performance for different memory technologies [21]. Further performance improvement in terms of power consumption, speed, and variability is under way.

So far, the metal oxide materials show promising properties for nonvolatile memory applications, their integration into the ReRAM architectures is at an early stage of development. Prior to any successful commercialization, any technology demands a robust and predictive understanding of its underlying mechanisms. Therefore, the switching mechanisms involved are still the subject of

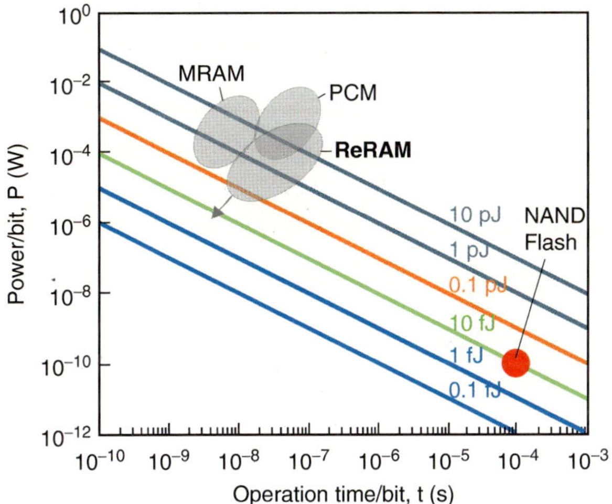

Fig. 7.1 Comparison of switching energy performance for ReRAM, NAND flash, MRAM and PCM technologies [21]

current research activities. Several physical models, i.e. the redox-based conductive filament model [18], Schottky barrier model [22], charge-trapping model [23], and electrochemical migration of point defects [24], have been reported to elucidate the resistive switching phenomenon. It is widely accepted that redox processes are mainly involved in the switching mechanisms. However, many underlying details of the switching are yet to be explained. To reveal all the mechanisms and to predict more accurate resistive switching behavior, a deeper understanding of defect chemistry and interactions of defects under the electrical field are required. For example, the effects of oxygen vacancy ordering on the energy band diagram and electron density of states are being studied in detail [25].

7.2 Physical Mechanism of Metal-Oxide ReRAM

Depending on the switching material and electrode type, the physical mechanism of the resistive switching is explained by various models. Nevertheless, the resistive switching operation in oxides is regarded as a toggling of the resistance states as a result of electric stimulus. This operation which changes the resistance of the device from the high resistance state (HRS) to the low resistance state (LRS) is known as the SET process, whereas the reverse process is referred to as the RESET process. Depending on the relationship of electrical polarity between the SET and the RESET processes, the resistive switching behavior can be divided into the unipolar and bipolar mode. Both switching modes are shown in Fig. 7.2.

Generally, the resistive switching is presumed to be the combination of physical, chemical, and thermal effects. A typical resistive switching based on the thermal

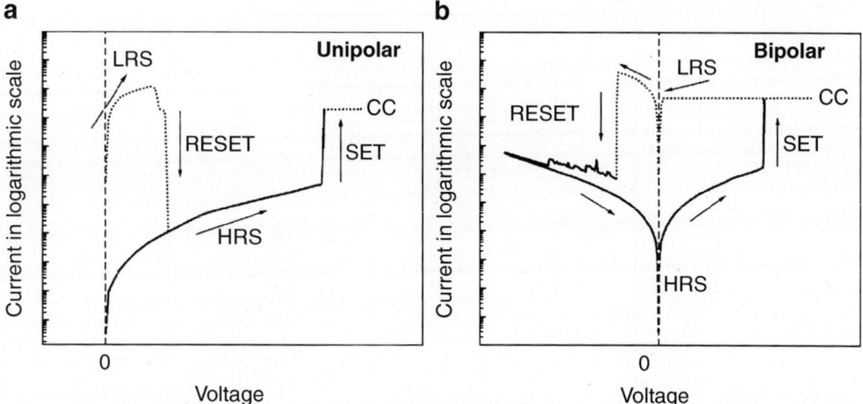

Fig. 7.2 Schematic demonstration of unipolar and bipolar resistive switching mode in binary metal oxides. Current compliance (I_{CC}) is used to avoid permanent breakdown [26]

effect shows the unipolar characteristic. In this mode, the switching direction does not depend on the polarity of the applied voltage. A controlled soft breakdown process initiated by an electric stimulus results in a conductive filament (CF) inside the dielectric film. The filament is generally composed of the electrode metal transported into the insulator or decomposed insulator material such as suboxides. During the reset transition, this CF is again disrupted by Joule heating. This type of conduction is referred to as *thermochemical memory* (*TCM*) [27] and has been verified in a symmetrical metal–insulator–metal (MIM) stack, e.g. Pt/TiO_2/Pt [28], Pt/NiO/Pt [29].

In contrast, the bipolar switching mode depends on the polarity of the applied voltage stimulus as shown in Fig. 7.2 and is commonly observed in asymmetrical structures such as Pt/TiO_2/Ti/Pt. In this type of device, the polarity of the reset voltage (V_{RESET}) is always opposite to the set voltage (V_{SET}). The bipolar switching relies on the migration of the anions/cations in the oxide as a result of oxidation and reduction of an electrochemically electrode.

Several studies have been conducted to elucidate the origin of the resistive switching mechanism in metal oxides. However, redox-based filamentary [18] and interface-type conduction models [22] are widely accepted mechanisms for the LRS conduction. According to the filamentary model, ionic transport and electrochemical redox reactions provide the essential switching mechanism, and current in the LRS mode flows through the confined local path in the insulating layer, whereas the current in the HRS mode flows uniformly through the film. The switching MIM systems consist of an active interface/electrode (AE) at which the switching takes place, a mixed ionic-electronic conducting (MIEC) layer and an ohmic counter electrode (OE). These systems generally require an initial electroforming process prior to any resistive switching behavior. This process typically needs a higher voltage/current than regular switching. As a first approximation, the forming voltage

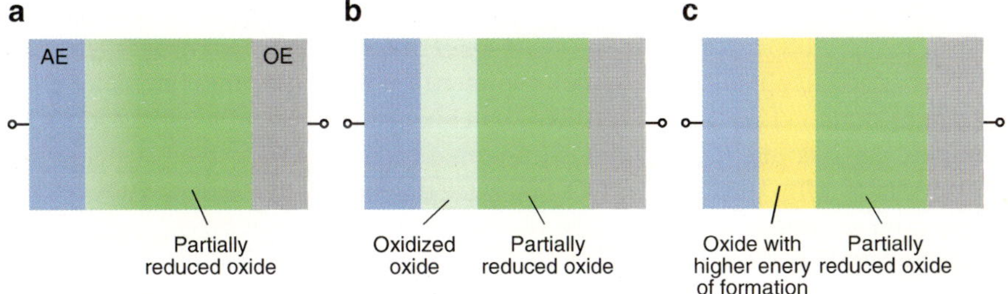

Fig. 7.3 Typical MIM stacks with the different concepts of insulating layer; (**a**) homogeneous monolayer, including the gradient of degree of reduction, (**b**) homogeneous bi-layer, (**c**) heterogeneous bilayer [6]

(V_{FORM}) is proportional to the insulator thickness. The forming process is described in detail in the following Sect. 7.4. The bipolar switching relies on the migration of anions, typically oxygen vacancies (V_o^{2+}), toward the cathode, and is referred to as the *valance-change memory (VCM)*, widely observed in transition metal oxides. Upon the application of a positive voltage bias to the OE, an oxygen-deficient region is created. If the cathode blocks ion exchange reactions, this region starts to expand towards the anode. Transition metal cations accommodate this deficiency by trapping electrons emitted from the cathode. In the case of TiO_2, this reduction reaction is equivalent to filling the Ti 3*d* band.

$$ne^- + Ti^{4+} \rightarrow Ti^{(4-n)+} \tag{7.1}$$

The reduced valence states of the transition metal cations, which are generated by this electrochemical process, typically turn the oxide into a metallically conducting phase, such as $TiO_{2-n/2}$, for approximately $n > 1.5$. This virtual cathode moves towards the anode and finally forms a conductive path [27]. This conductive path is terminated at a certain current/voltage compliance and the forming process is completed. During the electroforming process, the local redox reaction leaves an oxygen-deficient filament in the dielectric layer and the subsequent resistive switching takes place in the so-called disk region—an interface part of the filament and the high work function electrode [6]. Depending on the charge transfer during switching, the resistance of the system can be established at intermediate levels, which might help in creating multi-bit storage in future resistive memory cells. The VCM-type MIM system can be realized in several ways [30]. Typical VCM switching systems are shown in Fig. 7.3. In any case, the AE consists of low-oxygen-affinity material (e.g., Pt, Ir, TiN). A distinct feature of the filamentary model is that the resistance in the LRS state is independent of the device area, whereas the resistance in the HRS state is inversely proportional to the cell size. In this switching model, the CF size determines the ultimate scaling limit of the device. Szot et al. [15], however, reported bipolar resistive switching in individual dislocations of single-crystalline $SrTiO_3$ by applying the electrical voltage with a local-conduction AFM tip (LC-AFM).

In the interface-type resistive switching, the switching event occurs at one interface only, the rectifying non-ohmic interface, and the current is controlled by the barrier height at the interface. The interface barrier height is modulated by electrical stimuli. This type of resistive switching is observed in binary oxides and complex perovskites. Yang et al. [22] reported on the interface-type switching in $Pt/TiO_2/TiO_{2-x}/Pt$, where the Pt/TiO_2 interface is Schottky-like and the TiO_{2-x}/Pt is ohmic. Upon the application of negative voltage, the movement of oxygen vacancies from the TiO_{2-x} to the TiO_2 region can significantly modify the Schottky barrier height and can lower the local potential within the TiO_2 layer. With the opposite polarity, the vacancies drift back to the TiO_{2-x} layer and switch off the conduction. In perovskites, a mechanism concerning the charging effect at the interface is also proposed to explain the resistive switching [31]. The distribution of trapped charge is modulated by the voltages applied in forward or reverse directions, resulting in the modification of band lineup or tunneling probability at the interface barrier. Similarly, the Mott transition induced by the carriers doped at the interface is also classified as the interface-type resistive switching [26]. In the case of interface-type switching, the resistance of the LRS and the HRS is inversely proportional to the device size.

As a variant of the above-mentioned models, the bipolar resistive switching in HfO_x is also explained by a trap-assisted tunneling (TAT) model [32]. In this model, the forming event is presented as a movement of oxygen ions rather than oxygen vacancies, since the oxygen ion diffusion in crystalline HfO_2 is more efficient than the oxygen vacancy diffusion [33]. The oxygen dissociation is initiated by the electric field and elevated temperature. The former is effective due to the high polarizability of high-k dielectrics, while the latter is caused by electron transport through the existing oxygen vacancies, which is accompanied by energy dissipation that increases the local temperature. Due to the variation of trap-level energy (shallow or deeper), a nonuniform heat dissipation across the dielectric is enhanced and leads to a nonuniform temperature profile through the conductive path and promotes the formation of a CF. The electron transport through the CF is dominated by a hopping process. The nonuniformity of the CF seems to play a critical role in switching by enhancing the redox processes at the region of the CF with the highest resistance.

Another class of switching relies on migration of the cations as a result of oxidation and reduction of an electrochemically active electrode metal such as Ag and Cu upon the application of different voltage polarities. This type of cell consists of an insulating layer sandwiched between an electrochemically active electrode and an inert counter electrode. The insulating layer can either be a solid electrolyte containing a host cation such as Ag_2S, Cu_2S or an insulator such as SiO_2, WO_3, GeS, or GeSe doped with the cations. In this type of switching, the forming process is accompanied by the migration of the metal cations towards the cathode where they are reduced. The reduced metal atoms form a metal filament which grows towards the anode to turn on the cell. This process is generally referred to as the *electrochemical metallization memory* (*ECM*) effect [18]. This type of switching behavior was first demonstrated in 1976 in the lateral $Ag/As_2S_3/Mo$ structure, where the redox process of silver dendrites plays a key role in the formation and annihilation of the CF [34]. Later, Kozicki et al. [8] developed a vertical MIM system by using GeSe as the ion conductor and applied this in a nonvolatile memory device. In the case of the GeSe electrolyte system with silver electrodes, the following chemical reaction occurs at the anode and the cathode:

$$\text{At anode}: \quad Ag \rightarrow Ag^{+} + e^{-} \tag{7.2}$$

$$\text{At cathode}: \quad Ag^{+} + e^{-} \rightarrow Ag \tag{7.3}$$

Upon the application of an opposite bias voltage, Ag metal atoms start to dissolve at the edge of the metal filament, eventually annihilating the entire filament so that the cell is turned off. Typically, one filament dominates the growth and provides the contact for the ON state and the filamentary growth terminates at current compliance (I_{CC}). The diameter of the CF and strength of the electrical contact between the filament and active electrode are controlled by the I_{CC}. Generally, the V_{FORM} is significantly higher than the V_{SET} and increases linearly with the insulator layer thickness. However, the V_{SET} for all subsequent switching cycles is independent of the thickness [35]. This observation reveals that the remaining structural template of the dissolved filament after the first RESET serves as a fast transport and growth path for subsequent switching cycles. Both the ECM- and the VCM-type of resistive switching are generally bipolar in nature.

7.3 ReRAM Array Concepts and Device Fabrication

The simplest ReRAM device is composed of a metal–insulator–metal (MIM) stack and also exists in the form of a nano-crossbar array with and without a selection device. A number of selection devices are reported in the literature and can be classified into MOSFET transistor, diodes, and nonvolatile switches [36]. The choice of the selection device certainly impacts the ReRAM scaling limits. The nano-crossbar configuration is highly desirable as it offers cross-point architecture with an effective cell area of 4 F^2, where F is the minimum lithographic limit. Additionally, the ReRAM can be stacked in 3D integration in this configuration. Figure 7.4 shows an isolated MIM stack and a nano-crossbar architecture. However, the integrated nano-crossbar architecture leads to a parasitic current through non-selected memory devices. To avoid this undesirable effect, a selection device is therefore required and should be integrated with the memory elements.

Two-terminal selection devices, such as diodes and volatile switches, offer the scaling advantage as these devices could in principle be scaled down below 10 nm along with the memory element. A series connection of the diode at every cross-point, shown in Fig. 7.5a, allows all non-selected devices to be reverse biased. This can be arranged in a similar way to a conventional nano-crossbar memory array. However, a sufficient drive current with a large rectification factor and

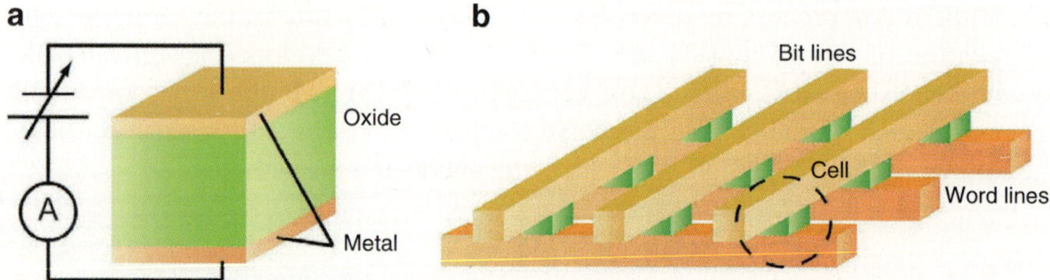

Fig. 7.4 Different ReRAM cell architectures (**a**) MIM isolated structure, (**b**) simple passive array nano-crossbar configuration [37]

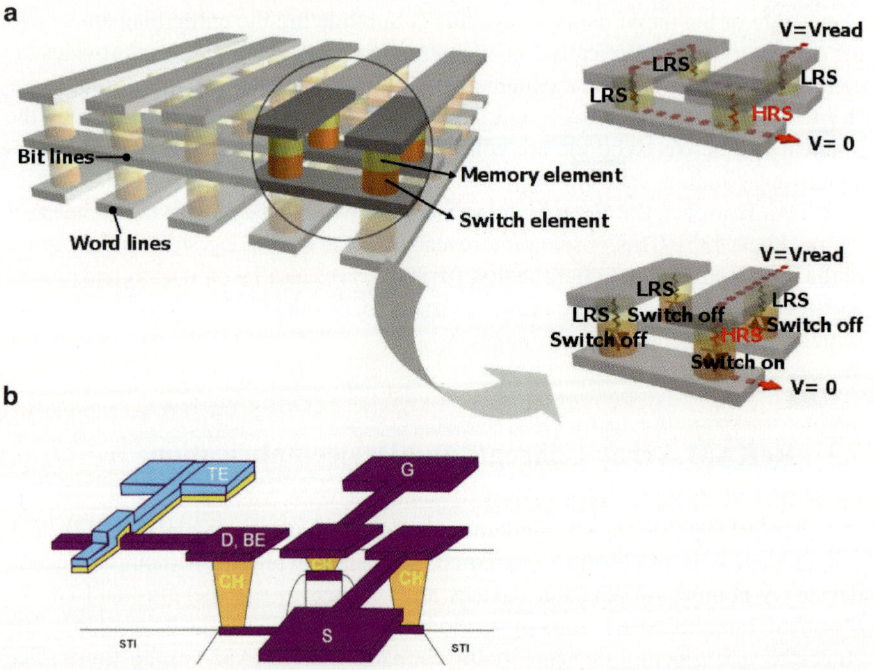

Fig. 7.5 (**a**) Nano-crossbar integration under 1D-1R configuration where one bit cell consists of a memory element and a switch element between bottom and top electrode [38]. (**b**) 1T-1R architecture of the ReRAM device [39]

poor current-control capabilities is still a major challenge in the development of these devices. Most diodes only work for the unipolar switching. For the bipolar switching, the MOS transistor (1T-1R) offers the best memory selection function in terms of on/off ratio, high drive current, and better current-controlling capability. The transistor can limit the switching current by gate voltage modulation. However, the larger footprint, high processing temperature, and additional processing steps are the main disadvantages. To avoid the high-temperature processing steps, the memory device may be integrated into the backend of the process line (BEOL) of an advanced CMOS process.

Another important requirement of the 1T-1R configuration is that the channel resistance of the selector transistor should be significantly lower than the memory element; otherwise, the effective memory window of the 1T-1R stack will be dominated by the channel resistance and memory operation will fail. In order to make a proper selection for the selector device, the on/off ratio, maximum drive current, scalability, speed, endurance, and manufacturability should be considered. For example, where the device size is less of a concern, the 1T-1R configuration is preferred. For 3D architectures, either 1D-1R or a self-selecting device could be the best candidate. Recently, a complementary resistive switch (CRS) concept [40] was proposed as a self-selecting structure where two resistive switching elements are connected anti-serially. The high resistance of the CRS cell helps to reduce the sneak-paths in the nano-crossbar array. The destructive read-out scheme is the main drawback of this concept. However, a capacitance-based nondestructive read-out scheme is under investigation [41]. An interesting two-terminal concept of the threshold type of resistive switching [42] is proposed and realized with Mott insulators such as VO_x, Nb_2O_5. The series connection of the Pt/VO_x/Pt resistor with Pt/NiO/Pt ReRAM showed a successful reduction of cross-talk at 10 ns programming speed [43]. Compared with the 1T-1R configuration, 1D-1R and 1S-1R configurations utilize a smaller area and are more suitable for 3D integration. However, there are certain requirements of the selector device such as high on/off current ratio, large forward current density, and low processing temperature [44, 45].

Depending on the selector device, the fabrication approach could vary. For example, when the diodes or the volatile switches are used as the selector device, the fabrication of the memory element is comprised of minimum of three steps; patterning the bottom electrode, deposition of switching and memory element, and patterning of the top electrode. In case of the MOS transistor as the selector device, the ReRAM devices are generally fabricated in the BEOL of the standard CMOS process. For the nano-crossbar architecture fabrication, a high throughput and low-cost approach is nanoimprinting lithography (NIL) [46]. This technology is very suitable for 3D integration. The basic types of the NIL are thermal nanoimprinting, which uses thermoplastic polymers as a resist, and UV nanoimprinting (UV-NIL), which uses ultraviolet light to crosslink a thin layer of liquid resist [47]. In most cases, the UV-NIL is the method of choice for the nano-crossbar ReRAM device fabrication because of its low operational pressure and temperature [48]. Figure 7.6 shows micrographs of the nano-crossbar architecture fabricated by the UV-NIL method [48].

Several research groups have demonstrated a successful integration of ReRAM devices at different technological node ranging from 180 to 32 nm. In the 1T-1R integrated configuration, the MOSFET transistor works as an ideal current limiter during the forming/SET process and avoids any possible overshoot in the ReRAM devices. Govoreanu et al. [21] successfully integrated a 10×10 nm^2 Hf/HfO_x ReRAM device in the BOEL of the 65 nm CMOS process. After the completion of the FOEL processing steps, a TiN bottom electrode is patterned in the BEOL of the CMOS process. The next steps were planarization of the bottom electrode, deposition of the resistive switching stack (Hf/HfO_x), and deposition and patterning of the TiN top electrode. All backend processing steps are performed at a lower

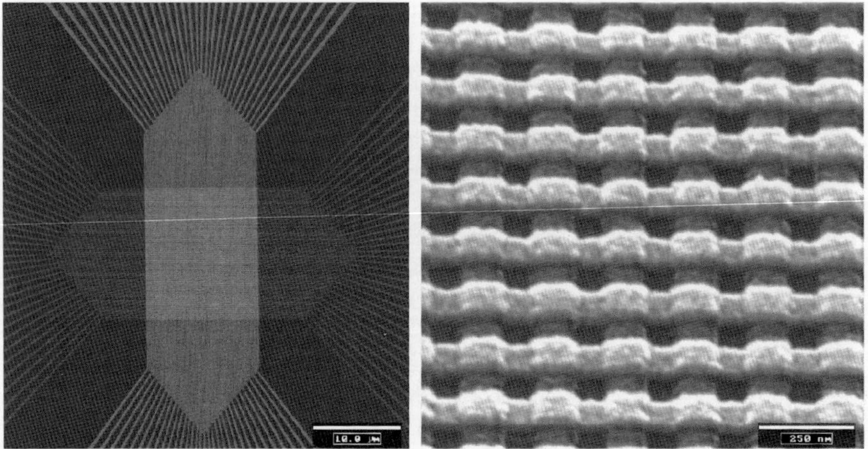

Fig. 7.6 Micrograph of the Pt/TiO_2/Ti/Pt nano-crossbar architecture [48]

thermal budget (∼400 °C) so that the MOSFET devices are not damaged. Lee et al. [49] also demonstrated a fully integrated HfO_x-based ReRAM with 0.18 μm CMOS technology in 1R and 1T-1R configuration. The integrated TiN/HfO_x/Ti/TiN ReRAM device showed excellent performance in terms of low operating current (<25 μA), high on/off resistance ratio (>10^3), fast switching speed (5 ns), endurance cycle (>10^6 cycles), and data retention for 10 years. The integration of ZrO_2, TiO_x, Ta_2O_{5-x}, AlO_x and WO_x in the 1T-1R configuration was successfully demonstrated in references [50, 51].

At the sublithographic scale, a different fabrication approach is adopted. One-dimensional (1D) ReRAM nanostructures can be synthesized based on bottom-up technology [52] by using various techniques, including vapor–liquid–solid (VLS) growth, chemical-vapor-deposition and atomic-layer-deposition. A variety of materials have been synthesized in the form of 1D nanostructure and demonstrated as the nonvolatile memory device. The resistive switching behavior has been demonstrated in NiO nanowires grown by using the VLS [53] and anodized aluminum oxide (AAO) [54] approach. With a similar approach, other metal oxide nanowires such as ZnO and TiO_2 are grown and used for the resistive switching applications [55, 56]. A forming process is generally required to realize a memory operation in the ReRAM devices in any configuration. This process will be described in detail in next section.

7.4 Forming Process

Forming process refers to the first CF formation in a pristine film and is to be equivalent to the nondestructive breakdown of the dielectric film. When a sufficiently large voltage is applied to the MIM system, a significant increase of

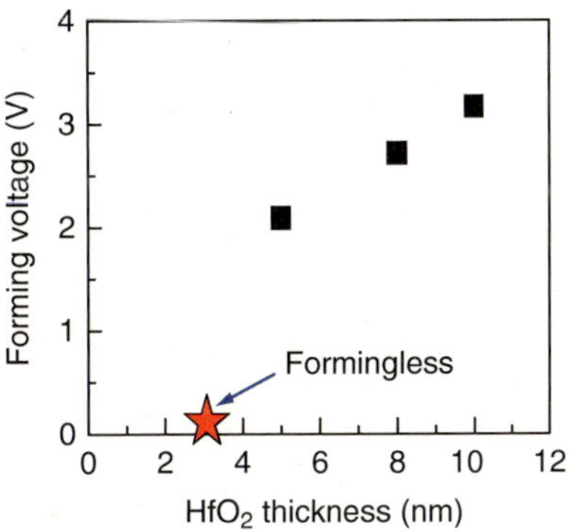

Fig. 7.7 Forming voltage dependence on the HfO_2 thickness in the TiN/HfO_x/Ti/TiN ReRAM [49]

the oxygen vacancy concentration near the cathode and a strong depletion of the oxygen vacancy region are generated. With the application of a higher voltage or longer time, this conducting cathodic region, so-called virtual electrode, propagates towards the anode. The O_2 ions will either oxidize the anode metal or discharge at the anode in the form of molecular oxygen (observable as gas bubbles). At the current compliance, the virtual cathode makes an electronic contact with the anode and the electroforming process is completed. The nature of the forming process can be explained as an electro-reduction and oxygen vacancy transport process towards the cathode caused by high electric fields and Joule heating [27]. This process is highly nonuniform and leads to morphological changes in dielectric films. As a rule of thumb, the forming voltage is higher than the V_{SET} and the V_{RESET} and causes extra time and power consumption in memory operation. The forming voltage (V_{FORM}), as shown in Fig. 7.7, is approximately linearly dependent on the dielectric thickness and, as a consequence, the as-fabricated device can even be made without the forming process as the thickness of HfO_2 film is thinned further to 3 nm [49].

In order to have the reproducible switching process, different forming approaches were studied by Nauenheim et al. [57]. The forming processes shown in Fig. 7.8 were performed on the Pt/TiO_2/Ti/Pt nano-crossbar memory device with the application of a sweeping voltage to the Ti/Pt electrode having either a positive/negative voltage or a current polarity with respect to the other electrode (Pt). The negative current controlled forming method is found to be most reliable for the switching process as the memory device is formed into the higher resistance/off state and avoids a complete breakdown of the device.

Generally, a high current overshoot phenomenon during the voltage-controlled forming process is observed and may lead to either permanent breakdown or high switching current. To overcome this shortcoming, an ideal current limiter such as a MOSFET transistor is integrated with the ReRAM device. In this configuration, the parasitic capacitance is minimized and the MOSFET transistor serves as a current limiter during the forming and set process. Kinoshita et al. [58] first demonstrated the effect of overshoot phenomena on a Pt/NiO/Pt memory device by integrating

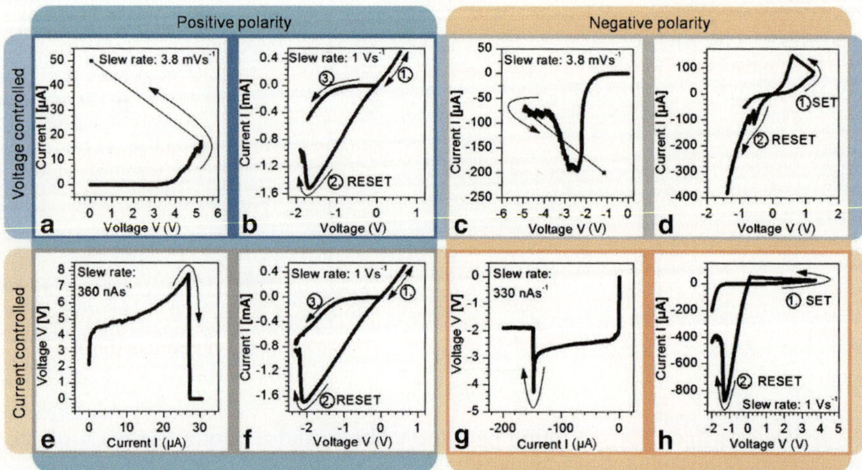

Fig. 7.8 Electroforming procedures with (**a**) positive voltage sweep and (**b**) subsequent operation, starting with a reset. (**c**) Forming with a negative voltage sweep resulting in the (**d**) HRS and a set step. (**e**) Positive current sweep into the (**f**) LRS and a reset operation, and (**g**) negative current sweep with the (**h**) subsequent set sweep [57]

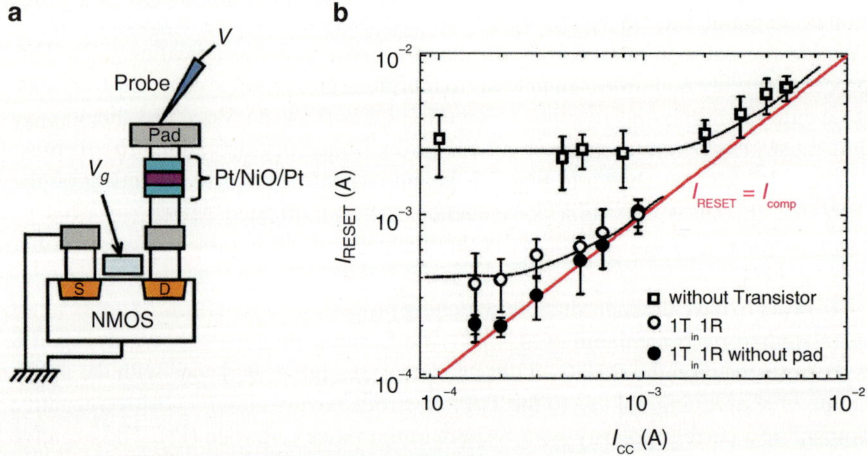

Fig. 7.9 I_{RESET} in 1T-1R configuration shows the function of the I_{SET} compliance (I_{CC}). This is due to avoidance of overshoot phenomena and minimizing the parasitic capacitance in NiO ReRAM [58]

the memory device in the 1T-1R configuration. In this configuration, the I_{CC} is a key parameter to define the switching characteristics and the I_{RESET} reduces as a function of the set I_{CC} as shown in Fig. 7.9. The relationship, $I_{RESET} = I_{CC}$ in the binary oxides is only valid when the 1T-1R configuration does not impose any parasitic capacitance on the memory cell.

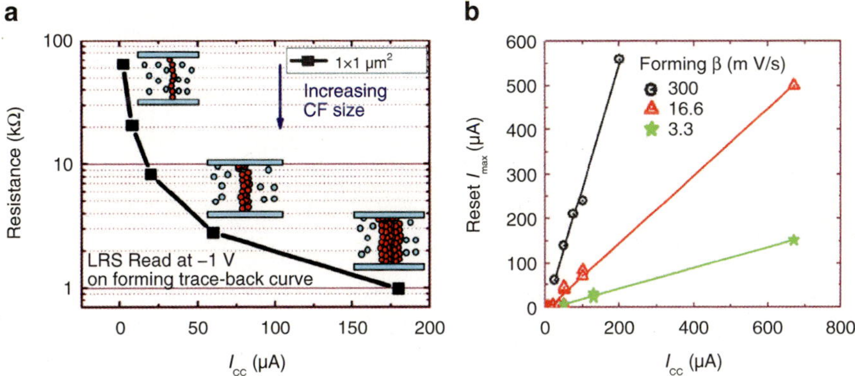

Fig. 7.10 (**a**) Cell resistance after forming as a function of I_{CC}. A decrease in resistance is postulated in terms of increase of CF diameter. (**b**) Dependence of RESET I_{max} on the forming current I_{CC} [59]

Generally, the maximum I_{RESET} during the first cycle of switching is assumed to be the figure of merit of the ReRAM memory device. It is directly related to the diameter or numbers of the CFs. As the I_{SET} compliance decreases, the diameter or numbers of CFs decrease resulting in a lower maximum I_{RESET} during the switching process [59], shown in Fig. 7.10.

From this result, it can be concluded that in order to achieve the minimum size of the CF, the compliance current should be reduced during the forming process. Recently, a sub-nA range of the I_{RESET} has been achieved in Al_2O_3 and HfO_2/Al_2O_3 memory devices [60, 61, 62]. After the forming process, the device can be successfully switched between the LRS and the HRS. The switching characteristics and related processes will be discussed in the next section.

7.5 Switching Characteristics

7.5.1 RESET Characteristics

In order to demonstrate the switching ability of the ReRAM device, the RESET process is generally carried out after the forming process. During the RESET process, the generated filament is ruptured and the oxide returns to the HRS state. The RESET kinetic in the unipolar switching device is described by the thermal dissolution model [63]. According to this model, local heating and current crowding effects are responsible for the rupture of the CF and the I_{RESET} is closely related to the ON resistance of the device. For the bipolar switching, the RESET mechanism is generally explained by the redox-reaction model [27], as described in the previous section. A local electrochemical redox process near the metal-oxide interface is

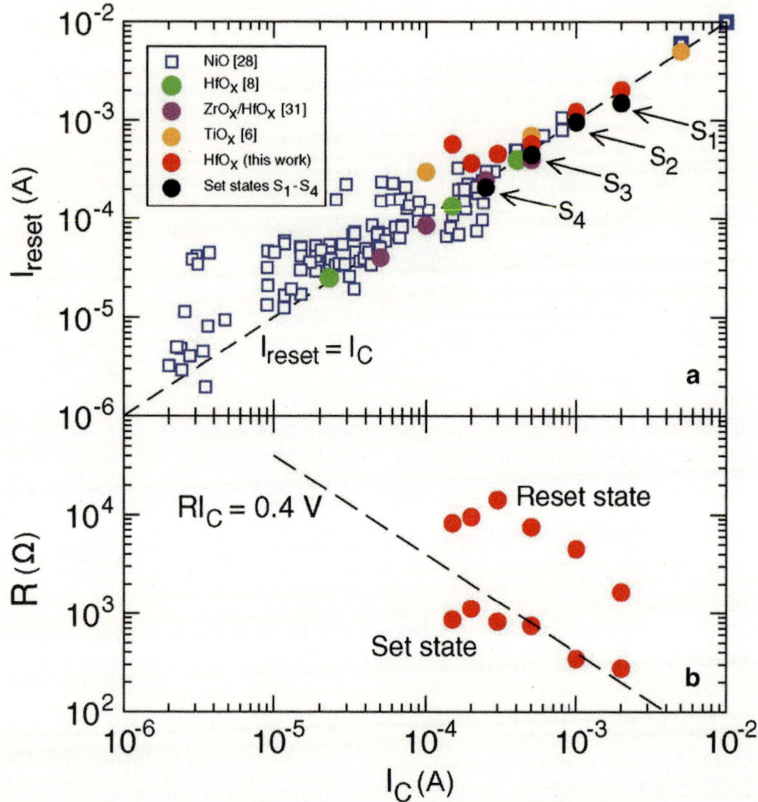

Fig. 7.11 Measured I_{RESET} for unipolar and bipolar switching oxides. The figure is redrawn from reference [65]

presumably responsible for the rupture of the CF during the RESET process. This creates a significant high-energy barrier and turns the device into the OFF state. For homogeneous systems, this corresponds to complete depletion of the oxygen vacancies, whereas for heterogeneous systems, there are either no charge carriers in the disk region or a certain density of negative ions (O^{2-}), which further increases the barrier height [6]. Indeed, other reports have assumed the RESET process to be the phase transition. The RESET operation in the TiO_2 might be achieved by locally melting the Magneli CFs with sufficient electric current (I_{RESET}) and subsequent cooling down [64].

The I_{RESET} in the various metal oxides increases linearly with the I_{CC}, shown in Fig. 7.11 and affects the resistance of the HRS in a similar way to the resistance of the LRS. The resistance of both states decreases at higher I_{CC}, whereas the resistance ratio between two levels remains almost unchanged. This can be understood from the fact that the I_{CC} controls the cross-sectional area of the CF. A higher I_{CC}, i.e. lower resistance of LRS, leads to a larger CF diameter, which results in the higher

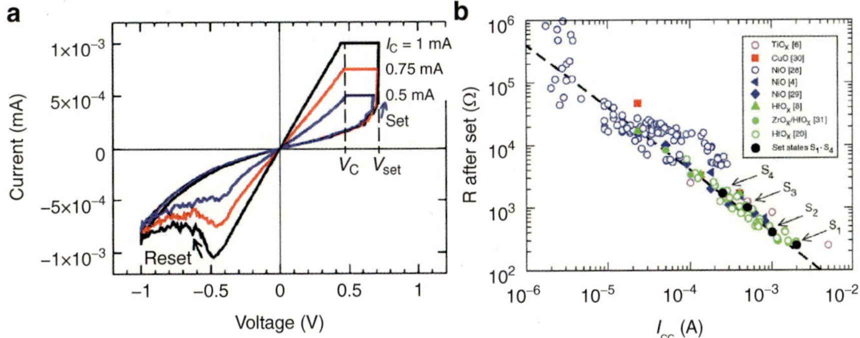

Fig. 7.12 (**a**) Measured I–V characteristics for a bipolar ReRAM with the TiN/HfO_x/TiN structure. (**b**) Set resistance is lowered with the I_{CC} for different binary oxides ReRAM [66]

I_{RESET} [65, 64]. The HRS resistance is mainly controlled by the disk region and the disk resistance is a function of the CF cross-section since the remaining tips of the CF serve as the virtual electrode. However, key questions such as the rupture location of the CF and influence of the CF shape on the RESET process are still under debate.

7.5.2 SET Characteristics

The SET process is closer to the forming process, except that the breakdown only occurs in the recovered region or in the gap between the broken CF and the metal electrode. During the SET process, the oxygen ions are removed from the disk region. For homogeneous systems, this is identical to the injections of the oxygen vacancies and is (at least partially) compensated by electrons. In the case of the heterogeneous systems, the overall process is the same while the vacancy picture might be more complicated and depends on the type of oxide. As for many other phase transitions, a local lattice arrangement might take place. In any case, the extraction of the oxygen ions will lead to a (chemical) reduction of the disk region, which in turn leads to a decrease in barrier height.

In the ON state, the cell resistance is determined by a series combination of the virtual electrode and disk region resistance. The ON-state resistance is a function of the I_{SET} compliance (I_{CC}) due to the fact of a lateral increase in the CF cross-sectional area at higher I_{CC} [65]. The lateral growth stage can be controlled by observing the dependence of the final set state R on the I_{CC}. Figure 7.12a shows the bipolar resistive switching in the TiN/HfO_x/TiN ReRAM device with different I_{CC}. A larger diameter of the CF is expected for the $I_{CC} = 1$ mA than that of 0.5 mA. However, the product of the resistance after the SET process and the corresponding I_{CC} remains constant $V_C = RI_{CC} = 0.4$ V. This characteristic voltage, V_C, represents the value of the voltage across the ReRAM device at the end of the SET transition, corresponding to the corner in the I–V characteristics of the ON state between the ohmic current increase and apparent constant current region where $I = I_{CC}$. This characteristic voltage is constant for other metal oxides, shown in Fig. 7.12b. This voltage controls both the electric field and the local temperature and is one of the controlling parameters of the CF growth [66]. Although there is a consensus among researchers that the resistive switching in the metal oxides is related to the creation and rupture of the CF, the recreation/regrowth of the filament after each switching cycle is still controversial and under debate.

7.6 Retention and Speed

Retention refers to the ability of a memory bit to retain its data state over long periods of time regardless of the power supply. The retention time is estimated based on accelerated stress methods and activation energy models [67, 68]. The speed of a ReRAM device is determined by its ability to toggle the LRS to the HRS or vice versa. Generally, a memory device with long data retention ($\sim$10 years) and high speed ($\sim$10 ns) is required for the applications. However, the combination of these requirements in the ReRAM devices leads to a voltage-time dilemma [69], which suggests a physical mechanism of the switching kinetics that needs to be extremely nonlinear. Based on the Nernst–Einstein model and oxygen vacancy diffusion data, the oxygen vacancies can hardly migrate for a distance of about 1 nm under normal conditions during the switching [15]. Therefore, the solution to the voltage-time dilemma must be caused by other factors.

Waser et al. [27] described this phenomenon in terms of the thermal activation and field enhancement of the oxygen vacancy transport. The temperature-accelerated drift of the oxygen vacancies leads to an exponential acceleration ($>10^{10}$) of the switching speed and is interpreted as the exponential dependence of the switching kinetics on the switching voltage. Furthermore, a lattice strain caused by a high oxygen vacancy concentration may increase the diffusion coefficient of the oxygen vacancies. Similar behavior has been observed by other researchers [70, 71]. Wei et al. [72] explained the voltage-time dilemma in a non-stoichiometric TaO_x-based memory device by the formation of another phase. The phase change from Ta_2O_5 to TaO_2 in the Pt/TaO_x/Pt memory device leads to a smaller band gap and a significantly higher conductivity than the fully oxidized Ta_2O_5 phase. The change in stoichiometry from an oxidized Ta_2O_5 phase (HRS) to the reduced TaO_2 phase (LRS) is explained by the redox reaction. In order to elucidate the origin of the solution to the voltage dilemma in the ReRAM device, the formation of a nanoscale phase and nucleation rate in different metal oxides needs to be experimentally investigated for the VCM systems. Recently, Hermes [70] experimentally demonstrated that the SET/RESET time in the Pt/TiO_2/Ti/Pt nano-crossbar memory device, shown in Fig. 7.13, exponentially decreases when the voltage amplitude is linearly increased. This provides experimental evidence of the solution of the voltage-time dilemma in metal-oxide-based ReRAM devices.

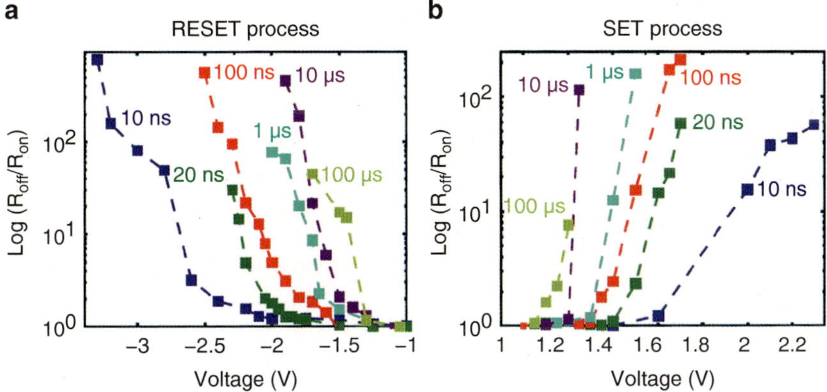

Fig. 7.13 (**a**) RESET, (**b**) SET event is a function of applied voltage for the Pt/TiO_2/Ti/Pt nano-crossbar memory device

The retention characteristics in the binary oxide ReRAM are explained by the oxygen diffusion model [73], and the HRS and the LRS characteristics depend on the CF size and the oxygen diffusion [74]. The data retention for 10 years at 85 °C has been reported on high-density Ir/$Ta_2O_{5-\delta}$/TaO_x/TaN ReRAM devices. The density of the oxygen vacancies in the CF is presumed to play a key role in ensuring data retention [75]. The retention property of the memory cell is considered to be a function of the I_{SET} compliance (I_{CC}) and degrades in the lower I_{SET} regime. This degradation is acknowledged to be caused by the weakening of the CFs. In the low-current regime, longer retention can only be achieved by scaling down the CF and increasing the density of the oxygen vacancies during the forming process [76].

As another important aspect of the memory; speed is defined as the ability to write/read the information. Mainstream nonvolatile memories, such as flash memory, exhibit write/read times in the order of tens of microseconds. This low speed hinders their use in high-performance applications. The metal-oxide-based ReRAM devices show a switching speed ranging from nanoseconds down to few hundred picoseconds. For example, a switching speed of 5 ns has been successfully achieved in the Nb-doped single-crystal $SrTiO_3$ [77] and the Pt/TiO_2/Ti/Pt nano-crossbar structure [78]. The ultra-fast speed of 300 ps, maintaining a sufficient memory window for 10^{10} switching cycles in the HfO_x-based resistive device, shown in Fig. 7.14 has been reported [79].

Recently, resistive switching in a tantalum oxide coplanar waveguide (CPW) structure was successfully achieved using the SET pulse of 105 ps and the RESET pulse of 120 ps [81]. The limiting factor for the switching speed in the binary metal oxides is to be the mobility of the oxygen vacancies. However, a significant temperature increase within the entire CF during the switching process overcomes this limit. A study shows that both the temperature and the electric field play an important role in the switching kinetics [80]. The published literature and ongoing research activities show that the binary metal-oxide ReRAMs are very suitable for high-speed memory applications.

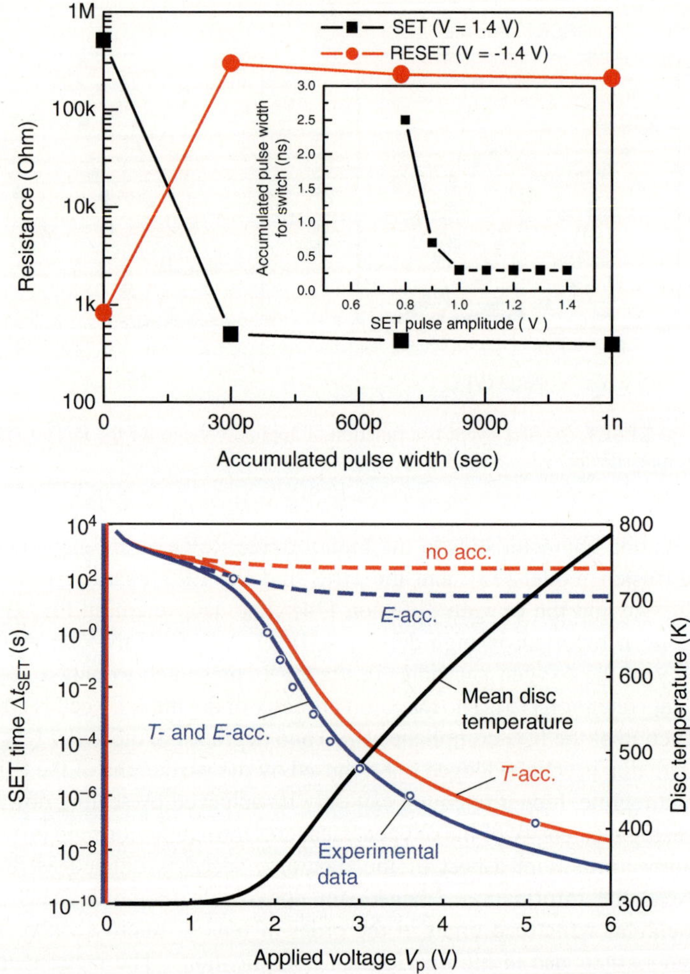

Fig. 7.14 (**a**) Toggling of the resistance states as a function of pulse width at constant amplitude. The device shows switching speed of 300 ps [79]. (**b**) SET time decreases exponentially with the applied voltage and filament temperature [80]

7.7 Reliability and Endurance

Reliability and endurance determine the number of switching cycles with a predefined resistance window. Both parameters depend on a number of factors such as device structure, material processing, and read/write scheme. However, mainly three types of failure behavior are observed in the ReRAM devices [82]; one is due to the loss of the R_{HRS}/R_{LRS} ratio accompanying the decreased R_{HRS} and increased R_{LRS}, the second is the sudden loss of the RESET capability, and the last is a gradual decrease of the R_{HRS} value with the switching cycle and approaches the low

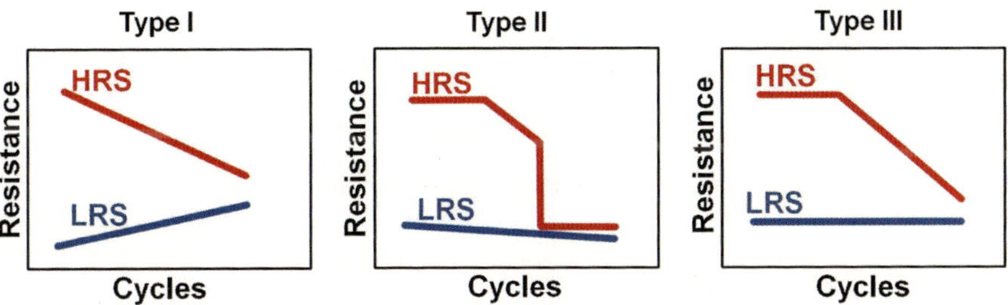

Fig. 7.15 Schematic views of endurance degradation mechanism. The scale on the Y axis is assumed to be logarithmic. Type I: interface oxide formation during the forming/SET process is mainly responsible for this degradation. Type II: electric-field-induced generation of extra oxygen vacancies leads to degradation of the HRS state. Type III: reduction in the recombination rate of the oxygen vacancies in the electrode layer leads to a reduced HRS state [82]

resistance state. These phenomena are shown in Fig. 7.15. Possible explanations for the above-mentioned phenomena are anode interface oxidation and extra vacancy generation and depletion of O^{2-} ions [82]. An additional impact could arise from the fact that all bipolar ReRAM act as batteries, i.e. they display a cell voltage in the OFF state due to an inherent emf voltage generation in the stack structure [83]. Doping in oxides, different device structures and novel read/write schemes have been proposed to improve the endurance of the memory cells.

Lee et al. [84] proposed the chemical mechanical polishing method to smoothen the bottom electrode and improve the endurance of HfO_x-based memory devices over 10^{10} cycles using 40 ns write/erase pulses in comparison with the method without a polished bottom electrode. Further, the reliability of the ReRAM device is improved by the forming and the switching in the long current regime. In this way, the diameter of the CF is scaled down while keeping a sufficient high density of oxygen vacancies in the filament. Ninomya et al. [75] demonstrated the scaling of the CF in TaO_x bipolar ReRAMs and achieved a long retention time of 1,000 h at 150 °C with a low SET current (~80 μA).

The reliability of the ReRAM devices is generally affected by the SET/RESET pulse amplitude/width. Unbalanced SET/RESET amplitude could lead to premature breakdown. For example, too strong SET pulses lead to an excess amount of the oxygen vacancies induced at the switching interface so that the RESET pulse is unable to rupture the CF. On the other hand, too strong RESET pulses result in the depletion of the oxygen vacancies at the interface so that the SET pulse is unable to reconstruct the CF. Researchers at IMEC [85] studied the failure mechanism in the Hf/HfO_2 1T-1R devices and observed that the unbalanced SET/RESET pulse amplitude/width influences the stability of the HRS/LRS states and causes premature device failure. By optimizing pulse conditions, the endurance of 40 nm Hf/HfO_2 1T-1R devices, shown in Fig. 7.16, can be extended to 10^{10} cycles in comparison with 10^6 cycles without optimized pulse conditions. In order to improve the endurance and the reliability of ReRAM devices, further optimization of the forming/read/write scheme, switching current, cell design, and material selection is required.

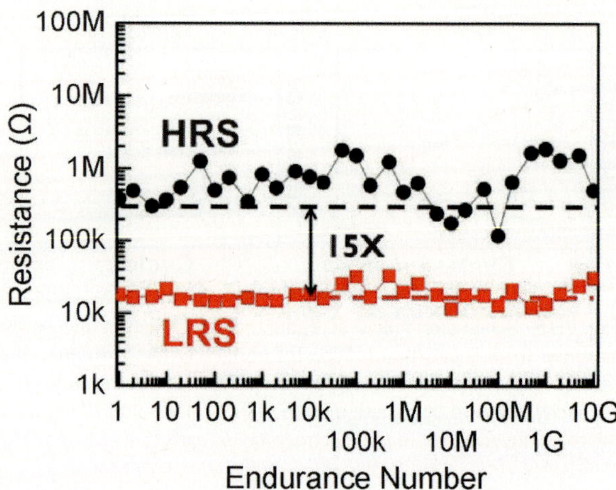

Fig. 7.16 Optimizing the SET and the RESET pulse, a higher endurance in 40-nm Hf/HfO_2 1T-1R devices is achieved [85]

7.8 Multiple-Bit Operation

Multiple level cell (MLC) operation exploits the phenomenon of storing more than one digital data per cell. For high-density memory applications, it is desirable to store more and more information in one cell. This phenomenon is considered to be due to the fact that the I_{SET} compliance modulates the diameter or number of the CFs. The main requirements for the MLC operation are a sufficiently large resistance window between the HRS and the LRS, high endurance of each state, and thermal stability of the stored data. Figure 7.17 shows the MLC operation in 40 nm W/WO_x/TiN ReRAM for more than 10 K cycle endurance with a verified programming algorithm [86]. In the programming algorithm, an incremental pulse amplitude is applied until the resistance state toggles to the target value. To write/erase the information, the first pulse is chosen to raise the cell resistance close to the desired value, and then the smaller incremental voltages are used to finish the MLC programming. Other oxide-based ReRAM materials such as CuO_x [87], HfO_x [49], ZrO_2 [50], TiO_x [88], Ta_2O_{5-x} [89], and AlO_x [13] have also shown the capability of the MLC operation. Yu et al. [90] utilized a field-driven oxygen ion migration model to realize the multiple-bit operation in the ReRAMs and proposed two pulse programming schemes. One depends on the linear increment in pulse amplitude and the other relies on an exponential increment in programming pulse width. Although both schemes are able to achieve the target resistance values the first scheme consumes considerable low energy during the operation.

As a variant, a constant signal pulse programming (CSPP) scheme is also proposed and is verified on a triple layer (base layer/oxygen exchange layer/barrier layer) TaO_x-based ReRAM [91]. In this algorithm, the constant reset pulses are repeatedly applied until the resistance level reaches the target range. With the CSPP algorithm, the switching endurance of the TaO_x ReRAM in 2 bit/cell operation

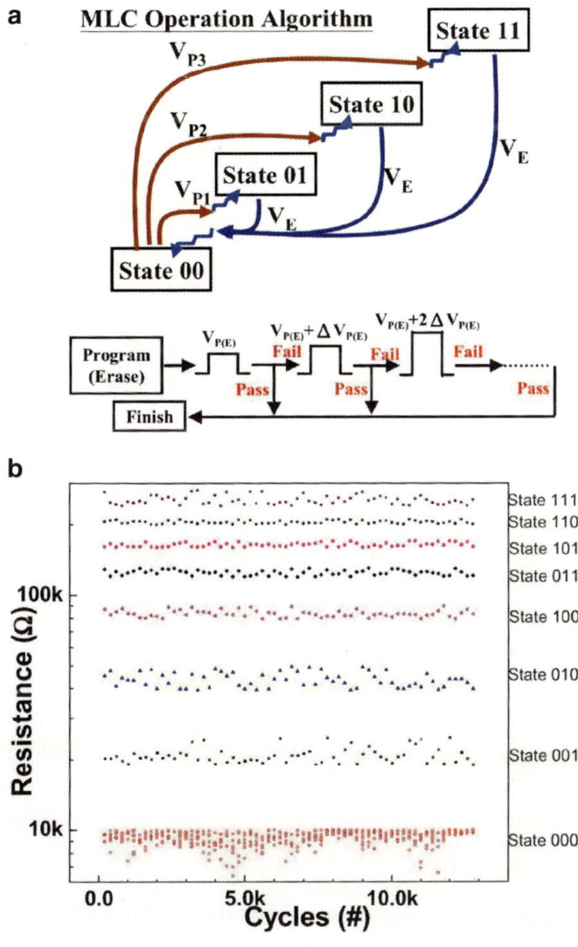

Fig. 7.17 (**a**) MLC operation scheme and verification algorithm. The MLC program starts from state 00 (lowest R). Different first RESET pulses are chosen depending on the target state. Then small increments in pulse voltage are applied until the cell resistance reaches the target value. (**b**) Multilevel bit operation in a $W/WO_x/TiN$ ReRAM. A 3 bit per cell operation is achieved by using a verified algorithm [86]

was extended to 10^7 cycles. In order to enhance further reliability and variability of multi-resistance states in the ReRAM devices, a better forming/set/reset/reading scheme is required.

7.9 Scalability of ReRAM Technology

Much of the interest in the ReRAM technology is due to its tremendous scaling potential down to the atomic regime. The smallest possible structure of the binary oxide that can potentially demonstrate the bipolar switching is a chain of ions in the crystal, changing their valence state. The occurrence of the redox process in such a crystal system seems to be the fundamental scaling limit of these devices.

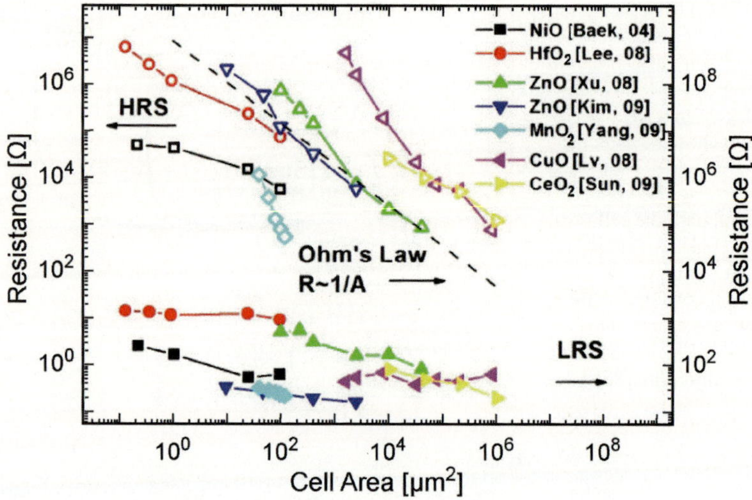

Fig. 7.18 Trends in HRS/LRS states with cell area in metal-oxide ReRAM. Data were taken from [68]

Local conductivity AFM studies on single-crystal $SrTiO_3$ shows the confinement of the bipolar switching within the 1–2 nm area [15]. For a more practical and realistic approach, we need to fulfill the requirement of distinguishable R_{OFF}/R_{ON} in order to implement the switching event and to charge/discharge the bit line capacity along with the switching operation. As a first order calculation in the TiO_2-based resistive system, a donor concentration of 3.10^{21} cm^{-3} is required for the $R_{OFF}/R_{ON} = 10$ [92]. This leads to the CF diameter of 4 nm.

For scaling and integration with the CMOS devices, conventional patterning methods such as optical lithography [93], nanoimprint lithography [22, 94], or e-beam lithography [95] are being used to fabricate the sub-nanometer ReRAM device. Recently, the HfO_x ReRAM [21] has been aggressively scaled down to 10 nm × 10 nm by the conventional photolithography process. Further, device size miniaturization has also been realized by a bottom-up approach [54]. Primarily, the resistance states (HRS and LRS) are affected by the ReRAM device scaling. The HRS resistance is often determined by the resistance of the residual filament and of the rest of the area, whereas the LRS typically shows no or only a weak dependence on the cell area since the conductance of the CF predominantly determines the cell resistance. The R_{OFF}/R_{ON} therefore tends to improve with the smaller cell area. The scaling trends of the HRS and the LRS with cell area from various metal-oxide memories are plotted in Fig. 7.18 [68].

Another important device parameter is the I_{RESET} that defines maximum power consumption during memory operation. The I_{RESET} does not show any dependence on the device size. However, it can be scaled down by using a smaller I_{SET} compliance (I_{CC}) during the set process. Ielmini et al. [96] demonstrated the scaling of the I_{RESET} with the I_{SET} compliance for various unipolar and bipolar ReRAM devices. In short, it can be concluded that the smaller device size scales down the I_{SET} leading to a decrease in the I_{RESET} while keeping the HRS/LRS window large. Therefore, the I_{RESET} can scale down with the device size. The ultimate limiting factor for the scaling of the ReRAM technology will be the tunneling distance between neighboring cells as well as the leakage current from the word and the bit lines [92]. Technologically, the scaling of the ReRAM devices will, however, be determined by the fabrication of efficient and reliable electrode contacts and interconnects within the memory matrix. Another obvious limit to scaling of the chip size will be the size of the periphery circuit and, for active matrices, the size of the access transistors within the memory matrix.

7.10 Future Outlook

In the past decade, the binary metal-oxide ReRAMs have emerged as high-density and ultra-fast nonvolatile memories which are compatible with conventional CMOS technology. The simplest way to integrate the ReRAM cells is the passive nano-crossbar matrix, which just connects the word and the bit lines at each node. However, this configuration has inherent problems of current sneak-paths and requires a selector device. Current research trends indicate that the bipolar switching devices (ECM- and VCM-type ReRAM) show better performance and lower variability than the unipolar switching devices (TCM-type ReRAM). This makes the search for a suitable cell selection device even more difficult as the selection device has to conduct current in both directions and the conventional reverse-bias blocking effect cannot be used. For this purpose, the MOSFET device is most suitable as a selector. However, this increases both the memory device area and the total processing cost.

Another major challenge is device uniformity. Device variation is a major barrier to using the ReRAMs in large memory arrays. To overcome the cycle-to-cycle and device-to-device variation of the device characteristics, circuit designers should innovate and implement new ideas. To make progress in this area, it is necessary to have a more complete understanding of the conduction and the resistive switching mechanism. Ultimately, the solution may come from a combination of materials engineering, device structure optimization, and innovations in addressing/read-out circuitry. Meanwhile, further understanding of the underlying physics of the ReRAM device has been obtained through progress in modeling from the atomistic level to the device level [25], although much work is still needed on this front. Compared with several other emerging memory concepts, the binary oxide ReRAM technology has the potential to be a universal memory, i.e. low power consumption, high integration density, and long retention. In addition, the redox-based resistive switching cells have also shown the potential to serve as logic devices. For example, the oxide resistive switches are suggested as the core elements in a CMOS-nano hybrid reconfigurable field-programmable gate array (FPGA) architecture [97]. Another huge emerging application field is hardware-based neuromorphic computing, where the redox-based resistive switches can serve as artificial synapse elements. Owing to their multilevel capability, they can be used like an analog memory emulating the function of plastic synapses in a neural network. Recently, TiO_x- [98], WO_x- [99], and HfO_x-based [100] synapses have been experimentally demonstrated for spike-timing-dependent plasticity [101]. Additionally, a low-temperature oxide-based ReRAM technology has been realized on flexible substrates [102].

In order to promote the practical applications of the ReRAM technology, a deeper understanding must be achieved of the microscopic mechanism of the switching, process and material optimization, the effects limiting the reliability and the retention time, the yield improvement, all aspects of the fabrication technology, and guidelines for the scaling. These unresolved questions currently represent an exciting hotspot for research in the fields of physics, electronics, and material science.

Acknowledgments The authors would like to thank Dr. Susanne Hoffmann-Eifert (FZ Jülich) for valuable feedbacks during the writing and Thomas Pössinger (RWTH Aachen) for editing and improving the figures in this chapter.

References

1. S. Lai, Flash memories: successes and challenges. IBM J. Res. Dev. **52**, 529–535 (2008)
2. G. Burr, B. Kurdi, J. Scott, C. Lam, K. Gopalakrishnan, R. Shenoy, Overview of candidate device technologies for storage-class memory. IBM J. Res. Dev. **52**, 449–464 (2008)
3. S.A. Wolf, J. Lu, M.R. Stan, E. Chen, D.M. Treger, The promise of nanomagnetics and spintronics for future logic and universal memory. Proc. IEEE **98**, 2155–2168 (2010)
4. A. Sheikholeslami, P.G. Gulak, A survey of circuit innovations in ferroelectric random-access memories. Proc. IEEE **88**, 667–689 (2000)
5. W. Zhao, S. Chaudhuri, C. Accoto, J.-O. Klein, D. Ravelosona, C. Chappert, P. Mazoyer, High density spin-transfer torque (STT)-MRAM based on cross-point architecture, in *2012 4th IEEE International Memory Workshop (IMW)* (2012), p. 4
6. R. Waser (ed.), *Nanoelectronics and Information Technology*, 3rd edn. (Wiley-VCH, Berlin, 2012)
7. A. Asamitsu, Y. Tomioka, H. Kuwahara, Y. Tokura, Current switching of resistive states in magnetoresistive manganites. Nature **388**, 50–52 (1997)
8. M.N. Kozicki, M. Yun, L. Hilt, A. Singh, Applications of programmable resistance changes in metal-doped chalcogenide. J. Electrochem. Soc. **99–13**, 298–309 (1999)
9. A. Beck, J.G. Bednorz, C. Gerber, C. Rossel, D. Widmer, Reproducible switching effect in thin oxide films for memory applications. Appl. Phys. Lett. **77**, 139–141 (2000)
10. B.J. Choi, D.S. Jeong, S.K. Kim, C. Rohde, S. Choi, J.H. Oh, H.J. Kim, C.S. Hwang, K. Szot, R. Waser, B. Reichenberg, S. Tiedke, Resistive switching mechanism of TiO_2 thin films grown by atomic-layer deposition. J. Appl. Phys. **98**, 33715-1–33715-10 (2005)
11. Y.M. Kim, J.S. Lee, Reproducible resistance switching characteristics of hafnium oxide-based nonvolatile memory devices. J. Appl. Phys. **104**, 114115 (2008)
12. I.G. Baek, M.S. Lee, S. Seo, M.J. Lee, D.H. Seo, D.-S. Suh, J.C. Park, S.O. Park, H.S. Kim, I.K. Yoo, U.-I. Chung, I.T. Moon, Electron Devices Meeting, Technical Digest. IEEE International, pp. 587–590, 13–15 dec. (2004)

13. Y. Wu, S. Yu, B. Lee, P. Wong, Low-power TiN/Al(2)O(3)/Pt resistive switching device with sub-20μA switching current and gradual resistance modulation. J. Appl. Phys. **110**, 94104/1–94104/5 (2011)
14. L. Chen, Q.Q. Sun, J.J. Gu, Y. Xu, S.J. Ding, D.W. Zhang, Bipolar resistive switching characteristics of atomic layer deposited Nb2O5 thin films for nonvolatile memory application. Curr. Appl. Phys. **11**, 849–852 (2011)
15. K. Szot, W. Speier, G. Bihlmayer, R. Waser, Switching the electrical resistance of individual dislocations in single-crystalline $SrTiO_3$. Nat. Mater. **5**, 312–320 (2006)
16. S.Q. Liu, N.J. Wu, A. Ignatiev, Electric-pulse-induced reversible resistance change effect in magnetoresistive films. Appl. Phys. Lett. **76**, 2749–2751 (2000)
17. D. Morgan, M. Howes, Electroforming and switching in copper oxide films. Phys. Status Solidi (A) Appl. Res. **21**, 191–195 (1974)
18. R. Waser, M. Aono, Nanoionics-based resistive switching memories. Nat. Mater. **6**, 833–840 (2007)
19. M.N. Kozicki, C. Gopalan, M. Balakrishnan, M. Park, M. Mitkova, Nonvolatile memory based on solid electrolytes, in *2004 Non-Volatile Memory Technology Symposium, Proceedings* (2004), pp. 10–17
20. S.-E. Ahn, M.-J. Lee, Y. Park, B.S. Kang, C.B. Lee, K.H. Kim, Write current reduction in transition metal oxide based resistance-change memory. Adv. Mater. **20**, 924 (2008)
21. B. Govoreanu, G.S. Kar, Y.-Y. Chen, V. Paraschiv, S. Kubicek, A. Fantini, I.P. Radu, L. Goux, S. Clima, R. Degraeve, N. Jossart, O. Richard, T. Vandeweyer, K. Seo, P. Hendrickx, G. Pourtois, H. Bender, L. Altimime, D.J. Wouters, J.A. Kittl, M. Jurczak, Electron Devices Meeting, Technical Digest. IEEE International, pp.31.6.1–31.6.4, 5–7 dec. (2011)
22. J.J. Yang, M.D. Pickett, X. Li, D.A.A. Ohlberg, D.R. Stewart, R.S. Williams, Memristive switching mechanism for metal/oxide/metal nanodevices. Nat. Nanotechnol. **3**, 429 (2008)
23. S. Yu, X. Guan, H.P. Wong, Conduction mechanism of TiN/HfOx/Pt resistive switching memory: a trap-assisted-tunneling model. Appl. Phys. Lett. **99**, 063507–063507 (2011)
24. I. Valov, R. Waser, J.R. Jameson, M.N. Kozicki, Electrochemical metallization memories-fundamentals, applications, prospects. Nanotechnology **22**, 254003/1–254003/22 (2011)
25. S.G. Park, B. Magyari-Koepe, Y. Nishi, Impact of oxygen vacancy ordering on the formation of a conductive filament in TiO_2 for resistive switching memory. IEEE Electron Device Lett. **32**, 197–199 (2011)
26. F. Pan, C. Chen, Z. Wang, Y. Yang, J. Yang, F. Zeng, Nonvolatile resistive switching memories-characteristics, mechanisms and challenges. Prog. Nat. Sci. Mater. Int. **20**, 1–15 (2010)
27. R. Waser, R. Dittmann, G. Staikov, K. Szot, Redox-based resistive switching memories – nanoionic mechanisms, prospects, and challenges. Adv. Mater. **21**, 2632–2663 (2009)
28. C. Rohde, B.J. Choi, D.S. Jeong, S.I. Choi, J. Zhao and C.S. Hwang, Identification of a determining parameter for resistive switching of TiO_2 thin films, Appl. Phys. Lett. **86**, 262907–09 (2005)
29. S. Seo, M.J. Lee, D.H. Seo, E.J. Jeoung, D.S. Suh, Y.S. Joung, I.K. Yoo, I.R. Hwang, S.H. Kim, I.S. Byun, J.S. Kim, J.S. Choi, B.H. Park, Reproducible resistance switching in polycrystalline NiO films. Appl. Phys. Lett. **85**, 5655–5657 (2004)
30. R. Waser, S. Menzel, R. Bruchhaus, *Nanoelectronics and Information Technology*, 3rd edn. (Wiley-VCH, Berlin, 2012)
31. A. Sawa, T. Fujii, M. Kawasaki, Y. Tokura, Interface resistance switching at a few nanometer thick perovskite manganite active layers. Appl. Phys. Lett. **88**, 232112-1–232112-3 (2006)
32. G. Bersuker, D.C. Gilmer, D. Veksler, P. Kirsch, L. Vandelli, A. Padovani, L. Larcher, K. McKenna, A. Shluger, V. Iglesias, M. Porti, M. Nafria, Metal oxide resistive memory switching mechanism based on conductive filament properties. J. Appl. Phys. **110**, 124518/1 (2011)
33. A. Foster, A. Shluger, R. Nieminen, Mechanism of interstitial oxygen diffusion in hafnia. Phys. Rev. Lett. **89**, 225901/1 (2002)

34. Y. Hirose, H. Hirose, Polarity-dependent memory switching and behaviour of Ag dendrite in Ag-photodoped amorphous As_2S_3 films. J. Appl. Phys. **47**, 2767–2772 (1976)
35. C. Schindler, G. Staikov, R. Waser, Electrode kinetics of Cu-SiO_2-based resistive switching cells: Overcoming the voltage-time dilemma of electrochemical metallization memories. Appl. Phys. Lett. **94**, 072109/1–072109/3 (2009)
36. A. Chen, V.V. Zhirnov, J.A. Hutchby, C. Michael Garner, ITRS chapter: emerging research devices. Future Fab. Special ITRS Focus (44) (2013)
37. A. Sawa, Resistive switching in transition metal oxides. Mater. Today **11**, 28–36 (2008)
38. M.-J. Lee, Y. Park, B.-S. Kang, S.-E. Ahn, C. Lee, K. Kim, W. Xianyu, G. Stefanovich, J.-H. Lee, S.-J. Chung, Y.-H. Kim, C.-S. Lee, J.-B. Park, I.-K. Yoo, Electron Devices Meeting, Technical Digest. IEEE International, pp. 771–774, 10–12 dec. (2007)
39. G. Kar, A. Fantini, Y. Chen, V. Paraschiv, B. Govoreanu, H. Hody, N. Jossart, H. Tielens, S. Brus, O. Richard, T. Vandeweyer, D. Wouters, L. Altimime, M. Jurczak, Process-improved RRAM cell performance and reliability and paving the way for manufacturability and scalability for high density memory application, in *Digest of Technical Papers – Symposium on VLSI Technology* (2012), pp. 157–158
40. E. Linn, R. Rosezin, C. Kügeler, R. Waser, Complementary resistive switches for passive nanocrossbar memories. Nat. Mater. **9**, 403–406 (2010)
41. S. Tappertzhofen, E. Linn, L. Nielen, R. Rosezin, F. Lentz, R. Bruchhaus, I. Valov, U. Böttger, R. Waser, Capacity based nondestructive readout for complementary resistive switches. Nanotechnology **22**, 395203/1–395203/7 (2011)
42. B.S. Kang, S.E. Ahn, M.J. Lee, G. Stefanovich, K.H. Kim, W.X. Xianyu, C.B. Lee, Y. Park, I.G. Baek, B.H. Park, High-current-density CuOx/InZnOx thin-film diodes for cross-point memory applications. Adv. Mater. **20**, 3066–3069 (2008)
43. M.-J. Lee, Y. Park, D.-S. Suh, E.-H. Lee, S. Seo, D.-C. Kim, R. Jung, B.-S. Kang, S.-E. Ahn, C.B. Lee, D.H. Seo, Y.-K. Cha, I.-K. Yoo, J.-S. Kim, B.H. Park, Two series oxide resistors applicable to high speed and high density nonvolatile memory. Adv. Mater. **19**, 3919–3923 (2007)
44. Q. Zuo, S. Long, Q. Liu, S. Zhang, Q. Wang, Y. Li, Y. Wang, M. Liu, Self-rectifying effect in gold nanocrystal-embedded zirconium oxide resistive memory. J. Appl. Phys. **106**, 73724/1–73724/5 (2009)
45. M.-J. Lee, S. Seo, D.-C. Kim, S.-E. Ahn, D.H. Seo, I.-K. Yoo, A low-temperature-grown oxide diode as a new switch element for high-density, nonvolatile memories. Adv. Mater. **19**, 73 (2007)
46. Q. Xia, J.J. Yang, W. Wu, X. Li, R.S. Williams, Self-aligned memristor cross-point arrays fabricated with one nanoimprint lithography step. Nano Lett. **10**, 2909–2914 (2010)
47. H. Lan, Y. Ding, *Nanoimprint Lithography* (InTech, 2010). Available from: http://www.intechopen.com/books/lithography/nanoimprintlithography
48. M. Meier, C. Nauenheim, S. Gilles, D. Mayer, C. Kuegeler, R. Waser, Nanoimprint for future non-volatile memory and logic devices. Microelectron. Eng. **85**, 870–872 (2008)
49. H.Y. Lee, P.S. Chen, T.Y. Wu, Y.S. Chen, C.C. Wang, P.J. Tzeng, C.H. Lin, F. Chen, C.H. Lien, M.J. Tsai, Low power and high speed bipolar switching with a thin reactive Ti buffer layer in robust HfO_2 based RRAM, in *IEEE International Electron Devices Meeting 2008, Technical Digest* (2008), pp. 297–300
50. M. Wu, Y. Lin, W. Jang, C. Lin, T. Tseng, Low-power and highly reliable multilevel operation in ZrO_2 1T1R RRAM. IEEE Electron Device Lett. **32**, 1026–1028 (2011)
51. W.C. Chien, Y.R. Chen, Y.C. Chen, A.T.H. Chuang, F.M. Lee, Y.Y. Lin, E.K. Lai, Y.H. Shih, K.Y. Hsieh, C.Y. Lu, A forming-free WOX resistive memory using a novel self-aligned field enhancement feature with excellent reliability and scalability, in *2010 International Electron Devices Meeting – Technical Digest* (2010)
52. Y. Dong, G. Yu, M. McAlpine, W. Lu, C. Lieber, Si/a-Si core/shell nanowires as nonvolatile crossbar switches. Nano Lett. **8**, 386–391 (2008)
53. K. Oka, T. Yanagida, K. Nagashima, T. Kawai, J. Kim, B. Park, Resistive-switching memory effects of NiO nanowire/metal junctions. J. Am. Chem. Soc. **132**, 6634–6635 (2010)

54. S.I. Kim, J.H. Lee, Y.W. Chang, S.S. Hwang, K.-H. Yoo, Reversible resistive switching behaviors in NiO nanowires. Appl. Phys. Lett. **93**, 033503–05 (2008)
55. Y. Chiang, W. Chang, C. Ho, C. Chen, C. Ho, S. Lin, T. Wu, J. He, Single-ZnO-nanowire memory. IEEE Trans. Electron Devices **58**, 1735–1740 (2011)
56. E. Herderick, J. Tresback, A. Vasiliev, N. Padture, Template-directed synthesis, characterization and electrical properties of Au-TiO_2-Au heterojunction nanowires. Nanotechnology **18**, 155204–09 (2007)
57. C. Nauenheim, C. Kuegeler, A. Ruediger, R. Waser, Investigation of the electroforming process in resistively switching TiO_2 nanocrosspoint junctions. Appl. Phys. Lett. **96**, 122902 (2010)
58. K. Kinoshita, K. Tsunoda, Y. Sato, H. Noshiro, S. Yagaki, M. Aoki, Y. Sugiyama, Reduction in the reset current in a resistive random access memory consisting of NiO_x brought about by reducing a parasitic capacitance. Appl. Phys. Lett. **93**, 033506 (2008)
59. B. Butcher, S. Koveshnikov, D. Gilmer, G. Bersuker, M. Sung, A. Kalantarian, C. Park, R. Geer, Y. Nishi, P. Kirsch, R. Jammy, High endurance performance of 1T1R HfOx based RRAM at low (20μA) operative current and elevated (150 °C) temperature, in *IEEE International Integrated Reliability Workshop Final Report* (2011), pp. 146–150
60. Y. Wu, B. Lee, H. Wong, Al_2O_3-based RRAM using atomic layer deposition (ALD) with 1-μA RESET current. IEEE Electron Device Lett. **31**, 1449–1451 (2010)
61. W. Kim, S. Park, Z. Zhang, Y. Yang-Liauw, D. Sekar, H. Wong, S. Wong, Forming-free nitrogen-doped AlOX RRAM with sub-μA programming current, in *Digest of Technical Papers – Symposium on VLSI Technology* (2011), pp. 22–23
62. L. Goux, A. Fantini, G. Kar, Y. Chen, N. Jossart, R. Degraeve, S. Clima, B. Govoreanu, G. Lorenzo, G. Pourtois, D. Wouters, J. Kittl, L. Altimime, M. Jurczak, Ultralow sub-500nA operating current high-performance TiN\$Al_2O_3$\$HfO_2$\Hf\TiN bipolar RRAM achieved through understanding-based stack-engineering, in *Digest of Technical Papers – Symposium on VLSI Technology* (2012), pp. 159–160
63. U. Russo, D. Ielmini, C. Cagli, A.L. Lacaita, Self-accelerated thermal dissolution model for reset programming in unipolar resistive-switching memory (RRAM) devices. IEEE Trans. Electron Devices **56**, 193–200 (2009)
64. K. Kim, D.S. Jeong, C.S. Hwang, Nanofilamentary resistive switching in binary oxide system; a review on the present status and outlook. Nanotechnology **22**, 254002 (2011)
65. F. Nardi, S. Larentis, S. Balatti, D. Gilmer, D. Ielmini, Resistive switching by voltage-driven ion migration in bipolar RRAM. Part I: Experimental study. IEEE Trans. Electron Devices **59**, 2461–2467 (2012)
66. D. Ielmini, Filamentary-switching model in RRAM for time, energy and scaling projections, in *2011 IEEE International Electron Devices Meeting – IEDM'11* (2011), pp. 17.2.1–17.2.4
67. S. Yu, H. Wong, A phenomenological model for the reset mechanism of metal oxide RRAM. IEEE Electron Device Lett. **31**, 1455–1457 (2010)
68. H.-S.P. Wong, H.-Y. Lee, S. Yu, Y.-S. Chen, Y. Wu, P.-S. Chen, B. Lee, F.T. Chen, M.-J. Tsai, Metal–oxide RRAM. Proc. IEEE **100**, 1951–1970 (2012)
69. H. Schroeder, V.V. Zhirnov, R.K. Cavin, R. Waser, Voltage-time dilemma of pure electronic mechanisms in resistive switching memory cells. J. Appl. Phys. **107**, 054517/1–054517/8 (2010)
70. C. Hermes, Interaction between redox-based resistive switching mechanisms. Forschungszentrum Jülich GmbH **25**, 134 (2013)
71. D.B. Strukov, R.S. Williams, Intrinsic constrains on thermally-assisted memristive switching. Appl. Phys. A Mater. Sci. Process. **102**, 851–855 (2011)
72. Z. Wei, Y. Kanzawa, K. Arita, Y. Katoh, K. Kawai, S. Muraoka, S. Mitani, S. Fujii, K. Katayama, M. Iijima, T. Mikawa, T. Ninomiya, R. Miyanaga, Y. Kawashima, K. Tsuji, A. Himeno, T. Okada, R. Azuma, K. Shimakawa, H. Sugaya, T. Takagi, R. Yasuhara, H. Horiba, H. Kumigashira, M. Oshima, Electron Devices Meeting, Technical Digest. IEEE International, pp. 1–4, 15–17 Dec. (2008)

73. B. Gao, J. Kang, H. Zhang, B. Sun, B. Chen, L. Liu, X. Liu, R. Han, Y. Wang, B. Yu, Z. Fang, H. Yu, D. Kwong, Oxide-based RRAM: physical based retention projection, in *2010 Proceedings of the European Solid State Device Research Conference, ESSDERC 2010* (2010), pp. 392–395
74. Z. Wei, T. Takagi, Y. Kanzawa, Y. Katoh, T. Ninomiya, K. Kawai, S. Muraoka, S. Mitani, K. Katayama, S. Fujii, R. Miyanaga, Y. Kawashima, T. Mikawa, K. Shimakawa, K. Aono, Retention model for high-density ReRAM, in *2012 4th IEEE International Memory Workshop, IMW 2012* (2012)
75. T. Ninomiya, T. Takagi, Z. Wei, S. Muraoka, R. Yasuhara, K. Katayama, Y. Ikeda, K. Kawai, Y. Kato, Y. Kawashima, S. Ito, T. Mikawa, K. Shimakawa, K. Aono, Conductive filament scaling of TaOx bipolar ReRAM for long retention with low current operation, in *Digest of Technical Papers – Symposium on VLSI Technology* (2012), pp. 73–74
76. Z. Wei, T. Takagi, Y. Kanzawa, Y. Katoh, T. Ninomiya, K. Kawai, S. Muraoka, S. Mitani, K. Katayama, S. Fujii, R. Miyanaga, Y. Kawashima, T. Mikawa, K. Shimakawa, K. Aono, Demonstration of high-density ReRAM ensuring 10-year retention at 85 °C based on a newly developed reliability model, in *Technical Digest – International Electron Devices Meeting, IEDM* (2011), pp. 31.4.1–31.4.4
77. X.T. Zhang, Q.X. Yu, Y.P. Yao, X.G. Li, Ultrafast resistive switching in $SrTiO_3$:Nb single crystal. Appl. Phys. Lett. **97**, 222117/1–222117/3 (2010)
78. C. Hermes, M. Wimmer, S. Menzel, K. Fleck, G. Bruns, M. Salinga, U. Boettger, R. Bruchhaus, T. Schmitz-Kempen, M. Wuttig, R. Waser, Analysis of transient currents during ultra fast switching of TiO_2 nanocrossbar devices. IEEE Electron Device Lett. **32**, 1116–1118 (2011)
79. H. Lee, Y. Chen, P. Chen, P. Gu, Y. Hsu, S. Wang, W. Liu, C. Tsai, S. Sheu, P. Chiang, W. Lin, C. Lin, W. Chen, F. Chen, C. Lien, and M. Tsai, Evidence and solution of over-RESET problem for HfO_X based resistive memory with sub-ns switching speed and high endurance, in *Technical Digest – International Electron Devices Meeting, IEDM* (2010), pp. 19.7.1–19.7.4
80. S. Menzel, M. Waters, A. Marchewka, U. Böttger, R. Dittmann, R. Waser, Origin of the ultra-nonlinear switching kinetics in oxide-based resistive switches. Adv. Funct. Mater. **21**, 4487–4492 (2011)
81. A.C. Torrezan, J.P. Strachan, G. Medeiros-Ribeiro, R.S. Williams, Sub-nanosecond switching of a tantalum oxide memristor. Nanotechnology **22**, 485203/1–485203/7 (2011)
82. B. Chen, Y. Lu, B. Gao, Y.H. Fu, F.F. Zhang, P. Huang, Y.S. Chen, L.F. Liu, X.Y. Liu, J.F. Kang, Y.Y. Wang, Z. Fang, H.Y. Yu, X. Li, X.P. Wang, N. Singh, G.Q. Lo, D.L. Kwong, Physical mechanisms of endurance degradation in TMO-RRAM, in *2011 IEEE International Electron Devices Meeting (IEDM)* (2011)
83. I. Valov, E. Linn, S. Tappertzhofen, S. Schmelzer, J. van den Hurk, F. Lentz, R. Waser, Nanobatteries in redox-based resistive switches require extension of memristor theory. Nat. Commun. **4**, 1771 (2013)
84. H.Y. Lee, Y.S. Chen, P.S. Chen, P.Y. Gu, Y.Y. Hsu, S.M. Wang, W.H. Liu, C.H. Tsai, S.S. Sheu, P.C. Chiang, W.P. Lin, C.H. Lin, W.S. Chen, F.T. Chen, C.H. Lien, M. Tsai, Evidence and solution of Over-RESET Problem for HfO_X based resistive memory with sub-ns switching speed and high endurance, in *2010 International Electron Devices Meeting – Technical Digest* (2010)
85. Y. Chen, B. Govoreanu, L. Goux, R. Degraeve, A. Fantini, G. Kar, D. Wouters, G. Groeseneken, J. Kittl, M. Jurczak, L. Altimime, Balancing SET/RESET pulse for $>10^{10}$ endurance in HfO_2 1T1R bipolar RRAM. IEEE Trans. Electron Devices **59**, 3243–3249 (2012)
86. W.-C. Chien, M.-H. Lee, F.-M. Lee, Y.-Y. Lin, H.-L. Lung, K.-Y. Hsieh, C.-Y. Lu, A multi-level 40nm WOX resistive memory with excellent reliability, in *2011 IEEE International Electron Devices Meeting – IEDM'11* (2011)
87. Y. Wang, Q. Liu, S. Long, W. Wang, Q. Wang, M. Zhang, S. Zhang, Y. Li, Q. Zuo, J. Yang, M. Liu, Investigation of resistive switching in Cu-doped HfO_2 thin film for multilevel non-volatile memory applications. Nanotechnology **21**, 45202/1–45202/6 (2010)

88. J. Park, K.P. Biju, S. Jung, W. Lee, J. Lee, S. Kim, S. Park, J. Shin, H. Hwang, Multibit operation of TiOx-based ReRAM by schottky barrier height engineering. IEEE Electron Device Lett. **32**, 476–478 (2011)
89. M. Terai, Y. Sakotsubo, S. Kotsuji, H. Hada, Resistance controllability of Ta_2O_5/TiO_2 stack ReRAM for low-voltage and multilevel operation. IEEE Electron Device Lett. **31**, 204–206 (2010)
90. S. Yu, Y. Wu, H. Wong, Investigating the switching dynamics and multilevel capability of bipolar metal oxide resistive switching memory. Appl. Phys. Lett. **98**, 103514/1–103514/3 (2011)
91. S. Lee, Y. Kim, M. Chang, K. Kim, C. Lee, J. Hur, G. Park, D. Lee, M. Lee, C. Kim, U. Chung, I. Yoo, K. Kim, Multi-level switching of triple-layered TaOx RRAM with excellent reliability for storage class memory, in *Digest of Technical Papers – Symposium on VLSI Technology* (2012), pp. 71–72
92. V.V. Zhirnov, R. Meade, R.K. Cavin, G. Sandhu, Scaling limits of resistive memories. Nanotechnology **22**, 254027/1–254027/21 (2011)
93. L. Goux, J.G. Lisoni, X.P. Wang, M. Jurczak, D.J. Wouters, Optimized Ni oxidation in 80-nm contact holes for integration of forming-free and low power Ni/NiO/Ni memory cells. IEEE Trans. Electron Devices **56**, 2363 (2009)
94. C. Nauenheim, Integration of resistive switching devices in crossbar structures, Phd thesis, Forschungszentrum Jülich GmbH (2009)
95. B. Lee, H. Wong, NiO resistance change memory with a novel structure for 3D integration and improved confinement of conduction path, in *2009 Symposium on VLSI Technology, Digest of Technical Papers* (2009), pp. 28–29
96. D. Ielmini, F. Nardi, C. Cagli, Universal reset characteristics of unipolar and bipolar metal-oxide RRAM. IEEE Trans. Electron Devices **58**, 1–8 (2011)
97. S. Tanachutiwat, M. Liu, W. Wang, FPGA based on integration of CMOS and RRAM. IEEE Trans. Very Large Scale Integration (VLSI) Syst. **19**, 2023–2032 (2011)
98. K. Seo, I. Kim, S. Jung, M. Jo, S. Park, J. Park, J. Shin, K.P. Biju, J. Kong, K. Lee, B. Lee, H. Hwang, Analog memory and spike-timing-dependent plasticity characteristics of a nanoscale titanium oxide bilayer resistive switching device. Nanotechnology **22**, 254023 (2011)
99. T. Chang, S. Jo, W. Lu, Short-term memory to long-term memory transition in a nanoscale memristor. ACS Nano **5**, 7669–7676 (2011)
100. S. Yu, Y. Wu, R. Jeyasingh, D. Kuzum, H.P. Wong, An electronic synapse device based on metal oxide resistive switching memory for neuromorphic computation. IEEE Trans. Electron Devices **58**, 2729–2737 (2011)
101. G.S. Snider, Spike-timing-dependent learning in memristive nanodevices, in *IEEE International Symposium on Nanoscale Architectures* (2008), pp. 85–92
102. H. Jeong, Y. Kim, J. Lee, S. Choi, A low-temperature-grown TiO_2-based device for the flexible stacked RRAM application. Nanotechnology **21**, 115203 (2010)

Chapter 8
Silicon Nanowire-Based Memristive Devices

Davide Sacchetto, Yusuf Leblebici, and Giovanni De Micheli

8.1 Introduction

Due to the natural limitations of materials, future nano-scale circuits will have to exploit more efficient ways for computation and memory storage. One possible scenario envisages an end of charge-based technologies, after which computation will rely on alternative, more power efficient state variable manipulation. A long list of fundamental state variables other than charge includes the spin, phase, multipole orientation, mechanical position, polarity, orbital symmetry, magnetic flux quanta, molecular configuration, and other quantum states [1]. Nevertheless, technologies using new state variables would have to be implemented within a completely new technological platform and cannot be seen as CMOS-compatible alternatives in the short term.

The physical realization of the memristor, whose behavior was postulated by Leon Chua [2] and generalized by Chua and Kang [3] for memristive devices and systems, offers a completely new set of possibilities for logic [4] and memory [5] applications. It is worth noting that a generalized model for memristive systems can be implemented under direct current, small signal, and sinusoidal excitation [3]. The implications of such modeling is linked with the observation of memristive functionalities over a broad range of technologies based on nanoelectronic and nanoionic behaviors.

One typical application targets standalone memories, and in this respect, the two-terminal memristive devices have a potential for very high density storage.

D. Sacchetto (✉) • Y. Leblebici
EPFL-STI-IEL-LSM, Bldg ELD, Station 11, CH-1015 Lausanne, Switzerland
e-mail: davide.sacchetto@epfl.ch; yusuf.leblebici@epfl.ch

G. De Micheli
EPFL-IC-ISIM-LSI, Building INF 3rd floor, Station 14, 1015 Lausanne, Switzerland
e-mail: giovanni.demicheli@epfl.ch

R. Tetzlaff (ed.), *Memristors and Memristive Systems*,
DOI 10.1007/978-1-4614-9068-5_8,

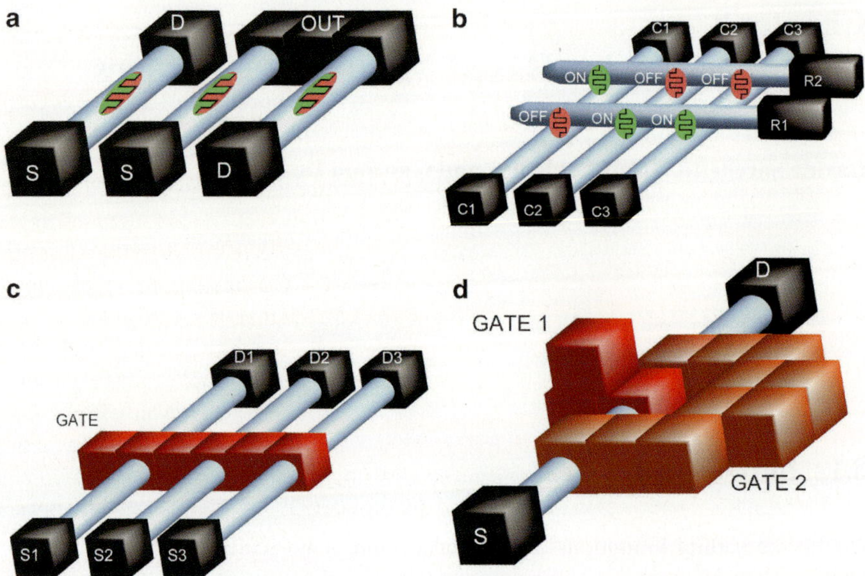

Fig. 8.1 (**a**) Parallel nanowire two-terminal memristive devices. (**b**) Crossbar array consisting of memristive cross-points (two-terminal). (**c**) Gate controlled three-terminal nanowire memristive device. (**d**) Double-gate four-terminal nanowire memristive devices

Complementary logic based on two-terminal memristive devices (see Fig. 8.1a) or ultra-dense crossbar arrays with memristive cross-points (see Fig. 8.1b) can dramatically improve device density up to 10^{11} bits per square centimeter [6]. Moreover, the use of memristive effects as new state variables for computation can be exploited to build new types of functional devices with three- or four-terminals (see Fig. 8.1c, d, respectively).

In the following sections, after a short review on the top-down fabrication methods of Si nanowire-based memristive devices (Sect. 8.2), two-terminal (Sect. 8.3), three-terminal (Sect. 8.4), and four-terminal (Sect. 8.5) memristive devices are presented.

8.2 Top-Down Fabrication Methods

The methods presented here are fully scalable by using a more advanced lithography and the authors demonstrated an alternative solution that skips the oxidation steps for the fabrication of ultra-dense vertically stacked nanowires that are below 30 nm diameter [7]. Nevertheless here is reported an inexpensive top-down processing solution that uses standard photolithography with 1 μm resolution applicable to both bulk-Si and SOI substrates.

8.2.1 Fabrication of Vertically Stacked Si Nanowire Arrays

The following top-down fabrication technique enables the structuring of Si nanowires at a pitch that is not limited by the lithographic resolution is based on the *Deep Reactive Ion Etching (DRIE)* technique. This approach is utilized to obtain arrays of vertically stacked *Si nanowires (SiNWs)*. While the density of horizontal strands is limited by the lithographic pitch, each strand can be composed of several vertically stacked nanowires by adjusting the number of cycles in the DRIE process.

In [8], the process begins by defining a photoresist line on a p-type silicon bulk wafer (see Fig. 8.2a). This mask will be used as protective layer for the successive DRIE technique. This technique that alternates a plasma etching with a passivation step has been optimized to produce a scalloped trench in silicon with high reproducibility. Etching time, passivation time, and plasma platen power can be changed in order to enhance the scalloping effect. The application of the DRIE technique gives a trench like the one depicted in Fig. 8.2b. The flexibility of the process allows us to change the number of scallops easily. After trench definition, a sacrificial oxidation step is carried out. The effect of oxidation results in the total Si consumption of the smaller portions of the trench. The wider parts of the trench leave vertically stacked Si nanowires embedded in the grown oxide (see Fig. 8.2c). Then the cavities produced by the Bosch process are filled with photoresist (Fig. 8.2d). After a combination of *Chemical Mechanical Polishing (CMP)* and buffered hydrofluoric acid dip, the wet oxide is removed around the nanowires (see Fig. 8.2e). After removal of the resist, caves with stacks of several nanowires are freestanding on a layer of thick wet oxide, which is left to isolate the substrate from the successive processes (Fig. 8.2f). Nanowires are oxidized in dry atmosphere, for a 10–20 nm higher quality oxide, as the dielectric for FET devices (Fig. 8.2g) as gate dielectric. Then between 200 nm and 500 nm of LPCVD polysilicon is deposited (Fig. 8.2h). The LPCVD polySi layer allows conformal coverage of the 3D structure, enabling the formation of *gate-all-around (GAA)* devices. The polysilicon gate is then patterned by means of a combination of isotropic and anisotropic recipes (see Fig. 8.2i). Depending on the structure, implantation, or metallization of the Si pillars can be carried out, so to produce MOSFETs or SBFETs, respectively.

Examples of fabricated structures demonstrating arrays having from 3 up to 12 vertically stacked Si nanowires are shown in Fig. 8.3. The obtained nanowires can be used to build gate-all-around field effect transistors (see Fig. 8.4) interconnected through Si pillars.

8.2.2 Si Nanowires with Double Independent Gates

The device consists of a 20 μm-long crystalline SiNW attached between two Si pillars on a SOI wafer (see Fig. 8.5a). The SiNW is then covered by two independent n++ polysilicon gates with this scheme: a main central gate (gate 1) of 7.5 μm

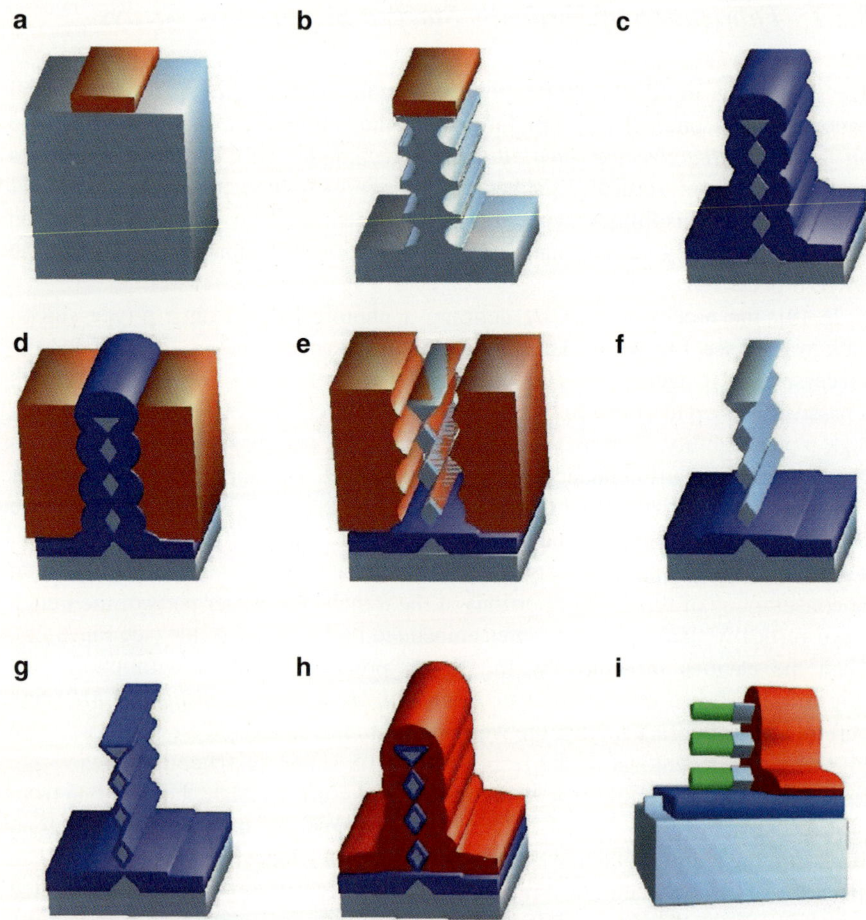

Fig. 8.2 Vertically stacked Si nanowire process steps. (**a**) Optical lithography; (**b**) Four steps DRIE etch; (**c**) Wet oxidation; (**d**) Cave filling with photoresist; (**e**) BHF oxide removal; (**f**) Photoresist removal; (**g**) Dry oxidation; (**h**) Conformal LPCVD polySi deposition; (**i**) PolySi patterning

length and a second gate (gate 2) that is used to control the SiNW portions between the main gate and source and drain regions, respectively. First, a low doping p-type ($N_A \approx 10^{15}$atoms/cm^2) SOI wafer with 1.5 μm device layer is spin coated. The photoresist is then patterned in 1.5 μm wide lines (see Fig. 8.5b) and used as mask for a next isotropic Si etching. The ICP SF_6 plasma etching recipe is tuned to form a triangular 75 nm wide SiNW lying on top of the *buried oxide (BOX)* layer (Fig. 8.5c). Then a 30 nm thick gate oxidation and a 150 nm polysilicon layer are

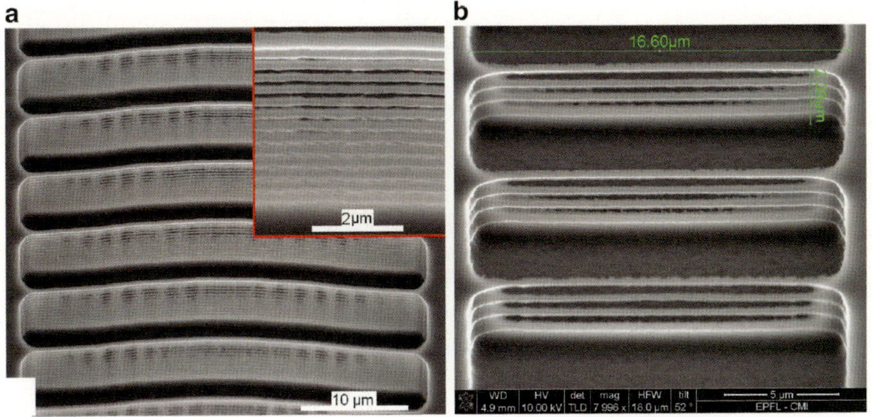

Fig. 8.3 Arrays of vertically stacked Si nanowires [8]: (**a**) Silicon nanowire arrays with 12 vertical levels. (**b**) Silicon nanowire arrays with three vertical levels

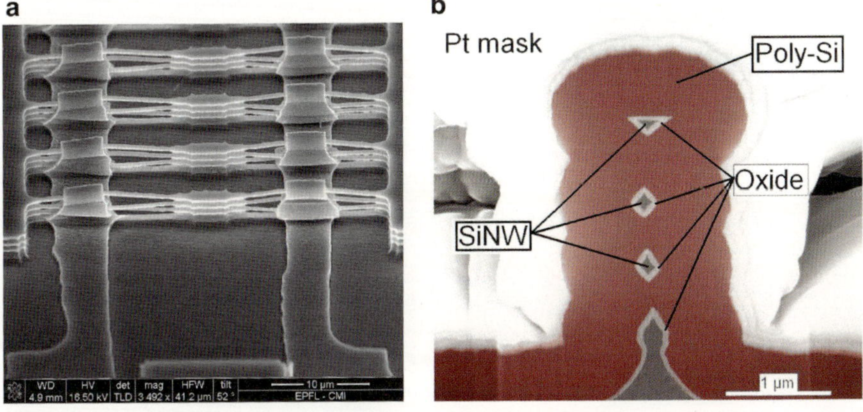

Fig. 8.4 Vertically stacked Si nanowire transistors [8]: (**a**) Three horizontal Si nanowire strands with two parallel polysilicon gates. (**b**) FIB cross-section showing triangular and rhombic nanowires embedded in a gate-all-around polysilicon gate

deposited with *low pressure chemical vapour deposition (LPCVD)* method to form a main gate with 7.5 μm length (gate 1, Fig. 8.5d). The main gate is then isolated by a 300 nm LPCVD low temperature oxide (LTO). A second 500 nm polysilicon layer is then deposited. Then a thick photoresist is spun over the wafer and planarized using a chemical mechanical polishing procedure similar to the one previously described by the authors in [8]. This method leaves a protective polymer layer that is used to etch a second polysilicon gate self-aligned within the cavity, thanks to the topography (see gate 2 in Fig. 8.5e). After standard cleaning steps, one additional

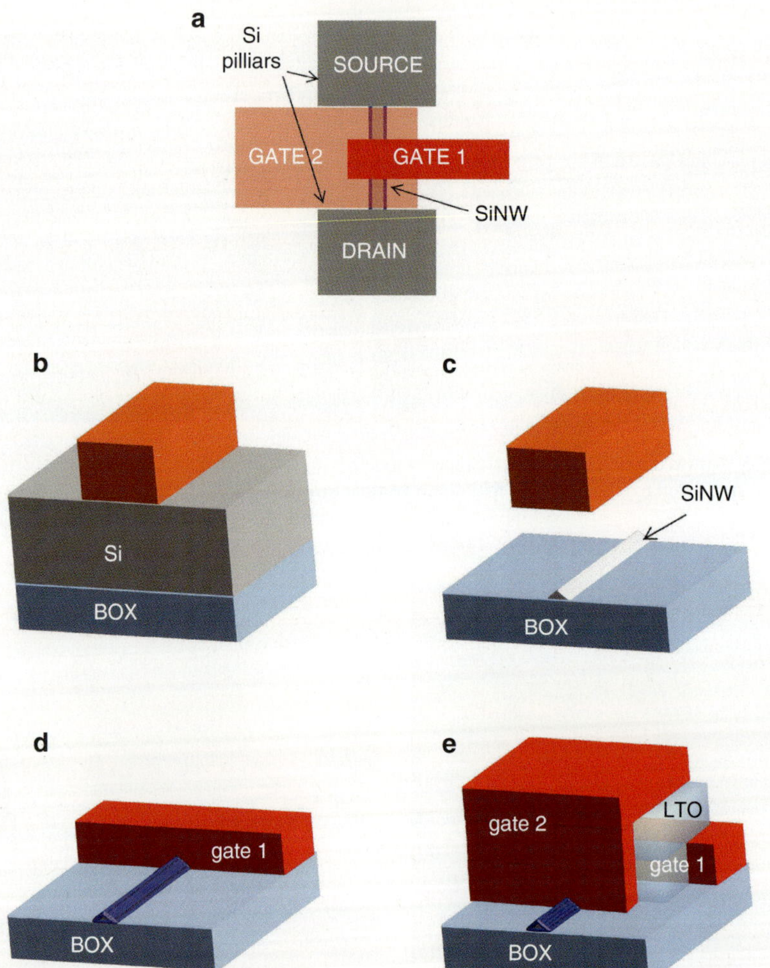

Fig. 8.5 Fabrication flow. (**a**) Top view of the dual-gate device. (**b**) Photoresist mask is patterned. (**c**) After isotropic etching of Si a triangular-shaped SiNW is formed. (**d**) Gate oxidation and LPCVD polysilicon are deposited and patterned to form the main, central gate stack. (**e**) An LTO inter-poly dielectric is deposited and a second polysilicon gate is made self-aligned with the nanowire

patterning of gate 2 is performed to remove the unnecessary polysilicon and to form areas for the contacts (see the top view of the device in Fig. 8.5a). In Fig. 8.6, a focused ion beam cross-section of the triangular SiNW channel with the double independent gate stack is shown. Then source/drain contacts are formed by means of NiSi silicidation in a horizontal wall furnace in forming gas at 400°C. Finally Al metal lines and pad area are defined for the electrical characterization.

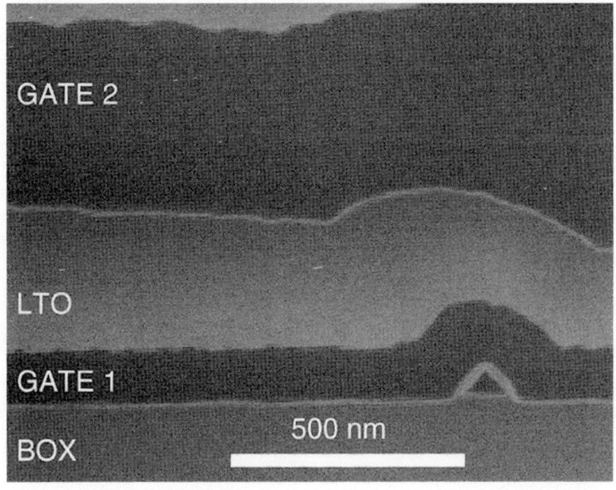

Fig. 8.6 Focused ion beam cross-section showing a triangular SiNW channel with two 75 nm sides and 100 nm base. The gate 1, gate 2 and the LTO inter-poly dielectric have 150 nm, 500 nm, and 300 nm thicknesses, respectively

8.2.3 Si Nanowire with Memristive Functionality

Bulk-Si wafers with low boron concentration ($N_A \sim 10^{15}$ atoms/cm^3) have been used as a substrate for the fabricated devices. Vertical stacks of Si nanowires are defined on the substrate by optical lithography (see Fig. 8.7a) without any constraint on the resolution limit (1 μm). The photoresist is then used as a mask for a *deep reactive ion etching (DRIE)* (Fig. 8.7b). The optimized Si etching technique, which uses an isotropic Si etching, defines scalloped trenches attached to Si pillars with high reproducibility. The enhanced scalloping effect produces vertical modulation of the trench width. A sacrificial oxidation is then carried out with the double purpose of eliminating the Si where the trench is thin, and also to reduce the surface roughness induced by the etching (Fig. 8.7c). A combination of CMP and buffered HF dip leaves an SiNW suspended on a thick layer of insulating oxide (Fig. 8.7d). The gate oxide is grown in a horizontal furnace with a dry atmosphere (Fig. 8.7e). The gate poly-silicon is conformally deposited and doped with phosphorous by means of a diffusion process and then patterned with a combination of isotropic and anisotropic plasma etching steps (Fig. 8.7f). The fabrication of SBFETs requires the use of metallic source and drain contacts, meaning source-to-body and drain-to-body Schottky junctions. We pattern a Cr/Ni bilayers (10 nm/50 nm) on top of the Si pillars, partially covering the SiNW at the anchor points (see Fig. 8.8). This leads to the silicidation of the nanowire starting from the Cr/Ni bilayer toward the gated region of the nanowire.

8.2.4 Ambipolar Si Nanowires for Memristive-Bio-Sensing

The fabrication method utilizes some of the steps that were previously reported [9] for memristive Schottky-barrier silicon nanowire field-effect-transistors. The process starts from low resistivity SOI substrates, with 1.5 μm device layer

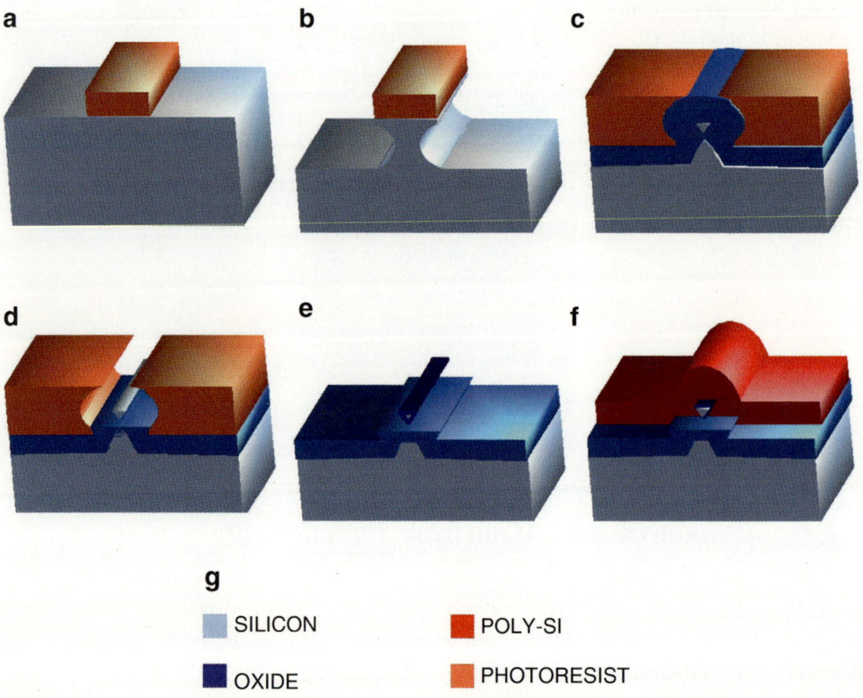

Fig. 8.7 GAA SiNW SBFET process flow. (**a**) A photoresist line determines the nanowire position. (**b**) DRIE etching forms a scalloped trench. (**c**) After wet oxidation, the Si trench reduces to a suspended nanowire. The caves are filled with photoresist and planarized with CMP. (**d**) Buffered HF oxide etch releases the SiNWs. (**e**) Gate oxidation. (**f**) Poly-silicon is deposited and patterned to form the gate. (**g**) Legend

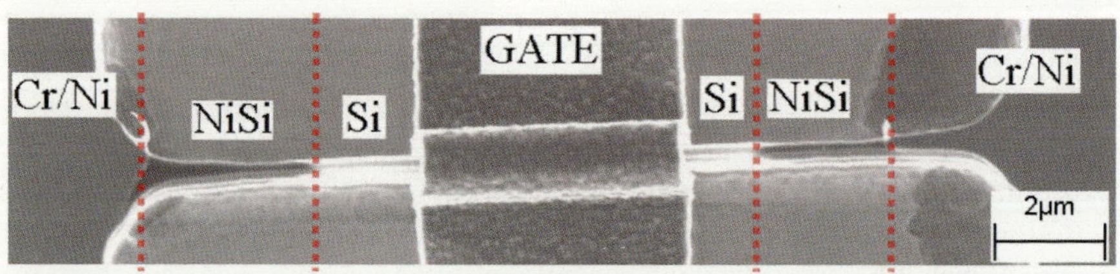

Fig. 8.8 GAA SiNW SBFET with Cr/Ni source/drain after the lift-off process. The change in contrast on the NW channel is attribute to a difference between Si and silicided regions

and 3 μm SiO_2 insulating layer. After standard lithography, the silicon nanowire is carved anchored at the top of two silicon pillars. Then, Ni is deposited on top of the pillars with overlap on the outer portions of the silicon nanowire. Hence, an annealing step at 450°C in a horizontal wall furnace forms NiSi contacts. Figure 8.9a shows a nano-fabricated memristive silicon nanowire with NiSi contacts.

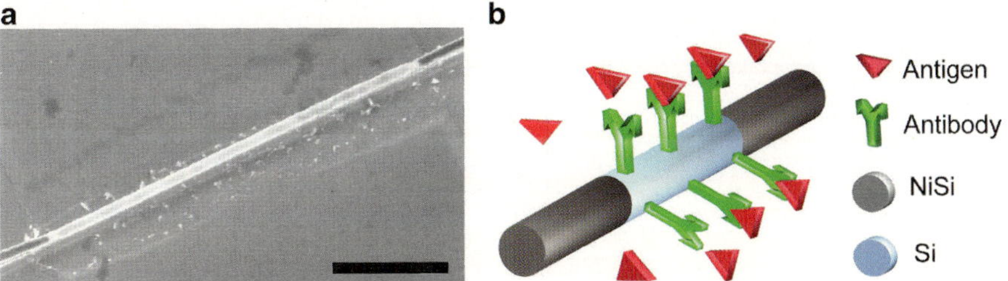

Fig. 8.9 (**a**) Suspended functionalized silicon nanowire with NiSi extremities. Scale bar is 4 μm. (**b**) Suspended functionalized Si nanowire with NiSi extremities. The functionalized layer is capable of trapping antigen molecules which in turn affects the memristive hysteresis behavior, giving a new method for bio-sensing [12]

The silicon nanowire surface was derivatized with 3-glycidoxypropyltrimethoxysilane GPTS [10] and functionalized by covalent attachment of anti-rabbit polyclonal antibodies (AB) [11]. Antigen (rabbit antibodies, AG) interacts with the functionalized wire as depicted in Fig. 8.9b.

8.3 Two-Terminal Memristive Devices

Two-terminal memristive devices can be based on metal/oxide switches, such as for SiO_2, Hf_2O [13], CuO [14], NiO [15], ZnO [16], Al_2O_3 [17], VO_2 [18], $SrTiO_3$ [19]. These devices behave as solid-state electrochemical switches, whose resistance is defined by a metallic filament formation mechanism related to the solid-state redox reactions stimulated by the polarity of the applied electric field [20]. One example is the CuO-based ReRAM of Dong et al. [14] that shows repeatable resistive switching at very low voltages (see Fig. 8.10).

The well-known TiO_2-based ReRAM [21, 22] seems to be based on a different mechanism, which is attributed to the vacancy/dopant diffusion in the oxide layer. The re-distribution of oxygen vacancies into the TiO_2 depends on the polarity of the applied voltage, and it causes the switching between a semiconductor state into a metallic one. Typical ReRAM functionality of the TiO_2-based ReRAM is shown in Fig. 8.11 [21].

Another type of two-terminal memristive device is the *phase change (PC)* RAM [23]. The main switching mechanism is based on phase transition between an amorphous and a crystalline type due to Joule heating dynamics controlled by a voltage pulse. For instance, Si nanowires can be engineered such that melting and solidification processes can be iterated, thus giving rise to alternate resistance states [24].

Another class is based on polymers [25]. Several memristive switches can be built by inter-posing a bio-molecule layer with properties ranging from

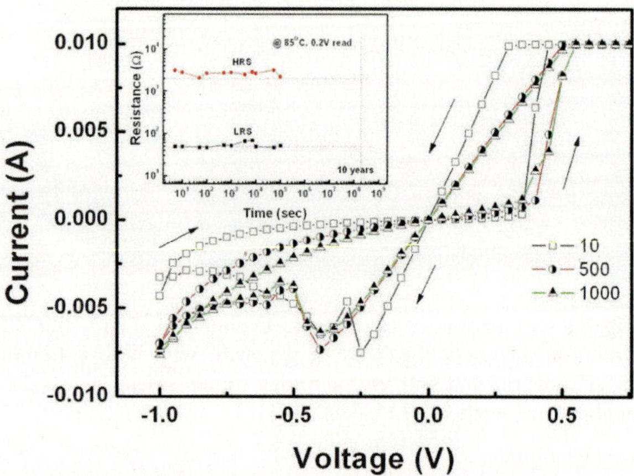

Fig. 8.10 Switching mechanism of a CuO-based ReRAM. The image is taken from [14]

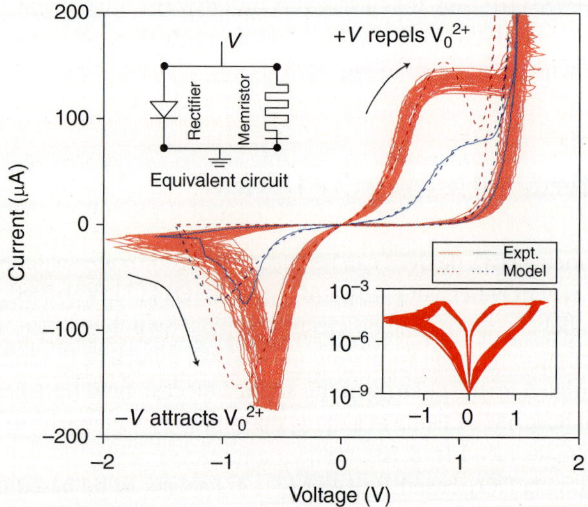

Fig. 8.11 Pinched hysteresis loop from a $Pt/TiO_2/Pt$ memristive junction. The image is taken from [21]

molecule-dependent switching, such as rotaxanes [6], or more in general on inter-locked molecules [26] but also on molecule-independent switching, where a filament formation mechanism through the molecular layer is involved [27].

A fifth class belongs to spintronics [28]. Pershin et al. demonstrated that electron-spin polarization controlled by the external voltage applied to a spintronic device, acts as a state variable that can be modeled as memristance (see Fig. 8.12 [29]).

In all these devices the amplitude and frequency of the input signal contribute in the formation of a so-called pinched hysteresis loop, whose salient feature is its zero crossing property [3], which is critical for ultra low power operation.

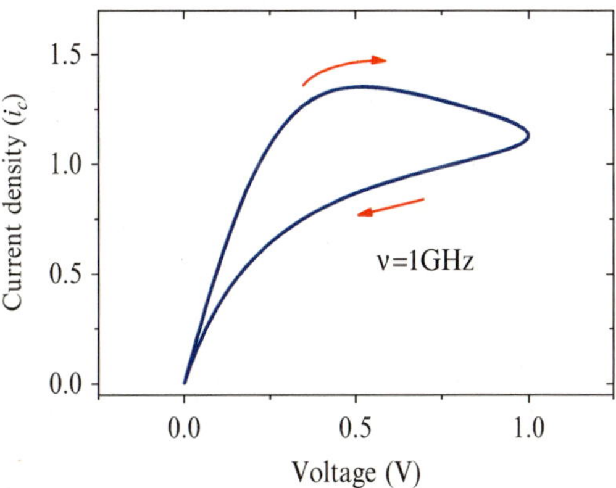

Fig. 8.12 Unipolar I-V simulation for a semiconductor spintronic system. The image is taken from [29]

8.4 Three-Terminal Nanowire Memristive Devices

Examples of three-terminal memristive devices are the electrochemical organic memristor [30], the solid-electrolyte nanometer switch [31], the ferroelectric FET [32], and the ambipolar *Si nanowire Schottky barrier FET (SiNW SBFET)* [8, 33].

A classification of three-terminal memristive devices can be based from the general concept of the FET structure (see Fig. 8.13a) in which memristive functionality can be inserted either by engineering the gate dielectric or by gating a memristive channel. For instance, trap charging dielectric layers inserted between the channel and the gate fall into the category of FET with capacitive memory storage (Fig. 8.13b). One example in this category can be the flash memory for which the trap charging into the gate dielectric influences the transconductance state of the channel. Thus a first category that exploits the operation of writing/erasing cycles into the gate dielectric will be a generalization of the flash memory concept, for which volatility of the charges that are injected into the trap charging layer can be tuned according to a desired frequency response.

A second category is the one of the gated memristors (Fig. 8.13c). A few examples are the electrochemical organic memristor [30], the bio-memristive nanowire [12], and the solid-electrolyte nanometer switch [31]. In the electrochemical organic memristor the gate potential is represented by the potential of the bath, which is used to transfer positively charged Rb^+ ions into a polyaniline (PANI) layer. The conductivity change can be iterated by switching the polarity of the bath potential, thus giving rise to a unipolar I_{ds}-V_{ds} curve that can be modeled as a memristor. In this case the device can be set into either memristive or diode functionality. Similarly, a novel method for bio-sensing that has been recently proposed exploits the memristive effect to detect low concentration of bio-molecules [12]. The device consists of a NiSi/Si/NiSi nanowire structure coated with antibody layer (see Fig. 8.14) shows memristive behavior. The hysteresis loop of this device has been demonstrated for detection of low concentrations of bio-molecules

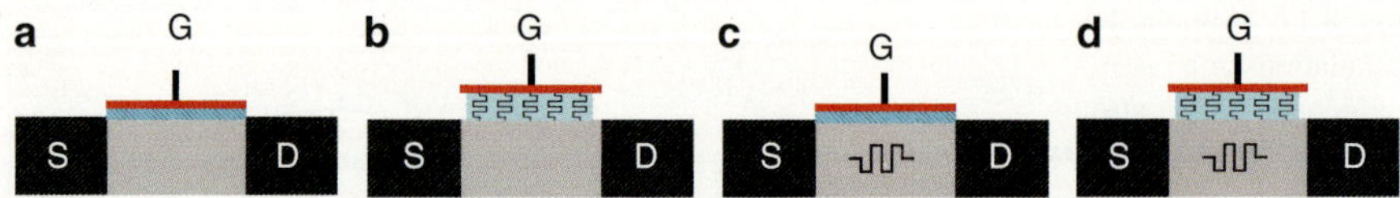

Fig. 8.13 Categorization of FETs with memristive functionalities: (**a**) Conventional FET. (**b**) FET with memristive gate dielectric. (**c**) Gated memristor (memristive channel). (**d**) Gated memristor with memristive dielectric

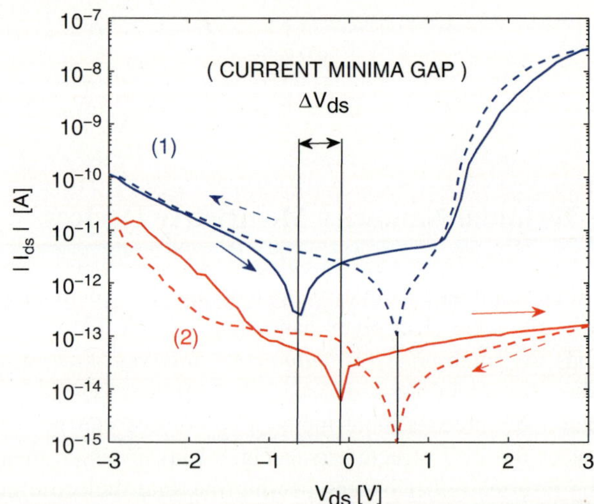

Fig. 8.14 A typical memristive hysteresis is observed; the *blue curve* (1) is measured after drying the sample from de-ionized water. The *red curve* (2) is measured after dipping in 5 pM antigen solution and drying. The measured ΔV_{ds} is proportional to the concentration of antigen [12]

(antigen) in a dry environment. Conversely, the three-terminal solid-state electrolyte nanometer switch shows a typical bistable resistance state but using 100 times less current than standard two-terminal operation [34]. This device is based on controlling the filament formation mechanism using the voltage of a gate terminal (Figs. 8.15, 8.16).

The ambipolar SB FET with SiNW channel reported in [9] (see Fig. 8.13d) falls in both categories of gated memristor and trap charging dielectric, as it shows dynamic trap charging mechanisms at the Schottky junctions and in the gate dielectric insulator. The result depicted in Fig. 8.17 shows a hysteretic behavior that is reminiscent of a two-terminal monolithic memristive device [35]. The hysteresis reflects the fact that the $I_{ds} - V_{ds}$ curve for forward V_{ds} sweep is not identical to the same curve for backwards V_{ds} sweep. It can be attributed to the presence of interface states at the metal/semiconductor junctions as reported in literature for Schottky diodes [36]. First, two-terminal measurements are performed. The drain-source current I_{ds} is measured vs. the drain-source voltage V_{ds} at constant $V_{gs} = 5\ V$. The device is equivalent to two back-to-back Schottky diodes. The two

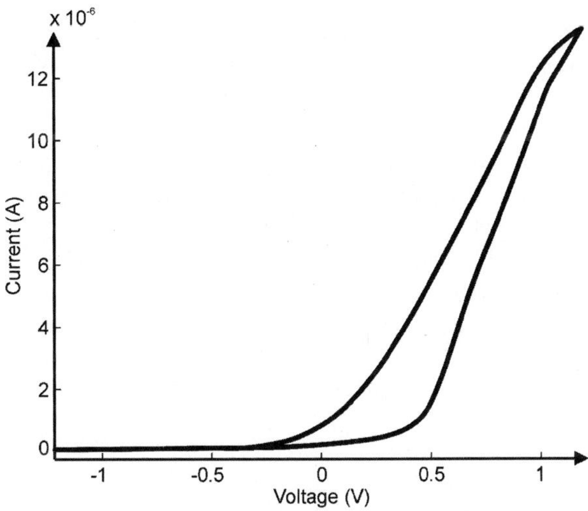

Fig. 8.15 Cyclic voltage measurement of an electrochemical organic memristor showing bistable memristive behavior. The image was taken from [30]

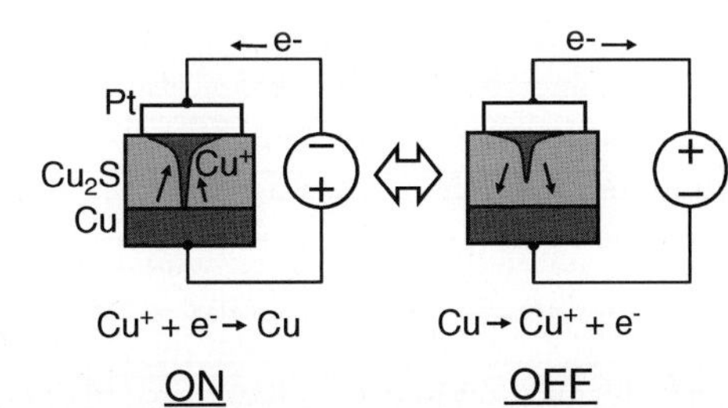

Fig. 8.16 Controlled filament formation of the three-terminal nanobridge device. The image was taken from [31]

diodes operate in opposite regimes: for negative V_{ds}, the source-to-channel diode is reversely biased while the drain-to-channel diode is forward biased. For positive V_{ds} both diodes invert their respective bias conditions. In either case, I_{ds} is limited by the current flowing in the reverse-biased diode. The reverse current of a metal-insulator-semiconductor diode has been observed to be very sensitive to charge trapping at the metal/semiconductor interface [17]. The large current value in the range of mA is most likely due to the large parallel parasitic structure in the bulk. In an ideal Schottky diode, the current is given by:

$$I = I_{\mathrm{S}} \cdot \mathrm{e}^{-\phi_{\mathrm{B}} q/kT} \left(\mathrm{e}^{Vq/kT} - 1\right) \tag{8.1}$$

with I and V the diode current and voltage, respectively, ϕ_{B} the Schottky barrier, k the Boltzmann constant, q the elementary charge, and T the absolute temperature. From the measured hysteretic behavior, it seems that the diode curve is modified as follows:

$$I = I_{\mathrm{S}} \cdot \mathrm{e}^{-\phi_{\mathrm{B}} q/kT} \left(\mathrm{e}^{(V - V_0(V))q/kT} - 1\right) \tag{8.2}$$

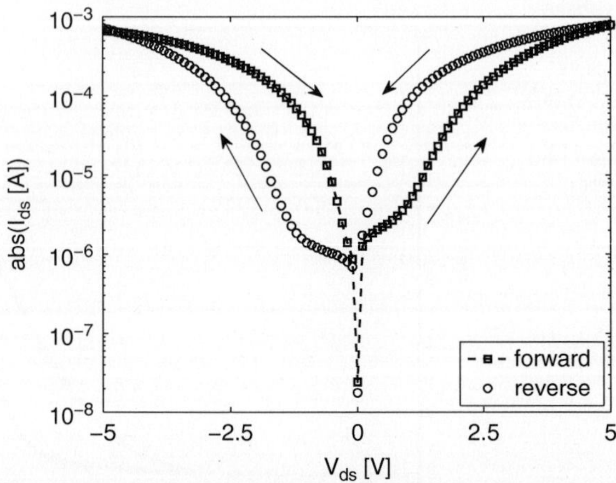

Fig. 8.17 $I_{ds} - V_{ds}$ characteristic showing the trapping/detrapping of charges at the metal/semiconductor junction. The device channel consists of ten SiNW in parallel. The forward sweep curve has a symmetrical correspondence with the reverse sweep curve, showing the respective Schottky barrier modulation. A current ratio of about 50 is found at either $V_{ds} = \pm 1V$. This behavior is typical of two-terminal memristive devices for ReRAM applications

with V_0 a built-in voltage at the Schottky contact that is positive for a positive V sweep and negative for a negative V sweep. As the measurement is performed at a speed which is comparable with the dynamic of interface states, a variation of the Schottky barrier height can be obtained by the variation of the image charge:

$$\Delta\phi_{\mathrm{B}} = \sqrt{\frac{q \cdot E_{app}}{4 \cdot \pi \cdot \epsilon_s}} \tag{8.3}$$

where E_{app} is the applied electric field across the Schottky junction and ϵ_s is the permittivity of the semiconductor. Moreover, for Schottky junctions, the barrier height can also be influenced by dipole lowering, which depends on the quality of the metal to semiconductor interface, and by the presence of impurities:

$$\Delta\phi_{\mathrm{B,dl}} \approx \alpha \cdot E_m \tag{8.4}$$

where α is the rate of change and E_m is the electric field perpendicular to the surface.

8.5 Four-Terminal Memristive Devices

Memristive functionality can be seen as state variable that can be used for more expressive logic gates [4]. The memristive behavior reported in Sect. 8.4 for the SB SiNW FETs can be tuned by operating on the polarity of the gate voltage [37].

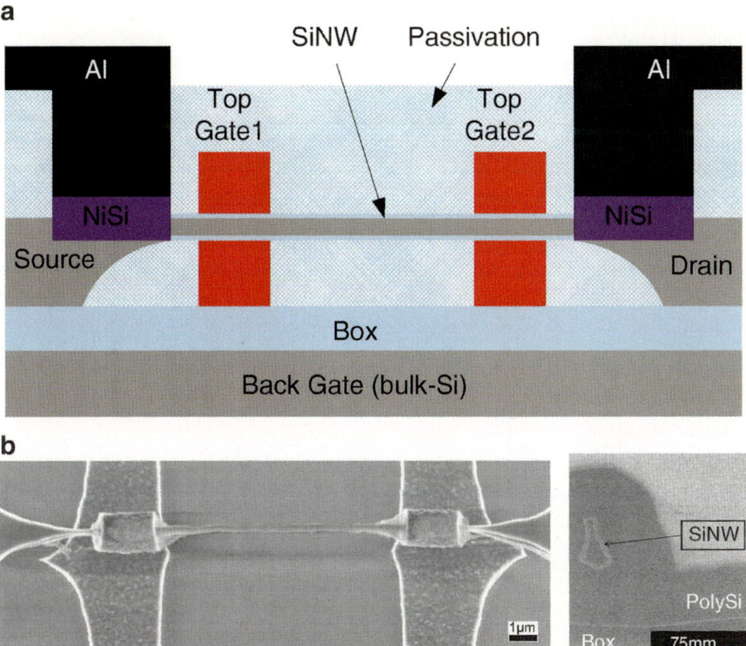

Fig. 8.18 (**a**) Schematic cross-section of a dual gate device with NiSi source and drain regions on SOI substrate. (**b**) A 20 μm long SiNW with two parallel GAA polysilicon gates having 4 μm gate lengths. (**c**) An FIB cut cross-section image showing the SiNW channel surrounded by a 500 nm polysilicon top gate

This type of behavior is linked with the double conductance, for holes and electrons. As described in [38] for SB *carbon-nanotube (CNT)* FETs, the ambipolarity can be controlled by using an additional control gate, such that it blocks one type of carrier conductance. Following this principle, four-terminal memristive SB SiNW FETs can be built (see Fig. 8.18). In the following, two modes of operation are reported, depending on the nature of the controlling signal applied at the Si nanowire channel.

8.5.1 Voltage-Controlled 4-T Memristive Device

A voltage-controlled four-terminal memristive Schottky barrier SiNW FET is obtained by using a dual gate configuration such that one of the two gates is controlling a portion of the channel that is between the source/drain contacts and the main gate. This configuration is exploited to control the ambipolarity imbalance, such as for CNT FETs [38]. Since the back-gate voltage modifies the ambipolar conductance, this fact can be used in ambipolar memristive devices to limit the current levels for one of the carriers. A fixed back-gate voltage $V_{bg} = +5$ V

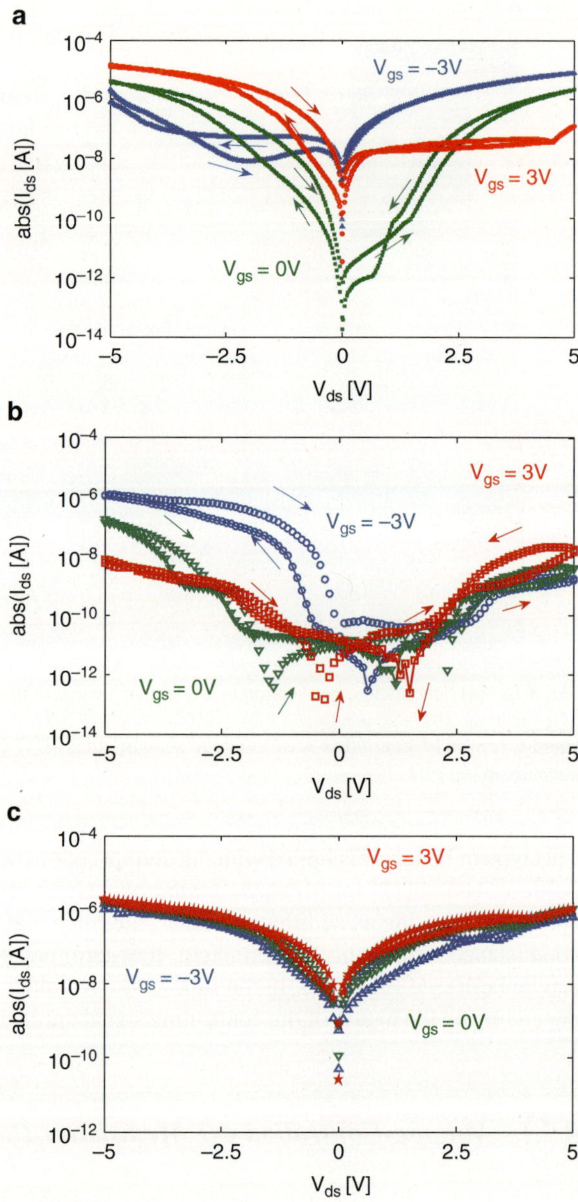

Fig. 8.19 (**a**) Controlled memristance for fixed top and back gate voltages $V_{bg} = +5$ V. (**b**) Controlled memristance for fixed top and back gate voltages $V_{bg} = -5$ V. (**c**) Controlled memristance for fixed top and back gate voltages $V_{bg} = 0$ V

(see Fig. 8.19a) leads to imbalanced bistable hysteresis loops under different V_{gs} voltages. By applying a negative $V_{bg} = -5$ V this imbalance is toggled to the negative side of the characteristics, giving a complementary effect (Fig. 8.19b). Finally, a $V_{bg} = 0$ V (Fig. 8.19c) levels off the conductances of electrons and holes, giving a fairly symmetric hysteresis.

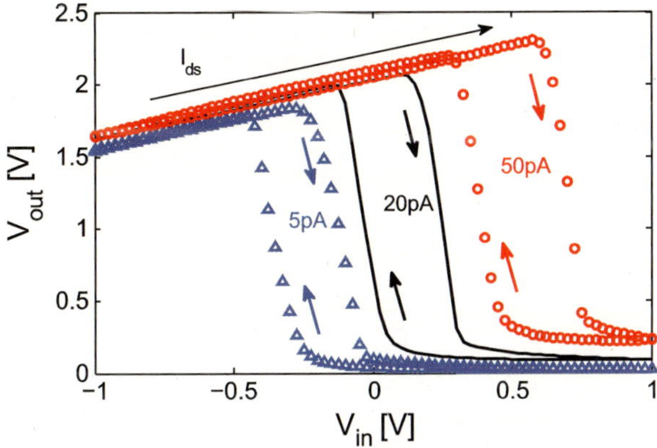

Fig. 8.20 Current-controlled memristive Schottky barrier SiNW FET hysteresis loop. For increasing I_{ds} current bias a V_{in} inversion threshold increases, suggesting applications for both memory, sensing (see Sect. 8.6.3) and threshold logic (see Sect. 8.6.4)

8.5.2 Current-Controlled 4-T Memristive Devices

A current-controlled version of the four-terminal memristive Schottky barrier SiNW FET is obtained by using a current I_{ds} bias instead of a V_{ds}. The output voltage is then compared with V_{gs} (Fig. 8.20). The obtained hysteresis can be used as a latch device, whose position in the $V_{out} - V_{in}$ plane can be adjusted by using a different value of the current bias. A similar behavior has been exploited with three-terminal Schottky-barrier polysilicon nanowire FETs circuits to build a new logic family based on precharge and evaluation scheme [39]. Moreover, a similar scheme has been demonstrated for DRAM type of memory [40] and for pA current and temperature detection [41]. Similarly, SB Si nanowire transistors fabricated with a low thermal budget process and biased in current-controlled mode show a similar hysteresis. Moreover, polycrystalline SiNWs SB FETs can give an hysteretic transfer characteristic (Fig. 8.21) very similar to the one reported for crystalline SiNW SB FETs fabricated with a low thermal budget process [42]. As it is shown in Fig. 8.21 the maximum output voltage in the transfer characteristics increases with the I_{ds} bias current. From Fig. 8.21a–c the sweep time is reduced. Similarly to what was discussed for the three-terminal SB SiNW FET memristive device, the sweeping time impacts on the amount of charge that traps at the gate oxide/channel interface, thus influencing the conductance state of the nanowire channel. A faster sweeping time outbalances the charge trapping/detrapping mechanism, resulting in lower output voltages (see Fig. 8.21a, b compared with Fig. 8.21c).

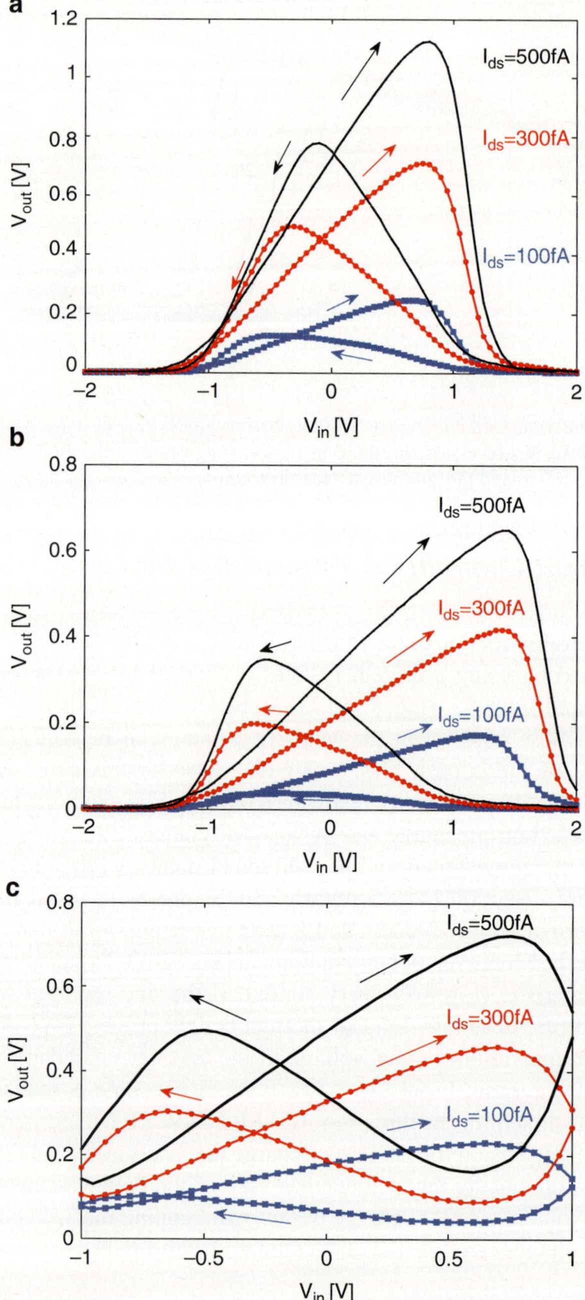

Fig. 8.21 V_{out} voltages for increasing with I_{ds} current bias. In all the figures I_{ds} bias values are 100 fA, 300 fA, and 500 fA. Notice the output voltage hysteresis narrows for increasing frequency sweep: (**a**) 24 s. (**b**) 6 s. (**c**) 0.5 s

8.6 Applications

Several applications of memristive devices can be envisaged, for instance: standalone memories, FPGA applications, synapse emulators for neuromorphic circuits, multi-value logic, physical and chemical sensors. In the following sections, examples of such applications will be showcased.

8.6.1 Field Programmable Gate Arrays

With the recent development of ReRAM technology, a number of novel *Field Programmable Gate Array (FPGA)* building blocks and architectures have been proposed in the past few years. For example, routing structures based on ReRAMs have shown promise. In [43], a cross point for switchboxes, using the ReRAMs as nonvolatile switches, is proposed to route signals through low-resistive paths, or to isolate them by means of high-resistive paths. The concept of routing elements based on ReRAM switches was then exploited in [44, 45] for timing optimization in FPGAs. In FPGAs, the programming customizable resources play a major role in the overall performance indexes and contribute to more than 80 % of the total delay and area. Instead of simply using ReRAMs as a replacement technology for configuration memories, we highlight an efficient opportunity by incorporating ReRAM-based MUXs into the routing architecture of an FPGA. With this respect, the authors proposed a solution to reduce the size of the building blocks by a factor of up to 3x compared to traditional flash memories [46]. The impact in FPGA design and we showed that area and critical path delay could be reduced by a factor of up to 28 % and 34 %, respectively, due to the compactness of ENVMs and the speed of ENVM-based switchboxes. Moreover, it is important to notice that the innovative architectures based on ReRAM devices can also be exploited in combination with other disruptive technologies, for instance 3D-stacking with *Through Silicon Via (TSV)* interconnects (see Fig. 8.22) [47].

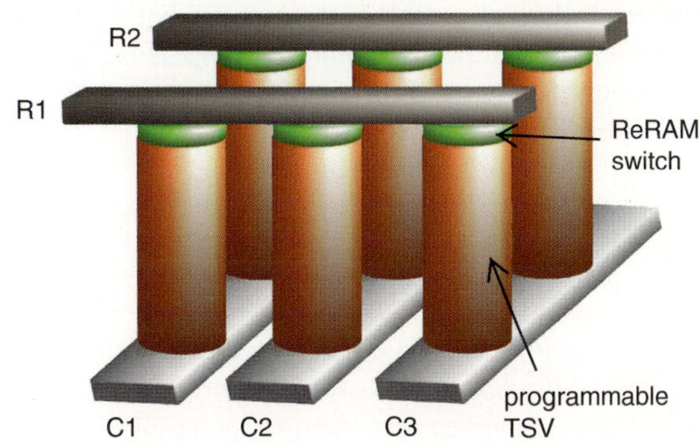

Fig. 8.22 A crossbar architecture made of ReRAM-TSV vertical interconnects

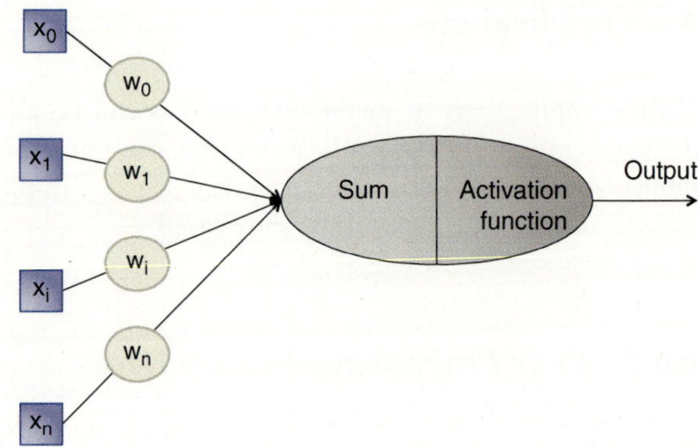

Fig. 8.23 Perceptron model of a neuron. Input signals x_i are weighted by programming each weight w_i and then summed. The output signal will be an activation function that depends on the summation

8.6.2 Neuromorphic Circuits

The nonvolatile property of the two-terminal memristive devices has a tremendous potential for neuromorphic circuits, in particular forming artificial synapses following the Hebbian rule of learning based on spike-rate-dependent plasticity (SRDP) as well as new building elements for hybrid CMOS/memristor circuits. For instance, when considering the perceptron model of the neuron (see Fig. 8.23), the weighted connections of the inputs to the summation element can be modeled with the properties of nonvolatile memristive devices.

With this respect, the Hudgkin-Huxley model can be mathematically described by first order differential equations. More specifically, Chua and Kang demonstrated that the H-H model of the potassium channel can be identified as a first-order time invariant voltage-controlled memristive one-port and that the sodium channel can be described as a second-order time invariant voltage-controlled memristive one-port. Since this representation is compatible with the mathematical representation of memristive devices, it is noteworthy to notice that memristive devices-based circuits can be built to emulate the behavior of biological systems, in this particular case emulating the potassium and sodium channels of the neurons. One example is an energy-efficient memristor-based integrate and fire neuron circuit which exploit the bistability of a ReRAM to model both the short time spike event and the refractory period [48]. Another example is the use of the analog programmability of the ReRAM devices that can be used to emulate the weighted connections of the perceptron model.

8.6.3 Multi-value Logic

Multiple-valued logic (MVL) is a candidate circuit design technique to replace traditional Boolean logic based on CMOS [49]. One-dimensional devices exploiting novel functionalities are the most credible candidates for future nanosystems [38, 50]. Recently, a fully CMOS compatible process flow has been proposed for

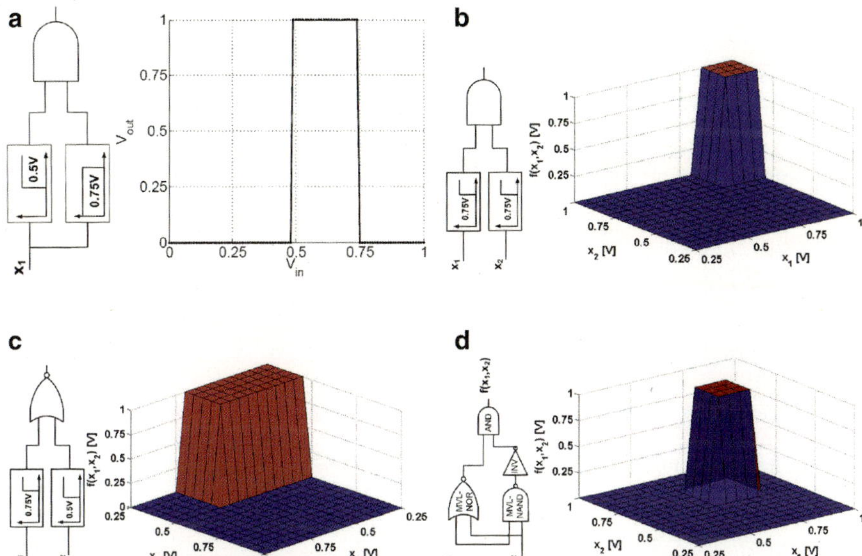

Fig. 8.24 (**a**) Opposite polarities TL nanowire gates having two different threshold voltages are used to construct a literal function of MVL input. (**b**) MVL-AND function obtained from two TL nanowire gates and a AND logic gate. (**c**) MVL NOR function obtained from two TL nanowire gates and a NOR logic gate. (**d**) Product of terms function obtained using a MVL-NOR gate and MVL-NAND in combination with a inverter and a Boolean logic AND gate

the fabrication of vertically stacked GAA SB SiNW transistors showing ambipolar behavior [8]. Ambipolarity has been envisaged as a way to build gates capable of exploiting denser logic functions [51], and here we originally propose to exploit the Schottky barrier SiNW ambipolarity for threshold logic (TL) gates, enabling the constructions of MVL functions.

Threshold logic (TL) gates are implemented exploiting the shift of the inversing point of the devices by changing the current bias. A set of three threshold gates having threshold voltages of 0.25 V, 0.5 V, 0.75 V are obtained by using bias currents of 20 pA, 35 pA, and 50 pA, respectively. A one-variable literal function using an AND gate and 2 TL gates having opposite polarities and thresholds at 0.5 V and 0.75 V is shown in Fig. 8.24a. A two-input MVL function can be obtained by connecting a two-input Boolean function with 2 TL gates. By using an AND gate we obtain a MVL-AND function (Fig. 8.24b), whereas an NOR gate is used to implement an MVL-NOR function (Fig. 8.24c). More complex combinations are possible, and a proposed MVL function is depicted in Fig. 8.24d, where MVL-NOR and MVL-NAND gates are combined with a Boolean INV and AND. The surface plot of Fig. 8.24d shows how a function can be mapped inside a two-variable input space. Some of the most commonly used functions in MVL are the MIN and MAX operators. A three-layer arrangement consisting of a first layer of TL gates, a second layer of Boolean logic gates, and a third layer of level recovery allows the

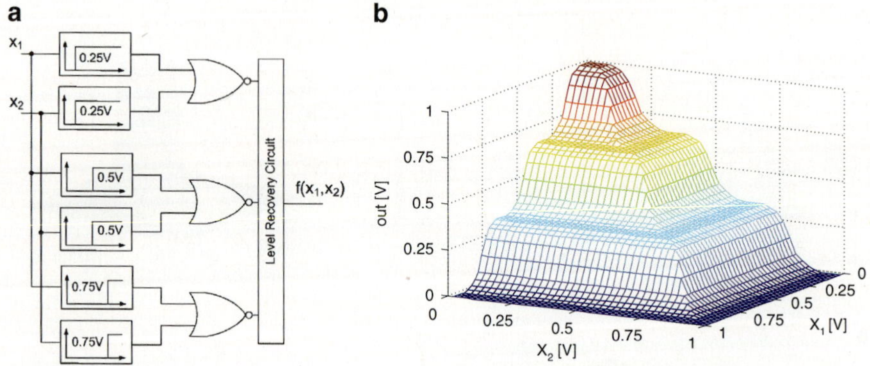

Fig. 8.25 (**a**) Circuit schematic of a NMAX logic function. Three MVL-NOR gates driving a level recovery circuit are used for mapping a two-variable input space into a one-variable output space. Note that the NMIN function can be implemented by changing the Boolean layer into a NAND one. (**b**) Output representation of the MVL-NMAX function implemented with nanowire based threshold gates

construction of the MVL-NMAX or MVL-NMIN operators (Fig. 8.25a). The 3D staircase output representation of the MVL-NMAX function is shown in Fig. 8.25b.

The proposed nanowire-based device proves to be a versatile TL block for the synthesis of several MVL functions. The excellent input/output voltage ranges observed experimentally enable cascaded operation of TL nanowire gates with Boolean gates. A reset signal is required to set the device threshold before using it for MVL logic circuit operation. Moreover, the threshold function related to the output characteristic of SB SiNW FET can also present an hysteretic behavior, which can further extend the functionality of the MVL circuits.

8.6.4 Current and Temperature Sensor

This paper reports on the fabrication and characterization of a pA current and temperature sensing device with ultra-low power consumption based on a Schottky barrier silicon nanowire transistor. Thermionic and trap-assisted tunneling current conduction mechanisms are identified and discussed on the base of the device sensitivity upon current and temperature biasing. In particular, very low current sensing properties are confirmed also with previously reported polysilicon-channel nanowire Schottky barrier transistors, demonstrating that these devices are suitable for temperature and current sensing applications. Moreover, the process flow compatibility for both sensing and logic applications makes these devices suitable for heterogeneous integration. A range of device operation conditions are investigated, showing how an ambipolar device can be used for different applications, the only requirement being the biasing condition.

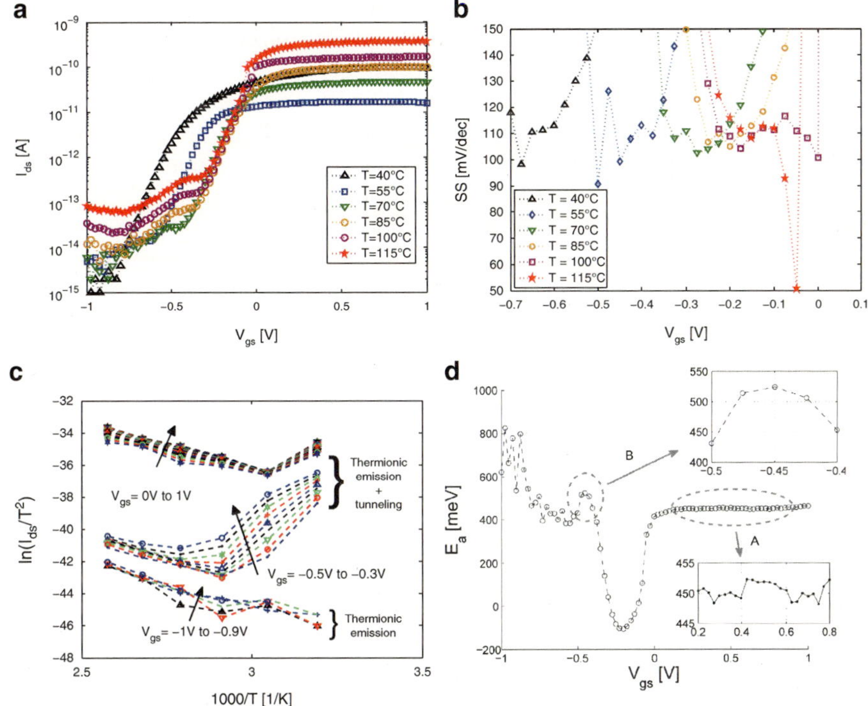

Fig. 8.26 (**a**) Effect of the temperature on the $I_{ds} - V_{gs}$ at $V_{ds} = 100\,mV$ (**b**) Subthreshold swings associated with the $I_{ds} - V_{gs}$. Very low swing minima are measured at 100 °C and 115 °C close to threshold voltages. Notice the voltage shift with temperature increase and the extremely low minima of 40 mV/dec for the highest temperature. (**c**) Arrhenius plot for different V_{gs} values showing both thermionic emission and tunneling mechanisms. The linear decreasing slopes are associated with thermionic emission regimes. (**d**) Extracted E_a over a large range of V_{gs}. Inset A shows constant $E_a \approx 450 \pm 5$meV. Inset B shows a maximum at 525meV which is taken as the value of the effective Schottky barrier height

The $I_{ds} - V_{gs}$ dependence with T is mainly attributed to the I_{th}; however; T also influences the Itunnel since hotter carriers pass through a narrower Schottky barrier, leading to an increasing current level [36]. The I_{OFF} current is increasing exponentially with temperature and its main contribution is a thermionic emission component. A different behavior has been observed for the I_{ON} current. Increasing the temperature makes the I_{ON} current to decrease until the temperature reaches 55 °C and then it rises exponentially with linear increase of T. At lower temperatures tunneling and trap-assisted tunneling are more important than thermionic emission. Rising T up to 70 °C makes the charges trapped into the gate oxide to un-trap, reducing the Itunnel component. A different behavior is observed for the I_{ON} currents for $70\,°C \leq T \leq 115\,°C$. In this range, the I_{ON} exponentially increases with T. This effect is evidence of two main current components, for which the I_{ON} changes from a tunneling to a thermionic emission dominated regime. A set of $I_{ds} - V_{gs}$ curves (Fig. 8.26a) taken at different temperatures at constant $V_{ds} = 100\,mV$ and $V_{bg} = 5\,V$

are used to extrapolate the Arrhenius plot (Fig. 8.26c). The constant $V_{bg} = 5\,V$ is used to set the device operation more favorable for electron conductance at low Vgs. Constant subthreshold swings$\approx 110\,mV/dec$ are observed independently from the temperature (see Fig. 8.26b). Low negative V_{gs} voltages ranging from $-1\,V$ to $0\,V$ show an almost linear slope with inverse of temperature and can be correlated with a thermionic-emission regime. However, for this V_{gs} range the current level is on the order of fAs, which is comparable to the background noise, and it cannot be used to extrapolate the Schottky barrier height. Another distinct regime is observed for $-0.3\,V \leq V_{gs} \leq -0.5\,V$, for which the slopes are greatly affected by tunneling. This regime shows a dominant tunneling component for the two lowest temperatures. Finally, an exponential dependence with T is observed again for $V_{gs} \geq 0V$ with the exception of the lower temperature. All these regimes demonstrate that the current in our device is mainly thermionic for $\geq 70°C$ and that the tunneling contribution is trap assisted. The slopes from the Arrhenius plot are then used to extract the effective Schottky barrier height ϕ_{Beff} with the activation energy E_a method. As shown in inset A of Fig. 8.26d, an average effective barrier height $E_a \approx 450 \pm 5meV$ is found over a large range of $V_{gs} \geq 0.2V$. However, these values cannot be taken as Schottky barrier height since in this regime the device has both tunneling and thermionic components. As suggested by Svensson et al. [51], a better evaluation of the Schottky barrier height can be taken at the maximum of Ea for low current levels. As shown in the inset B of Fig. 8.26d, this maximum corresponds to $V_{gs} = -0.45\,V$ and gives a $\phi_{Beff} = 525\,meV$, confirming the mid-gap Schottky barrier height.

8.6.4.1 pA Current Sensing

Current biasing the devices with a constant Ids current makes the device to behave as a pseudo-inverter configuration with hysteretic transfer function. Thanks to the ambipolarity, the Vout-Vin curves shift linearly with the applied current bias. For instance in Fig. 8.27a, low pA current levels can be either read from the high-to-low or the low-to-high transition voltage with sensitivities of 17 mV/pA. A similar biasing scheme for polysilicon nanowires has been previously characterized by the authors show a similar trend. In Fig. 8.27b, forward and reverse threshold voltages for currents between 100 fA and 500 fA show a linear increase with current (adapted from [52]).

8.6.4.2 Temperature Sensing

Another application is temperature sensing. Upon application of increasing temperature of operation, the hysteresis window observed in pseudo-inverter biasing scheme shrinks. The crystalline Si nanowire Schottky barrier FET shows different sensitivities at different temperature regimes, depending on which mechanism dominates the conductance. Since the hysteresis is attributed to the storage of charges in either gate oxide and/or at the Schottky barrier junctions [9], an increased

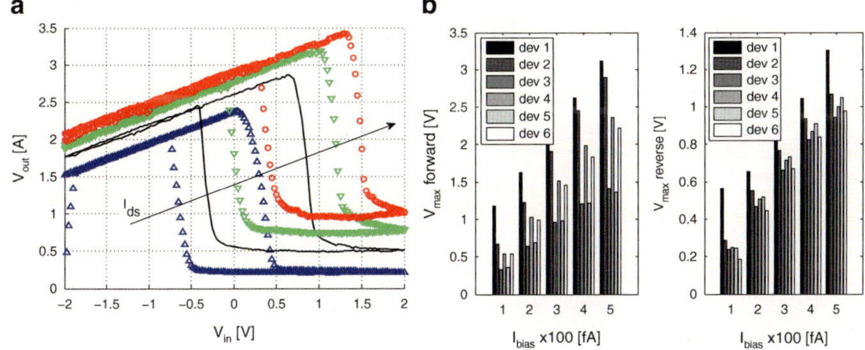

Fig. 8.27 (**a**) Measured input–output transfer characteristics of a hysteretic inverter based on a single Si nanowire FET with low current bias, showing current-dependent thresholds. (**b**) Forward and reverse threshold voltages for polysilicon Schottky barrier FETs under constant current biasing from 100 fA up to 500 fA (adapted from [52])

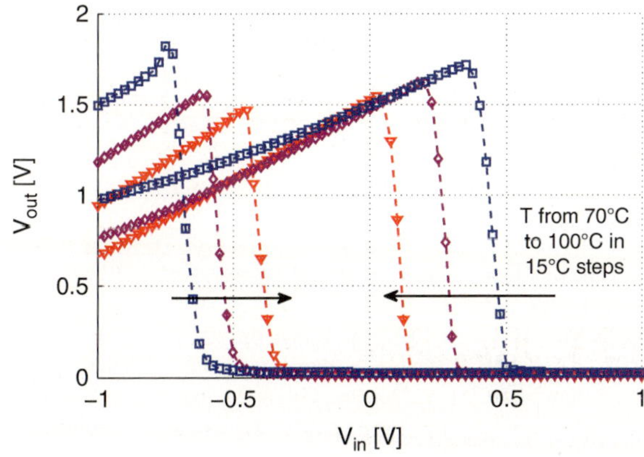

Fig. 8.28 The hysteresis window shrinks with increasing temperature. Within this T range, the temperature sensitivity of 10 mV/°C is related to the thermionic current regime

hysteresis window is expected for the lowest temperatures. The highest sensitivity of 40 mV/°C is found in the T range around 40 °C at which the trap tunneling mechanism dominates. For temperatures higher than 55 °C the sensitivity tends to saturate according to the dominance of thermionic current contribution, leading to lower sensitivity of 10 mV/°C. In Fig. 8.28 the hysteresis window shrinks for increasing T when 70 °C $\leq$ T $\leq$ 100 °C.

8.7 Conclusions

An overview on Si nanowire-based multi-terminal memristive devices has been given, showcasing several potential disruptive applications touching the area of logic, memory and sensing. In addition to the very broad application potential offered by the memristive Schottky barrier SiNW FETs, the ease of fabrication and top-down CMOS process compatibility paves the way to novel circuit architectures implementing the memristive functionality.

Acknowledgements The authors would like to thank Dr. Sandro Carrara of LSI at EPFL for useful discussions. Moreover, the authors thank the CMI staff of EPFL for help with the micro- and nano-fabrication.

References

1. URL http://www.itrs.net
2. L. Chua, IEEE Trans. Circuit Theory **18**(5), 507 (1971)
3. L. Chua, S.M. Kang, Proc. IEEE **64**(2), 209 (1976)
4. J. Borghetti, G.S. Snider, P.J. Kuekes, J.J. Yang, D.R. Stewart, R.S. Williams, Nature **464**, 873 (2010)
5. D.B. Strukov, G.S. Snider, D.R. Stewart, R.S. Williams, Nature **453**, 80–83 (2008)
6. J.E. Green, J. Wook Choi, A. Boukai, Y. Bunimovich, E. Johnston-Halperin, E. Deionno, Y. Luo, B.A. Sheriff, K. Xu, Y. Shik Shin, H. Tseng, J.F. Stoddart, J.R. Heath, Nature **445**, 414 (2007). doi:10.1038/nature05462
7. M. Zervas, D. Sacchetto, G. De Micheli, Y. Leblebici, Microelectron. Eng. **88**(10), 3127 (2011). doi:10.1016/j.mee.2011.06.013. URL http://dx.doi.org/10.1016/j.mee.2011.06.013
8. D. Sacchetto, M. Ben-Jamaa, G. De Micheli, Y. Leblebici, in *Solid State Device Research Conference, 2009. ESSDERC '09. Proceedings of the European*, , pp. 245–248, Athens, 14–18 September 2009
9. D. Sacchetto, M. Ben-Jamaa, S. Carrara, G. De Micheli, Y. Leblebici, in *Circuits and Systems (ISCAS), Proceedings of 2010 IEEE International Symposium on*, Paris, 30 May 2010–2 June 2010, pp. 9–12. doi:10.1109/ISCAS.2010.5537146
10. D. Kim, H. Lee, H. Jung, S. Kang, Bull. Korean. Chem. Soc. **28**(5), 783 (2007)
11. W. Kusnezow, A. Jacob, A. Walijew, F. Diehl, J. Hoheisel, Proteomics **3**(3), 254 (2003)
12. D. Sacchetto, M.A. Doucey, G. De Micheli, Y. Leblebici, S. Carrara, BioNanoScience **1**, 1 (2011). URL http://dx.doi.org/10.1007/s12668-011-0002-9. doi: 0.1007/s12668-011-0002-9
13. Y. Kim, J. Lee, J. Appl. Phys. **104**(11), 114115 (2008)
14. R. Dong, D.S. Lee, W.F. Xiang, S.J. Oh, D.J. Seong, S.H. Heo, H.J. Choi, M.J. Kwon, S.N. Seo, M.B. Pyun, M. Hasan, H. Hwang, Appl. Phys. Lett. **90**(4), 042107 (2007)
15. K. Tsunoda, K. Kinoshita, H. Noshiro, Y. Yamazaki, T. Iizuka, Y. Ito, A. Takahashi, A. Okano, Y. Sato, T. Fukano, M. Aoki, Y. Sugiyama, in *Electron Devices Meeting, 2007. IEDM 2007. IEEE International*, pp. 767–770, Washington, DC, 10–12 December, 2007. doi:10.1109/IEDM.2007.4419060
16. W.Y. Chang, Y.C. Lai, T.B. Wu, S.F. Wang, F. Chen, M.J. Tsai, Appl. Phys. Lett. **92**(2), 022110 (2008)
17. W. Zhu, T. Chen, Y. Liu, M. Yang, S. Zhang, W. Zhang, S. Fung, IEEE Trans. Electron Devices **56**(9), 2060 (2009)
18. T. Driscoll, H.T. Kim, B.G. Chae, M.D. Ventra, D.N. Basov, Appl. Phys. Lett. **95**(4), 043503 (2009)
19. J. Blanc, D.L. Staebler, Phys. Rev. B **4**(10), 3548 (1971)
20. R. Waser, M. Aono, Nat. Mater. **6**, 833 (2007)
21. J.J. Yang, M.D. Pickett, X. Li, O.A. A., D.R. Stewart, R.S. Williams, Nat Nanotechnol. **3**(7), 429 (2008)

22. C. Kugeler, C. Nauenheim, M. Meier, A. Rudiger, R. Waser, *Non-volatile Memory Technology Symposium, 2008. NVMTS 2008. 9th Annual*, pp. 1–6, Pacific Grove, CA, 11–14 November 2008. doi:10.1109/NVMT.2008.4731195
23. M.H.R. Lankhorst, B.W.S.M.M. Ketelaars, R.A.M. Wolters, Nat. Mater. **4**, 347 (2005) doi:10.1038/nmat1350
24. A. Cywar, G. Bakan, C. Boztug, H. Silva, A. Gokirmak, Appl. Phys. Lett. **94**(7), 072111 (2009)
25. D.R. Stewart, D.A.A. Ohlberg, P.A. Beck, Y. Chen, R.S. Williams, J.O. Jeppesen, K.A. Nielsen, J.F. Stoddart, Nano Lett. **4**(1), 133 (2004)
26. A.R. Pease, J.O. Jeppesen, J.F. Stoddart, Y. Luo, C.P. Collier, J.R. Heath, Acc. Chem. Res. **34**(6), 433 (2001). doi:10.1021/ar000178q. URL http://pubs.acs.org/doi/abs/10.1021/ar000178q. PMID:11412080
27. C.N. Lau, D.R. Stewart, R.S. Williams, M. Bockrath, Nano Lett. **4**(4), 569 (2004). doi:10.1021/nl035117a. URL http://pubs.acs.org/doi/abs/10.1021/nl035117a
28. X. Wang, Y. Chen, H. Xi, H. Li, D. Dimitrov, IEEE Electron Device Lett. **30**(3), 294 (2009)
29. Y.V. Pershin, M. Di Ventra, Phys. Rev. B **78**(11), 113309 (2008)
30. T. Berzina, S. Erokhina, P. Camorani, O. Konovalov, V. Erokhin, M. Fontana, ACS Appl. Mater. Interfaces **1**(10), 2115–2118 (2009)
31. T. Sakamoto, N. Banno, N. Iguchi, H. Kawaura, S. Kaeriyama, M. Mizuno, K. Terabe, T. Hasegawa, M. Aono, in *Electron Devices Meeting, 2005. IEDM Technical Digest. IEEE International*, pp. 475–478, Washington, DC, 5–7 December 2005. doi:10.1109/IEDM.2005.1609383
32. Y. Kaneko, H. Tanaka, M. Ueda, Y. Kato, E. Fujii, in *Device Research Conference (DRC), 2010*, pp. 257–258, South Bend, Indiana, USA, 21–23 June 2010. doi:10.1109/DRC.2010.5551971
33. M. Haykel Ben Jamaa, S. Carrara, J. Georgiou, N. Archontas, G. De Micheli, in *Nanotechnology, 2009. IEEE-NANO 2009. 9th IEEE Conference on*, pp. 152–154, Genoa, 26–30 July 2009
34. M. Bawedin, S. Cristoloveanu, D. Flandre, Solid State Electron. **51**, 1252 (2007). doi:10.1016/j.sse.2007.06.024
35. J.J. Yang, F. Miao, M.D. Pickett, D.A.A. Ohlberg, D.R. Stewart, C.N. Lau, R.S. Williams, Nanotechnology **20**(21), 215201 (9pp) (2009)
36. S.M. Sze, K.K. Ng, *Physics of Semiconductor Devices*, John Wiley & Sons, Inc., Hoboken, New Jersey, 3rd edn. (Wiley, New York, 2007)
37. D. Sacchetto, Y. Leblebici, G. De Micheli, IEEE Electron Device Lett. **33**(2), 143 (2012)
38. J. Appenzeller, J. Knoch, M. Bjork, H. Riel, H. Schmid, W. Riess, IEEE Trans. Electron Devices **55**(11), 2827 (2008)
39. S. Ecoffey, M. Mazza, V. Pott, D. Bouvet, A. Schmid, Y. Leblebici, M. Declereq, A. Ionescu, *Electron Devices Meeting, 2005. IEDM Technical Digest. IEEE International*, pp. 269–272, Washington, DC, 2–5 December 2005. doi:10.1109/IEDM.2005.1609325
40. S. Ecoffey, Ultra-thin nanograin polysilicon devices for hybrid CMOS-NANO integrated circuits. Ph.D. thesis, Lausanne (2007). doi:10.5075/epfl-thesis-3722. URL http://library.epfl.ch/theses/?nr=3722
41. D. Sacchetto, G. De Micheli, Y. Leblebici, in *The 16th International Conference on Solid-State Sensors, Actuators and Microsystems*, Beijing, China, 5–9 June 2011
42. D. Sacchetto, V. Savu, G.D. Micheli, J. Brugger, Y. Leblebici, Microelectron. Eng. **88**(8), 2732 (2011)
43. P. Gaillardon, M. Ben-Jamaa, G. Beneventi, F. Clermidy, L. Perniola, in *Electronics, Circuits, and Systems (ICECS), 2010 17th IEEE International Conference on*, pp. 62–65, IEEE, Athens, 12–15 December 2010
44. S. Tanachutiwat, M. Liu, W. Wang, IEEE Trans. VLSI Syst. **19**(11), 2023 (2011)
45. J. Cong, B. Xiao, in *Nanoscale Architectures (NANOARCH), 2011 IEEE/ACM International Symposium on*, pp. 1–8, IEEE, San Diego, CA, 8–9 June 2011
46. P. Gaillardon, D. Sacchetto, G.B. Beneventi, M.H. Ben Jamaa, L. Perniola, F. Clermidy, I. O'Connor, G. De Micheli, Design and Architectural Assessment of 3-D Resistive Memory Technologies in FPGAs, *Nanotechnology, IEEE Transactions on*, **12**(1), 40–50 (2013). doi:10.1109/TNANO.2012.2226747

47. D. Sacchetto, M. Zervas, Y. Temiz, G. De Micheli, Y. Leblebici, IEEE Trans. Nanotechnol. **PP**(99), 1 (2011). doi:10.1109/TNANO.2011.2160557
48. S. Shin, D. Sacchetto, Y. Leblebici, S.M. Kang, in *Cellular Nanoscale Networks and Their Applications (CNNA), 2012 13th International Workshop on*, pp. 1–4, Turin, Italy, 29–31 August 2012. doi:10.1109/CNNA.2012.6331427
49. A. Schmid, Y. Leblebici, IEE Proc. Comput. Digit. Tech. **151**(6), 435 (2004). doi:10.1049/ip-cdt:20041099
50. H. Iwai, Microelectron. Eng. **86**(7–9), 1520 (2009)
51. J. Svensson, A. Sourab, Y. Tarakanov, D. Lee, S. Park, S. Baek, Y. Park, E. Campbell, Nanotechnology **20**(17), 175204 (2009)
52. D. Sacchetto, V. Savu, G.D. Micheli, J. Brugger, Y. Leblebici, Microelectron. Eng. **88**(8), 2732 (2011). doi:10.1016/j.mee.2010.12.117. URL http://www.sciencedirect.com/science/article/pii/S0167931711000062. <ce:title>Proceedings of the 36th International Conference on Micro- and Nano-Engineering (MNE)</ce:title><xocs:full-name>36th International Conference on Micro- and Nano-Engineering (MNE)</xocs:full-name>

Chapter 9
Spintronic Memristor as Interface Between DNA and Solid State Devices

Yiran Chen, Hai Li, and Zhenyu Sun

9.1 Introduction

Magnetic sensing is widely used in various modern bio-medical devices since many physiological functions (e.g., nerve impulses) generate electrical currents that create magnetic field [24]. Monitoring such signals by detecting magnetic field is less invasive and more reliable than implanting electrodes to sense the electronic signals. Generally, magnetic sensors can be utilized to detect the changes or disturbances of magnetic field, i.e., the strength and/or direction of magnetic flux. For example, magnetic sensors with high sensitivity have been widely used in heart disease monitor by detecting the bio-magnetic signals from heart (known as *magnetocardiography*, or MCG) [8]. The magnetic sensors in bio-medical applications are required to detect the low-field signals that are much lower than the Earth's magnetic field ($< 0.5\,\text{Oe}$) [17].

Compared to other low-field sensing techniques, such as search coil, flux gate, and *superconducting quantum interference detectors* (SQUID) [14, 15, 18], solid state sensors have demonstrated many unique advantages, including small size ($< 0.1\,\text{mm}^2$), low power consumption, high sensitivity ($\sim 0.1\,\text{Oe}$), and good compatibility with CMOS technology, etc. [17]. A solid state magnetic sensor usually converts a magnetic field into the change of resistance that can be easily detected by applying a sense current/voltage.

The *giant magnetoresistance* (GMR) spin valve sensor and *tunnel magnetoresistance* (TMR) sensor are two major solid state magnetic sensor technologies used for low magnetic field detection. In both technologies, the relative angle of the magnetization directions of two ferromagnetic layers changes in the presence of magnetic field, leading to the resistance variation of the sensor. The GMR and TMR

Y. Chen (✉) • H. Li • Z. Sun
University of Pittsburgh, 1140 Benedum Hall, Pittsburgh, PA 15261, USA
e-mail: yic52@pitt.edu; hal66@pitt.edu; zhs25@pitt.edu

R. Tetzlaff (ed.), *Memristors and Memristive Systems*,
DOI 10.1007/978-1-4614-9068-5_9,

technologies have been successfully utilized in the recording head of *hard disk drive* (HDD) for almost two decades [11]. Thanks to the advance of integrating GMR/TMR sensors with CMOS VLSI circuitry, fabricating a large TMR/GMR micro-array became economically feasible to realize some complicated sensing systems. For example, a GMR sensor array was designed for DNA assay [6]: The DNA samples being detected are pre-labeled (tagged) by magnetic nanoparticles. Under an external magnetic field, the GMR sensor array can capture the magnetic responses of the nanoparticles, which reflect the density and distributions of DNA samples.

On the one hand, the utilization of nanoparticles with a diameter of $100 - 1000A$ is attracting in magneto-nano biochip [12]. Such a dimension range is comparable to that of the target DNA molecules. In addition, the magnetic field for nanoparticle excitation can be modulated to reduce the impact of $1/f$ noise, enabling useful signal detection in a limited frequency band. On the other hand, detecting such tiny magnetic nanoparticles with limited physical volume is extremely challenging because the induced magnetic moments are very low and constrained by external factors such as noise and thermal turbulence [10]. Thus, ultra high sensitive detector is required in mass-produced magnetic microarray.

Besides TMR and GMR devices-based sensor, another emerging nonvolatile device—memristor is also very promising in the nanoparticle detection. Especially, the recent proposed domain wall motion-based spintronic memristor [22] can change its memristance (resistance) when responding to its external electrical or magnetic excitation. Unlike GMR and TMR sensors, the resistance states of memristor still maintain even the external excitations are removed. Due to the different magnetic and electrical characteristics among TMR, GMR, and spintronic memristor, operating mechanisms of these devices are also different. In this work, we present a novel spintronic memristor-based sensing mechanism which has different relative resistance change, noise, sensitivity, reliability compared to TMR and GMR sensor. Also corresponding optimization philosophy will be discussed here.

The rest of this chapter is organized as follows. Section 9.2 gives the fundamentals of GMR/TMR sensors and spintronic memristor, as well as the application of magnetic sensors in DNA assay. The proposed magnetic sensing mechanism of spintronic memristors will be described in Sect. 9.3. We then discuss the design tradeoffs of different devices in DNA assay application in Sect. 9.4 and analyze the noise endurability of the proposed readout scheme in Sect. 9.5. At last, Sect. 9.6 concludes our work.

9.2 Preliminary

9.2.1 GMR and TMR Sensors

Figure 9.1a, b illustrate the structures of a typical GMR spin valve sensor and a typical TMR device, respectively. Both structures are in the form of *magnetic*

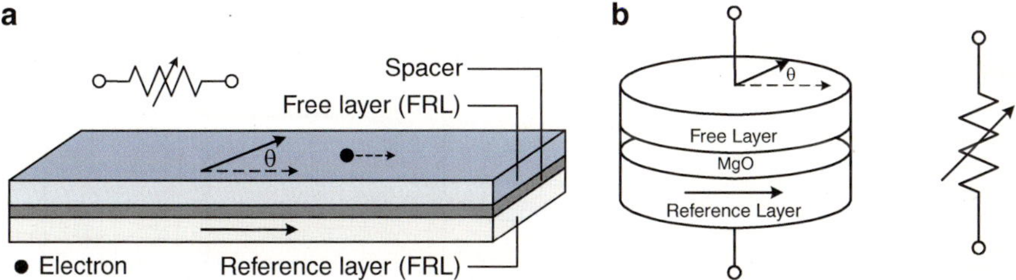

Fig. 9.1 (**a**) GMR spin valve sensor structure and (**b**) TMR MTJ structure

tunneling junction (MTJ) consisting of two ferromagnetic layers, namely, *reference layer* (RL) and *free layer* (FL). The two ferromagnetic layers are separated by a thin non-ferromagnetic spacer, e.g., AlO in GMR and MgO in TMR devices. The *magnetization direction* (MD) of RL is fixed by coupling to a pinned magnetic layer while the MD of FL can be changed in the presence of magnetic field. The MTJ resistance is determined by the relative angle between the MD's of RL and FL: when the MD of RL is parallel or anti-parallel to that of FL, the MTJ reaches its minimum or maximum resistance value represented by R_L or R_H, respectively. The magnetoresistance ratio, therefore, is defined as $MR = (R_H - R_L)/R_L$.

As illustrated in Fig. 9.1a, the electrons in a GMR spin valve device travel with their spins orientated parallel to the ferromagnetic layers. In contrast, we can eject electrons through the oxide barrier to change the MD of FL in a TMR device, as shown in Fig. 9.1b.

In the presence of a magnetic field, the resistance of a GMR or TMR sensor vary between R_L and R_H. The exact resistance value that is determined by the direction and amplitude of the magnetic field can be detected by applying a small sensing current through the device and measuring the voltage across it. Once the sensed magnetic field is removed, the sensor will resume back to its stable state, at which the MD of FL is parallel or anti-parallel to that of RL.

9.2.2 Spintronic Memristor

As early as year 1971, Professor L. Chua predicted the existence of memristor based on the completeness of circuit theory: besides resistor, capacitor, and inductor, there must be a fourth fundamental passive circuit element to bridge the electrical charge (q) and the magnetic flux (ϕ) [4]. The element was named as memristor. The corresponding physical variable—memristance (M) is uniquely determined by the historic profile of the current/voltage through it.

The memristor can be realized by many different materials, such as Mn-doped ZnO films [23], Pt/$BiFeO_3$/Nb-doped $SrTiO_3$ [7], or even carbon nanotube [13]. Figure 9.2 shows the spintronic memristor proposed by Wang et al. [22]. The

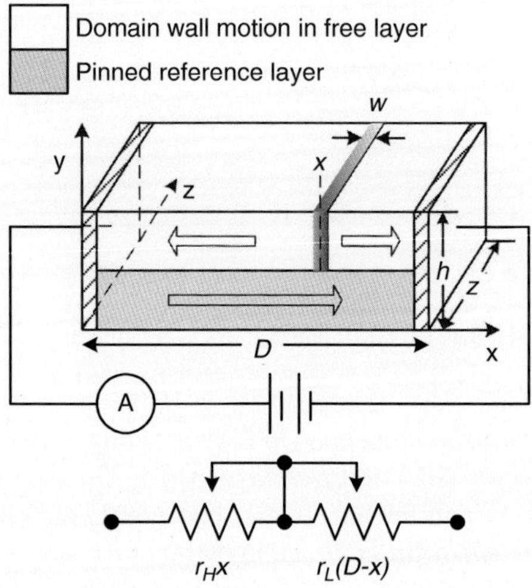

Fig. 9.2 A spintronic memristor based on magnetic domain wall motion

structure is very similar to the GMR sensor except that its free layer is divided by a magnetic domain wall into two segments with opposite magnetization directions. When applying a current through the spintronic memristor, the domain wall could move along the longitudinal direction. Because of the difference in resistance per unit length of the two segments, the overall memristance varies from R_L to R_H, determined by the domain wall position $X(t)$ at time t. In general, the device resistance can be calculated as:

$$R(t) = R_H - (R_H - R_L)X(t)/D, \tag{9.1}$$

where D represents the length of memristor.

We note that the domain wall movement happens only when the applied current density (J) is above the critical current density (Jcr). Also, the domain wall velocity v is proportional to the current density J. Besides the spin torque excitation generated by electrical current, the domain wall mobility is also affected by the thermal fluctuation and the applied magnetic field [21]. Under certain conditions, the amplitude of magnetic field can be sensed as the resistance of the spintronic memristor (or the domain wall location). As an obvious advantage of spintronic memristor, the device state, i.e., the domain wall location, will maintain after the sensed magnetic field is removed.

9.2.3 Magnetic Sensor Microarray for DNA Assay

The principle of magnetic sensor microarray-based DNA assay is shown in Fig. 9.3 [20]: Single-stranded DNA receptors (or known as DNA probes) are immobilized on the surface of a magnetic sensor microarray. The unknown

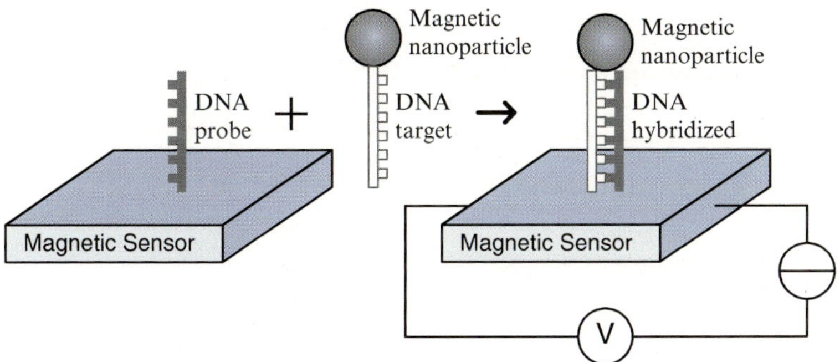

Fig. 9.3 Magnetic sensor microarray for DNA assay

DNA fragments (targets) are labeled by high moment magnetic nanoparticles (nano-tags) with some binding technique, e.g., biotin-streptavidin chemistry. During DNA detection, the tagged DNA fragments (targets) are captured by the complimentary DNA probes. To detect the density and the distribution of the DNA targets, nanoparticles are excited by applying an external magnetic field and their corresponding magnetic responses are sensed by the magnetic sensor array.

9.3 Spintronic Memristor Based Magnetic Sensing

The spin torque-induced domain wall motion at finite temperature can be described by stochastic Landau–Lifshitz–Gilbert equation [2] with a spin torque term [1, 16]. Using rigid wall approximation [5, 19], the domain wall motion is expressed in terms of magnetization spherical angle θ and φ as

$$\theta(x,t) = \theta_0(x - X(t)), \varphi(x,t) = \varphi_0(t) \tag{9.2}$$

where x is the position of the domain wall along the length of the memristor. $\theta_0(x)$= arccos[tanh(x/w)] is the function of domain wall shape. w is the domain wall thickness. $X(t)$ is the domain wall position. Domain wall velocity $v = dX(t)/dt$. The domain wall position $X(t)$ satisfies following stochastic differential equations [5, 19]:

$$\frac{d\phi}{dt} + \frac{a}{w}\frac{dX}{dt} = \gamma \cdot H + \eta_\phi, \frac{1}{w}\frac{dX}{dt} - \alpha\frac{d\phi}{dt} = w_0 sin(2\phi) + \frac{V_s}{w} + \eta_x. \tag{9.3}$$

where $w_0 = \gamma H_p/2$ (δ is gyro-magnetic ratio); is the damping parameter; H is the magnetic field including external unbalanced field and intrinsic domain wall pinning field. $v_s = PJ\mu_B/eMs$ is the spin torque excitation strength; P is the polarization efficiency; B is the Bohr magneton; and e is the elementary electron charge. $\eta\varphi(t)$ and $\eta X(t)$ are the φ and X components of the thermal fluctuation fields, respectively.

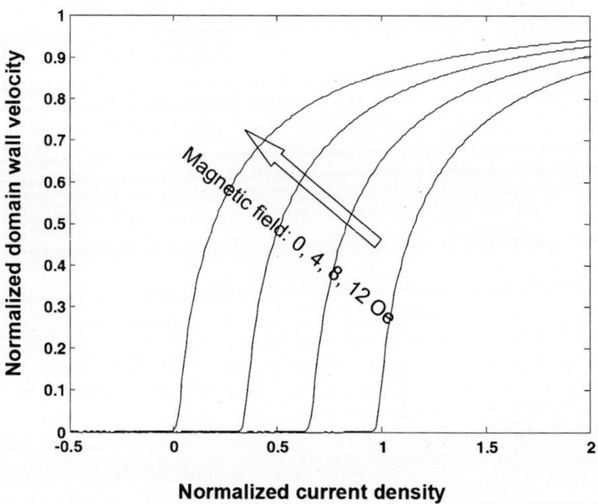

Fig. 9.4 The normalized domain wall velocity as a function of the normalized current density under the different magnetic fields at 300 K

Spintronic memristors sense the magnetic field change with magnetic nanoparticles relative to that without magnetic nanoparticles [3]. Figure 9.4 demonstrates the normalized domain wall velocity as a function of the normalized current density under different magnetic fields at 300 K. The magnetic field is the additional field due to the existence of the bound magnetic nanoparticles. When the applied current density is close to the critical one (*Jcr*), the domain wall velocity becomes very sensitive to the amplitude of the applied external magnetic field.

During the sensing operation, a current with the density slightly below the critical current density *Jcr* is applied to spintronic memristor for certain duration. The domain wall will move to the different locations under the different magnetic fields excited by the bound magnetic nanoparticles.

Figure 9.5 shows the spintronic memristor resistance as a function of time under the different number of magnetic nanoparticles at 300 K. Based upon [3], the amplitude of magnetic field generated by nanoparticles is approximately proportional to nanoparticle numbers. 500 nanoparticles in Fig. 9.5 corresponds to a magnetic field around 10 Oe at the spintronic domain wall center.

9.4 Design Space Exploration

9.4.1 Biosensor Array Architecture

Similar biosensor array architecture shown in Fig. 9.6 can be used for GMR, TMR, and spintronic memristor. In our design, an N×N array of biosensor cells form a DNA spot. The whole on-chip biosensor is composed of 44 DNA spots with the corresponding control bus. To increase the throughput of the biosensor array,

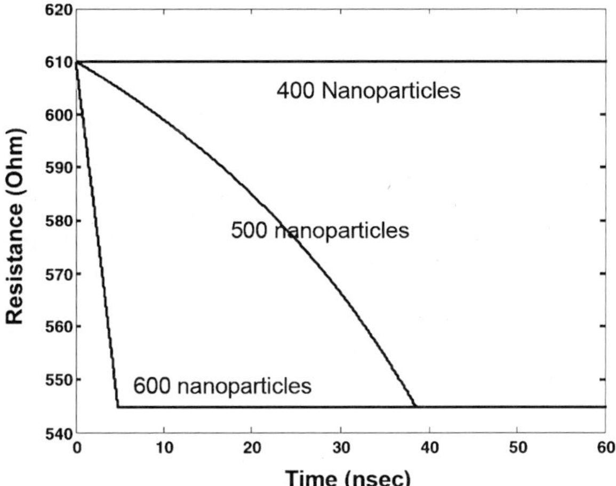

Fig. 9.5 Spintronic memristor resistance as a function of time under different number of magnetic nanoparticles at 300 K

both frequency division multiplexing (FDM) and time division multiplexing (TDM) are utilized [5]. In this architecture, every four DNA spots share one physical link with four carrier frequencies to realize FDM. Then all the four FDM channels are connected to one 4-to-1 multiplexer to achieve TDM. The final output is sent to off-chip signal processing system to carry out Fourier Transform and frequency spectrum analysis.

9.4.2 Circuit Design

Figure 9.7 shows the circuit schematic of the readout channel for biosensor. The readout channel includes a read current source, a frequency divider, a mixer, a low noise differential amplifier, a programmable operational amplifier, and a transmission gate-based multiplexer. The readout circuit is designed at PTM 90 nm technology node [3]. All circuit level simulations are conducted under Cadence Spectre Analog environment. The device models of GMR, TMR, and spintronic memristor were developed by using Verilog-A language.

During the readout operation, current source injects a small (i.e., 0.1 mA) read current into both the sensor cell (Act) and the reference resistance (Ref) to incur Act and Ref voltages, respectively. These two voltages are connected to the inputs of the mixer, which multiplies the readout voltage by sampling pulse (carrier frequency) generated by the frequency divider with a master clock of 20 kHz. As shown in Fig. 9.7, each DNA spot within one readout channel has its own carrier frequency. The low noise differential amplifier (DA) is used to amplify the voltage difference between the Act and Ref voltages. Furthermore, it can also serve as an analog adder by swapping the Act and Ref voltages at the inputs of the mixer. Thus, at the end of the each readout channel, the sensing signals (difference between Act and Ref

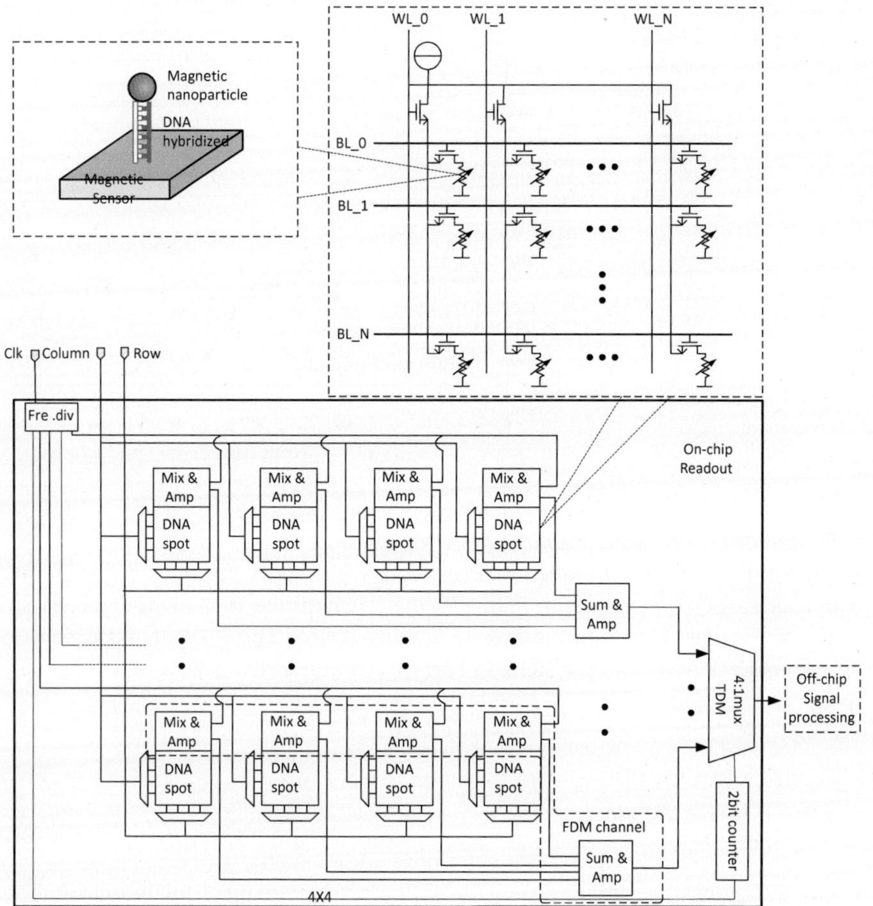

Fig. 9.6 Biosensor architecture design

voltages) of all the four DNA spots have been summed up to share one physical channel. In our design, FDM is realized by modulating the signals with different carrier frequencies, which can be extracted by off-chip bandpass filter. FDM can also reduce 1/f noise effectively. A programmable operational amplifier (POA) with a gain range from 10 to 60 is designed to maximize the dynamic range of the output signal. Instead of pushing the amplifier gain even higher, we use multi-level amplifiers to reduce design complexity and improve circuit reliability.

Figure 9.9 shows the simulated output signal of one readout channel. The main clock frequency is set as 20 kHz. The four carrier frequencies are 20/6 kHz, 20/8 kHz, 20/10 kHz, and 20/12 kHz. To guarantee cycle integrality of all signals, the minimal time slot should cover integral multiples of all the carrier clock period,

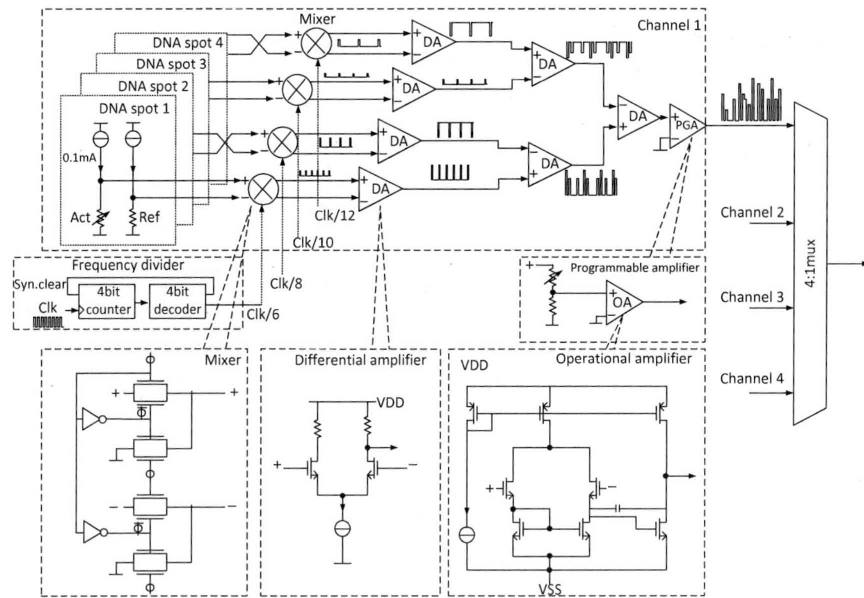

Fig. 9.7 Schematic of readout channel

Table 9.1 Baseline circuit design parameters

Differential amp gain	2 or 3
Programmable operation amp gain	10–60
Output swing of operation amp (mV)	300
Main clock frequency and limitation (kHz)	20
Carrier frequencies for FDM (kHz)	3.33, 2.5, 2, 1.67
One time slot of TDM (ms)	6

e.g., 6 ms in our configuration. The outputs of the four readout channels are connected to a 4-to-1 multiplexer. Each channel is assigned to a time slot to achieve TDM function. The baseline design parameters are summarized in Table 9.1.

9.4.3 Comparison Among Different Devices

GMR, TMR, and spintronic memristor are all spintronic sensors. However, these devices demonstrate the different responses to extraneous magnetic field disturbances of nanoparticles in terms of the resistance value changing (R) and the resistance transition time. Accordingly, the circuit design should be adjusted for each specific device as shown in Table 9.2. The MR ratio of a typical GMR device is around 10 % [9]. The minimum resistance gap before and after absorbing nanoparticles could be only 100 m [10], which results in only 10 V voltage difference at the

Table 9.2 Circuit design parameters specific to three device

Device	R (m)	DA gain	PGA gain	V (mV)
GMR	100–400	333	60	16.2–64.8
TMR	1,000–5,000	222	10	8–40
Memristor	10,000	222	10	80

output of DNA spot. Thereby, the overall voltage gain of the amplifiers need to be relatively high. Usually, a TMR device has an MR ratio of 100 % or higher [11]. Hence, the resistance gap and voltage difference could be significantly improved.

Unlike spintronic memristor, neither GMR nor TMR sensors can maintain the resistance states after removing the external magnetic fields. Note that the resistance gap (R) of a memristor sensor can be adjusted by controlling the time period of apply external magnetic field as shown in figure. We assume the magnetic field applied for 10 ns, and then the resistance gap could be more than 10.

As shown in Table 9.2, the gain of differential amplifier and programmable operation amplifier can be adjusted accordingly. In GMR, due to the small R (from 100 m to 400 m), we select the maximum gain for both DA with 3 and PGA with 60. Therefore the output voltage can be speculated by R(DA gain)3 (PGA gain). If DA gain and PGA gain is 3 and 60, respectively, the output voltage could range from 16.2 to 64.8 mV. Nevertheless, if all circuit parameters keep same but substituting GMR by TMR device, the maximum voltage will be 648 mV which is 10 times of GMR sensor. Under above situation, voltage signals of four DNA spots are summed up in FDM so that the total output could surpass VDD (1 V). Hence, the PGA gain and DA gain should be adjusted to a lower value. On the other hand, we have another constraint that the sum up of output voltages of four DNA spots cannot beyond the voltage swing (300 mV) of PGA; otherwise, the amplifier does not work at linear region. Moreover, the DA gain may need to be redesigned when different device is used (i.e., TMR, memristor). Decreasing in DA gain can not only address voltage swing issue but also save chip area. For example, the MOS transistor width of a DA with gain of 2 is 5 times smaller than that of a DA with gain of 3 (Fig. 9.7).

9.5 FDM Signal Processing and Noise Analysis

9.5.1 Modulation Theory and Implementation

In the previous section, we have discussed about the combination of FDM and TDM scheme by using pulse sampling to achieve high throughput and low noise. In this section, we will further improve the efficiency and leverage the noise endurability of the FDM modulation technology by using cosine signal which is originally developed for radio telecommunication. In its basic form, a signal with power concentrated at the carrier frequency will be produced. In modulation, message

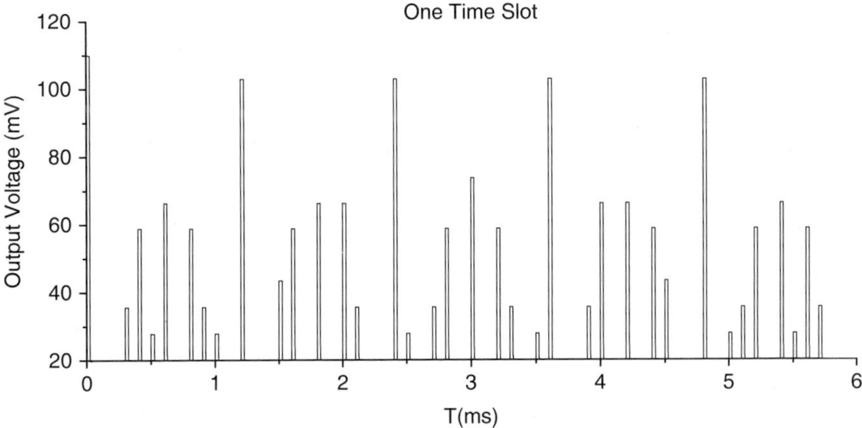

Fig. 9.8 Simulation result of one channel at output of PGA

signal is multiplied by carrier signal which has a much higher frequency than the message signal and is defined as

$$f(t) = m(t) \cdot cos(2\pi \cdot f_0 \cdot n/f_s). \tag{9.4}$$

where $m(t)$ is usually referred to as the message signal and f_0 is the carrier frequency. As shown in above equation, the modulation consists of multiplying thc message signal $m(t)$ by the carrier $cos(2\pi \cdot f_0 \cdot n/f_s)$. Therefore, we can use the modulation theorem of Fourier Transforms to obtain the spectrum $F(f)$ in frequency domain by calculate:

$$F(f) = \frac{1}{2}[M(f - f_0) + M(f + f_0)]. \tag{9.5}$$

where $M(f)$ is the Fourier Transform of $M(f)$. According to the property of Fourier Transform, multiplication of signals in time domain will result in their convolution in frequency domain. And the Fourier Transform of cosine function is shifted Dirac delta function, so the wave of message signal will be shifted to $-f_0$ and f_0. In the implementation of our biosensor circuit, the message signal is the voltage variation caused by the memristance changing of the microarray spintronic memristor detector. The voltage signal of each detection point will be multiplied by the carrier frequency through the analog mixer. We have four DNA spots in one FDM channel, so totally four cosine signals with four different carrier frequencies are needed. They can be expressed by $f(t) = \sum_{i=1}^{4} m_i cos(2 \cdot \pi \cdot f_i \cdot t)$. We still use the frequencies as shown in Table 9.1. The benefit by using this FDM modulation technology is that most of the power of the signal will concentrate at the carrier frequency; therefore, we can use bandpass filter to obtain the useful information and eliminate most of the noise at the same time.

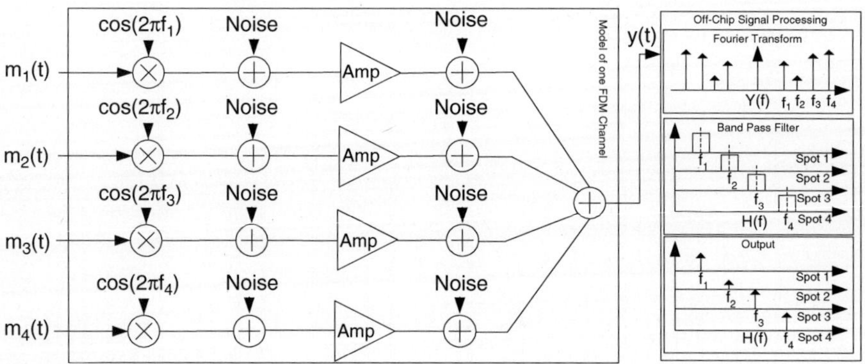

Fig. 9.9 FDM channel model

We use simplified model to illustrate the FDM principle of circuit as shown in Fig. 9.9. Three-stage amplifiers in readout circuit of Fig. 9.7 are simplified into one ideal amplifier added with a noise source. And the mixer also brings in certain noise. The target signals $m_1(t)$ to $m_4(t)$ are four voltage levels (v_1 to v_4) generated by the memristor which is applied by a read current. As we have mentioned before, the output $y(t)$ will be TDM to off-chip signal processing, hinting that it is impossible for the time slot of each FDM channel to be infinite. With limited time period, the spectrum of the output signal in frequency domain cannot be ideal Dirac delta function as shown in Fig. 9.9. In the following sections, we will show the simulation results with different lengths of time slots and different types of noise in order to measure the noise endurability of the design.

9.5.2 Noise Source

The noise behavior of the bio-chip is dominated primarily by two noise sources: thermal noise and pink or the so-called $1/f$ noise. Thermal noise is approximately white, meaning that the power spectral density is nearly equal throughout the frequency spectrum. Additionally, the amplitude of the signal is similar with Gaussian probability density function. In the signal process simulation, we use white noise to approximately model the thermal noise. We can simulation stationary, continuous-time random process $x(t)$: t is real with constant mean μ and covariance function

$$K_x(\tau) = E(x(t_1) - \mu))(x(t_2) - \mu)^*. \tag{9.6}$$

where $\tau = t_1 - t_2$ and power spectral density

$$S_x(f) = \int_{-\infty}^{\infty} K_x(\tau) e^{-jw\tau} d\tau. \tag{9.7}$$

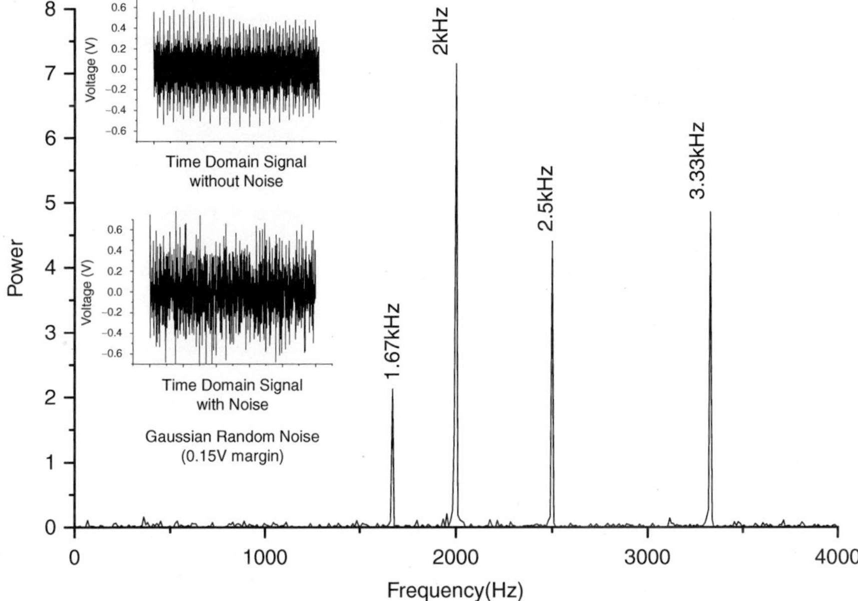

Fig. 9.10 Output power after Fourier Transform with gaussian white noise

Pink noise or $1/f$ noise is a signal with a frequency spectrum that the power spectral density is inversely proportional to the frequency

$$N_f = \frac{\alpha}{N \cdot f}(I \cdot R)^2. \tag{9.8}$$

where α is the dimensionless (sometimes field dependent) Hooge constant, N is the total number of conduction electrons in the sensor (often taken as the number of atoms in the active area of the sensor), f is the frequency, I is the read current, and R is the resistance of micro sensor. [14] gave α values ranging from 6.7×10^{-3} to 2.8×10^{-1}.

We will show Fast Fourier Transform (FFT) of FDM voltage output mixed with the above-mentioned two noise. According to Nyquist sampling theory, sampling frequency must be at least twice bigger than the message signal. The highest carrier frequency is 3.33 kHz, so we define the sampling frequency f_s to be 10 kHz in our simulation.

9.5.3 Simulation Result

Figure 9.10 shows the FFT of FDM output with gaussian random white noise. The main curve is in frequency domain. As we can see from the main curve, the four message signals are clearly centered at 1.67 kHz, 2 kHz, 2.5 kHz, and 3.33 kHz

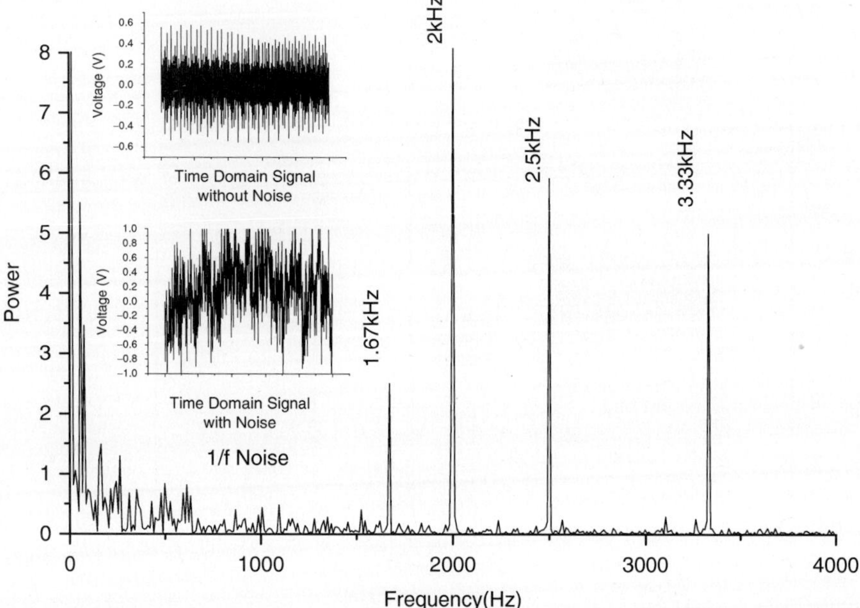

Fig. 9.11 Output power after Fourier Transform with 1/f noise

along with the frequency coordinate. Subgraphs shows the FDM output before FFT in time domain with and without white noise. As we can see, the impact of white noise in the time domain is very larger, while it is much more trivial in the frequency domain. Once we use bandpass filter to extract the useful information, the impact of the white noise will be even smaller. We will measure the signal-to-noise ratio (SNR) later to estimate the benefit brought by FDM in the bio-circuit readout. Figure 9.11 shows the FFT of FDM output with $1/f$ noise. The power density of $1/f$ noise is gathered around low frequency region and rolls off with the increasing of frequency. When the frequency is beyond 1 KHz, the impact of $1/f$ noise turns to be very small. Subgraphs shows the FDM output before FFT in time domain with and without $1/f$ noise. Obviously, the impact of $1/f$ noise in time domain is much greater than that in the frequency domain that further proves the efficiency of FDM in eliminating noise influence.

In order to leverage the endurability of our scheme, signal-to-noise (SNR) is used as a leveraging metric. Under the impact of gaussian random white noise, the SNR can be calculated by

$$SNR_{white} = 10log\left(\frac{\sum_{i=1}^{4}(I\triangle R_i \times cos(2\pi f_i t))^2}{\int_{-\infty}^{\infty} K_x(\tau)e^{-jw\tau}d\tau}\right). \tag{9.9}$$

where I is the read current, $\triangle R_i$ is the resistance change of each detection device.

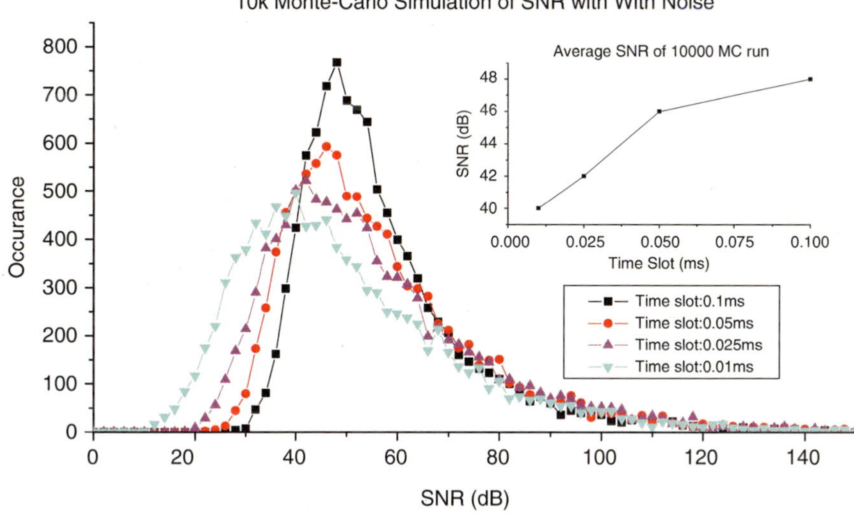

Fig. 9.12 Monte-Carlo simulation of SNR with gaussian random noise

Figure 9.12 shows the 10^4-run Monte-Carlo simulation of SNR with gaussian random white noise. Four lengths of time slot are compared in the simulation. The larger the time slot is, the higher the SNR will be. That is because larger time slot implies the closer the Fourier Transform will be to the Dirac delta function at frequency domain. Because we use ideal bandpass filters whose center frequencies are 1.67 kHz, 2 kHz, 2.5 kHz, and 3.33 kHz with bandwidth of 200 Hz. Therefore, the power of the output signal physically concentrates more at certain frequency point resulting in high SNR after passing the ideal bandpass filter. However, larger time slot means low throughput, so trade-off between SNR and throughput has to be balanced according to requirement. Because the resistance changing of GMR and TMR device is relatively small, so the impact of noise will become bigger in return. When GMR and TMR are used as bio-sensor interface, we will give more priority to SNR. For spintronic memristor, the resistance changing is much larger than GMR or TMR device, so the SNR is reasonably high as we can see from the simulation. When the time slot is 0.1 ms, the SNR can approach almost 50 dB. That means the power of useful information is 316.22 times of the power of noise. If we want to improve the throughput, we can tune the time slot to be 0.01 ms. In other words, we can obtain information of four detection spots in every 10 ms, and the corresponding SNR also stays as high as 40 dB.

Under the impact of 1/f noise, the SNR can be calculated by

$$SNR_{1/f} = 10log\left(\frac{\sum_{i=1}^{4}(I\triangle R_i \times cos(2\pi f_i t))^2}{\alpha/(N_f\triangle f)I^2R^2}\right) = 20log\left(\frac{\sum_{i=1}^{4} I\triangle R_i \times cos(2\pi f_i t)}{\sqrt{\alpha/(N_f\triangle f)}IR}\right) \tag{9.10}$$

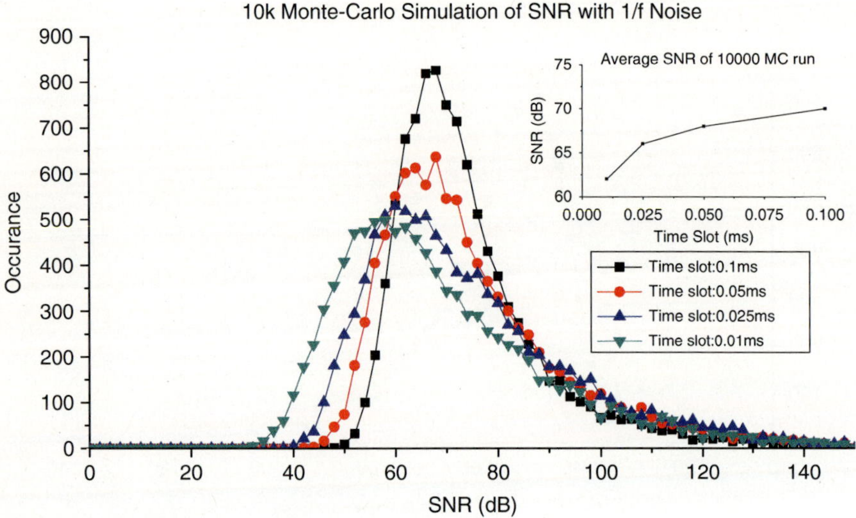

Fig. 9.13 Monte-Carlo simulation of SNR with 1/f noise

Figure 9.13 shows the 10^4-run Monte-Carlo simulation of SNR with 1/f noise. When the time slot is selected as 0.1 ms, the SNR can reach 70 dB. The reason behind is when frequency surpasses 1 kHz, the 1/f noise levels off to zero. We also compared SNR before bandpass filter and after bandpass filter. The bandpass filter can help improve SNR_{white} and $SNR_{1/f}$ by 3.4 and 5.83 times, respectively, when one time slot period is 0.01 ms. Explicitly, the reason $SNR_{1/f}$ has been improved more than SNR_{white} is the power of 1/f noise is gathered at low frequency region and most of them is filtered by the bandpass filter.

9.6 Conclusion

In this work, we discuss a possible magnetic field sensing mechanism of spintronic memristors, which can be utilized in DNA hybridization detection. The circuit implementation of a biosensor array based on spintronic memristors is also proposed. We compare the differences between spintronic memristors and the other two popular spintronic devices—GMR and TMR, in terms of working mechanism, magnetic and electric parameters, etc. Our analysis shows that besides the nonvolatility (the sensed value is kept in the device after the sensed magnetic field is removed), spintronic memristors also show a higher sensing signal amplitude than TMR and GMR devices. We also propose a on-chip readout scheme with a telecommunication method-frequency division multiplexing (FDM) technique that can efficiently transmit useful information and filtrate the noise. We can achieve high SNR up to 70 dB by using proposed scheme.

Acknowledgments This work is supported in part by NSF grants CNS-1253424 (CAREER) and ECCS-1202225.

References

1. L. Berger, Emission of spin waves by a magnetic multilayer traversed by a current. Phys. Rev. B **54**(13), 9353–9358 (1996)
2. W.F. Brown, Thermal fluctuations of a single-domain particle. Phys. Rev. **130**(5), 1677–1686 (1963)
3. Y. Cao, T. Sato, M. Orshansky, D. Sylvester, C. Hu, New paradigm of predictive MOSFET and interconnect modeling for early circuit simulation. In *IEEE Proceedings of Custom Integrated Circuits Conference (CICC)*, pp. 201–204, Orlando, FL, 21–24 May 2000
4. L. Chua, Memristor-the missing circuit element. IEEE Trans. Circuit Theory **18**(5), 507–519 (2002)
5. R.A. Duine, A.S. Nunez, A.H. MacDonald, Thermally assisted current-driven domain-wall motion. Phys. Rev. Lett. **98**(5), 56605 (2007)
6. S.J. Han, L. Xu, H. Yu, R.J. Wilson, R.L. White, N. Pourmand, S.X. Wang, CMOS integrated DNA microarray based on GMR sensors. In *International Electron Devices Meeting (IEDM)*, pp. 1–4, San Francisco, CA, 11–13 December 2006
7. Z. Hu, Q. Li, M. Li, Q. Wang, Y. Zhu, X. Liu, X. Zhao, Y. Liu, S. Dong, Ferroelectric memristor based on pt/bifeo3/nb-doped srtio3 heterostructure. Appl. Phys. Lett. **102**(10), 102901–102901 (2013)
8. S. Lau, R. Eichardt, L. Di Rienzo, J. Haueisen, Tabu search optimization of magnetic sensor systems for magnetocardiography. IEEE Trans. Magn. **44**(6), 1442–1445 (2008)
9. G. Li, V. Joshi, R.L. White, S.X. Wang, J.T. Kemp, C. Webb, R.W. Davis, S. Sun, Detection of single micron-sized magnetic bead and magnetic nanoparticles using spin valve sensors for biological applications. J. Appl. Phys. **93**, 7557 (2003)
10. G. Li, S. Sun, R.J. Wilson, R.L. White, N. Pourmand, S.X. Wang, Spin valve sensors for ultrasensitive detection of superparamagnetic nanoparticles for biological applications. Sens. Actuators A Phys. **126**(1), 98–106 (2006)
11. S. Mao, Y. Chen, F. Liu, X. Chen, B. Xu, P. Lu, M. Patwari, H. Xi, C. Chang, B. Miller, et al., Commercial TMR heads for hard disk drives: characterization and extendibility at 300 gbit/in 2. IEEE Trans. Magn. **42**(2), 97–102 (2006)
12. Q.A. Pankhurst, J. Connolly, S.K. Jones, J. Dobson, Applications of magnetic nanoparticles in biomedicine. J. Phys. D Appl. Phys. **36**, R167 (2003)
13. A. Radoi, M. Dragoman, D. Dragoman, Memristor device based on carbon nanotubes decorated with gold nanoislands. Appl. Phys. Lett. **99**(9), 093102–093102 (2011)
14. A. Rhouni, G. Sou, P. Leroy, C. Coillot, A very low 1/f noise and radiation-hardened cmos preamplifier for high sensitivity search coil magnetometers. IEEE Sens. J. **13**, 159–166 (2012)
15. L. Rovati, S. Cattini, Contactless two-axis inclination measurement system using planar fluxgate sensor. IEEE Trans. Instrum. Meas. **59**(5), 1284–1293 (2010)
16. J.C. Slonczewski, Current-driven excitation of magnetic multilayers. J. Magn. Magn. Mater. **159**(1–2), L1–L7 (1996)
17. C.H. Smith, R.W. Schneider, Low Magnetic Field Sensing with GMR Sensors, Part 1: The Theory of Solid-State Magnetic Sensing. Sensors Magazine, 1999
18. M. Takemoto, T. Akai, Y. Kitamura, Y. Hatsukade, S. Tanaka, Hts rf-squid microscope for metallic contaminant detection. IEEE Trans. Appl. Supercond. **21**(3), 432–435 (2011)
19. G. Tatara, H. Kohno, Theory of current-driven domain wall motion: Spin transfer versus momentum transfer. Phys. Rev. Lett. **92**(8), 86601 (2004)
20. S.X. Wang, G. Li, Advances in giant magnetoresistance biosensors with magnetic nanoparticle tags: Review and outlook. IEEE Trans. Magn. **44**(7), 1687–1702 (2008)

21. X. Wang, Y. Chen, Y. Gu, H. Li, Spintronic memristor temperature sensor. IEEE Electron Device Lett. **31**(1), 20–22 (2010)
22. X. Wang, Y. Chen, H. Xi, H. Li, D. Dimitrov, Spintronic memristor through spin-torque-induced magnetization motion. IEEE Electron Device Lett. **30**(3), 294–297 (2009)
23. X.L. Wang, Q. Shao, C.W. Leung, A. Ruotolo, Non-volatile, reversible switching of the magnetic moment in mn-doped zno films. J. Appl. Phys. **113**, 17C301 (2013)
24. S. Yamada, High-spatial-resolution magnetic-field measurement by giant magnetoresistance sensor—applications to nondestructive evaluation and biomedical engineering. Int. J. Smart Sen. Intell. Syst. **1**(1), 160–175 (2008)

Part IV
Reconfigurable Logic Circuits and Neuromorphic Systems

Chapter 10
Memristor-Based Resistive Computing

Sung-Mo Steve Kang and Sangho Shin

10.1 Introduction

In 1971, Chua published a seminal paper on memristor as a missing basic circuit element by explaining the constitutive relationship between electrical charge q and flux π linkage [1]. Chua demonstrated that the memristor can be physically realized by using other passive and active circuit elements and predicted that inherently passive memristors would be found. In 1976, Chua and Kang published a paper that defined a large class of devices and systems which they named memristive devices and systems to broaden thc domain of useful nonlinear devices with memristive characteristics substantially and showed that many physical systems can be categorized as memristive devices and systems [2]. In 2008, almost 40 years later, Stan Williams and his research team at HP Labs unveiled a two-terminal titanium dioxide nanoscale device that exhibited memristor characteristics in a restricted operating range [3].

Continuing demands for more complex information processing require future systems integration to overcome various physical limitations of traditional CMOS technologies, including the lithographical limitation and the power density limit [4]. Such barriers to Moore's Law call for revolutionary approaches to systems integration. To meet the increasingly difficult technological requirements, the emergent nanoscale resistive memory device technologies [3] have received much attention in order to help overcome the limitations of CMOS technologies. Memristive devices [2, 5] have been realized in a form of bipolar voltage-actuated nanoscale switches such that they can be used to build ultra-dense resistive memory arrays. Nanoscale memristor devices can be reconfigured into nonvolatile

S.-M.S. Kang (✉) • S. Shin
Department of Electrical Engineering, Jack Baskin School of Engineering, University of California, Santa Cruz, CA, USA
e-mail: kang@soe.ucsc.edu; sshin@soe.ucsc.edu

R. Tetzlaff (ed.), *Memristors and Memristive Systems*,
DOI 10.1007/978-1-4614-9068-5_10,

memories, logic gates, programmable interconnects with a very high integration density and, more importantly, with CMOS compatibility [6, 7]. Thus, memristors-based nanocomputing architectures offer promises for low-power and high-density computing applications, pushing Moore's Law far beyond the present silicon roadmap horizons.

A myriad of research opportunities have been opened by the memristive technologies. Besides the ultra-dense nonvolatile memory applications, they include self-adaptable analog/digital electronics [8–12], resistive signal processors [13], and synaptic neuromorphic networks [14]. One of the most plausible architectures is the CMOS/Molecular hybrid (CMOL) [6] which can be implemented by using Field Programmable Nanowire Interconnect (FPNI) [7, 15]. This is based on hybrid circuits composed of a conventional CMOS layer connected to multiple crossbar layers that contain memristive switches. These architectures have been proposed by primarily taking advantage of their regularity and compatibility with CMOS processes for the ease of fabrication of nanoelectronic circuits into the single-digit nanometer scale. Such "molecular electronics and computing" enabled by a nanoscale memristive devices technology can extend Moore's Law well beyond 2020. High-density vertical interconnects such as through silicon vias (TSVs) can provide high-bandwidth, low-latency communication between stacked layers without sacrificing area in each layer for vias. FPNI leverages on a Field-Programmable Gate Array (FPGA) architecture by lifting configuration bits and associated components out of the CMOS plane and replacing them with nonvolatile switches in interconnects. In FPNI, logic gates and flip-flops which form the most crucial part of the FPGA reside in the CMOS layer.

When the properties of nanoscale memory devices are exploited together with CMOL/FPNI architecture, replacing CMOS transistors with memristive devices for latching or switching circuits, new analog/digital functions can be developed to significantly reduce the form factor, manufacturing cost, and active/leakage power consumption. Recently, stateful logic operations for which the memristive switches function as gates and latches have been demonstrated [16–26]. Since these memristive switches can implement memory, programmable interconnects, logic operation, and latches, with higher density and lower power consumption, the nanowire crossbar layers of memristive devices have become an intriguing technology. In such stateful logic gates, the voltage-actuated bipolar memristive devices offer their unique properties of nonvolatile memory function and conditional set/reset capability. And such stateful logic gates functioning NAND or NOR operation can serve as basic elements to construct general-purpose large-scale stateful logics and low-power field programmable logic arrays [21, 22]. In addition resistive signal processing units for nonvolatile arithmetic computation such as signal multiplication have been reported [24–26].

This chapter introduces a high-density resistive computing architecture that can perform as general logic gates and signal multipliers, and as base technologies of future nanocomputing. The resistive multipliers, which can be imbedded in hybrid integrated circuits, are consisted of bistable memristive devices and CMOS switches. Using nonvolatile memristors with a conditional set/reset, these resistive multipliers can execute, even with passive unipolar resistances, exclusive OR (XOR) and exclusive NOR (XNOR) equivalent bipolar multiplications, in two and three phases, respectively. The function of the resistive multiplier is demonstrated for XOR/XNOR operation with inherent register capability. In particular, a massively parallel array of the resistive multiplier, each of which functions as either XOR or XNOR, is constructed to compute complex multiplication tasks, such as correlation computing.

10.2 Memristors and Memristive Devices

The memristive devices realized in a form of nanoscale bipolar voltage-actuated devices use "resistance" as a physical state variable [1, 5]. Since nanoscale memristor devices can be reconfigured into nonvolatile memories, logic gates, programmable interconnects with high integration density and, more importantly, with CMOS compatibility, the memristor technology together with CMOS is a formidable technology candidate that can advance Moore's Law beyond the present silicon roadmap horizons. A multitude of new opportunities can be provided by memristive nanotechnologies. Due to their potential for inexpensive manufacturing and ultrahigh density, various forms of memristors are being explored for many potential applications. Besides the ultra-dense nonvolatile memory applications, other opportunities may reside in highly programmable and self-adaptable analog/digital electronics, resistive nanocomputing architectures, and synaptic neuromorphic networks, among many others. A hybrid type of integration of CMOS circuits and memristive nano devices [6, 7] is considered promising for diverse next generation applications such as the implantable low-power biological sensor application.

10.2.1 Properties and Modeling of Memristive Devices

The resistances of memristors and memristive devices change depending on their current state and the voltages applied across them and/or currents driven through them [1–3]. When operated with low voltages over short time intervals, they behave as analog devices displaying a controllable hysteresis in their current–voltage characteristics. However, when overdriven by large voltages, they become essentially two-state digital switches or latches which can be opened or closed by applying opposite polarity voltages.

Among the several properties of the memristive systems which were identified in [2], the most distinct property is that the output voltage (or current) is always zero whenever the input current (or voltage) is zero regardless of the internal state variables and the momentary resistance (or conductance). This property is manifested in the pinched hysteretic *voltage–current* (v–i) characteristics for a

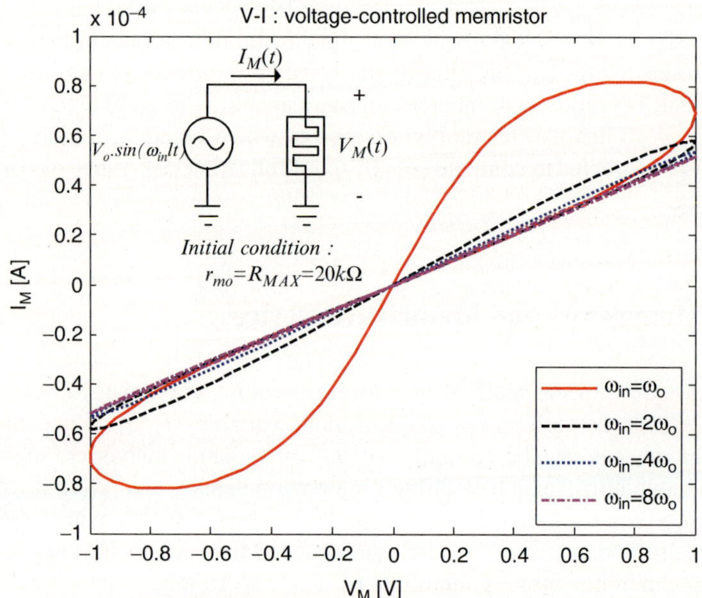

Fig. 10.1 Momentary *v–i* behaviors of memristors

periodic balanced input. Also it is noteworthy that the memristor behaves as a linear resistor in the limit of infinite frequency at which there is no time allowed to respond and produce resistance change during the short input period.

The *v–i* characteristics of memristors under a periodic sinusoidal input, i.e., $v_M(t) = V_O \cdot \sin(_{in}t)$, are shown in Fig. 10.1, where the pinched hysteretic loops appear at low frequencies. They are collapsed to that of a linear resistor at high frequencies. For a fully balanced periodic input, the memristance is also periodic over the input period, since the injected amount of net charge or flux over each period is equal to zero. Also it can be noted that the range of memristance change is strongly dependent on the input frequency and the amplitude, because the amplitude of the flux pattern is linearly proportional to the input voltage amplitude and almost inversely proportional to the input frequency. For voltage input with balanced polarity, memristors have negligible memristance changes at high frequencies and thus can be treated as static linear resistors. However, if the input polarity is unbalanced, the memristance will change even at high input frequencies, since the net flux will be accumulated over time and eventually will affect the memristance.

Unlike the theoretical memristors or memristive devices [1–3], the practical bistable memristive devices, e.g., bistable Resistive Random Access Memory (RRAM) devices, exhibit three distinguishable features: a finite number of threshold voltages exist based on which the device function can be set either for read or write mode; the device switches its resistance state in a linear or nonlinear switching rate; and such device behaves with its unique boundary assurance when the device is in

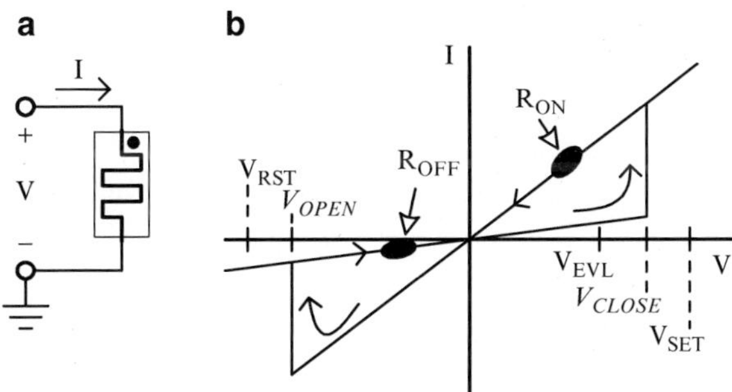

Fig. 10.2 (**a**) Memristive device, and (**b**) its low-frequency voltage–current characteristics exhibiting bistable resistance of R_{ON} for the close state and R_{OFF} for the open state

a resistance state. Such bistable memristive devices which exhibit two distinctive resistance states, namely R_{ON} for their on-state resistance and R_{OFF} for off-state, are considered in this chapter [16–26]. Their resistance is assumed to switch in a short switching duration (T_S) when a voltage with magnitude higher than their switching thresholds (V_{CLOSE} for off-to-on and V_{OPEN} for on-to-off switching) is applied across them for a sufficiently long time duration. In other words, the device is assumed to change its resistance after a fixed time duration (T_{hold}), and the transition will be completed after attaining the other resistance [19–21]. For simplicity, we assume that the resistance switching rate is constant, zero for smaller voltage amplitude than the thresholds and nonzero constant for larger amplitudes, the resistance switching behaviors are described in a Verilog-A behavioral model.

Figure 10.2 shows low-frequency voltage–current characteristics of a bistable memristive device that exhibit distinctive two resistance states, the low-resistance of R_{ON} (closed) and the high-resistance of R_{OFF} (open). A long enough positive voltage pulse with its level higher than the voltage threshold V_{CLOSE} closes the device switching from R_{OFF} to R_{ON}, while a negative pulse with magnitude larger than V_{OPEN} opens the device for long enough voltage applications. On the other hand, for the applied voltage with amplitude smaller than V_{CLOSE} or V_{OPEN}, the device remains in its previous state with no change of the resistance state. With the thresholds of V_{CLOSE} and V_{OPEN}, the device can be set (or reset) by applying larger voltage amplitude of V_{SET} (or V_{RST}) than the thresholds, while the resistance is held for the readout voltage (V_{EVL}) with smaller amplitude than the thresholds.

10.3 Memristive Stateful Logics

Stateful logics is an intriguing important application of memristive devices which function for both logic and latch operations [16, 17]. Since these memristive devices can implement memory, programmable interconnects, and logic operations and latches, with ultra-high integration density together with CMOS devices, the memristive device technology structured in a nanowire crossbar array has emerged as an intriguing technology that can allow Moore's Law to continue overcoming the

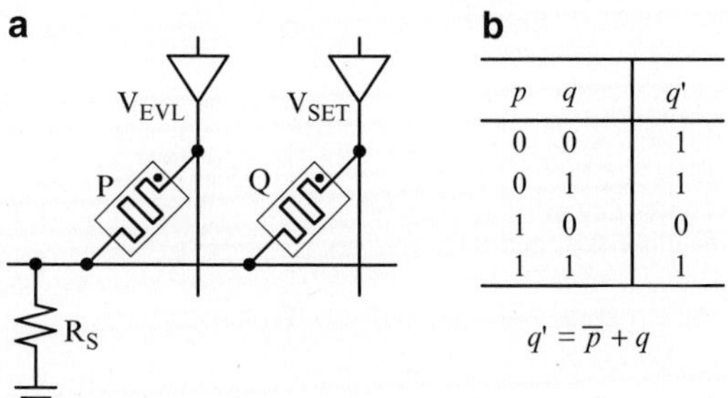

p	q	q'
0	0	1
0	1	1
1	0	0
1	1	1

Fig. 10.3 Basic operations in stateful logic: (**a**) material implication performed by two simultaneous voltage pulses V_{EVL} and V_{SET}, applied to memristive switches P and Q, respectively, and (**b**) a truth table for material implication

barriers of CMOS technologies due to lithographical limitation and power density limit among others. One of the most plausible architectures for hybrid integration of memristive nano devices and CMOS devices is the CMOS/Molecular hybrid (CMOL) [6], which can be implemented by using Field Programmable Nanowire Interconnect (FPNI) [7]. This is based on hybrid circuits composed of a conventional CMOS layer connected to multiple crossbar layers that contain memristive devices. With its inherent data-latching property, stateful logic can effectively implement low-cost fully pipelined digital systems [20–22], and therefore holds potential for implementing high performance next generation nano computers. Also, the latches with embedded memristors can provide a power shutoff mode with non-volatility against power failures.

However, the memristive stateful logic structure holds several technical issues that must be resolved to effectively complement CMOS logic. The challenging issues are on power consumption, operation speed, multiple fan-in/fan-out capability, and two- or three-dimensional integration for large-scale computations [21]. This section discusses the feasibility of nanowire crossbar memristive devices as candidate devices to implement a general-purpose computation.

10.3.1 Basic Operations of Stateful Logic

Based on the memristive devices' conditional set/reset operation and their nonvolatile memory property, a stateful NAND gate has been reported [19]. Figure 10.3a illustrates the base material implication operation in the stateful logic, where two memristive devices P and Q are connected by a common horizontal nanowire to a load resistor R_S. The logic values stored in the devices P and Q as their resistance status are represented by p and q, respectively, and logic 0 (1) corresponds to the open (closed) state of the switch. A switch is set to logic 0 (1) by applying a negative (positive) voltage V_{RST} (V_{SET}) pulse through its corresponding voltage driver. A V_{EVL} pulse that is applied to a switch which is previously set to logic 1 (0) prevents (allows) a state change of the other switch which is concurrently driven by a V_{SET} pulse. This conditional set operation enables the circuit to function as a gate implementing material implication as shown in Fig. 10.3b.

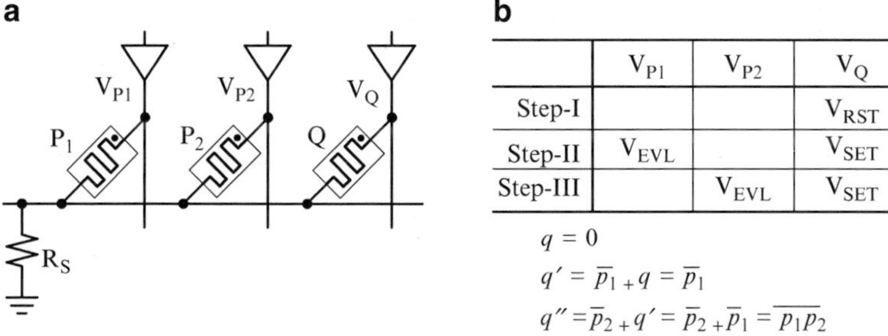

	V_{P1}	V_{P2}	V_Q
Step-I			V_{RST}
Step-II	V_{EVL}		V_{SET}
Step-III		V_{EVL}	V_{SET}

$q = 0$

$q' = \overline{p_1} + q = \overline{p_1}$

$q'' = \overline{p_2} + q' = \overline{p_2} + \overline{p_1} = \overline{p_1 p_2}$

Fig. 10.4 Stateful NAND gate: (**a**) circuit configuration, and (**b**) a schedule to execute a NAND operation

A stateful NAND gate is shown in Fig. 10.4a of which operation is executed through three sequential steps as illustrated in Fig. 10.4b [19]. The inputs are logic levels of p_1 and p_2 stored in switches P_1 and P_2, and the output is the logic value q'' accumulated in switch Q. Initially, a V_{RST} pulse is applied to switch Q to execute $q = 0$. The second step $q' = \overline{p_1} + q = \overline{p_1}$ is performed by applying a V_{EVL} pulse to V_{P1} concurrently with a V_{SET} pulse to V_Q. Finally, the operation $q'' = \overline{p_2} + q' = \overline{p_2} + \overline{p_1} = \overline{p_1 p_2}$ is executed by concurrently applying V_{EVL} and V_{SET} pulses to V_{P2} and V_Q, respectively. Since the NAND operation is known to be universal, any Boolean logic operation can be constructed by a network of NAND gates [19].

10.3.2 Issues for Large-Scale Logic Integration

However, several technical issues arise in designing memristive devices-based stateful logic architecture [21]: Firstly the stateful NAND gate requires deep pipeline steps for logic execution, as shown in an example where three dedicated steps are required for a two-input NAND operation shown in Fig. 10.4b. The pipeline steps are not dependent on the circuit size, but determined by the maximum fan-in of the constituent NAND gates. If this dependency is removed, and the multiple implication operations can be executed concurrently in a single step, the performance—the data latency as well as the pipeline period—can be significantly reduced.

Secondly, a simultaneous execution of multiple logic calculations is a basic required capability of large-scale logic arrays. When there are functionally disjoint logic cells, they should be executed in parallel so that the total required time for computation can be reduced. However in the stateful NAND gate of Fig. 10.4a with shared common horizontal nanowire and load resistance (R_S), every one-to-one implication process should be non-overlapped in time and thus very deep pipeline steps are needed for large-scale logics. A proper switch-based array configuration with isolated logic gates can perform concurrently multiple logic operations.

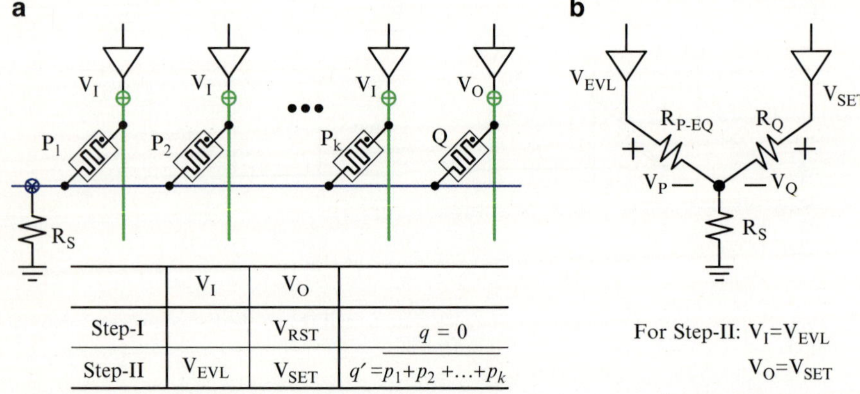

	V_I	V_O	
Step-I		V_{RST}	$q = 0$
Step-II	V_{EVL}	V_{SET}	$q' = \overline{p_1 + p_2 + \ldots + p_k}$

Fig. 10.5 (**a**) A configuration of *k*-input NOR operation that executes multiple implications in a single-step (Step-2), and (**b**) its equivalent circuit for Step-2

In stateful logic, resistance of a memristive device is used as the physical state variable. Also, the memristive switches involved in an operation need to be isolated from those in other operations. This constraint prohibits a logic state from fanning out. Therefore, each logic state should be explicitly duplicated by a dedicated operation. In order to integrate a complex logic function with high density, two- or three-dimensional circuit configurations that can be mapped onto the FPNI fabric are favored instead of the one-dimensional devices array [21].

10.3.3 Reconfigurable Stateful NOR Gate

In consideration of aforementioned technical issues, a CMOS/memristors hybrid structure of NOR function was reported to execute multiple implications in a single step [21]. Any Boolean logic operation can be constructed by a network of NOR gates since NOR operation is also universal. Figure 10.5a shows the circuit configuration for two-step NOR operations where $k + 1$ number of memristive devices are connected in parallel with a shared common horizontal nanowire and a load resistor R_S. Devices P_1, P_2, ..., P_k are input memristive switches with logic states p_1, p_2, ..., p_k, respectively, and the Q is the output device with its logic state q. Instead of composing a network of multiple two-input NAND gates to form a k-input NOR operation, this circuit is devised to execute $q\prime = \overline{p_1 + p_2 + \cdots + p_k}$ in two steps. After q is cleared by applying a V_{RST} pulse to V_O in Step-I, a single-step simultaneous execution of multiple implication processes is made enabled in Step-II by applying properly conditioned V_{EVL} pulse simultaneously to all input switches and concurrently applying V_{SET} to the device Q. V_{EVL} and V_{SET} for the conditional set operations of the single-step NOR executions can be analyzed by using an equivalent resistor network model shown in Fig. 10.5b [21].

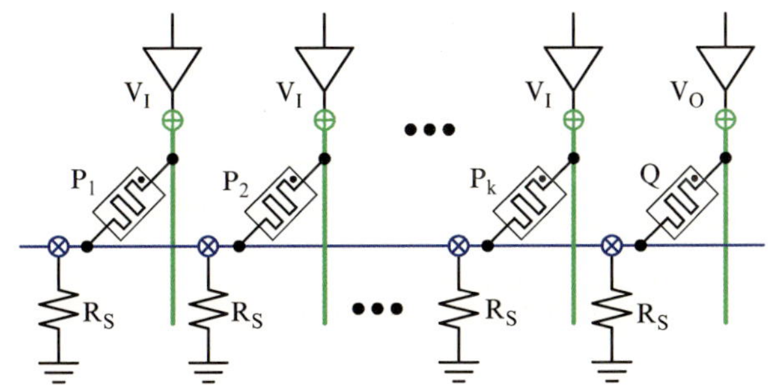

Fig. 10.6 A circuit modification for multiple fan-in capability, where load resistors are dedicated to all participating logic devices so that the equivalent load resistance can be automatically scaled by the number of fanning in

However, since more than two parallel connected memristive switches are involved in the NOR operation, the load resistance needs to be scaled with the number of inputs so that the voltages across the memristive switches are properly maintained within the valid ranges for all possible input fanning-ins. The scaled equivalent load resistance will lower the required levels of conditional set voltages (V_{SET} and V_{EVL}) and desensitize the dependency of the voltage conditions on the input fan-in. The circuit construction becomes simple if a load resistor is attached to each memristive switch as shown in Fig. 10.6. Here, the effective load resistance is inversely proportional to the number of inputs, and the logic computation always uses all load resistors unlike that the sequential implications in [18] use only one shared load resistor.

Based on the logic structure of Fig. 10.6, a generalized hybrid type of two-dimensional circuit structure is devised as shown in Fig. 10.7 so that a group of the logic units can be easily mapped onto the FPNI fabric [21]. CMOS and nanowire inter-layer vias to blue (red) horizontal (vertical) input (output) nanowires are marked by $\otimes$ ($\oplus$). Green wires are connected to voltage drivers through the vias marked by $\oplus$. It is composed of four different devices groups: memristive logic devices (P_1, P_2, ..., P_k, and Q) for input and output; memristive interconnect switches (I_{11}, I_{21}, ..., and I_{k1}) for logic configuration; CMOS switches (S_I and S_O) for implication scheduling; dedicated load resistors (R_S). Each load resistor can be implemented by using either correctly biased CMOS transistors or multiply stacked nanowires. Note that the functionality can be reconfigured solely by the interconnect crossbar switches on the nanowire layer and no reconfiguration is required in the CMOS layer. Since every logical value is latched in memristive logic devices, this architecture can operate reliably as a pipeline [20]. In other words, the pipelined architecture which can execute the corresponding computation at a constant rate independent of the system size can be devised by using the inherent data-latching property of memristive switches. The pipeline architecture can be easily mapped to the fabric as a two-dimensional array of columns. In a column, a tri-state voltage driver (V_I) is shared by multiple unit cells each of which consists of a memristive switch, a load resistor, and a CMOS control switch. The units in a column of the pipeline alternate between the write mode in which a new state is registered, and the read mode in which the state is read to determine the state at the next column. Any two neighboring columns are executed in different modes each other.

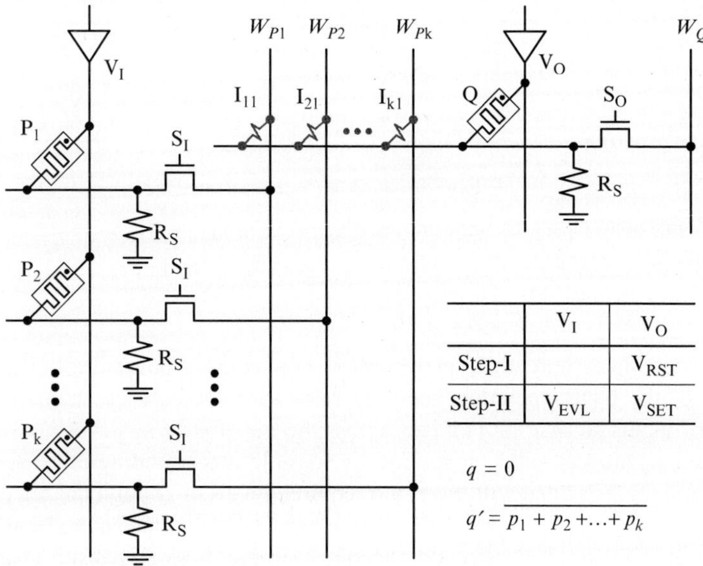

	V_I	V_O
Step-I		V_{RST}
Step-II	V_{EVL}	V_{SET}

Fig. 10.7 Two-dimensional stateful logic construction of k-input NOR operation for large-scale logic integrations

The single-step NOR execution is validated for large input fan-in by using a Verilog-A compact model for the memristive devices. For voltage pulse levels higher than V_{OPEN} (V_{CLOSE}), devices' resistance states switch with the minimum pulse width of 10 ns. Nine transient analyses for an 8-input NOR gate of Fig. 10.7, each with $n = 0, 1, \ldots, 8$, respectively, were performed, with devices' switching thresholds of $V_{OPEN} = 1$ V and $V_{CLOSE} = -1$ V. The analyzed conditional set voltages of $(V_{SET}, V_{EVL}) = (1.7 \times V_{CLOSE}, 1.2 \times V_{CLOSE})$ and $V_{RST} = 1.6 \times V_{OPEN}$ with an adaptive β-ratio (two for Step-2 and 200 for Step-1) are used. The pulse widths of voltage drivers are all set at 20 ns which is twice the minimum requirement for a state switching.

Figure 10.8a shows the applied voltages for Steps-I and -II where thin blue and thick red pulses, respectively, display V_I and V_O pulses. All results for Step-I are displayed in dotted lines while those for Step-II are in solid waveforms. The input states of p_1 through p_k are programmed before each Step-I process such that $n = 0, 1, 2, \ldots,$ and 8, sequentially. However for clarity, only the transient results of Step-I and Step-II are compiled into the graphs. As shown the equivalent input resistance ($R_{P\text{-}EQ}$) in Fig. 10.8b, the programmed input states are well maintained, not affected by the subsequent conditional set processes. This can be validated by the well confined small enough voltage drops across the input devices (V_P) in between V_{CLOSE} and V_{OPEN}, shown in thin blue lines in Fig. 10.8d, for all possible cases. While the voltage drops across the memristive interconnecting devices, shown in Fig. 10.8c, are also maintained with less than 120 mV far below the V_{OPEN}, and that of the output device (V_Q) exceeds V_{CLOSE} only when $n = 0$ as shown in

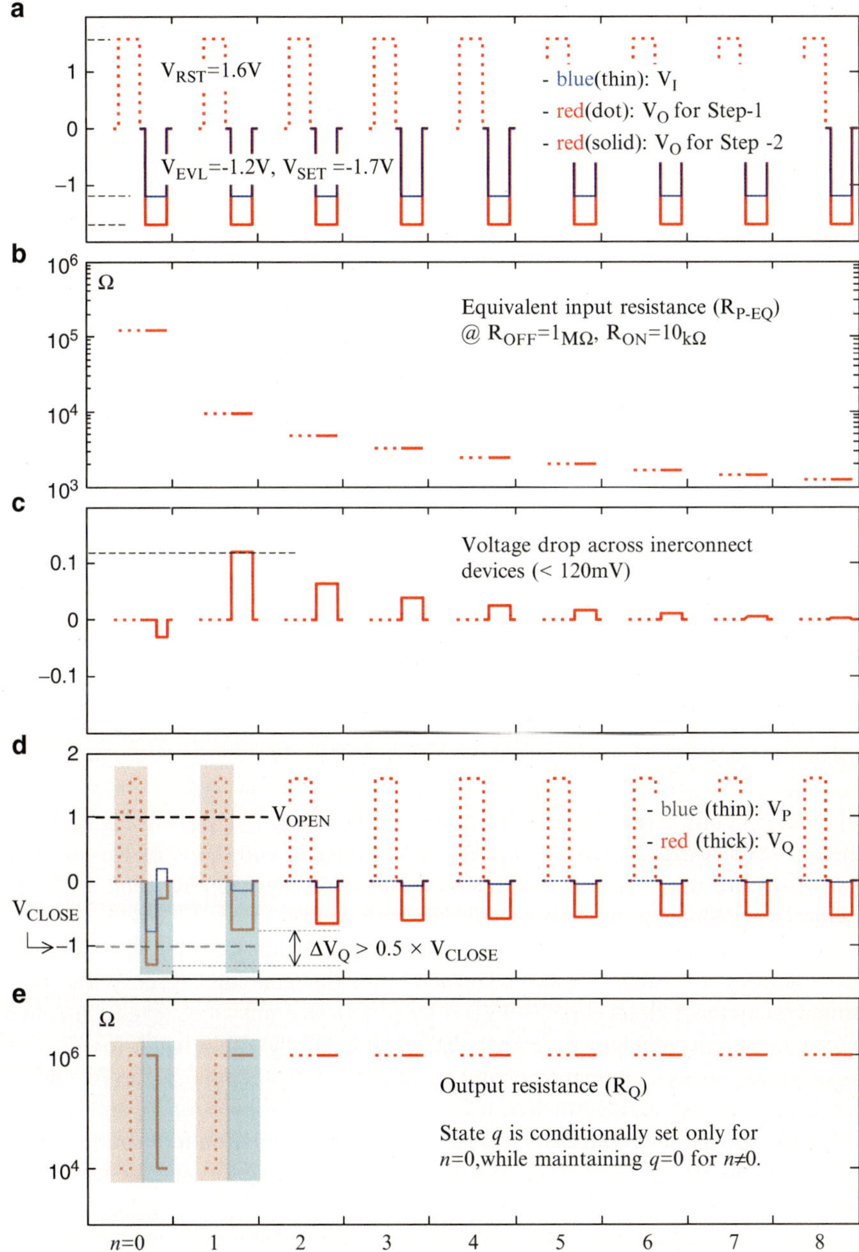

Fig. 10.8 Transient simulation results of an 8-input NOR gate for all possible input state combinations: (**a**) voltage pulses applied to V_I and V_O, (**b**) equivalent resistance of the input memristive switch network, (**c**) voltage across the interconnecting memristive devices, (**d**) voltages across the input and the output memristive switches, and (**e**) resistance of the output memristive switch (R_Q)

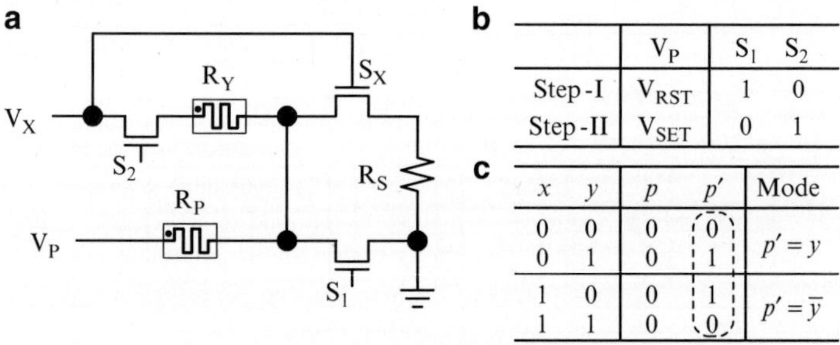

	V_P	S_1	S_2
Step-I	V_{RST}	1	0
Step-II	V_{SET}	0	1

x	y	p	p'	Mode
0	0	0	0	$p' = y$
0	1	0	1	
1	0	0	1	$p' = \overline{y}$
1	1	0	0	

Fig. 10.9 (**a**) Memristive XOR circuit, (**b**) its two-step computing phases, and (**c**) the truth table for XOR operations

Fig. 10.8d, where the smallest V_Q difference between the cases of $n = 0$ and $n \neq 0$ is larger than $0.5 \times V_{CLOSE}$. The conditional V_Q transition successfully yields, with a large margin, the 8-input NOR operation in a single-step by setting $q' = 1$ only for $n = 0$ from its initial state of $q = 0$. The resulting transient output resistance (R_Q) behaviors are shown in Fig. 10.8e, where the resistance switches to R_{ON} from R_{OFF} only for $n = 0$.

10.3.4 Memristive XOR Gate Functioning as a Resistive Multiplier

Recently [23] reported that the memristors enable bipolar signal multiplications with inherent property of output registering, in two computing phases. Figure 10.9a shows the reported resistive multiplier circuit which is composed of two bistable memristive devices (R_Y and R_P) together with a load resistor (R_S) and a CMOS switch (S_X). While V_X is a two-level digital input voltage, V_P is a fixed bias voltage chosen to guarantee memristors' conditional set operation together with the logic level of V_X. The switch S_X is turned on and off by the input logic level. When S_X is turned on by the high level of V_X ($V_X = V_{EVL} > 0$), R_S sinks the currents supplied by the memristive devices R_P and R_Y in parallel. On the other hand, for $V_X = 0$, S_X is turned off by the input level and R_S becomes isolated, thus a current path is formed through R_P and R_Y in series. By choosing proper levels of V_P and V_X, the two operation modes controlled by V_X can perform a bipolar multiplication, where the two inputs are the level of V_X and the resistance state of R_Y, and the multiplied output is registered as the resistance state of R_P [23].

With the variables x, y, and p, respectively, representing the logic values of V_X, R_Y, and R_P, where the logic level "1" stands for $V_X = V_{EVL}$, $R_Y = R_{ON}$, and $R_P = R_{ON}$, and the logic "0" for $V_X = 0$, $R_Y = R_{OFF}$, and $R_P = R_{OFF}$, the computing sequence is summarized in Fig. 10.9b. As shown for the evaluation sequence in Fig. 10.9c for all possible combinations of x and y, the evaluated XOR output after two steps is stored as the last state of R_P (p'), where the R_P is initially reset to "0" ($p = 0$).

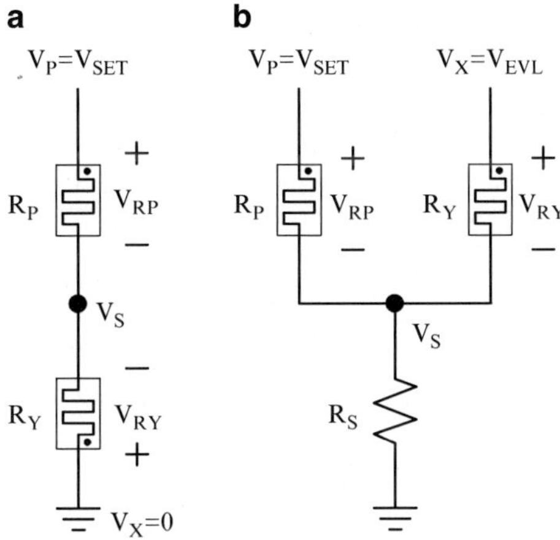

Fig. 10.10 Equivalent circuits of Fig. 10.5: (**a**) Step-II with $x = 0$, and (**b**) Step-II with $x = 1$

In order for the conditional set operations which are the basic function of the XOR computation to be done securely, the bias voltage of V_{SET} and the logic level V_{EVL} need to be carefully chosen [23, 25]. Figure 10.10a, b shows equivalent circuits of Fig. 10.9a, respectively, for Step-II with $x = 0$ and $x = 1$, where the V_P bias of V_{SET} is normally higher than V_{CLOSE} for these conditional set operations. If $x = 0$ in Step-II, the voltage drop across R_P, V_{RP}, which is initially in off state ($p = 0$) is proportional to y, and thus V_{RP} can be higher than V_{CLOSE} only for $y = 1$. Under such condition, the state of R_P registers a duplication of y, i.e., $p' = y$. On the other hand, if $x = 1$, V_{RP} becomes inversely related to y, since the smaller R_Y state yields the higher value of V_S. When the V_{SET} and V_{EVL} are chosen to have V_{RP} higher than V_{CLOSE} only for $y = 0$, the circuit can function as an inverter where the inverted version of y is registered in the memristive device R_P, i.e., $p' = \bar{y}$. The bias constraints for the conditional set operations in Step-III are exactly the same as for the case with $x = 1$ for Step-II.

With parameters of α and β, respectively, for off-to-on resistance ratio (R_{OFF}/R_{ON}) and the ratio of R_{OFF}/R_S, aforementioned conditional set operations for Steps-II ($x = 0$ and $x = 1$) eventually arrive at the following set of four bias constraints:

$$V_{SET} < 2 \times V_{CLOSE}, \tag{10.1}$$

$$V_{SET} \geq \left(\frac{1}{1+\beta}\right) \times V_{EVL} + \left(1 + \frac{1}{1+\beta}\right) \times V_{CLOSE}, \tag{10.2}$$

$$V_{SET} > (1+\beta) \times V_{EVL} - (2+\beta) \times V_{CLOSE}, \tag{10.3}$$

$$V_{SET} < \left(\frac{\alpha}{\alpha+\beta}\right) \times V_{EVL} + \left(1 + \frac{1}{\alpha+\beta}\right) \times V_{CLOSE}, \tag{10.4}$$

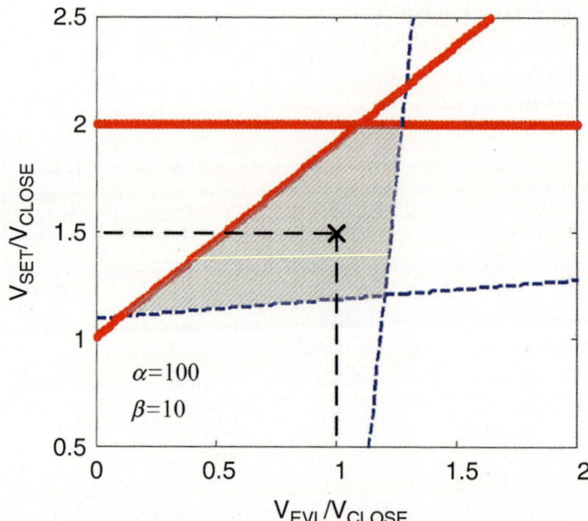

Fig. 10.11 Example of an available voltage space (*shaded*) of (V_{SET}, V_{EVL}) satisfying conditional set operations for XOR functions

where (10.1) and (10.4) offer two upper bounds, and (10.2) and (10.3) show the lower bounds of available bias space for V_{EVL} and V_{SET}.

Figure 10.11 depicts an example solution space for a case with $\alpha = 100$ and $\beta = 10$, where the solid and dashed lines, respectively, plot the upper and lower bounds. Within the wide range of available biases, we chose $V_{EVL} = V_{CLOSE}$ and $V_{SET} = 1.5 \times V_{CLOSE}$ as a demonstrative case. The solution space satisfying the constraints always exists for$\alpha > \beta > 1$. In consideration of process variations that can happen to the bistable resistance levels (R_{ON} and R_{OFF}) and the threshold voltages (V_{CLOSE} and V_{OPEN}), it is important to choose the bias voltages of V_{SET} and V_{EVL} around the center of the solution space.

10.3.5 Extension to XNOR Operation

Based on the resistive XOR, we introduce a resistive multiplier circuit that operates as an XNOR gate. As the circuit diagram shown in Fig. 10.12a, the memristive XNOR is composed of three bistable memristive devices (R_Y, R_P, and R_Q), five CMOS switches (S_1–S_4, and S_X), and a load resistor (R_S) [25]. Similar to the resistive XOR circuit of Fig. 10.9a, its computing sequence is composed of 3 phases, Steps-I through Step-III, where the devices R_P and R_Q are simultaneously reset in Step-I, and the $x \oplus y$ is computed and registered in R_P during Step-II, $p' = x \oplus y$, whereas the registered p' is then inverted and the complementary state of R_P is registered in R_Q during the Step-III, $q'' = \overline{x \oplus y}$. The three computing steps allow the circuit to register the logic values of $x \oplus y$ and $\overline{x \oplus y}$, respectively, as the resistance states of R_P and R_Q. The registered outputs can be accessed anytime later on by reading the nonvolatile state of R_P and R_Q. The computing sequence and switch controls are summarized in Fig. 10.12b, and its truth table is shown in Fig. 10.12c.

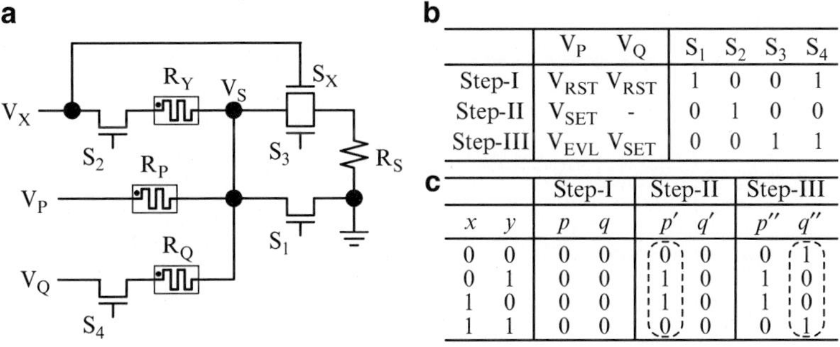

	V_P	V_Q	S_1	S_2	S_3	S_4
Step-I	V_{RST}	V_{RST}	1	0	0	1
Step-II	V_{SET}	-	0	1	0	0
Step-III	V_{EVL}	V_{SET}	0	0	1	1

		Step-I		Step-II		Step-III	
x	y	p	q	p'	q'	p''	q''
0	0	0	0	0	0	0	1
0	1	0	0	1	0	1	0
1	0	0	0	1	0	1	0
1	1	0	0	0	0	0	1

Fig. 10.12 Memristors-based resistive multiplier that registers an evaluated XNOR output in the state of R_Q (q), in three computing phases. (**a**) circuit configuration, (**b**) XNOR execution phases, and (**c**) truth table for the 3-step XNOR execution

Table 10.1 Readout configuration

Mode	V_P	V_Q	S_1	S_2	S_3	S_4
XOR	V_{EVL}	0	0	0	1	1
XNOR	0	V_{EVL}	0	0	1	1

Consisting of memristive devices and CMOS switches, the resistive multipliers of Figs. 10.9a and 10.12a can execute, even with passive unipolar resistances, bipolar multiplications in two or three computing phases. Compared to the CMOS XOR/XNOR gate, the memristors-based resistive XOR/XNOR take less chip area and hence increase the packing density, since only three or five transistors and a resistor will be needed on the CMOS layer and the memristors can be stacked on top of the CMOS circuit through hybrid integration. Other required peripherals for Step control and voltage drivers can be shared with other circuits in the system. In addition to the superior packing density, the proposed memristors-based XOR and XNOR inherently register the computed outputs in nonvolatile memristors, and thus can save the costly registers, especially for massively parallel deep pipeline computing applications [27, 28].

After completing the Steps-I through III, either of evaluated logic values p or q can be read by sensing V_S under a readout voltage appliance to the device to be read, since both the evaluated XOR and XNOR outputs are registered as the state of R_P and R_Q. Since the R_P and R_Q are basically complementary each other, i.e., $q = \overline{p}$, a sensed voltage window can be maximized by configuring the switches and readout voltages as summarized in Table 10.1. With the configurations, V_S becomes proportional to XNOR output (q) if $V_Q = V_{EVL}$ and $V_P = 0$, whereas it is proportional to XOR output (p) for $V_Q = 0$ and $V_P = V_{EVL}$. Here the switch S_3 in the XNOR gate is turned on together with S_4 for this readout duration in order to prevent potential V_S dependency on V_X. When the switch-induced parasitics are ignored, V_S for a readout of XNOR output can be represented with parameters of

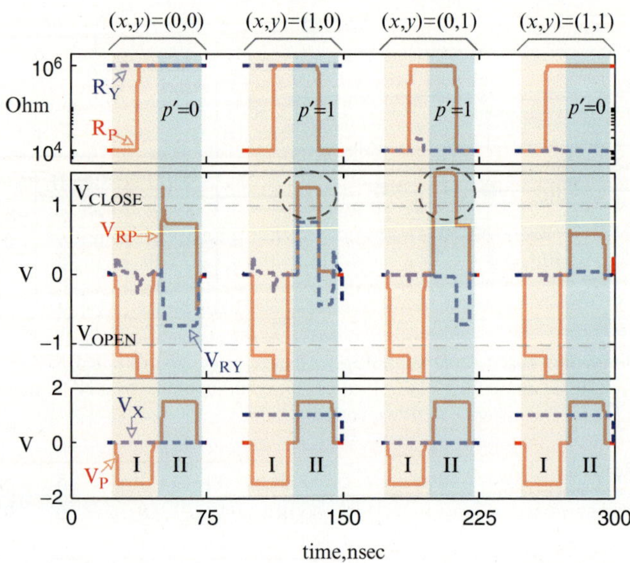

Fig. 10.13 XNOR execution. Simulation results for the stateful two-input XOR gate

α and β, respectively, for off-to-on resistance ratio (R_{OFF}/R_{ON}) and the ratio of R_{OFF}/R_S, as:

$$V_S = \left(\frac{R_P//R_S}{R_Q + R_P//R_S}\right) \times V_{EVL} = \left\{\frac{1}{\alpha+\beta+1} + \left(\frac{\alpha-1}{\alpha+\beta+1}\right) \cdot q\right\} \times V_{EVL}.$$

For a case with $\alpha = 100$ and $\beta = 10$, the sensed voltage V_S for $q = 0$ and $q = 1$ are, respectively, $0.09 \times V_{EVL}$ and $0.9 \times V_{EVL}$, of which the voltage difference is wide enough to identify the stored information.

10.3.6 Simulations of Resistive XOR/XNOR

The circuits in Figs. 10.9a and 10.12a were designed for the 0.18um CMOS process, and the behaviors of memristive devices were described in a Verilog-A model. Circuit simulations were performed with the circuit parameters of $\beta = 10$, $V_{EVL} = V_{CLOSE}$, $V_{SET} = 1.5 \times V_{CLOSE}$, and $V_{RST} = 1.5 \times V_{OPEN}$. Assuming 10 ns for the devices' hold time (T_{hold}) before resistance switching, pulse widths of V_P and switch control signals were chosen to be $2 \times T_{hold}$, 20 ns, so that the memristive devices can safely be switched when all other constraints are met.

While the switches S_1 and S_X in the XOR gate (Fig. 10.9a) are sized to the minimum width, $W_n/L_n = 0.22$ um/0.18 um, to minimize the parasitics, S_2 is sized to $W_n/L_n = 1$ um/0.18 um in order for the channel resistance not to significantly shift the conditional set constraints [23]. Figure 10.13 shows the simulated results

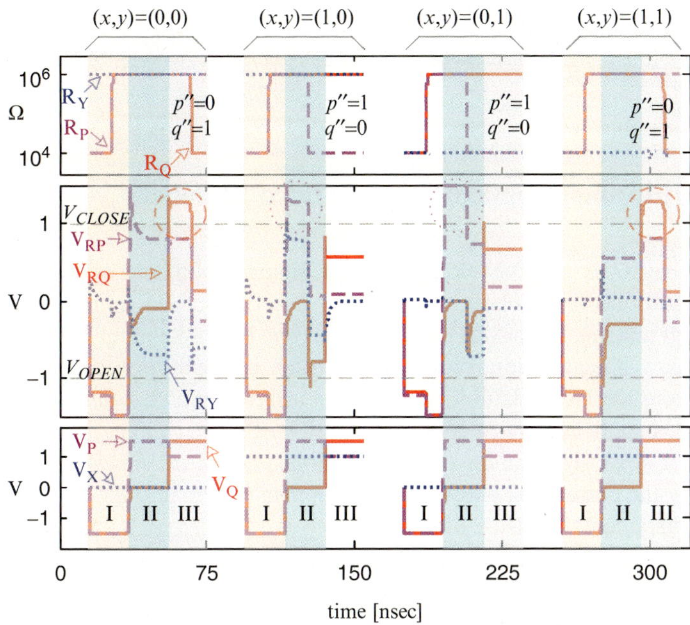

Fig. 10.14 XNOR circuit simulation results

of the resistive XOR circuit for all possible four input combinations, $x \in \{0,1\}$ and $y \in \{0,1\}$. Each simulation started from the initial R_P state of "1" ($R_P = R_{ON}$). In Steps-I, the initial R_P was reset to R_{OFF}, $p = 0$, and then the XOR function was executed in Steps-II registering the evaluated results (p'). The transient R_P behaviors clearly show that the R_P device successfully registers the evaluated output in its resistance state, $p' = x \oplus y$, for all input combinations. This can be confirmed by the observation that V_{RP} higher than V_{CLOSE} is sustained for longer duration than T_{hold} only for the case $x \neq y$. For other cases of $x = y$, V_{RP} in Steps-II remains in a range lower than V_{CLOSE}, with a margin larger than 250 mV. Also it can be noted that the state of R_Y (y) always holds its value, since V_{RY} is well confined within the range $[V_{OPEN}, V_{CLOSE}]$ with a margin larger than 250 mV [23].

Figure 10.14 shows the simulated results of the two-input resistive XNOR circuit shown in Fig. 10.12a, for all possible input combinations. In Steps-I, the initial R_P and R_Q were reset to R_{OFF}, $p = 0$ and $q = 0$, and then the XOR function was executed in Steps-II registering the evaluated results in R_P (p'). Finally in Step-III, the inverted version of p' is registered in R_Q, ($q'' = \overline{p'}$). The transient behaviors of R_P and R_Q clearly show that the devices successfully register the evaluated outputs in their resistance states, $p' = p'' = x \oplus y$ and $q'' = \overline{x \oplus y}$, for all input combinations. This can be confirmed since V_{RP} (V_{RQ}) of higher than V_{CLOSE} is sustained for longer duration than T_{hold} only for the case $x \neq y$ ($x = y$). For other case of $x = y$ ($x \neq y$), V_{RP} (V_{RQ}) in Steps-II and -III remains in a range lower than V_{CLOSE}, with more than

200 mV margin. It can be noted that the state of R_Y, y, holds its value, since V_{RY} is well confined within the range [V_{OPEN}, V_{CLOSE}] with larger margin than 200 mV. The evaluated p' also holds its value for Step-III [25].

10.4 Multiplier Application to Pattern Matchers

The resistive XOR and XNOR function as signal multipliers with negative and positive multiplication polarities, of which one of the two inputs is the stored information in a memristor (R_Y) and the other one being the input logic level (V_X). Among many potential applications where such multiplication capability can be used, we consider a correlation-based pattern matcher as a demonstrative example. In numerous applications of pattern recognition of voice or image, estimation and detection of noisy signal in the communication systems, the signal similarity more specifically the correlation between the input data pattern and the stored reference data is a fundamental measure of the matching as a function of a time-lag applied to one of them. Applications also include scientific data analysis of climate, astronomy, or bioinformatics, as well as national security applications including cyber security, surveillance, and forensics.

In the linear digital signal processing, this is commonly known as a sliding inner product. Since the quantity of correlation between two inputs is equivalent to the inner product of the two signals, a simple correlator can be constructed with a number of multipliers and one adder as shown its diagram in Fig. 10.15a. At every computation, an L-number of input bit sequence $\mathbf{X} = [x_0, x_1, \ldots, x_{L-1}]$ is, respectively, multiplied with the same number of reference sequence $\mathbf{Y} = [y_0, y_1, \ldots, y_{L-1}]$, and then summing the all products together becomes the quantity of signal correlation (c), i.e., $c = \sum_{i=0}^{L-1} x_i \cdot y_i$. Similar to the linear digital correlator,

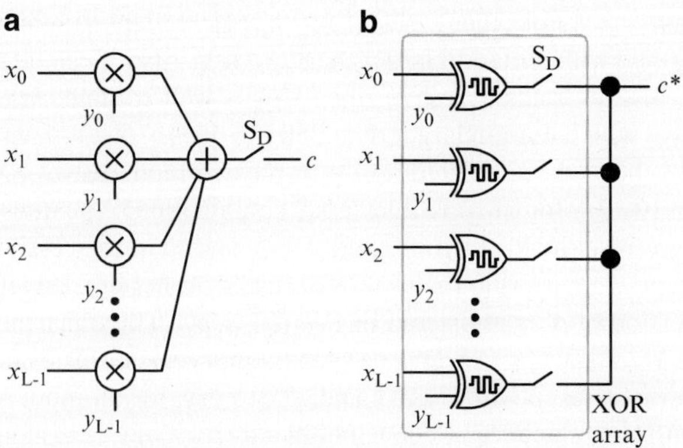

Fig. 10.15 (**a**) Linear digital correlator of which output (c) is dumped during ON durations of the dump switch (S_D), (**b**) nonlinear pattern matcher implemented by an array of resistive XORs

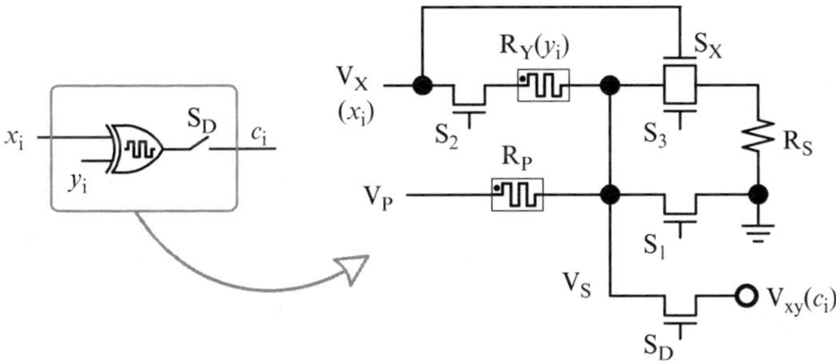

Fig. 10.16 Resistive XOR cell implementation for a pattern matcher. After registering XOR output, the S_D is enabled to evaluate and dump the signal similarity output

Table 10.2 Pattern matcher computing sequence

Step	V_P	V_Q	S_1	S_2	S_3	S_D	Mode
I	V_{RST}	V_{RST}	1	0	0	0	Reset
II	V_{SET}	–	0	1	0	0	XOR
III	V_{EVL}	–	0	0	1	1	Correlation

a nonlinear pattern matcher for L-bit width vector comparisons is constructed by a parallel network composed of L-number of memristive XOR multipliers, as the diagram shown in Fig. 10.15b [26].

Individual resistive XOR cell is comprised as the circuit in Fig. 10.16. Here the reference vector **Y** is assumed to be registered as the resistance states of devices (R_Y). During the XOR evaluation periods of Steps-I and -II as its sequence described in Table 10.2, each XOR cell computes the bit-wise signal multiplication between x_i and y_i with the dump switches (S_D) turned off. Then in the similarity evaluation period of Step-III, the S_D switches are turned on to connect all the V_S nodes of participating XOR cells while applying evaluation voltage V_{EVL} to V_P. The switch S_3 of each XOR gate is turned on for this period in order to prevent potential V_S dependency on V_X which cause undesirable inaccuracy of the evaluated output (c^*).

If the S_D switches are in off states under the bias voltage of $V_P = V_{EVL}$, each XOR gate operating in a readout mode produces its V_S node output ($V_{SO.i}$), when the parasitics are ignored, as:

$$V_{SO.i} = \left(\frac{R_S}{R_{P.i} + R_S}\right) \times V_{EVL} = \left(\frac{1 + z_i(\alpha - 1)}{1 + z_i(\alpha - 1) + \beta}\right) \times V_{EVL},$$

where $R_{P.i} = \frac{R_{OFF}}{1+z_i(\alpha-1)}$ and $z_i = x \oplus y$. It produces $V_{SO.i} = \left(\frac{1}{1+\beta}\right) \times V_{EVL}$ for matched bits ($z_i = 0$), and $V_{SO.i} = \left(\frac{\alpha}{\alpha+\beta}\right) \times V_{EVL}$ for mismatched bits ($z_i = 1$).

For the similarity evaluation of two L-bit width vectors with closed S_D switches, the output voltage at c^* node in Fig. 10.15b, V_{XY}, becomes:

$$V_{XY} = \left(\frac{R_{S.EQ}}{R_{P.EQ} + R_{S.EQ}}\right) \times V_{EVL},$$

where $R_{P.EQ} = \frac{R_{OFF}}{\sum_{i=0}^{L-1} \{1 + z_i(\alpha - 1)\}}$ and $R_{S.EQ} = \frac{R_S}{L} = \frac{R_{OFF}}{\beta L}$. The output voltage, V_{XY}, can be recast as a function of the number of matched bits (k) out of L-bits, as:

$$V_{XY} = \left(\frac{1 + \alpha(L - k)}{1 + \alpha(L - k) + \beta L} \right) \times V_{EVL}. \tag{10.5}$$

As the evaluated similarity output, V_{XY} is inversely related to the correlation between the vectors **X** and **Y**, ranging from the largest V_{XY} of $\left(\frac{1+\alpha L}{1+(\alpha+\beta)L}\right) \times V_{EVL}$ for $k = 0$, to the smallest V_{XY} of $\left(\frac{1}{1+\beta L}\right) \times V_{EVL}$ for $k = L$. Here it can be noted that the V_{XY} is nonlinear to the number of matched bits. As the k approaches to L, the V_{XY} is reduced much more rapidly compared to cases with less number of matched bits. Such nonlinear V_{XY} behavior to the signal correlation significantly enhances the matcher performance, since a small reduction of the correlation from the perfect matching case will cause significant amount of V_{XY} difference. It is also noteworthy that, even though the unit resistive multiplier individually may exhibit relatively slow computing speed, pattern matchers comprised of a massive array of the resistive multipliers can be much faster than conventional CMOS correlators by performing multi-bit calculations in a single step. In other words, the computation speed is independent of L, implying that the circuit can complete the computing for any complex sources in 3-steps, 2-steps for XOR evaluation, and 1-step for evaluation and readout of similarity output [26].

For the 0.18 um CMOS process, the resistive multiplier-based pattern matcher of Fig. 10.15b was designed for a simple demonstrative pattern matcher case. For an example case with $L = 8$, a stored reference code (**Y**) and nine different input code words (**X**) were generated such that the ideal correlation values are equally distributed within $[-1, 1]$. For the same device and circuit parameters that used in Sect. 10.3.6, Fig. 10.17a shows the correlation-dependent equivalent parallel impedances of $R_{P.EQ}$ and $R_{S.EQ}$, and the output similarity values of V_{XY} are shown in Fig. 10.17b. Induced by the nonlinear $R_{P.EQ}$ whose value is highly sensitive for the close matching cases, the matcher output (V_{XY}) sensitively identifies the pattern similarity [26].

In order to validate the matcher's functionality, we constructed an 8-bit memristive XOR-based pattern matcher with 8-different branches as shown in Fig. 10.18 [26]. The input symbol stream was encoded into 8-bit binary code words at a rate of $f_{\text{symbol}} = 1.67$ MHz ($T_{\text{symbol}} = 600$ ns), and then was serialized with paddings of two zeros in between the symbols at a rate of $f_{\text{bit}} = 16.7$ MHz ($T_{\text{bit}} = 60$ ns). The serialized bit stream was then pushed into the 8-bit shift registers whose outputs were fed to the 8 matcher branches in parallel. Here the matchers were clocked by $f_{CK} = 3 \times f_{\text{bit}} = 50$ MHz ($T_{CK} = 20$ns), so that the 3-Steps of matcher operations (Steps-I and II for XOR executions, and Step-III for the similarity evaluation) can be done in every T_{bit} duration.

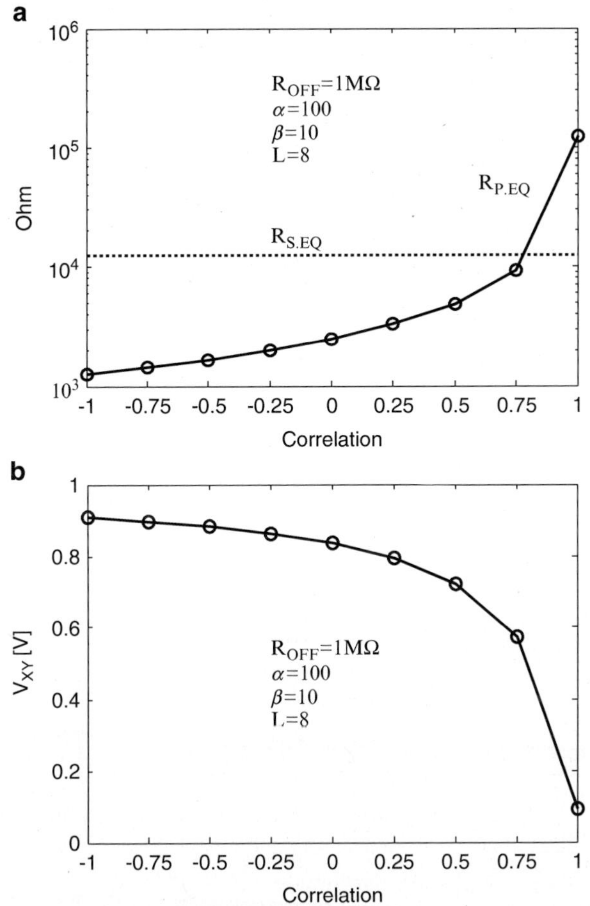

Fig. 10.17 Simulation results of 8-bit pattern matcher: (**a**) equivalent combined resistance after the bit-wise XOR evaluations, and (**b**) the signal similarity output as a function of correlation quantity of two inputs. Displayed samples were picked during every readout period from transient V_{XY} waveform

For a demonstrative application example, each of the matcher branches was set to store one of the eight binary encoded characters of "ACFILNOR," and the character stream of "CALIFORNIA" was fed to the input. Figure 10.15 depicts simulated matcher branch outputs (V_{XY}) captured for the similarity evaluation period (Steps-III) of every input bit cycle, showing the matcher successfully detects the matched input pattern [26]. As desired, each branch produces the low level V_{XY} only if the input bit stream is perfectly aligned and matched to its stored reference vector **Y** (Fig. 10.19).

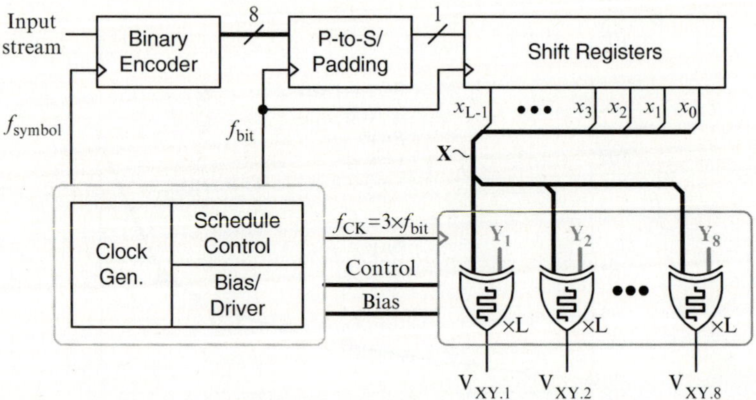

Fig. 10.18 Sturcture of L-bit width memristive-XOR-based pattern detector with 8-different branches of pattern matcher, where the binary encoded and serialized input bit stream (**X**) is fed to the eight matcher branches

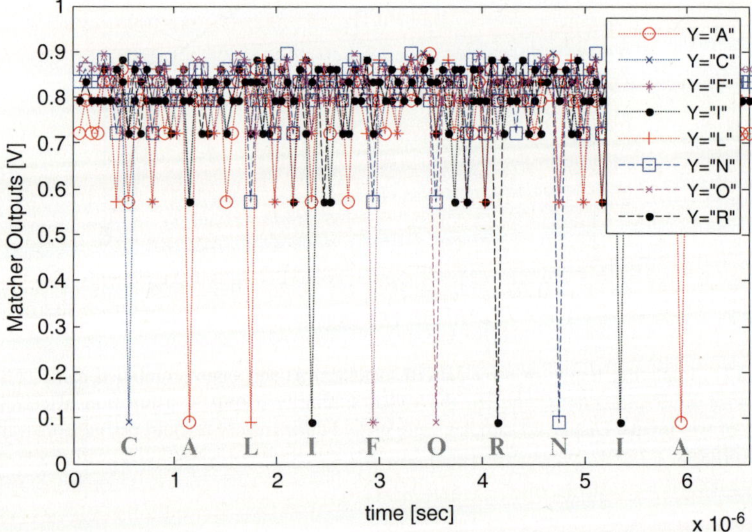

Fig. 10.19 Detection performance of 8-bit matcher with 8-different branches each of which stores one of the eight binary encoded characters of "ACFILNOR." As the input bit stream, binary encoded character stream of "CALIFORNIA" was fed to the matcher with zero paddings in between the characters

10.5 Summary and Future Memristive Electronics

In this chapter we discussed recently reported memristor applications to resistive stateful logic gates and pattern matchers. Memristive technology is anticipated to yield new revolutionary analog/digital functions by exploiting the nonlinear properties of memristors, and supplement and replace transistors with memristors for latching or switching circuits, such as FPGAs, which will significantly reduce the form factor, manufacturing cost, and active/leakage power consumption. One potential example is the dependable memristive logic array, for which basic architectures and appropriate circuit compositions for reconfigurable resistive logic array can be investigated with particular attention to the technical barriers such as difficulties for multiple fan-in/fan-out capability and functional robustness. The memristive logic array can effectively implement low-cost fully pipelined digital systems, and therefore holds great potential for implementing low-power and high performance next generation resistive nanocomputers. The memristive-technology-enabled low-cost, nonvolatile, solid-state circuit technology will significantly reduce the energy cost associated with all other electronic systems.

It is also anticipated that a reliable low-power hardware platform for truly dependable nanocomputing systems will help Moore's Law to be extended far beyond the barriers currently observed in CMOS technologies. In addition to the barriers to Moore's Law, such as the scaling limit and the power wall, future computers for complex and advanced information processing must overcome the severe connectivity issues. Since the power consumption for communication can take more than 90 % of the entire computing power, the computer architectures with co-location of memories and processors on a chip becomes preferable. Placing the massive memristive nano devices into local logic and processing cores will be a key to truly co-locate memory and processor on a chip. Since the co-located memory devices can basically function as nonvolatile local memories, stateful logic gates, and also as reconfigurable interconnects, all with very high integration density and low power, the hybrid hardware structure can become a leading technology, significantly relaxing the issue related with scaling and leakage power, and enabling instant resiliency at power shutoff.

Recently demonstrate dare several key technologies for high-density nonvolatile memory architectures; reconfigurable resistive logic gates and nonvolatile CMOS logic structure; resistive computation structures; and many others, all of which will contribute to building a fundamental basis for truly flexible nonvolatile resistive nanocomputers. The related research would open a new electronics and computing paradigm, and directing the research community to a completely new way.

References

1. L.O. Chua, Memristor-the missing circuit element. IEEE Trans. Circuits Theory **18**, 507–519 (1971)
2. L.O. Chua, S.M. Kang, Memristive devices and systems. Proc. IEEE **64**, 209–223 (1976)
3. D.B. Strukov et al., The missing memristor found. Nature **453**, 80–83 (2008)
4. ERD/ERM Final Report, ITRS (2010)
5. M. Di Ventra, Y.V. Preshin, L.O. Chua, Circuit elements with memory: memristors, memcapacitors, and meminductors. Proc. IEEE **97**(10), 1717–1724 (2009)
6. D.B. Strukov, K.K. Likharev, CMOLFPGA: a reconfigurable architecture for hybrid digital circuits with two-terminal nanodevices. Nanotechnology **16**, 888–900 (2005)
7. G.S. Snider, R.S. Williams, Nano/CMOS architectures using a field programmable nanowire interconnect. Nanotechnology **18**, 035204 (2007)
8. S. Shin, K. Kim, S.M. Kang, Memristor-based fine resolution resistance and its applications, in *ICCCAS 2009* (2009), pp. 948–951
9. S. Shin, K. Kim, S.M. Kang, Memristor application to programmable analog ICs. IEEE Trans. Nanotechnol. **10**(2), 266–270 (2011)
10. D. Varghese, G. Gandhi, Memristor based high linear range differential pair, in *ICCCAS 2009* (2009), pp. 935–938
11. Y.V. Pershin, M.D. Ventra, Practical approach to programmable analog circuits with memristors. IEEE Trans. Circuits Syst. I **57**(8), 1857–1864 (2010)
12. S. Shin, K. Kim, S.M. Kang, Analysis of passive memristive devices array: data-dependent statistical model and self-adaptable sense resistance for RRAMs. Proc. IEEE **100**(6), 2021–2032 (2012)
13. B. Mouttet, Proposal for memristors in signal processing. Nano-Net **2008**, 11–13 (2009)
14. S.H. Jo et al., Nanoscale memristor device as synapse in neuromorphic systems. Nano Lett. **10**(4), 1297–1301 (2010)
15. D.B. Strukov, R.S. Williams, Four-dimensional address topology for circuits with stacked multilayer crossbar arrays. Proc. Natl. Acad. Sci. USA **106**(48), 20155–20158 (2009)
16. P.J. Kuekes et al., The crossbar latch: logic value storage, restoration, and inversion in crossbar circuits. J Appl. Phys. **97**, 034301 (2005)
17. P. Kuekes, Material implication: digital logic with memristors, in *Memristor and Memristive Systems Symposium* (November 21, 2008)
18. J. Borghetti et al., A hybrid nanomemristor/transistor logic circuit capable of self-programming. Proc. Natl. Acad. Sci. USA **106**(6), 1699–1703 (2009)
19. J. Borghetti et al., 'Memristive' switches enable 'stateful' logic operations via material implication. Nature **464**, 873–875 (2010)
20. K. Kim, S. Shin, S.M. Kang, Stateful logic pipeline architecture. IEEE Int. Symp. Circuits Syst. **2011**, 2497–2500 (2011)
21. S. Shin, K. Kim, S.M. Kang, Reconfigurable stateful NOR gate for large-scale logic array integrations. IEEE Trans. Circuits Syst. II **58**(7), 442–446 (2011)
22. K. Kim, S. Shin, S.M. Kang, Field programmable stateful logic array. IEEE Trans. Comput.-Aided Design **30**(12), 1800–1813 (2011)
23. S.M. Kang, S. Shin, Energy-efficient memristive analog and digital electronics. Adv. Neuromorphic Memristor Sci. Appl. Springer Ser. Cogn. Neural Syst. **4**, 181–209 (2012)
24. S. Shin, K. Kim, S.M. Kang, Memristive XOR for resistive multiplier. Electron. Lett. **48**(2), 78–80 (2012)
25. S. Shin, K. Kim, S.M. Kang, Memristive computing-multiplication and correlation. IEEE Int. Symp. Circuits Syst. **2012**, 1608–1611 (2012)

26. S. Shin, K. Kim, S.M. Kang, Resistie Computing: Memristors-Enabled Signal Multiplication. IEEE Trans. Circ. Syst. I **60**(5), 1241–1249 (2013)
27. G. Gill, J. Hansen, M. Singh, Loop pipelining for high-throughput stream computation using self-timed rings, in *Proceedings of ICCAD* (2006), pp. 289–296
28. J. Hennessy, D.A. Patterson, *Computer Architecture: A Quantitative Approach*, 2nd edn. (Morgan Kaufmann, San Mateo, 1996)

Chapter 11
Memristor Device Engineering and CMOS Integration for Reconfigurable Logic Applications

Qiangfei Xia

11.1 Introduction

In the past few decades the integrated circuits (IC) industry has successfully followed Moore's Law in delivering more and more powerful computer chips with reduced cost per transistor [1]. However, CMOS (complementary metal oxide semiconductors) scaling is approaching a physical and economical limit. On the one hand, it becomes more and more challenging and expensive to build smaller transistors that are packed into a very small area. On the other hand, the leakage current associated with smaller transistors will deteriorate or even destroy the device. To sustain rapid progress in information technology, there are intensive efforts to go beyond Moore's Law in research areas including new devices/materials, new technologies, and new architectures and algorithms.

The memristor (memristive device) emerges as one of the most promising devices for the post-CMOS era. As a nonvolatile, two-terminal electronic device, the memristor has variable resistance that changes with the polarity and amplitude of the applied voltage [2–4]. High and low resistance states instead of charge storage are used to represent the logic "1" and "0" in these devices. The typical structure of a memristor consists of a layer of switching material sandwiched between two electrodes. With a cross-point structure, these devices offer great scalability since the junction area is dependent solely on the width of the two nanowires. Materials that are widely used for the switching layer include binary or ternary transition metal oxides, perovskites, and solid-state electrolytes [5]. Depending on the device switching mechanism, the electrode materials are usually inert metals such as Pt, W, or active metals such as Ag, Cu, etc.

Q. Xia (✉)
Nanodevices and Integrated Systems Laboratory, Department of Electrical and Computer Engineering, University of Massachusetts, Amherst, MA 01003, USA
e-mail: qxia@ecs.umass.edu

R. Tetzlaff (ed.), *Memristors and Memristive Systems*,
DOI 10.1007/978-1-4614-9068-5_11,

Although the fundamental physics for the switching mechanism is not thoroughly understood, there has been significant progress in memristive device research and development. Particularly, superior device performance has been achieved, such as high endurance higher than 10^{12} cycles [6], faster than 1 ns switching speed [7], sub-100 fJ energy consumption per switching event [8] and an extrapolated date retention time longer than 10 years at room temperature [9]. Because of the achievement in device performance, memristors have been proposed and demonstrated for applications in nonvolatile memory [10–12], reconfigurable switches [13], nonvolatile logic [14], and bio-inspired neuromorphic computing [15–17].

This section focuses on the memristor device engineering and integration for reconfigurable logic circuits applications. First, the requirements in particular device performance parameters are reviewed. Next, device engineering approaches used to achieve these metrics are discussed. These include the fabrication techniques, the choice of switching materials, the multilayer devices, and the device geometry engineering. Finally, hybrid memristor/CMOS circuits with reconfigurable logic functions are demonstrated.

11.2 Performance Requirements

Depending on the application, specific requirements for the device's performance are different. For reconfigurable logic applications, the metrics of the switches can be characterized by the following parameters.

11.2.1 Power Consumption

Resistive switching effect can be implemented through a number of physical mechanisms including phase change, electrochemical reaction, and the creation and migration of charge carriers (such as oxygen vacancies). However, widespread application of the resistive switching is limited by practical power consumption concern. For example, resistive switching based on the amorphous/crystalline phase change phenomenon has been widely studied and was proposed for memory and neuromorphic computing. But one potential barrier for such device is the high power consumption in heating the volume of the materials in the cell. For reconfigurable switches in a larger array, it is important that each device consumes a minute amount of power so that the switch network is not going to be the bottleneck of power consumption. Equally important is the power dissipated in the nanowire network for a memristor crossbar array. The demonstrated sub-100 fJ energy consumption per switch event in oxide-based memristor devices is encouraging, but the power dissipation in the metal nanowire electrodes is yet to be explored.

From the circuit's point of view, the power consumption of the memristors dictates the CMOS compatibility of the devices, as discussed in the following sections.

11.2.2 Endurance

In some applications, such as solid-state drives or as electrical synapses for neuromorphic computing, the devices might experience much more reading than writing. Consequently, the endurance requirement of memristors for those applications is not as demanding. However, for memristors as reconfigurable switches, this is not the case. The devices are programmed every single time when a reconfiguration is necessary. The frequent change of the device states imposes extremely high standards for the device's endurance. Previously, up to 10^{10} cycles of endurance were demonstrated for a TaO_x device at HP Labs [18], and this record was soon broken by Samsung using a bilayer TaO_{2-x}/Ta_2O_{5-x} device that exhibited 10^{12} cycles of endurance [6]. The common feature of their devices is that both used a TaO_x-based oxide as the switching layer. Although the best endurance was demonstrated with the TaOx-based devices, one drawback of such devices is their low ON/OFF ratios. It is desirable that the endurance of the memristor device be comparable to that of DRAM, that is, on the order of 10^{15} cycles.

11.2.3 Uniformity

The temporal (cycle-to-cycle) and spatial (device-to-device) variation in device performance has been identified as one of the biggest obstacles for the wide adoption of memristor devices. There are several fundamental origins for the nonuniformity. First of all, the number and location of conductive filaments in a junction are random, especially for a device with a relatively large junction area. Second, the junction geometry such as the thickness uniformity and the line edge roughness of the electrode wire are not the same from device to device. Third, during the RESET process, the interface between the conductive filament and the oxide is not deterministically controllable.

For logic computing purposes, some of the nonuniformity issues can be addressed by adopting new computing algorithms, such as statistical computing [19] that uses the stochastic nature of the memristive devices and use a bit stream from several devices rather than one to represent the logic state. However, this is not practical for reconfigurable switches since each junction serves as a switch. It is hence important that the switching voltages are uniform from device to device and from cycle to cycle so that proper drive circuits can be used for these devices. The best demonstrated device-to-device uniformity was 5 % variation in the programming voltages [20].

11.2.4 Switch Speed

How fast the device can SET (to low resistance or ON state) or RESET (to high resistance or OFF state) is critical for many applications. For reconfigurable switches, this will affect the bandwidth of the logic circuits. The measurement of the switch speed is practically challenging especially down to the nanosecond region. This is partly due to the lack of ultrahigh speed measurement equipment, as well as the fact that shorter electrical pulses are hard to reach the junction area due to significant reflectance loss on the metal wires. To couple ultra-short electrical pulses into the device with reduced reflection, a tapered coplanar waveguide (CPW) test structure was introduced and the best reported switch speed data was on the order of 100 ps [7]. It was also discovered that there is a speed-voltage trade-off for the memristor devices [21]. Higher switching voltage is usually needed to switch the devices if shorter electrical pulses are used.

11.2.5 CMOS Compatibility

In order for the memristor to be used for reconfigurable logic circuits, it is often used together with silicon-based transistors in a hybrid fashion. It is straightforward that the materials and process used in building the circuits should be compatible with state-of-the-art CMOS technology. Moreover, the programming voltage should be compatible with the CMOS driving circuits. Memristors can be built with standard IC fabrication techniques such as lithography, deposition, and etching. While there is a broad spectrum of materials that have exhibited memristive behavior, including standard materials used in the IC industry such as silicon, silicon oxides are considered advantageous for the wide adoption of memristors without significant alteration in the fabrication infrastructure. To further improve the device's behavior, memristors with electrodes embedded into the substrates have demonstrated much improved endurance and reduce series resistance [22]. This can be implemented in the industry with a process similar to dual Damascene process that is currently widely used for Cu interconnection in the IC industry.

Voltage compatibility with CMOS imposes high demand in the device's performance, especially the programming voltages. The current generation of transistors operates at a voltage lower than 1 V. To make memristors compatible with CMOS, it is hence important to (1) develop memristors that do not require an electroforming process (electroforming is a process in which a higher voltage/current is used to create a conductive path (or filament) by softly breaking down the switching layer for a wide variety of transition metal oxides), and (2) develop memristors with ultralow programming voltages.

11.2.6 I–V Nonlinearity

When memristors are connected in a cross bar, all the cells in a row are connected to each other. This inevitably creates a problem in selectively programming a particular cell because the neighboring cells could be affected. This is the so-called sneak path problem. When a crossbar array becomes bigger, this problem becomes more complicated. Several selective devices such as transistors, diode, etc. have been proposed to solve this problem, but they have their own problems as well. For example, diode is better for a unipolar switching device, while a transistor will increase the area of each cell, limiting the scalability of the cross bar.

For reconfigurable switches, it is hence desirable to use devices that exhibit intrinsic I–V nonlinearity without introducing external selector devices to precisely program a particular device. I–V nonlinearity of memristive devices will minimize the sneak path current with no overhead in the cell area.

11.2.7 ON/OFF Conductance Ratio

The ON/OFF ratio is important for high-density memory applications, especially for multi-level cells. Storing more than one single bit of information in a memory cell is one of the most promising approaches to effectively increase the information density. To this end, a big enough ON/OFF ratio will enable big enough resistance window between any resistance states. However, the requirement in the ON/OFF ratio for the reconfigurable switches is much relaxed. As a binary device, as long as the device can be reliably turned to stable ON or OFF states, it should serve the purpose for a switch. Given the discussion on the endurance requirements, it appears that TaO_x-based memristor devices are one of the best candidates for reconfigurable switch applications.

To meet all the aforementioned requirements, tremendous efforts should be devoted to device engineering by adopting new switching materials, novel device geometry, reliable fabrication techniques, and electrode nanostructure engineering.

11.3 Device Engineering

11.3.1 Fabrication Techniques

For memristors with a cross-point structure, a key issue in device fabrication is the patterning of the electrodes. Since the width of the electrode wires determines the junction size, nanopatterning techniques such as deep ultraviolet (DUV) photolithography, electron beam lithography (EBL), and nanoimprint lithography (NIL)

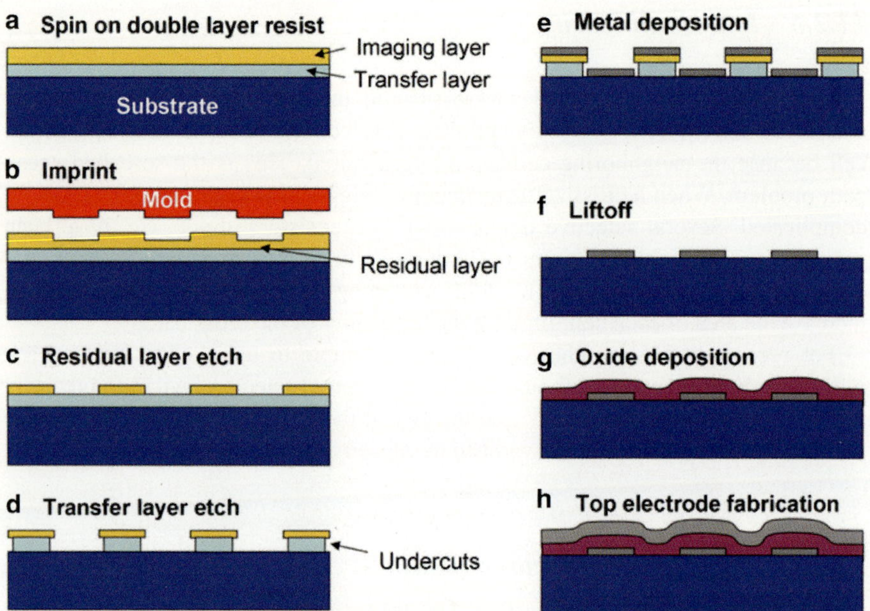

Fig. 11.1 The "standard" fabrication process flow for cross bar structures using nanoimprint lithography (NIL). Two NIL steps, one for the bottom electrodes and the other for the top electrodes are used in this scheme [26]

are candidates for making nanoscale memrsitor devices. Among these techniques, NIL has been widely used for fabricating memristor crossbar arrays over a large area due to its resolution, patterning area, and cost advantages.

Usually ultraviolet nanoimprint lithography (UV–NIL) [23–25] is the preferred method because it operates at low pressure and at room temperature. A key component for NIL is the imprint mold. A master mold (usually made of Si with a thin layer of thermally grown SiO_2) with nanoscale (nanowires, nanoscale fan outs, etc.) and microscale features (contact pads, alignment marks, microscale fan outs, etc.) is first made by EBL, photolithography, and reactive ion etching (RIE). These features are then duplicated using NIL and RIE onto a quartz (QZ) wafer. QZ is chosen because it is transparent to the UV light used for NIL, chemically compatible with the UV resist, and mechanically strong to achieve high resolution nanopatterning. The fabricated QZ molds are then treated with trichlorosilane to form an anti-sticking monolayer before being used for imprint. It is worthwhile pointing out that in order to achieve smaller devices, the feature sizes on the master SiO_2/Si mold can be further shrunk using simple techniques such as wet etching with diluted hydrofluoric (HF) acid [21].

Figure 11.1 schematically illustrates the procedure of memristor crossbar fabrication using NIL to pattern both the bottom and the top electrodes [27, 28]. The procedure is as follows. First, double layer resists consisting of an underlayer and a

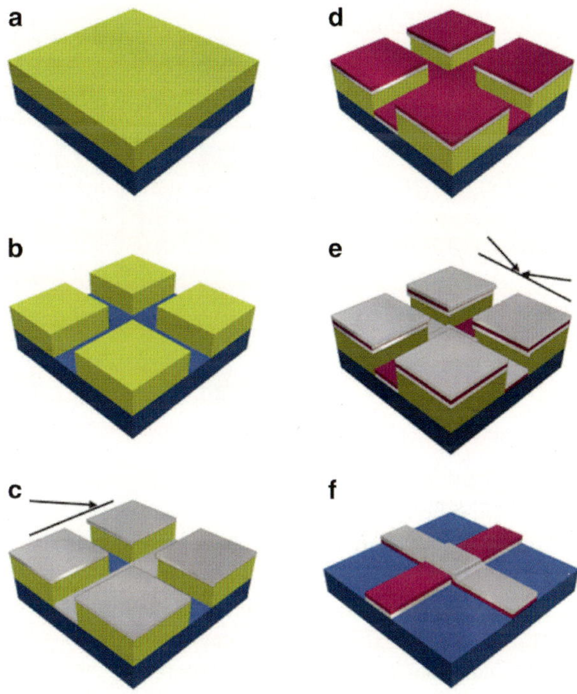

Fig. 11.2 Schematic of the self-aligned fabrication approach with one NIL step. The cross-shaped trenches are patterned by one NIL step and RIE in resists; after depositing the three layers, a lift-off process concludes the fabrication. The metal electrodes were deposited using angle evaporation so that the metals reach the bottom of one trench but not the other [29]

UV-curable top layer are spin coated onto the substrate sequentially. The underlayer is necessary if lift-off is used for electrode fabrication, and it can be dissolved in a solvent such as acetone. The UV-curable layer is a liquid that can be turned into a solid upon exposure to a UV source. Circuit patterns on the QZ mold are then transferred to the UV-curable layer during NIL. The residual cross-linked UV resist and the underlayer are etched by RIE using fluorine and oxygen-based etching chemistry, respectively. During the second RIE process, undercuts in the transfer layer are intentionally made by over etching to get better lift-off properties. The bottom electrodes are deposited onto the sample followed by a lift-off process in a solvent. A thin layer of switching material is then coated onto the sample with sputter or atomic layer deposition (ALD) equipment. Following a second set of NIL, RIE, metallization, and lift-off processes, the top electrodes are made on top of the oxide thin film with the nanowires oriented orthogonal to the bottom wires.

The fabrication process in Fig. 11.1 has been widely used as a "standard" process for memristor crossbar fabrication. Although nanoscale devices with excellent electrical performances have been routinely fabricated, the "standard" process can be optimized to improve the device performances. Recently, an alternative approach that employs only one lithography step was proposed and demonstrated (Fig. 11.2) [29]. In this new technique, an NIL mold with crossbar structures instead of nanowire arrays was used to pattern crossbar-shaped trenches in the resist stack. Metal evaporation with an oblique angle to the sample surface was then carried out

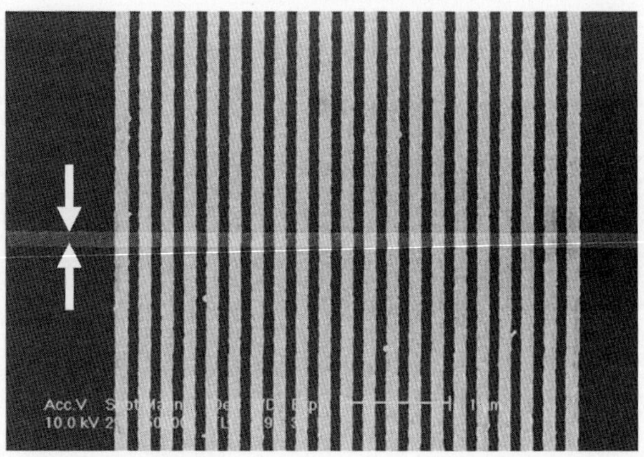

Fig. 11.3 Images of a 1 × 21 array of memristors fabricated using one NIL step. The junction area for each device is 100 × 100 nm^2 [29]

to make the bottom electrodes. Due to shadow effects, metal entered into one trench but not the other. The switching layer was immediately deposited in a sputter, and the top electrodes were made by shadow evaporation along the direction of the other trenches.

The advantages of the new fabrication technique are as follows. First, the critical interfaces between the switching layer and the two electrodes are protected from contaminants such as wet chemicals, resists, and charged particles during RIE. Cleaner interfaces could lead to better device performance. Second, with only one lithography step, the device fabrication process is self-aligned. As a result, the requirement on accurate overlay alignment between the top and bottom electrode is relaxed. The fabrication time and cost are reduced and the throughput will be increased. Third, as NIL is a contact lithography technique, the defectivity of the devices is increased if multiple NIL steps are used. Reducing to one NIL step means fewer defects and higher device yield.

Figure 11.3 shows a typical scanning electron microscope (SEM) image of a 1 × 21 TiO_2 device array with the single horizontal wire at the bottom fabricated with one NIL. The junction area of these devices is 100 nm × 100 nm. The device stack has 12.5 nm of Pt as the top electrode, 13 nm thick TiO_2 as the switching layer, and 9 nm of Pt and 3.5 nm of Ti as the bottom electrode. Compared with devices of the same geometry but fabricated with the "standard" two NIL steps, devices fabricated with this technique exhibited forming-free nonvolatile switching behavior with an ON/OFF ratio over 1,000. The switching current was at nA scale, much lower than usually observed for similar devices fabricated using a multistep lithography process (μA) (Fig. 11.4).

The forming-free behavior and the low programming current of devices fabricated with one NIL step were due to the rougher surface created by angle evaporation. Atomic force microscope (AFM) surface characterization showed that a smaller deposition angle results in a rougher surface. For a 90° incidence, the root-mean-square (rms) roughness was 0.50 nm, while that for a 20° incidence angle was 0.98 nm. There were also a certain amount of metal spikes on the film

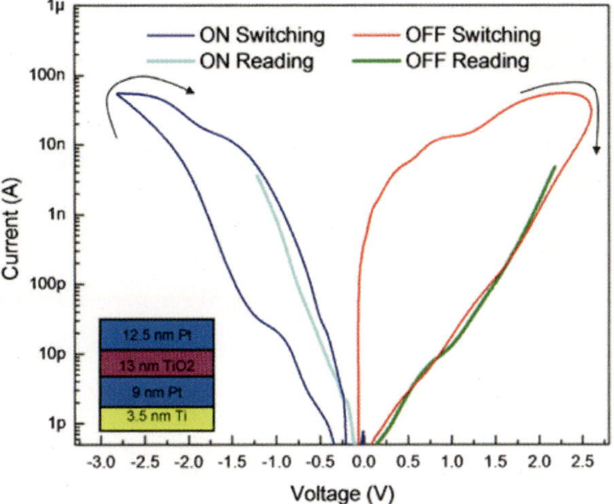

Fig. 11.4 Typical I–V curve of a forming-free device with nA operational current. The device exhibits nonvolatile switching behavior with an ON/OFF ratio over 1000 at 1 V [29]

surface. Figure 11.5 illustrates the electric field distribution on the surface of metal electrodes deposited at different angles. As the numerical simulation results showed, the electric field was concentrated at the metal spikes that served as "artificial filaments." For 3 V applied voltages, the calculated maximum electrical field for (a) and (b) are 2.8×10^8 and 6.4×10^8 V/m, respectively. Consequently, the devices were able to be switched at lower applied biases without the necessity of an electroforming process.

The discovery can be further engineering by precise control of the "artificial filaments" over a large area with high uniformity. The will lead to forming-free devices with better performance uniformity. It is also possible to design a deposition system so that all the critical layers (metal electrodes and switching materials) can be deposited in the same system without breaking the vacuum. Such an instrument will offer even better protection of the critical interfaces between the metal electrode and the switching material.

11.3.2 Switching Materials

First reported nearly 50 years ago in anodic oxide thin film [30], the signature hysteresis I–V curves have been discovered in different systems. A wide variety of materials have been demonstrated with memristive behavior recently and they can be roughly categorized as transition metal oxides (such as NiO [31], CuO [32], MoO_x [33], ZrO_2 [34], TiO_2 [4, 35, 36], TaO_x [18, 37]), perovskites [38, 39], and solid-state electrolytes [40] and even Si and SiO_2, etc. [20, 41–43]. The switching behavior has been widely studied for these materials and the reported device behavior can be best summarized in Table 11.1 [5].

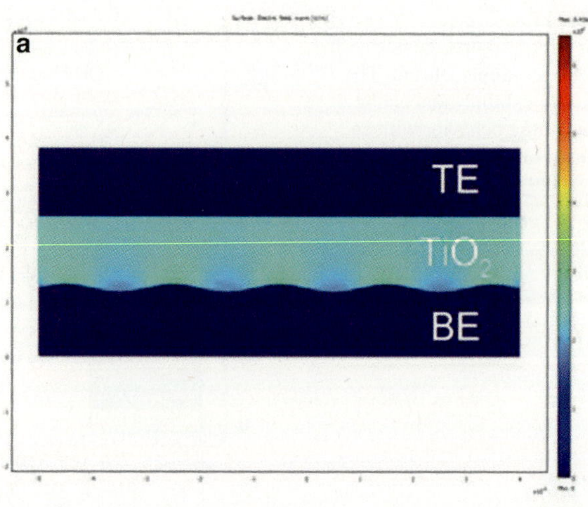

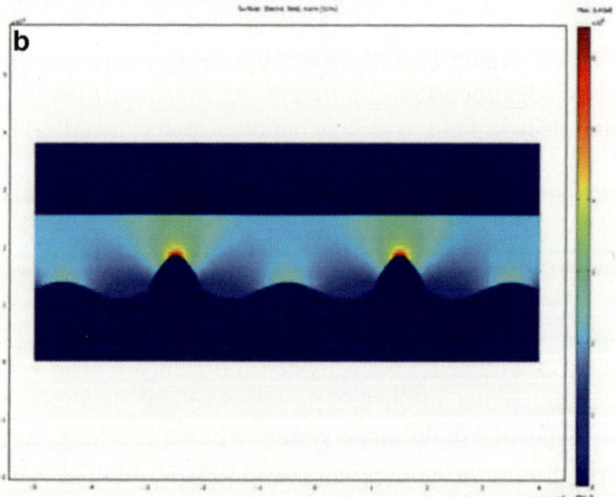

Fig. 11.5 Simulation of the effect of bottom electrode surface roughness on the electric field distribution. Sinusoidal surface profiles with 10 nm grain size are assumed for the bottom electrodes. The rms roughnesses are 0.5 and 0.98 nm for (**a**) and (**b**), respectively, and the metal pillars in (**b**) are 6 nm high, corresponding to the experimental observation. With a rougher surface, the electric field is more localized at the tips of the metal pillars [29]

Although most memristive devices use materials such as transition metal oxide and perovskites as the functional switching layer, not all of these materials are compatible with current IC industry infrastructure. There has recently been a focused interest in amorphous silicon and silicon oxide. The resistive switching phenomenon in these materials was actually discovered a long time ago [44]. Recent studies using mostly silicon oxide made by different technologies such as plasma enhanced chemical vapor deposition (PECVD) have made great progress [43, 45–47]. However, despite the fact that the devices are compatible with CMOS technologies regarding materials, the reported programming voltages for silicon oxide-based devices were too high (from 3.5 to 13 V). As a result, significant efforts have to be devoted to dramatically lower the programming voltages in order for the devices to be truly compatible with the state-of-the-art low-power CMOS

Table 11.1 List of oxides used for memristive devices and the corresponding device behavior (adapted from [5])

Oxide	TE-BE	ON/OFF ratio	Switching speed	Retention time(s)	Endurance
Binary, bipolar					
CoO	Ta–Pt	10^3	20 ns	–	100
Cu_xO	Ti/TiN–Cu	10^2	50 ns	10^5 @ 90°C	600
$HfLaO_x$	TaN–Pt	10^6	10 ns	10^4 @ 27°C	10^4
HfO_x/TiO_x	TiN–TiN	10^3	5 ns	10^4 @ 200°C	10^5
TaO_x	Pt–Pt	10^1	10 ns	10^7 @ 150°C	10^9
TiO_2	Pt–TiN	10^3	5 ns	10^6 @ 85°C	10^6
ZrO_2	TiN–Pt	10^1	1 μs	10^4 @ 27°C	10^3
Binary, unipolar					
Gd_2O_3	Pt–Pt	10^6	–	10^5 @ 85°C	60
HfO_2	Pt–Pt	10^2	–	10^6 @ 27°C	140
Lu_2O_3	Pt–Pt	10^3	30 ns	10^6 @ 27°C	300
NiO	Pt–Pt	10^2	5 μs	10^7 @ 27°C	10^6
TaO_x	Cu–Pt	10^2	80 ns	10^6 @ 27°C	100
TiO_x	Pt–Pt	10^4	–	–	25
WO_x	TiN–W	4	300 ns	10^4 @ 100°C	10^7
ZnO	Pt–Pt	10^4	–	–	100
Perovskite, bipolar					
$Cr{:}Ba_{0.7}Sr_{0.3}TiO_3$	$Pt–SrRuO_3$	4	0.2 s	10^4 @ 27°C	10^4
$Pr_{0.7}Ca_{0.3}MnO_3$	$Ag–YBa_2Cu_3O_{7-x}$	10^2	8 ns	–	10^5
$Pr_{0.7}Ca_{0.3}MnO_3$	Al–Pt	10^2	20 μs	10^4 @ 125°C	10^3
$Cr{:}SrTiO_3$	Au–Au	10	1 ms	8×10^4 @ 27°C	10^3
$Nb{:}SrTiO_3$	Pt	10^2	50 μs	10^8 @ 125°C	10^7
$Cr{:}SrZrO_3$	$Au–SrRuO_3$	20	100 ns	10^7 @ 27°C	–
$Cr{:}SrZrO_3$	$Al–LaNiO_3$	10^2	500 μs	10^3 @ 85°C	–

transistors. One promising approach demonstrated recently [48] involves the use of chemical method to create an ultrathin layer of silicon oxide (~1 nm) on the silicon wafer surface, and the demonstrated switch voltage was as low as 0.6 V.

The choice of materials in memristor devices is truly dependent on the application. For example, the TaO_x device has the best demonstrated endurance, but the ON/OFF ratio is less than ten. If the resistance window is not a concern, this materials system is a great choice for many applications. On the other hand, it has to be pointed out that even with the same materials systems and electrodes, slight change in device fabrication condition could lead to dramatically different device behavior. For example, the ON/OFF conductance ratio for a TiO_2 device can be modulated by changing the switching layer deposition condition. We have demonstrated that with a small amount of oxygen flow introduced into the sputter chamber, the ON/OFF ratio of a $Pt/TiO_2/Pt/Ti$ device read at 0.5 V was increased from below 100 to more than 10,000. This suggests that there is plenty of room in switching materials engineering in order to optimize the device performances.

11.3.3 Multilayer Devices

While devices with a single layer of switching materials are easy to fabricate, multilayer oxides are widely used to achieve properties that cannot be implemented with a single layer oxide. The multilayers can be homogeneous, meaning they are from the same origin but part of the original single layer is modulated by creating extra amount of charge carries such as oxygen vacancies. In reality, most of the TiO_2-based memristors consist of a conductive TiO_{2-x} layer and a semiconducting TiO_2 layer. The in situ doping was a result from the diffusion of the Ti adhesion layer through grain boundaries in the Pt bottom electrode, which reacts with TiO_2 and creates abundant oxygen vacancies [37]. Heterogeneous multilayer devices were also widely used and they are usually created by deposition of different oxide materials in sequence during device fabrication. In some rare cases, intermediate metal layers were also introduced in between the multi-level switching layers, effectively creating interconnected memristor device stacks at one junction.

The benefits of multilayer devices have been testified with novel properties such as electroforming free switching behavior, nonlinearity in the switching curves, improved switching uniformity, etc. For example, by intentionally creating a thick layer of oxygen vacancy reach layer (TiO_{2-x}) in the TiO_2 switching layer, forming-free behavior has been observed in the device [49]. Devices with HfO_2/TiO_2 double oxide layer showed much improved switching uniformity [50]. It is also expected that devices with both high endurance and high ON/OFF ratio are possible by introducing a heterogeneous oxide layer to the TaO_x stack.

Particularly, multilayer devices are attractive for memristor crossbar arrays to solve the sneak path problem. Excellent nonlinearity has been achieved using TaO_x/TiO_{2-x} heterostructure [51]. Complementary resistive switching (CSR) was proposed by stacking two memristors head to head in one junction [52] and was also implemented using a single device with Ta_2O_{5-x}/TaO_y bilayer structure [53].

11.3.4 Device Geometry Engineering

When fabricated on a flat substrate, the bottom electrodes of a crossbar array form a set of "ribs" on the substrates. These ribs are preserved after the conformal deposition of the thin layer of the switching material. As a result, the top electrodes have to climb up one side and down the other of each rib, leading to four kinks in the top electrode metal wire at each junction. These kinks are the mechanically and electrically weakest parts of the devices and can be detrimental to the device's performance, especially endurance (Fig. 11.6 and 11.7).

In order to avoid the kink structure in the junction, a planar device geometry was proposed and demonstrated. In this structure, the bottom electrodes are embedded into trenches in the substrate, the switching layer and the top electrode are built on flat surfaces. Figure 11.7 shows the schematic of the device geometry and

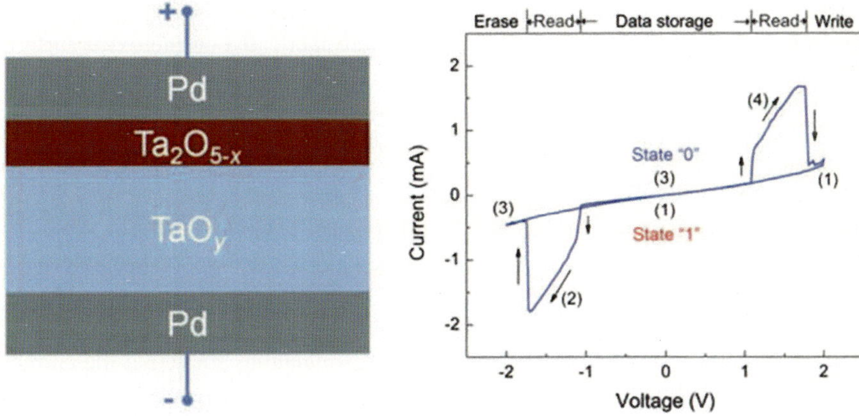

Fig. 11.6 (*Left*) Device structure of aPd/Ta_2O_{5-x}/TaO_y/Pd memristor and the I–V curve (*right*) if programmed with a voltage range of (−2, 2) V. This demonstrates a typical complementary resistive switching behavior [53]

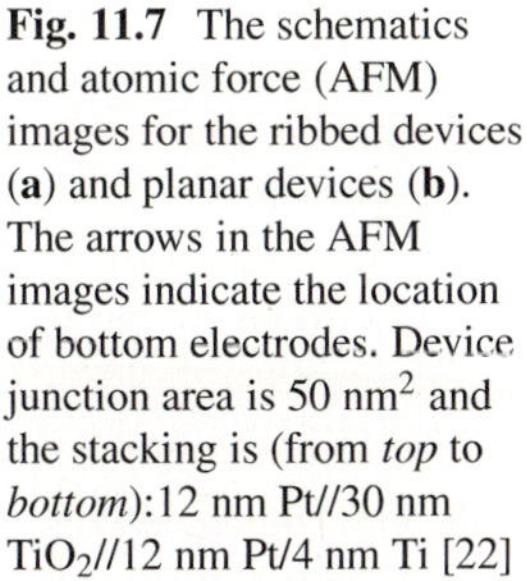

Fig. 11.7 The schematics and atomic force (AFM) images for the ribbed devices (**a**) and planar devices (**b**). The arrows in the AFM images indicate the location of bottom electrodes. Device junction area is 50 nm^2 and the stacking is (from *top* to *bottom*):12 nm Pt//30 nm TiO_2//12 nm Pt/4 nm Ti [22]

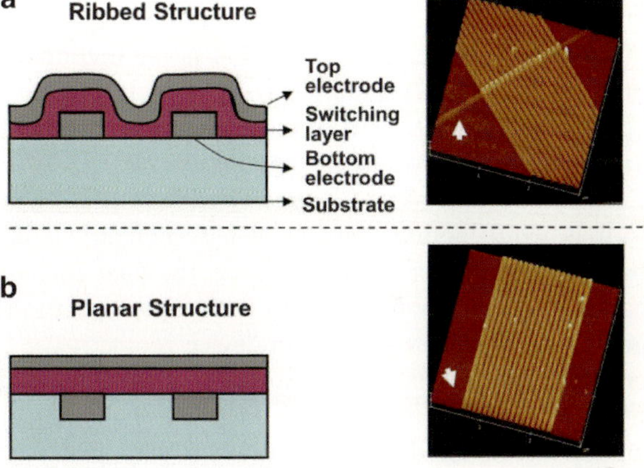

AFM images of traditional ribbed devices and the new planar devices. In contrast to the ribbed devices that have a bump at each cross point, the planar device structure does not have kinks in the top electrodes. There are two advantages for the new device geometry. First, without the kinks, the top electrode metal nanowires are less susceptible to electromigration, a popular failure mechanism for crossbar devices. As a result, the device's endurance is expected to be extended. Second, the electrodes can be made thicker to reduce the series resistance. This is extremely important as the nanowire electrodes become narrower and narrower for smaller devices.

To fabricate the devices, nanoscale trench arrays were first patterned in the substrate. With precise control on the subsequent RIE and metal deposition processes, these trenches were filled with the bottom electrode metals to the substrate surface level, achieving a flat surface for further device fabrication. After deposition of the oxide switching layer, the top electrodes were fabricated on the oxide surface without forming any protrusions.

The planar devices demonstrated four times improved endurance for a 50 nm^2 device with 12 nm Pt//30 nm TiO_2//12 nm Pt/4 nm Ti structures. The median endurance for the ribbed devices in this test was 2900 cycles, while the planar devices were 12,000 cycles (both without passivation and packaging) (Fig. 11.8). During the endurance measurements, electrical pulses with 100 μs pulse width was used to switch the device ON or OFF, and a much lower voltage (100 mV) was used to read the resistance states between switching steps.

The poorer device endurance for the ribbed devices was caused by the defects related to the kinks in these devices. Numerical simulations show that the current density and Joule heating are localized at the corners, especially at the kinks for the ribbed structure. As a result, the ribbed devices are more likely to fail before the planar structure. For example, for devices with the aforementioned structure, the induced total currents at the corners of the bottom electrodes were nearly at the same level (2.56×10^5 and 2.35×10^5 A/m^2 for the ribbed and planar structures, respectively) (Fig. 11.9). While those at the top electrode/oxide interfaces were significantly different (1.5×10^5 vs. 13.8 A/m^2). The four orders of magnitude higher induced current in the traditional ribbed devices has resulted in much more heat generation at the top electrode/oxide interfaces during device operation (Fig. 11.9). This simulation is supported by experimental observation that the popular failure mode of the crossbar devices is the breaking down of the top electrodes at the kinks.

It is worth noting that the planar geometry is compatible with the standard IC fabrication process. For example, the dual Damascene process in IC industry has enabled narrow metal wires with high aspect ratios for interconnection. This indicates that the new memristor device geometry can be effectively integrated with other components or CMOS substrates.

On the other hand, this is just one example of how the device geometry greatly affects the device performance, particularly endurance. It is expected that with nanostructure engineering, other aspects of the device performance such as the uniformity and power consumption can be significantly improved as well. Furthermore, the device geometry engineering is likely more powerful if used in combination with other device engineering approaches, such as multilayer oxide, new materials, etc.

11.3.5 Device Scaling

Extending the device to new physical dimensions is significant for memristor crossbar arrays. Memristor scaling will not only dramatically increase the packing density but also significantly enhance their electrical performance and deepen our understanding in fundamental device physics. Previously, electrical behavior of

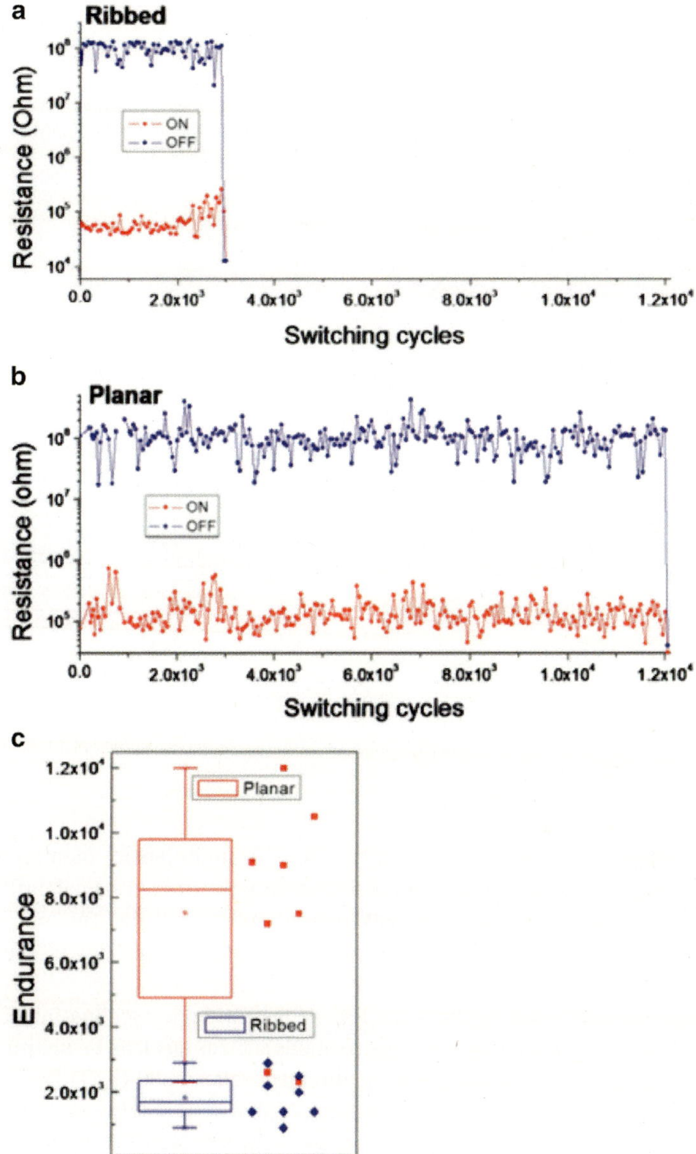

Fig. 11.8 Typical endurance test curves for (**a**) a "ribbed" memristive device and (**b**) a planar memristive device. (**c**) Statistical data of the endurance test for different types of devices. The median cycle number for the ribbed and planar devices are 1,700 and 8,300, respectively [22]

memristor crossbar devices with $50 \times 50\text{nm}^2$ junction area or larger was intensively studied [40, 54]. Single or discrete memristive devices of 10 nm size were reported recently, but they were not built in densely packed and addressable arrays [55, 56].

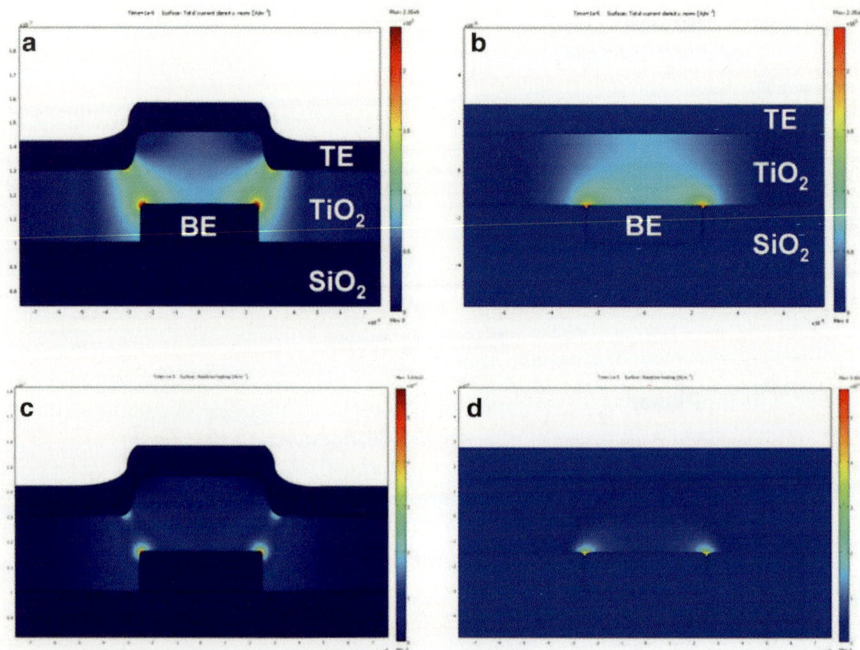

Fig. 11.9 Total current density (**a**, **b**) and resistive heating (**c**, **d**) simulation for ribbed and planar devices in the first 10 μs electric pulse of 2 V using the COMSOL multiphysics model. The localization of current density and heat at the kinks for the ribbed structure is likely the cause of lower device endurance [22]

To adopt the memristor devices for applications in high-density memories, high-efficiency reconfigurable logics and high-performance mixed-signal information processing, large area arrays of densely packed sub-10 nm devices are required. In other words, the scaling in this context means both "scaling down" the feature size and "scaling up" the area of the arrays.

Several approaches that can be used for memristor device scaling include some traditional and disruptive nanopatterning methods such as electron beam lithography (EBL), nanoimprint lithography (NIL), directed self-assembly (DSA) of block copolymers, and state-of-the-art optical lithography with resolution enhancement techniques (RETs). One particular interesting approach is to use EBL or DSA to make a master imprint mold and use NIL to make the densely packed memristor crossbar arrays over large area. We have successfully fabricated crossbar arrays of memristors that have a junction area of 8×8 nm^2 [21].

Devices with smaller junction areas exhibited much lower power consumption during the switching. For example, the programming current for the 8×8 nm^2 device was on the order of 600 pA, five orders of magnitude lower than that for a 25×25 nm^2 device [21]. On the other hand, the cycle-to-cycle switching uniformity was greatly improved as well. The improved device uniformity was a result from the small junction area that limited the location and dimension of the conductive filament (CF). The diameter of a CF in a TiO_2 device was proved to be smaller than 10 nm [57]. Hence it is reasonable to believe that in the $8 \times 8\ nm^2$ device, there was only a single CF that has participated in the switching event every time. Different from larger devices that have multiple CFs with different diameter, length and unpredictable locations, a single small CF will lead to a smaller programming current and much more uniform switching behavior.

11.4 Hybrid Memristor/CMOS Circuits

To get more computing per transistor on a chip, one approach is the CMOL (CMOS/molecule) architecture proposed by of Strukov and Likharev [58]. In this architecture, nanoscale reconfigurable molecular switches are used to connect CMOS transistors for computing. This architecture was later modified by Snider and Williams [59] by adapting larger area contact pads to improve its manufacturability and to separate the routing and computing functions. This scheme was called field-programmable nanowire interconnect (FPNI). In FPNI, the data routing network is separated from the Si CMOS layer using field-programmable memristor crossbar arrays. Since the data routing layer is placed directly above the CMOS circuit, the chip footage area is smaller and the communication distance between the transistors are shorter (which means shorter RC delay or higher information processing speed). With the cross bar structure, alternative routes can be used if defects are found in the data routing layer, resulting in a highly defect-tolerant circuit. The hybrid architecture of nanowire crossbar/CMOS has several advantages over conventional transistor-based field-programmable gate arrays (FPGA) in terms of density, functionality, and power consumption. Numerical simulations showed that this type of architecture can dramatically increase the logic density of an FPGA-like chip without degrading power dissipation or speed, even in the presence of large numbers (up to 20 %) of defective components [59].

The memristor was chosen for the FPGA-like hybrid chip for a number of reasons. First, a single memristor is capable of implementing functions that need several transistors in a CMOS circuit, namely, a configuration-bit flip-flop and associated data routing multiplexer. Second, memristors are nonvolatile devices, which mean they do not require power to refresh their states; hence, the hybrid chip is power efficient. Third, the simple device structure enables the feasibility of integration with CMOS substrates with cost-effective technology such as nanoimprint lithography.

The concept of the memristor-CMOS hybrid circuits is schematically shown in Fig. 11.10. The memristor crossbar layers (nanowire layer 1, switching material layer, nanowire layer 2) are fabricated on top of a CMOS substrate and connected to two sets of tungsten vias coming up from the CMOS through an area interconnect (pads).

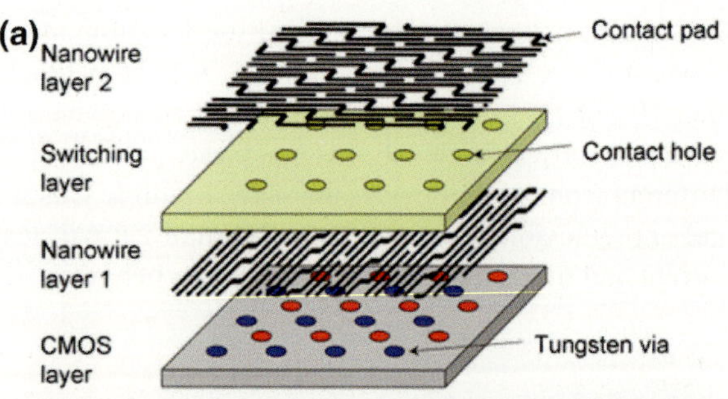

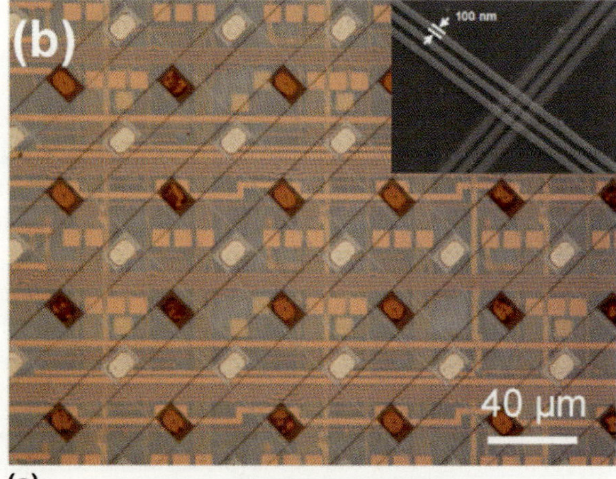

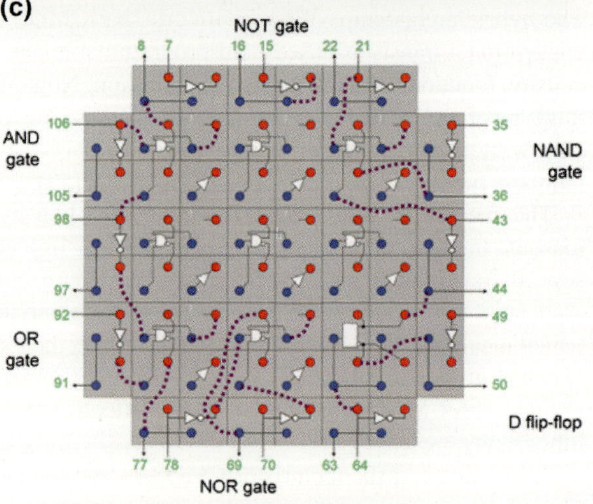

Fig. 11.10 (**a**) Conceptual illustration of the memristor-CMOS hybrid architecture. Tungsten vias (*red* and *blue*) on the CMOS substrate are designed as the area interface between the transistors and the memristors. (**b**) The hybrid chip with memristor crossbars built on top. The *inset* shows an SEM image of a fragment of the memristor crossbar array. (**c**) CMOS layer fabric on a die and possible wiring for digital circuits using configured memristors (*dashed purple lines*). This figure shows one of the many possible signal routings for some basic logic gates. The reading voltage across the memristors is 0.5–1.7 V and the CMOS logic voltage is 3.3 V. The size of each cell is 50 µm × 50 µm [13]

A milestone was recently achieved to physically implement the memristor/CMOS hybrid circuits with demonstrated FPGA-like functionality by integrating titanium dioxide memristor crossbars on top of a CMOS substrate using nanoimprint lithography.

In this particular demonstration, the CMOS substrates were fabricated in a commercial foundry using a high-voltage 0.5 μm technology. A 1 μm thick layer of silicon dioxide was deposited onto the completed wafers using tetraethyl orthosilicate (TEOS), and the resulting surface was planarized by chemical mechanical polishing (CMP) to expose the tungsten vias. This step resulted in surfaces with ∼50 nm deep depressions above the tungsten vias, which were planarized by using a thin layer of liquid NIL resist that was flattened with a polished quartz plate and cured with UV radiation. These memristors connected to the CMOS devices had 2 nm Ti/9 nm Pt as the bottom electrode, 36 nm thick TiO_2 as the switching layer, and 12 nm thick Pt as the top electrode. The junction area was 100 nm × 100 nm. To make alignment feasible between the CMOS and crossbar layers, interconnects between the memristor layer and the CMOS layer are implemented using larger contact pads, or "flags" that connected the nanowires to the tungsten vias in the CMOS substrate. Images of the CMOS substrate and the finished hybrid chip are shown in Fig. 11.10, with part of the memristor arrays shown in the inset.

In circuits operation, certain memristors are configured by the CMOS circuitry to be closed, thus connecting components in the CMOS layer to synthesize particular logic circuits in a field-programmable gate array (FPGA) fashion, as shown in Fig. 11.10. For example, configuring the memristor between channel 15 and 16 would result in a NOT gate. Similarly, other basic logic gates such as OR, AND, NAND, NOR, and a positive-edge triggered D flip-flop can be configured in this hybrid circuit (Fig. 11.11). The successful operation of the logic gates implies that (1) the memristors were configured correctly by the CMOS circuitry, (2) the transistors in the CMOS layer were successfully connected using the configured memristors and could communicate to perform higher level functions, and (3) the fabrication processes for building the memristor crossbars did not disturb the underlying CMOS.

More importantly, the memristors' states can be changed by applying the proper voltages, making the hybrid circuits reconfigurable. For example, a memristor initially in the OFF state was turned ON using a negative voltage and then turned back OFF using a positive voltage (Fig. 11.12). As a result, no signal could be passed through, thus decommissioning the NOT gate and making the resources available for another purpose. The reconfigurability provides significant flexibility in data routing, and it can also reduce power consumption by disconnecting components that are not needed for a particular task.

From a circuit-fabrication point of view, our current work incorporated technologies from different generations to implement a novel architecture. It may be possible and even necessary in the future to extend Moore's Law without CMOS scaling. To the best of our knowledge, this is the first demonstration of NIL on an active CMOS substrate that was fabricated in a commercial semiconductor fabrication facility. The successful integration opens up opportunities in other areas such as NV-RAM and non-Boolean neuromorphic computing.

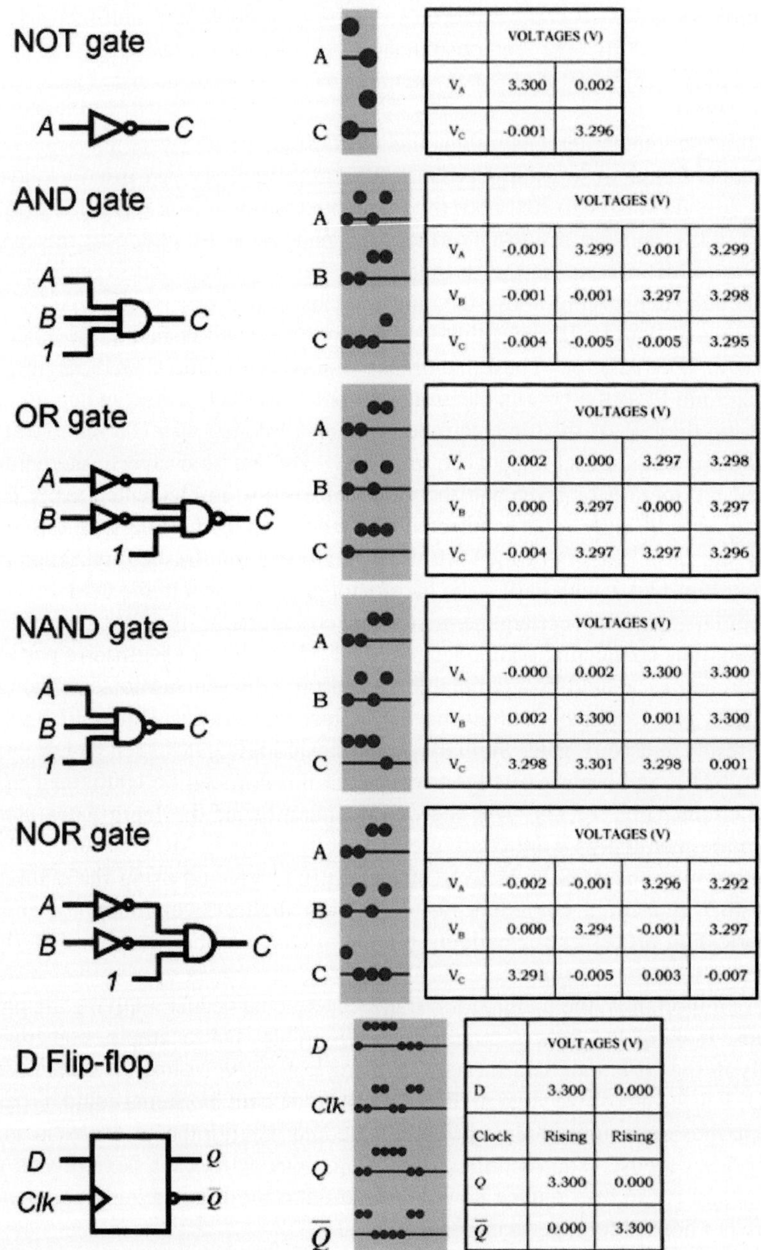

	VOLTAGES (V)	
V_A	3.300	0.002
V_C	-0.001	3.296

	VOLTAGES (V)			
V_A	-0.001	3.299	-0.001	3.299
V_B	-0.001	-0.001	3.297	3.298
V_C	-0.004	-0.005	-0.005	3.295

	VOLTAGES (V)			
V_A	0.002	0.000	3.297	3.298
V_B	0.000	3.297	-0.000	3.297
V_C	-0.004	3.297	3.297	3.296

	VOLTAGES (V)			
V_A	-0.000	0.002	3.300	3.300
V_B	0.002	3.300	0.001	3.300
V_C	3.298	3.301	3.298	0.001

	VOLTAGES (V)			
V_A	-0.002	-0.001	3.296	3.292
V_B	0.000	3.294	-0.001	3.297
V_C	3.291	-0.005	0.003	-0.007

	VOLTAGES (V)	
D	3.300	0.000
Clock	Rising	Rising
Q	3.300	0.000
$\overline{Q}$	0.000	3.300

Fig. 11.11 Equivalent computing circuits, visualized digital results and the measured truth tables for the logic gates in the hybrid circuits. In the visualization results, the lower blue dots are logic 0 and the upper dots are logic 1. The clock frequency for the flip-flop was 50 Hz [13]

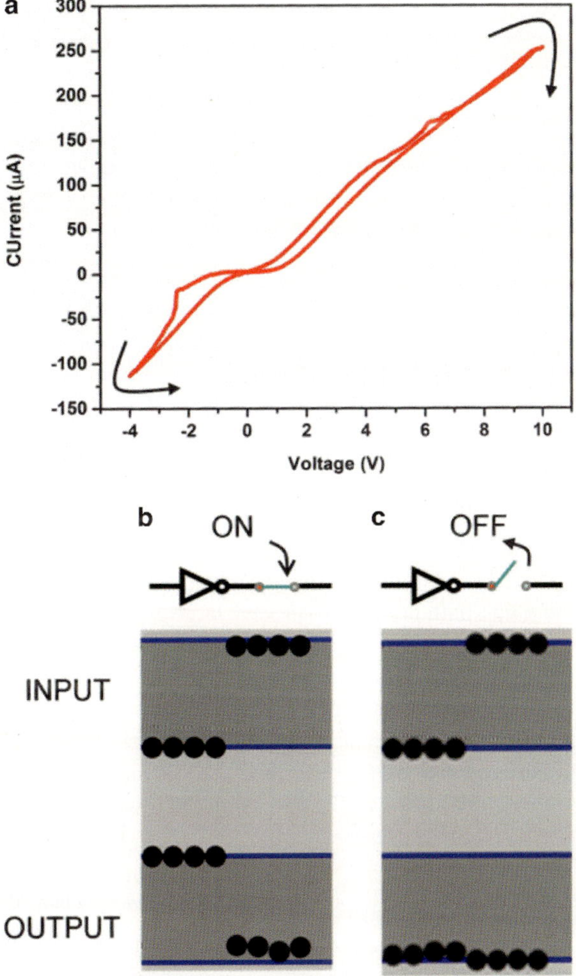

Fig. 11.12 (**a**) A memristor initially in the OFF state was turned ON using a negative voltage and then turned back OFF using a positive voltage. Note that the voltage scale is large because this is a two-wire measurement and most of the voltage dropped on the metal nanowire electrodes. (**b**) and (**c**) are the computation results for a NOT gate when the connecting memristor is at different states. A string of 00001111 was inverted by the NOT gate when the memristor is ON (**b**) while the signal did not pass through when it is OFF (**c**). The reading voltage across the memristors is 0.5 V [13]

However, in the first demonstration, the electrical performance of the memristor devices was not optimized. For instance, one particular problem was that the programming voltages were relatively high so that a separate high voltage circuitry was used to configure the memristors. The high programming voltage arose from the high series resistance of the thin metal nanowire electrodes, which took the majority of the applied voltage. There was a practical limit in the thickness of the metal nanowires for the devices. When the crossbar density increases, the nanowire widths need to be narrower, and the series resistance problem will become more severe. This will lead to even higher programming voltages that limit the compatibility of memristive devices with lower voltage CMOS circuits.

To solve this problem, planar crossbar memristor devices were integrated with CMOS substrates for reconfigurable logic. As discussed earlier in this chapter, with bottom electrodes embedded into the substrate, a planar memristive device exhibited greatly enhanced endurance and lower series resistance. By carefully designed the process parameters, much thicker bottom electrodes were used and were embedded into the TEOS passivation layer of the CMOS substrate [60]. As a result, the series resistances, and consequently the programming voltages, were lowered down. Furthermore, with the planar geometry, the devices also exhibited much improved switching uniformity. The integration of planar memristors with CMOS substrates opened the road to hybrid circuits of high-density memristor devices and low-power CMOS circuits.

11.5 Concluding Remarks

The memristor has a broad spectrum of applications, among which the reconfigurable switch has a small but important share. Although significant progress in memristor research and development has been made, challenges still remain for the wider applications. Particularly, there have yet to be demonstrations on practical approaches to dramatically improve the device performance uniformity and to avoid the sneak path problem over very large array of devices.

From the hybrid circuit point of view, increasing the memristor density to achieve more powerful hybrid chips is of great importance. This requires close collaboration between circuit designers, device engineers, integration experts, and computer architects. With recent progress in device engineering, further improvements of the integration process (as described later in this chapter), and some emerging computer architecture such as 4D addressing [61], we should be optimistic for the future of this bourgeoning research area.

Acknowledgments The author thanks the financial support from the U. S. Air Force Office of Scientific Research (AFOSR) (FA9550-12-1-0038), the Defense Advanced Research Program Agency (DARPA) (N66001-12-1-4217), and the U. S. National Science Foundation (NSF) (ECCS-1253073) for his research at UMass Amherst. Work done at Hewlett-Packard Labs was under partial sponsorship from the US Government's Nano-Enabled Technology Initiative. The author would like to acknowledge help from current students at UMass Amherst and former colleagues at HP Labs.

References

1. G.E. Moore, Cramming more components onto integrated circuits. Electron. Mag. **38**, 114–117 (1965)
2. L.O. Chua, Memristor—missing circuit element. IEEE Trans. Circuit Theory **18**, 507–519 (1971)
3. L.O. Chua, S.M. Kang, Memristive devices and systems. Proc. IEEE **64**, 209–233 (1976)
4. D.B. Strukov, G.S. Snider, D.R. Stewart, R.S. Williams, The missing memristor found. Nature **453**, 80–83 (2008)

5. S.D. Ha, S. Ramanathan, Adaptive oxide electronics: a review. J. Appl. Phys. **110**, 071101 (2011)
6. M.-J. Lee et al., A fast, high-endurance and scalable non-volatile memory device made from asymmetric Ta_2O_{5-x}/TaO_{2-x} bilayer structures. Nat. Mater. **10**, 625–630 (2011)
7. A.C. Torrezan et al., Sub-nanosecond switching of a tantalum oxide memristor. Nanotechnology **22**, 485203 (2011)
8. M.D. Pickett, R.S. Williams, Nanotechnology **23**, 215202 (2012)
9. C. Nauenheim, C. Kuegeler, S. Trellenkamp, A. Ruediger, R. Waser, in *Proceedings of 10th International Conference on Ultimate Integration on Silicon* (2009), p. 135
10. I.G. Baek, D.C. Kim, M.J. Lee, H.-J. Kim, E.K. Yim, M.S. Lee, J.E. Lee, S.E. Ahn, S. Seo, J.H. Lee, J.C. Park, Y.K. Cha, S.O. Park, H.S. Kim, I.K. Yoo, U.-I. Chung, J.T. Moon, B.I. Ryu, Multi-layer cross-point binary oxide resistive memory (OxRRAM) for post-NAND storage application, in *Proceedings of the Electron Device Meeting*, San Francisco, CA (December 2005), pp. 750–753
11. A. Chen, S. Haddad, Y.-C. Wu, T.-N. Fang, Z. Lan, S. Avanzino, S. Pangrle, M. Buynoski, M. Rathor, W. Cai, N. Tripsas, C. Bill, M. VanBuskirk, M. Taguchi, Non-volatile resistive switching for advanced memory applications. IEDM Tech. Dig., 765–768 (2005)
12. X.M. Chen, G.H. Wu, D.H. Bao, Resistive switching behavior of Pt/Mg0.2Zn0.8O/Pt devices for nonvolatile memory applications. Appl. Phys. Lett. **93**, 093501 (2008)
13. Q.F. Xia et al., Memristor-CMOS hybrid integrated circuits for reconfigurable logic. Nano Lett. **9**, 3640–3645 (2009)
14. J. Borghetti et al., 'Memristive' switches enable 'stateful' logic operations via material implication. Nature **464**, 873–876 (2010)
15. G.S. Snider, Self-organized computation with unreliable, memristive nanodevices. Nanotechnology **18**, 365202 (2007)
16. Y.V. Pershin, S. La Fontaine, M. Di Ventra, Memristive model of amoeba's learning. Phys. Rev. E **80**, 021926 (2009)
17. S.H. Jo, T. Chang, I. Ebong, B.B. Bhadviya, P. Mazumder, W. Lu, Nano Lett. **10**, 1297 (2010)
18. J.J. Yang et al., High switching endurance in TaO_x memristive devices. Appl. Phys. Lett. **97**, 232102 (2010)
19. S. Gaba, P. Sheridan, J. Zhou, S. Choi, W. Lu, Stochastic memristive devices for computing and neuromorphic applications. Nanoscale **5**, 5872–5878 (2013)
20. S.H. Jo, K.-H. Kim, W. Lu, High-density crossbar arrays based on a Si memristive system. Nano Lett. **9**, 870–874 (2009)
21. S. Pi, P. Lin, Q. Xia, Cross point arrays of ultralow power 8 nm by 8 nm memristive devices fabricated with nanoimprint lithography. J. Vac. Sci. Technol. B, (2013), accepted. doi: 10.1116/1.4827021
22. Q.F. Xia, M.D. Pickett, J.J. Yang, M.–.X. Zhang, J. Borghetti, X. Li, W. Wu, G. Medeiros-Ribeiro, R.S. Williams, Impact of geometry on the performance of memristive nanodevices. Nanotechnology **22**, 254026 (2011)
23. S.Y. Chou, P.R. Krauss, P.J. Renstrom, Imprint lithography with 25-nanometer resolution. Science **272**, 85–87 (1996)
24. M. Colburn, S. Johnson, M. Stewart et al., Step and flash imprint lithography: a new approach to high-resolution patterning. Proc. SPIE **3676**, 379–389 (1999)
25. J. Haisma, M. Verheijen, K. vanden Heuvel, J. vanden Berg, Mold-assisted nanolithography: a process for reliable pattern replication. J. Vac. Sci. Technol. B **14**, 4124–4128 (1996)
26. Q.F. Xia, Nanoscale resistive switches: devices, fabrication and integration. Appl. Phys. **A102**, 955–965 (2011)
27. W. Wu, G.Y. Jung, D.L. Olynick et al., One-kilobit cross-bar molecular memory circuits at 30-nm half-pitch fabricated by nanoimprint lithography. Appl. Phys. A **80**, 1173–1178 (2005)
28. G.Y. Jung, S. Ganapathiappan, D.A.A. Ohlberg et al., Fabrication of a 34×34 crossbar structure at 50 nm half-pitch by UV-based nanoimprint lithography. Nano Lett. **4**, 1225–1229 (2004)

29. Q.F. Xia, J.J. Yang, W. Wu, X. Li, R.S. Williams, Self-aligned memristor cross point arrays fabricated with one nanoimprint lithography step. Nano Lett. **10**, 2909–2914 (2010)
30. M.T. Hickmott, Low-frequency negative resistance in thin anodic oxide films. J. Appl. Phys. **33**, 2669–2682 (1962)
31. D.C. Kim, S. Seo, S.E. Ahn, D.-S. Suh, M.J. Lee, B.-H. Park, I.K. Yoo, I.G. Baek, H.-J. Kim, E.K. Yim, J.E. Lee, S.O. Park, H.S. Kim, U.-I. Chung, J.T. Moon, B.I. Ryu, Electrical observations of filamentary conductions for the resistive memory switching in NiO films. Appl. Phys. Lett. **88**, 202102 (2006)
32. T.-N. Fang, S. Kaza, S. Haddad, A. Chen, Y.-C. Wu, Z. Lan, S. Avanzino, D. Liao, C. Gopalan, M. Choi, S. Mahdavi, M. Buynoski, Y. Lin, C. Marrian, C. Bill, M. van Buskirk, M. Taguchi, Copper oxide resistive switching for non-volatile memory applications, in *Proceedings of the International Conference on Memory Technology and Design*, Giens, France (May 2007), pp. 143–146
33. D. Lee, D.-J. Seong, I. Jo, F. Xiang, R. Dong, S. Oh, H. Hwang, Resistance switching of copper doped MoOx films for nonvolatile memory applications. Appl. Phys. Lett. **90**, 122104 (2007)
34. W. Guan, S. Long, Q. Liu, M. Liu, W. Wang, Nonpolar nonvolatile resistive switching in Cu doped ZrO2. IEEE Electron Device Lett. **29**, 434–437 (2008)
35. H. Schröder, D.S. Jeong, Resistive switching in a Pt/TiO2/Pt thin film stack—a candidate for a nonvolatile ReRAM. Microelectr. J. **84**, 1982–1985 (2007)
36. S.-G. Park, B. Magyari-Köpe, Y. Nishi, Electronic correlation effects in reduced rutile TiO2 within the LDA + U method. Phys. Rev. B **82**, 115109 (2010)
37. J.J. Yang, J.P. Strachan, Q.F. Xia et al., Diffusion of adhesion layer metals controls nanoscale memristive switching. Adv. Mater. **22**, 4034–4038 (2010)
38. C.-H. Yang, J. Seidel, S.Y. Kim, P.B. Rossen, P. Yu, M. Gajek, Y.H. Chu, L.W. Martin, M.B. Holcomb, Q. He, P. Maksymovych, N. Balke, S.V. Kalinin, A.P. Baddorf, S.R. Basu, M.L. Scullin, R. Ramesh, Electric modulation of conduction in multiferroic Ca-doped BiFeO3 films. Nat. Mater. **8**, 485–493 (2009)
39. J. Blanc, D.L. Staebler, Electrocoloration in SrTiO3—vacancy drift and oxidation-reduction of transition metals. Phys. Rev. B **4**, 3548–3557 (1971)
40. S. Kim et al., Effect of scaling WO_x-based RRAMs on their resistive switching characteristics. IEEE Electron Device Lett. **32**, 671–673 (2011)
41. A. Avila, R. Asomoza, Switching in coplanar amorphous hydrogenated silicon devices. Solid State Electron. **44**, 17–27 (2000)
42. C. Schindler, S.C.P. Thermadam, R. Waser, M.N. Kozicki, Bipolar and unipolar resistive switching in Cu-doped SiO2. IEEE Trans. Electron Devices **54**, 2762–2768 (2007)
43. J. Yao, Z.Z. Sun, L. Zhong, D. Natelson, J.M. Tour, Resistive switches and memories from silicon oxide. Nano Lett. **10**, 4105–4110 (2010)
44. J.G. Simmons, R.R. Verderber, Proc. R. Soc. London Ser. A **301**, 77–102 (1967)
45. J. Yao, L. Zhong, D. Natelson, J.M. Tour, J. Am. Chem. Soc. **133**(4), 941–948 (2011)
46. A. Mehonic, S. Cueff, M. Wojdak, S. Hudziak, C. Labbe, R. Rizk, A.J. Kenyon, Nanotechnology **23**(45), 455201 (2012)
47. Y.-F. Chang, P.-Y. Chen, B. Fowler, Y.-T. Chen, F. Xue, Y. Wang, F. Zhou, J.C. Lee, J. Appl. Phys. **112**(12), 123702 (2012)
48. C. Li, H. Jiang, Q. Xia, Low voltage resistive switching devices based on chemically produced silicon oxide. Appl. Phys. Lett. **103**, 062104 (2013)
49. C.-I. Hsieh, et al., Forming-free resistive switching of TiOx layers with oxygen injection treatments, in *2011 International Symposium on VLSI Technology, Systems and Applications (VLSI-TSA)*. doi:10.1109/VTSA.2011.5872234
50. H. Jiang, Q. Xia, Improved switching uniformity for TiO_2/HfO_2 bi-layer memristive devices, in *The 57th International Conference on Electron, Ion, and Photon Beam Technology & Nanofabrication (EIPBN'13)*, Nashville, TN (28–31 May 2013)
51. J.J. Yang et al., Engineering nonlinearity into memristors for passive crossbar applications. Appl. Phys. Lett. **100**, 113501 (2012)

52. E. Linn, R. Rosezin, C. Kugeler, R. Waser, Complementary resistive switches for passive nanocrossbar memories. Nat. Mater. **9**, 403–406 (2010)
53. J. Yang, P. Sheridan, W. Lu, Complementary resistive switching in tantalum oxide-based resistive memory devices. Appl. Phys. Lett. **100**, 203112 (2012)
54. J.J. Yang et al., Memristive switching mechanism for metal/oxide/metal nanodevices. Nat. Nanotechnol. **3**, 429–433 (2008)
55. J. Park, et al., Quantized conductive filament formed by limited Cu source in sub-5nm era, in *2011 IEEE International Electron Devices Meeting (IEDM)* (2011), pp. 3.7.1–3.7.4. doi:10.1109/IEDM.2011.6131484
56. B. Govoreanu, et al., 10×10 nm^2 Hf/HfO$_x$ crossbar resistive RAM with excellent performance, reliability and low-energy operation, in *2011 IEEE International Electron Devices Meeting (IEDM)* (2011), pp. 31.6.1–31.6.4. doi:10.1109/IEDM.2011.6131652
57. D.H. Kwon et al., Atomic structure of conducting nano-filaments in TiO2 resistive switching memory. Nat. Nanotechnol. **5**, 148–153 (2010)
58. D.B. Strukov, K.K. Likharev, CMOL FPGA: a reconfigurable architecture for hybrid digital circuits with two-terminal nanodevices. Nanotechnology **16**, 888–900 (2005)
59. G.S. Snider, R.S. Williams, Nano/CMOS architectures using a field-programmable nanowire interconnect. Nanotechnology **18**, 035204 (2007)
60. P. Lin, S. Pi, Q. Xia, 3D integration of planar crossbar memristive devices with CMOS substrates. IEEE. Trans. Nanotechnol. (2013)
61. D.B. Strukov, R.S. Williams, Four-dimensional address topology for circuits with stacked multilayer crossbar arrays. Proc. Natl. Acad. Sci. USA **106**, 20155–20158 (2009)

Chapter 12
Spike-Timing-Dependent-Plasticity in Hybrid Memristive-CMOS Spiking Neuromorphic Systems

Teresa Serrano-Gotarredona and Bernabé Linares-Barranco

12.1 Introduction

Neuromorphic engineering "http://en.wikipedia.org/wiki/Neuromorphic" is an interdisciplinary discipline that takes inspiration from biology, physics, mathematics, computer science, and engineering to design artificial neural systems, such as vision systems, head-eye systems, auditory processors, and autonomous robots, the physical architecture and design principles of which are based on those of biological nervous systems. The term neuromorphic was coined by Carver Mead, in the late 1980s [2] to describe very large-scale integration (VLSI) systems containing electronic analog circuits that mimic neuro-biological architectures present in the nervous system. In recent times the term neuromorphic has been used to describe both analog, digital or mixed-mode analog/digital VLSI systems that implement models of neural systems (for perception, motor control, or sensory processing) and also software algorithms. A key aspect of neuromorphic engineering is understanding how the morphology and signal processing of neurons, circuits, and overall architectures creates desirable computations, influences robustness to damage, incorporates learning and development, and facilitates evolutionary change.

It is obvious that interdisciplinary research broadens our view of particular problems yielding fresh and possibly unexpected insights. This is the case of

---

This chapter is a reproduction (with minor changes and with permission from IEEE) of publication [1] with IEEE copyright.

T. Serrano-Gotarredona • B. Linares-Barranco (✉)
Instituto de Microelectrónica de Sevilla IMSE-CNM (CSIC and Universidad de Sevilla)
e-mail: bernabe@imse-cnm.csic.es

R. Tetzlaff (ed.), *Memristors and Memristive Systems*,
DOI 10.1007/978-1-4614-9068-5_12,

neuromorphic engineering, where technology and neuroscience cross-fertilize each other. One example of this is the recent impact of fabricated memristor devices [3–6], postulated since 1971 [7–9], thanks to research in nanotechnology electronics. Another is the mechanism known as Spike-Time-Dependent-Plasticity (STDP) [10–24] which describes a neuronal synaptic learning mechanism that refines the traditional Hebbian synaptic plasticity proposed in 1949 [25]. These are very different subjects from relatively unrelated disciplines (nanotechnology, biology, and computer science), which have nevertheless been drawn together by researchers in neuromorphic engineering [26–30]. STDP was originally postulated as a family of computer learning algorithms [10–12] and is being used by the machine intelligence and computational neuroscience community [15–24]. At the same time its biological and physiological foundations have been reasonably well established during the past decade [31–38]. If memristance and STDP can be related, then (a) recent discoveries in nanophysics and nanoelectronic principles may shed new light on the intricate molecular and physiological mechanisms behind STDP in neuroscience, and (b) new neuromorphic-like computers built out of nanotechnology memristive devices could incorporate biological STDP mechanisms, yielding a new generation of self-adaptive ultra-high-dense intelligent machines. Here we explain how by combining memristance models with the electrical wave signals of neural impulses (spikes) converging from pre- and post-synaptic neurons into a synaptic junction, STDP behavior emerges naturally [28–30]. This helps us to understand how neural and memristance parameters modulate STDP and may offer new insights to neurophysiologists searching for the ultimate physiological mechanisms responsible for STDP in biological synapses. At the same time, it also provides a direct means of incorporating STDP learning mechanisms into a new generation of nanotechnology computers employing memristors. Here we focus on this second aspect.

In this chapter we first describe some interesting concepts and properties behind spiking neural networks, also called event-driven neural systems, focusing on vision sensing and processing. Then we quickly review STDP (Sect. 12.3) and memristor (Sect. 12.4) concepts. Afterwards, in Sect. 12.5, we explain how the memristance mechanism, and one particular formulation of it, can explain the experimental characterization of the STDP phenomena in biological synapses. We will see how the shape of action potentials is a crucial component which influences and defines the mathematical learning of STDP, and how by changing action potential shapes the STDP learning rule can be modulated and changed. Section 12.5 also proposes circuit techniques for achieving STDP learning neural systems using memristors as synapses. In Sect. 12.6 we describe how by exploiting present day AER (Address Event Representation) technology it is feasible to build hybrid CMOS/memristive scalable and reconfigurable neural systems capable of assembling in the order of 10^8 neurons in one printed circuit board (PCB). Finally, Sect. 12.7 discusses practical limitations.

12.2 Spiking Neural Networks for Event-Driven Sensing and Processing

For the sake of clarity we will illustrate spike-based or event-driven (ED) sensing and processing within the domain of vision, but the principles described here extend to all other sensory domains and to new asynchronous ED computing paradigms. State of the art in artificial vision is based on video streams, by capturing sequences of images at a given "frame rate" and processing them frame after frame by computational algorithms. Frame-by-frame processing is CPU-hungry and always includes the latencies of sensing, transmitting, and processing each frame. On the other hand, biological vision is frame-free: nor the eyes nor the brain have a clue of what a video frame is. In biology, retina cells (pixels) respond to external stimulation asynchronously sending action potential ("spikes" or "events") to the brain through the optical nerve fibers. Cells in the brain process these spikes through complex hierarchical structures to achieve, for example, shape size and position invariant object recognition. There are no frames, but a continuous flow of events from the retina through the cortical brain structures. Each neuron autonomously decides when to send out an event depending on the spatio-temporal collection of the received events. This asynchronous frame-free sensing and processing is what we call here "event-driven," ED, (as opposed to "frame-driven"). In humans, object recognition can be performed as quickly as in about 150 ms [39], giving time to each neuron in the ventral stream hierarchy to fire just one spike [39], revealing a highly efficient timing-domain signal encoding in the brain. Based on these observations, neuromorphic researchers worldwide have developed in the last 10 years a collection of ED sensor [40–48] and processor [49–52] chips. For example, Fig. 12.1 illustrates the basic principle of ED sensing and signal encoding, by showing the characteristics of one of the most popular ED vision sensors, the "Dynamic Vision Sensor" (DVS) [42–45] developed first during project CAVIAR [53] and used in a variety of applications [29, 54–70]. Whenever a pixel senses a change of light above a threshold it sends out of the chip an "event" in the form of a digital word (x,y,p) representing its address (x,y) and a polarity bit p (positive for a dark to bright change and negative for a bright to dark change) that typically needs fractions of a micro-second to be communicated. The DVS output consists of an asynchronous flow of events, known as AER "Address Event Representation." Figure 12.1b shows the output event flow produced by a DVS chip when observing the 500 Hz spiral shown in Fig. 12.1a for 6 ms. Each event is represented as (x,y,p,t), where the polarity is denoted with color (blue for negative events and red for positive ones). The observed dynamic scene in Fig. 12.1a is thus represented by the frame-free event flow in Fig. 12.1b. As can be seen, the ED DVS sensor provides very rich temporal information within these 6 ms with sub-microsecond precision. Furthermore, such high temporal resolution was captured with the ambient light produced by the oscilloscope [42]. Physically, inter-chip AER communication typically uses a high speed digital multi-bit parallel bus, where one bit is used for 'p' (polarity) and the rest for (x,y), together with handshaking signals

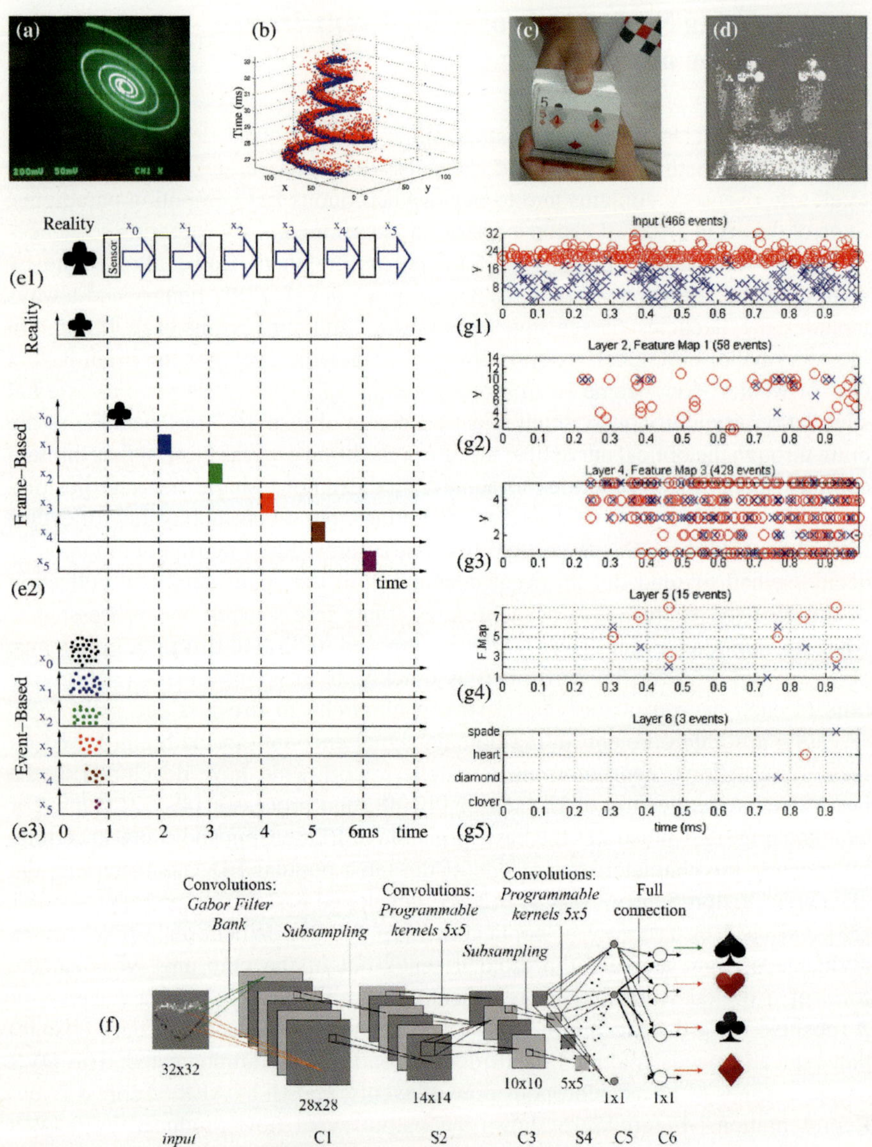

Fig. 12.1 Illustration of ED (event-driven) sensing and processing. (**a**) 500 Hz spiral displayed on an analog oscilloscope, and (**b**) events (x,y,t) captured from an ED artificial retina (DVS) during 6 ms. (**c**) Poker card deck browsed quickly, and (**d**) events captured during 3 ms with a DVS and displayed in the (x,y) plane. (**e**) Comparison of frame-driven vs. event-driven vision sensing and processing: (**e.1**) a symbol is flashed during 1 ms, captured by a sensor, and processed by five sequential stages; (**e.2**) in a frame-driven system with 1 ms frame-time, the sensor needs 1 ms to capture reality, and each processing stage requires a frame-time delay (assumed also 1 ms) for processing; (**e.3**) in an ED sensing and processing system the sensor generates events as it observes reality, and these events are processed by the first stage "as they flow"; the output events

for asynchronous communication. Alternatively, serial AER schemes have also been proposed where a differential microstrip communicates (x, y, p) events bit-serially and asynchronously [71–73].

The availability of ED sensing and processing chips has allowed the implementation of first ED sensory systems [50, 51, 55] that show the unique pseudo-simultaneity property, where the input and output event flows of a processing stage are (in practice) simultaneous or coincident in time. This is illustrated in Fig. 12.1(e–g) where a 5-layer structure of a feed-forward Convolutional Neural Network (ConvNet) [74] typically used for size and pose invariant object recognition, handwritten character recognition, scene recognition for robots, etc., is used. Figure 12.1e1, f show schematically this 5-layer ConvNet. If this ConvNet is implemented using traditional frame-driven sensing and image processing computing hardware [75], each stage has to wait until the output image from the previous stage is available. Figure 12.1e2 shows the latencies in a frame-driven system when a symbol is quickly flashed (in 1 ms) to the vision sensor. A total of six frame delays (each 1 ms) are needed for recognition. On the contrary, Fig. 12.1e3 shows the situation for an ED implementation. An ED processor module processes events as they flow in, with a delay typically in the 100 ns range per event [51]. The system does not need to wait for collecting image frames, but output events are emitted while the input events are processed as soon as enough input events are received, as is in cortical circuits. For orientation extraction, a 2D Gabor filter can produce an output event after just 4–6 correlated input events, signalling the presence of an oriented edge in that location at that time, producing an output that is almost simultaneous to the input event flow (with the delay of a few events). We call this the *pseudo-simultaneity* property between input and output event flows in an ED processing system. Thanks to the pseudo-simultaneity property the output events of all stages are available concurrently to the sensor output event flow (which is concurrent to reality), and correct object recognition is feasible while the sensor is still producing events. This pseudo-simultaneity property has already been verified experimentally with cascades of available ED convolution chips [51], with large arrays of ED convolution modules implemented within high-end FPGAs [54], and has been verified by simulations of full feed-forward ConvNets processing high speed DVS recordings, achieving symbol recognition with 1–2 ms delays [55], as shown in Fig. 12.1g. Figure 12.1g shows the detailed ED processing during a 1 ms flash of poker symbol "club" for the ConvNet in Fig. 12.1f, displaying the individual positive (red circles) and negative (blue crosses) events at the retina output, at internal layers 2, 4, and 5, and at the output category layer. As can be seen, correct recognition is available 0.84 ms after stimulus onset.

Fig. 12.1 (continued) of each stage are also processed by the next stage "as they flow," making it possible to achieve recognition while the sensor is still capturing the 1 ms flash; (**f**) 5-layer ED Convolutional Neural Network (ConvNet) for recognizing poker card symbols when browsing a card deck; (**g**) Simulation results of the ConvNet in (**f**) using DVS recorded data representing 1 ms events and describing each convolution unit using parameters from real convolution chips

Consequently, spiking (or ED) neural sensing and processing systems present interesting features inhereted from their biological counterparts. However, for proper recognition, neural networks need to be trained and correctly learn the intended application. Spiking neural networks can be trained using the Spike-Timing-Dependent-Plasticity learning rule, which can be implemented using memristors. This is the subject of the rest of this chapter.

12.3 Spike-Timing-Dependent Plasticity

Spike-timing-dependent plasticity (STDP) is a family of learning mechanisms originally postulated in the context of artificial machine learning algorithms (or computational neuroscience), exploiting spike-based computations (as in brains) with great emphasis on the relative timings of spikes. Gerstner started to report the first spike-timing-dependent learning algorithms [10, 11] in 1993. STDP has been shown to be better than Hebbian correlation-based plasticity at explaining cortical phenomena [23, 24] and has been proven successful in learning hidden spiking patterns [20] or performing competitive spike pattern learning [21]. Astonishingly, experimental evidence of STDP has been reported by neuroscience groups during the past decade [31–38], so today we can state that the physiological existence of STDP has been reasonably well established. For a historical overview on how STDP research evolved independently among computational and experimentalist groups, please refer to the Section on STDP history by Gerstner elsewhere [12]. However, the full implications of the molecular and electro-chemical principles behind STDP are still under debate [76]. Before describing STDP mathematically, let us first explain how neurons interchange information and what the synaptic connections are.

Figure 12.2 illustrates two neurons connected by a synapse. The pre-synaptic neuron is sending a pre-synaptic spike $V_{mem-pre}(t)$ through one of its axons to the synaptic junction. Neural spikes are membrane voltages from the outside of the cellular membrane V_{pre^+} with respect to the inside V_{pre^-}. Thus $V_{mem-pre} = V_{pre^+} - V_{pre^-}$ and $V_{mem-pos} = V_{pos^+} - V_{pos^-}$. The "large" membrane voltages during a spike (in the order of a hundred mV) cause a variety of selective molecular membrane channels to open and close allowing many ionic and molecular substances to flow, or preventing them from flowing through the membrane. At the same time, synaptic vesicles inside the pre-synaptic cell containing "packages" of neurotransmitters fuse with the membrane in such a way that these "packages" are released into the synaptic cleft (the inter cellular space between both neurons at the synaptic junction). Neurotransmitters are collected in part by the post-synaptic membrane, contributing to a change in its membrane conductivity. The cumulative effect of pre-synaptic spikes (coming from this or other pre-synaptic neurons) will eventually trigger the generation of a new spike at the post-synaptic neuron. Each synapse is characterized by a "synaptic strength" (or weight) w which determines the efficacy

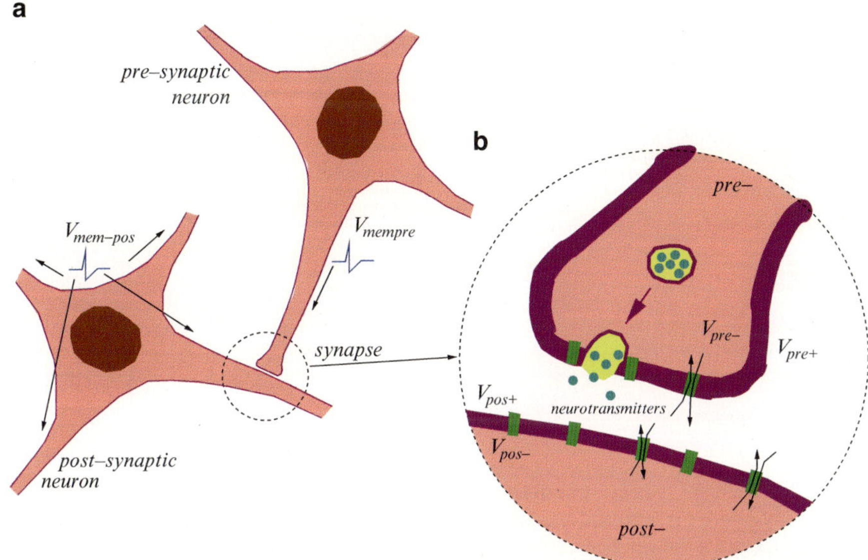

Fig. 12.2 Illustration of synaptic action. (**a**) A synapse is where a pre-synaptic neuron "connects" with a post-synaptic neuron. The pre-synaptic neuron sends an action potential $V_{mem-pre}$ traveling through one of its axons to the synapse. The cumulative effect of many pre-synaptic action potentials, generates a post-synaptic action potential at the membrane of the post-synaptic neuron, which propagates through all the neuron's terminations. (**b**) Detail of synaptic junction. The cell membrane has many membrane channels of varying nature which open and close with changes in the membrane voltage. During a pre-synaptic action potential vesicles containing neurotransmitters are released into the synaptic cleft

of a pre-synaptic spike in contributing to this cumulative action at the post-synaptic neuron. This weight w could well be interpreted as the size and/or number of neurotransmitter packages released during a pre-synaptic spike. However, for our analyses, we will interpret w more generally as some kind of structural parameter of the synapse (like the amount of one or more metabolic substances) that directly controls the efficacy of this synapse per spike. The synaptic weight w is considered to be nonvolatile and analog in nature, but it changes in time as a function of the spiking activity of pre- and post-synaptic neurons. This phenomenon was originally observed and reported in 1949 by Hebb, who introduced his hebbian learning postulate [25]: "*When an axon of cell A is near enough to excite a cell B and repeatedly or persistently takes part in firing it, some growth process or metabolic change takes place in one or both cells such that A's efficiency, as one of the cells firing B, is increased.*" Traditionally, this has been described by computational neuroscientists and machine learning computer engineers as producing an increment in synaptic weight Δw proportional to the product of the mean firing rates of pre- and post-synaptic neurons. STDP is a refinement of this 1949 rule which takes into account the precise relative timing of individual pre- and post-synaptic spikes, and

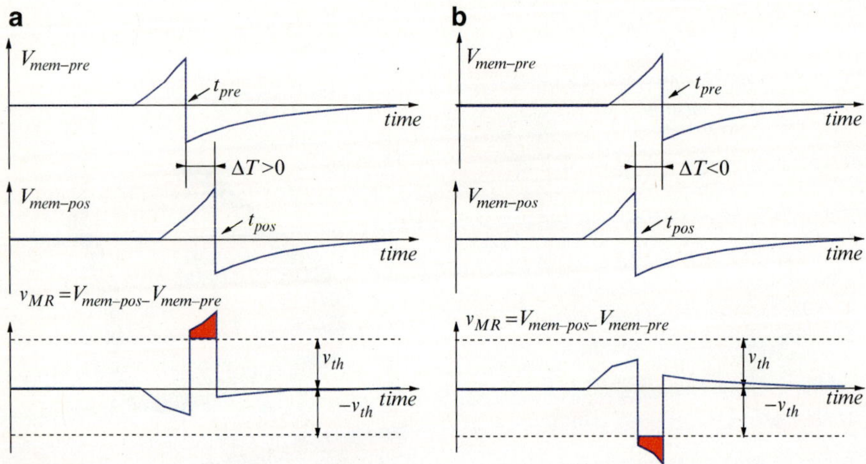

Fig. 12.3 Membrane voltage waveforms. Pre- and post-synaptic membrane voltages for the situations of positive ΔT (**a**) and negative ΔT (**b**). Voltage v_{MR} is the difference between the post-synaptic membrane voltage $V_{mem-pos}$ and the pre-synaptic membrane voltage $V_{mem-pre}$

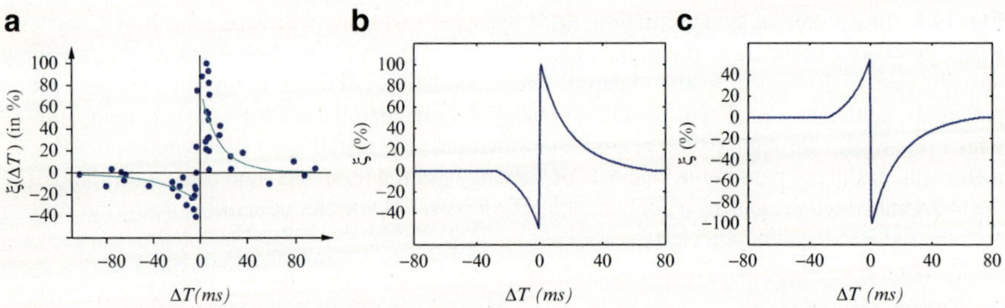

Fig. 12.4 (**a**) Experimentally measured STDP function $\xi(\Delta T)$ on biological synapses (data from Bi and Poo [32, 33]). (**b**) Ideal STDP update function used in computational models of STDP synaptic learning. (**c**) Anti-STDP learning function for inhibitory STDP synapses

not their average rates over time. In STDP the change in synaptic weight Δw is expressed as a function of the time difference between the post-synaptic spike at t_{pos} and the pre-synaptic spike at t_{pre} (see Fig. 12.3). Specifically, as is shown in Fig. 12.4, $\Delta w = \xi(\Delta T)$, with $\Delta T = t_{pos} - t_{pre}$. The shape of the STDP function ξ can be interpolated from experimental data from Bi and Poo as shown in Fig. 12.4a [33]. For positive ΔT (that is to say, the pre-synaptic spike has a highly relevant role in producing the post-synaptic spike) there will be a potentiation of synaptic weight $\Delta w > 0$, which will be stronger as $|\Delta T|$ reduces. For negative ΔT (that is to say, the pre-synaptic spike is highly irrelevant for the generation of the post-synaptic spike), there will be a depression of synaptic weight $\Delta w < 0$, which will be stronger as $|\Delta T|$ reduces. Bi and Poo concluded that they had observed an asymmetric critical window for ΔT of about $\pm$40–80 ms for synaptic modification to take place.

Mathematically, this $\xi(\Delta T)$ STDP learning function is described by computational neuroscientists as

$$\xi(\Delta T) = \begin{cases} a^{+}e^{-\Delta T/\tau^{+}} & if\ \Delta T > 0 \\ -a^{-}e^{\Delta T/\tau^{-}} & if\ \Delta T < 0 \end{cases} \tag{12.1}$$

The STDP learning function $\xi(\Delta T)$ as defined in Fig. 12.4a, b is useful for synapses with positive weights. In these cases, weight w is strengthened if it is increased ($\Delta w > 0$) when $\Delta T > 0$, and weakened if it is decreased ($\Delta w < 0$) when $\Delta T < 0$. However, if the weight is negative ($w < 0$), as in some inhibitory synapse implementations, the STDP learning function in Fig. 12.4b is not appropriate because an increase in weight ($\Delta w > 0$) would weaken the strength of the synapse, and vice versa. For negative weight synapses an STDP learning function with a shape similar to that shown in Fig. 12.4c [77] is required. In this case, the synapse is strengthened by decreasing its weight ($\Delta w < 0$), which should happen for $\Delta T > 0$. Let us call this an Anti-STDP synaptic update or learning function. Other more exotic shapes for $\xi(\Delta T)$ are also possible, as we will discuss later in Sect. 12.5.

Most of the present day literature on STDP presents a learning function ξ which depends on ΔT but not on the actual weight value w. This type of weight-independent STDP learning rule is usually known as "additive STDP." Additive STDP requires the weight values to be bounded to an interval because weights will stabilize at one of their boundary values [78,79]. On the other hand, in multiplicative STDP (mSTDP) [78–80] the learning function is also a function of the actual weight value $\xi_m(w, \Delta T)$. Furthermore, there usually appears a weight-dependent factor which multiplies the original additive STDP learning function ξ_a, and which may generally be different for the positive ($\Delta T > 0$) and negative ($\Delta T < 0$) sides

$$\xi_m(w, \Delta T) = F(w, sign(\Delta T))\xi_a(\Delta T) \tag{12.2}$$

In mSTDP weights can stabilize to intermediate values inside the boundary definitions. Thus, it is often not even necessary to enforce boundary conditions for the weight values [78]. Normally, factor $F(w, sign(\Delta T))$ is considered proportional to w. However, one may consider it to be proportional to w^a. As we will see later in Sect. 12.5, some memristors yield a multiplicative type of STDP with power $a = 2$ (quadratic STDP), while other memristors result in plain additive STDP (with $a = 1$).

12.4 Memristance

Memristance was postulated in 1971 by Chua [7] based on circuit theoretical reasonings. According to circuit theoretical fundamentals, there are four basic electrical quantities [9]: (1) voltage difference between two terminals "v," (2) current

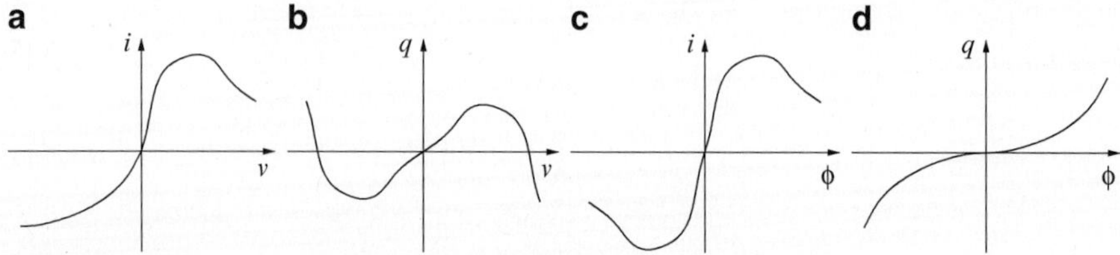

Fig. 12.5 Description of the four canonical two-terminal devices. (**a**) A resistor is defined by a static relationship between a device's voltage and current. (**b**) A capacitor is defined by a static relationship between a device's charge and voltage. (**c**) An inductor is defined by a static relationship between a device's current and flux. (**d**) And a memristor is defined by a static relationship between a device's charge and flux

flowing through into a device terminal "*i*," (3) charge flowing through a device terminal or integral of current $q = \int i(\tau)d\tau$, and (4) flux or integral of voltage $\phi = \int v(\tau)d\tau$. A two-terminal device is said to be canonical [9] if either two of the four basic electrical quantities are related by a static[1] relationship, as shown in Fig. 12.5. A resistor has a static relationship between terminal voltage v and device current i, as shown in Fig. 12.5a. A capacitor shows a static relationship between charge q and voltage v, as shown in Fig. 12.5b. An inductor has a static relationship between its current i and flux v, as shown in Fig. 12.5c. These three devices have been very well known since the origins of Electronics and Electricity. However, there are other possibilities for combining the four basic electrical quantities: (q,i), (v,ϕ), and (q,ϕ). Ignoring the combinations of a quantity with its time derivative leaves us with one single additional possibility: (q,ϕ). This reasoning led Chua to postulate the existence of a fourth basic two-terminal element, which he called the *Memristor*. The memristor would show a static relationship between charge q and flux ϕ, as shown in Fig. 12.5d. If the q vs. ϕ relationship is linear, the *memristor* degenerates into a linear resistor. *Memristors* behave as resistances in which the resistance changes through some of the basic electrical quantities and is somehow *memorized*. Although none of the so-far reported memristors can be described by a static constitutive relationship in the (q,ϕ) plane (and thus, strictly speaking, the 1971 fourth canonical element is still missing), they all fall within Chua's 1976 generalization of *Memristive Systems* [8]. From here on we will use the term *memristor* for Chua's 1976 definition of *memristive system*.

Memristance has recently been demonstrated in nanoscale two-terminal devices, such as certain titanium-dioxide [3, 4, 81, 82] and amorphous Silicon [6] cross-point switches. However, memristive devices were reported earlier by other groups [83–85]. Memristance arises naturally in nanoscale devices because small voltages

[1]By "static" we mean it is not altered by changes of the above electrical quantities, or by their history, integrals, derivatives, etc. These "static" curves can, however, be time-varying if the change is caused by an external agent. For example, a motor-driven potentiometer would have a "static" i/v curve that is time varying.

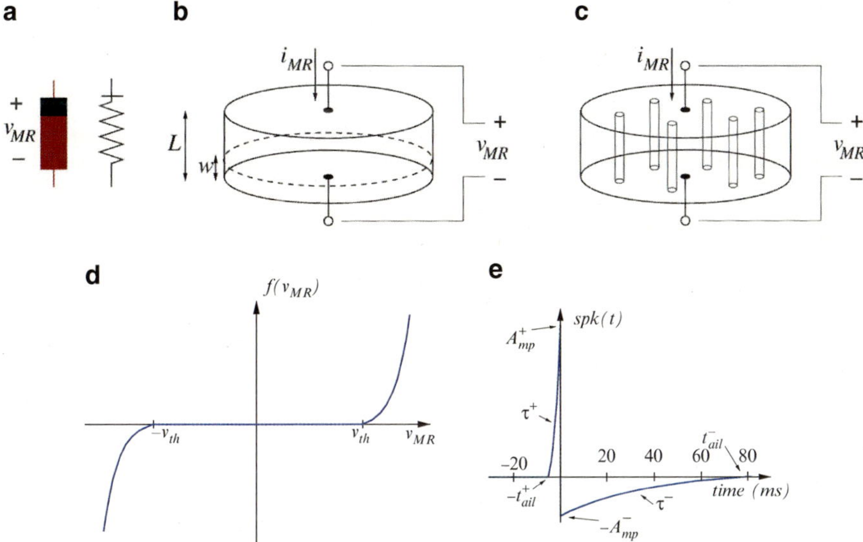

Fig. 12.6 (**a**) Memristor asymmetric symbols. (**b**) Illustration of moving wall model describing memristor operation as two variable resistors in series. (**c**) Illustration of filament formation/annihilation model describing memristor operation as two variable resistances in parallel. (**d**) Shape of memristor weight update function $f(v_{MR})$, (**e**) spike shape waveform

can yield enormous electric fields that produce the motion of charged atomic or molecular species, changing structural properties of a device (such as its doping profile) while it operates. Memristors are asymmetric two-terminal passive devices. Consequently, their circuit symbol must indicate somehow their polarity. Figure 12.6a shows two possible symbols. By definition, memristors can be either voltage/flux driven or current/charge driven. Voltage/flux-driven memristors can be described by [8]

$$i_{MR} = G(w, v_{MR}, t)v_{MR} \tag{12.3}$$

$$\dot{w} = f(w, v_{MR}, t) \tag{12.4}$$

while current/charge-driven memristors would be described as [8]

$$v_{MR} = R(w, i_{MR}, t)i_{MR} \tag{12.5}$$

$$\dot{w} = f(w, i_{MR}, t) \tag{12.6}$$

Here w represents some *structural* property parameter of the memristor. For example, in the 2008 HP paper [3] the operation of the reported memristor was postulated as described by the *moving wall model* depicted in Fig. 12.6b. In this simplified model a memristor of height L, sandwiched between two electrodes, has a low resistance region of height w and a high resistance region of height $L-w$. The memristor is considered to be divided into two regions. Both regions are separated by a boundary wall at position w, which moves up and down with the amount of charge that has flown through the memristor (in the case of being current/charge driven) or the accumulated flux (in case of being voltage/flux driven). The memristor would behave as two variable resistors in series. The total effective resistance of the memristor would be described by

$$R = R_{ON}\frac{w}{L} + R_{OFF}\left(1 - \frac{w}{L}\right) \tag{12.7}$$

This *moving wall model* can approximate phenomena like migration of oxygen ions [86] and vacancies [87], the lowering of Schottky barrier heights by trapped charge carriers at interfacial states [88], and the phase-change in some PCM (phase change materials) devices [89].

However, resistive switching effects in dielectric-based devices have normally been assumed to be caused by conducting filament formation across the electrodes, although the understanding and modeling of these phenomena remains controversial. As a matter of fact, some researchers are observing the formation and annihilation of nanoscale width conducting filaments in memristors [90, 91]. However, let us here propose the following very simplified view to approximate this physical mechanism. Figure 12.6c illustrates schematically a memristor with several conducting filaments between the two electrodes. The number of filaments or their cross-sectional area would increase or decrease with memristor operation. Let us call now w the total cross sectional area of the effective conducting filaments at a given instant in time, and S the total cross section area of the memristor. The filaments present high conductivity (low resistivity), while the bulk presents much lower conductivity (high resistivity). All formed parallel filaments behave as one effective resistance of low resistance, while the rest of the bulk behaves as another higher resistivity resistor. Therefore, now the memristor behaves as two variable resistors in parallel. Consequently, its total conductance (inverse of resistance) could be described as

$$G = G_{ON}\frac{w}{S} + G_{OFF}\left(1 - \frac{w}{S}\right) \tag{12.8}$$

where G_{ON} is the conductance per effective cross section area of the filaments, and G_{OFF} is the conductance per effective cross section area of the filament-less bulk material. Parameter w would change from 0 to w_{max}, the maximum possible effective cross section area of total conducting filaments ($w_{max} \leq S$).

This changing cross section description approximates not only filament formation/annhilation phenomena but also some other gradual cross section area variations observed in some phase-change or ferroelectric-domains-based materials [92].

As we will highlight later in Sects. 12.4 and 12.5, whether a memristor is better described by the moving wall model or the filament formation/annihilation model impacts severely on the resulting type of STDP learning mechanism. The latter yields an additive type of STDP, while the former results in a quadratic type STDP. Note that a memristor can be either voltage/flux or current/charge driven, independently of whether it is a "wall" or a "filament" memristor.

12.5 Combining Memristors and CMOS Neurons for STDP

The STDP learning rule, as modeled by (12.1), can be implemented by [28–30] (a) using a particular type of voltage/flux-driven memristor [5] whose operation might be approximated by (12.4–12.5) with (see Fig. 12.6d)

$$f(v_{MR}) = \begin{cases} I_o \ sign(v_{MR}) \left[e^{|v_{MR}|/v_o} - e^{v_{th}/v_o} \right] & if \quad |v_{MR}| > v_{th} \\ 0 & otherwise \end{cases} \tag{12.9}$$

and bounded synaptic strength $w \in [w_{min}, w_{max}]$, while (b) providing appropriately shaped pre- and post-synaptic spikes available at both synapse (memristor) electrodes [29]. For example, consider a pair of identical pre- and post-synaptic spikes with a shape resembling that of biological spikes, with an on-set duration $|t_{ail}^{+}|$ and a tail of duration $|t_{ail}^{-}|$, as shown in Fig. 12.6e,

$$spk(t) = \begin{cases} A_{mp}^{+} \frac{e^{t/\tau^{+}} - e^{-t_{ail}^{+}/\tau^{+}}}{1 - e^{-t_{ail}^{+}/\tau^{+}}} & if \ -t_{ail}^{+} < t < 0 \\ -A_{mp}^{-} \frac{e^{-t/\tau^{-}} - e^{-t_{ail}^{-}/\tau^{-}}}{1 - e^{-t_{ail}^{-}/\tau^{-}}} & if \ 0 < t < t_{ail}^{-} \\ 0 & otherwise \end{cases} \tag{12.10}$$

Under these circumstances, memristor voltage is $v_{MR}(t, \Delta t) = \alpha_{pos} spk(t) - \alpha_{pre} spk(t + \Delta t)$ and from (12.4) and (12.9) synaptic strength update can be computed as

$$\Delta w(\Delta T) = \int f(v_{MR}(t, \Delta T))dt = \xi(\Delta T) \tag{12.11}$$

which has been shown to result in the same shape illustrated in Fig. 12.4b [29]. Furthermore, by reshaping the spike waveform one can fine-tune or completely alter the STDP learning function $\xi(\Delta t)$, as illustrated in Fig. 12.7 [93]. This way, by building neurons with a given degree of shape programmability, it is possible to change the STDP learning function at will, depending on the application, or make it evolve in time as learning progresses.

Figure 12.8a shows a way of interconnecting memristors and CMOS neurons for STDP learning. Triangles represent the neuron soma, being the flat side its input (dendrites) and the sharp side the output (axon). Dark rectangles are memristors, representing each one synaptic junction. Each neuron controls the voltage at its input (V_{post} in Fig. 12.8b) and output (V_{pre} in Fig. 12.8b) nodes. When the neuron is not

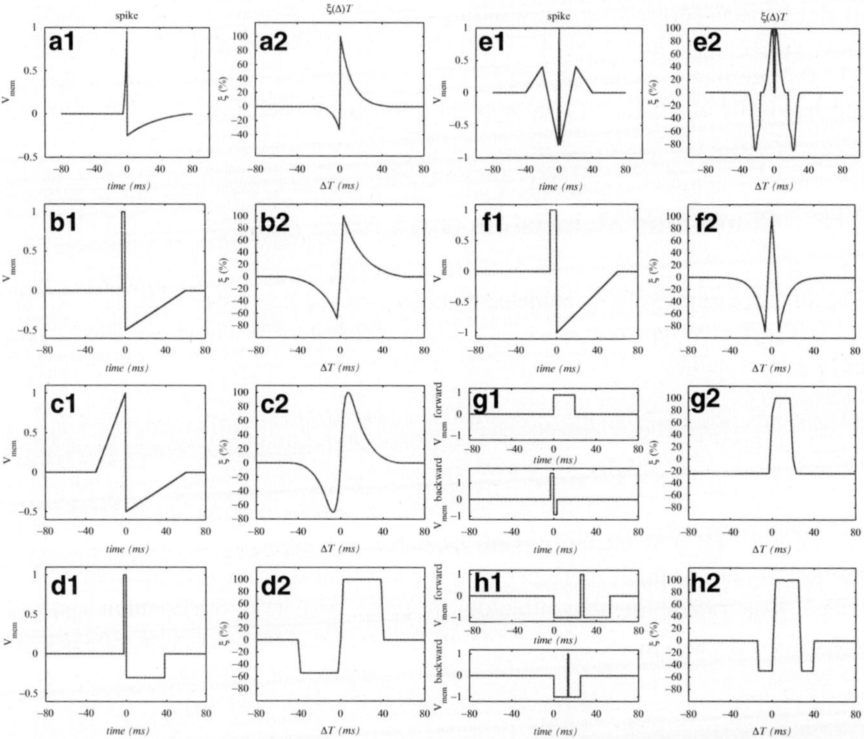

Fig. 12.7 Illustration of influence of action potential shapes on the resulting STDP memristor weight update function $\xi(\Delta T)$. Memristor upper and lower thresholds are normalized to amplitudes +/-1.0. From **(a1)**–**(a2)** to **(e1)**–**(e2)** the same spike waveform travels forward and backward. In **(f1)**–**(f2)** the forward and backward waveforms are the same but have opposite polarity. In **(g1)**–**(g2)** to **(h1)**–**(h2)** the forward and backward waveforms are different. In **(g1)**–**(g2)**, the positive pulse of the backward waveform exceeds amplitude +1.0, thus producing negative STDP update whenever there is a post-synaptic spike alone **(g2)**; otherwise if pre- and post-synaptic spikes happen within a given time window, there will be positive STDP update

spiking it forces a constant voltage at both nodes, while collecting through its input node the sum of input synaptic spike currents coming from the memristors, which contribute to changing the neuron internal state. When the neuron spikes, it sets a one-spike waveform at both input and output nodes. This way, they send their output spikes forward as pre-synaptic spikes for the destination synaptic memristors, but also backward to preceding synaptic memristors as post-synaptic spikes. Zamarreño et al. showed extensive simulations on these concepts, and how one can change from STDP to anti-STDP by switching polarities of spikes or memristors [29]. For example, Fig. 12.7f1, f2 illustrate the case where forward and backward spikes have opposite polarities, resulting in a symmetric STDP update function $\xi(\Delta T)$. Figure 12.7g1, g2 illustrate an example where forward and backward spikes are different, with the backward spike such that its positive part exceeds the positive memristor threshold ($v_{th} = 1.0$). This produces LTD (long-term depression) or

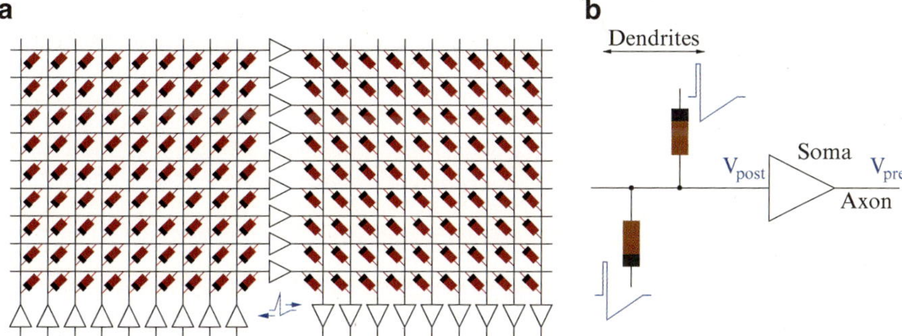

Fig. 12.8 (**a**) Example of Memristors and CMOS neuron circuits arrangement for achieving STDP learning: feed-forward neural system with three layers of neurons and two fully connecting synapse crossbars. (**b**) Details of parts around one post-synaptic neuron. While a neuron is silent, it sets a constant DC voltage at its input (V_{post}) and output (V_{pre}) nodes. When a neuron is sending a spike, it sets a voltage spike at both nodes. (**c**) Implementation of single spike STDP: Block diagram of CMOS neuron together with single memristor synapse connected between pre- and post-synaptic neurons, (**d**) example spike waveform with negative square neural activation shape, and (*e*) example spike waveform with positive more biological neural activation shape

negative STDP update whenever there is a post-synaptic spike sufficiently apart from a pre-synaptic one; and produces LTP (long-term potentiation) if pre- and post-synaptic spikes happen within a given time window [70, 94]. Figure 12.7h1, h2 illustrate a similar STDP update behavior, except that update (whether positive or negative) is restricted to a constraint time window.

If the system is structured into neural layers (for example, Fig. 12.8a shows a 3-layer neuron system) with memristive synapses in between, then for each layer all pre-synaptic neurons should have the same forward spike shape and all post-synaptic neurons should have the same backward shape. This way, all memristive synapses between these two neural layers will have the same STDP function $\xi(\Delta T)$.

In all these circuits, synaptic strength is the conductance G of the memristor: the higher the conductance of a memristor G is (or the lower its resistance $R = 1/G$ is) the stronger the synaptic efficiency will be, as it will let more current through and thus affect more strongly the destination neuron state. Therefore, if the memristors used obey a "*moving wall*" model (see (12.7)), then STDP update $\Delta w = \xi(\Delta T)$ changes wall position w, which from (12.7) is directly proportional to resistance

$$\Delta R(\Delta T) = (R_{ON} - R_{OFF})\frac{\Delta w(\Delta T)}{L} = \rho\xi(\Delta T) \tag{12.12}$$

where ρ is a constant. Consequently, synaptic strength $G = 1/R$ will change as

$$\Delta G(\Delta T) = -\frac{\Delta R(\Delta T)}{R^2} = -G^2\Delta R(\Delta T) \propto -G^2\rho\xi(\Delta T) \tag{12.13}$$

This means that synaptic strength update would follow a quadratic STDP learning rule.

If the memristor physics is better represented by the inter-electrode filament formation/annihilation model, then synaptic update would change parameter w of (12.8), which is now directly proportional to memristor conductance (synaptic strength),

$$\Delta G(\Delta T) = \frac{G_{ON}}{S}\Delta w(\Delta T) = \gamma\xi(\Delta T) \tag{12.14}$$

where γ is a constant. Therefore, synaptic update would be independent of actual weight (conductance) and the resulting STDP update rule is said to be of additive type. Note that (12.9)–(12.11) and the resulting functions $\xi(\Delta T)$ in Fig. 12.7 are common for both "wall" and "filament" models.

12.6 Proposal for Scalable Spiking Neural systems with STDP Learning Capability

Using present-day AER (Address Event Representation) technology it is quite feasible to build hybrid CMOS-memristor systems with many million neurons, once one could assemble reliably dense arrays of memristors on top of CMOS. This is illustrated in Fig. 12.9. On the top left we show a printed circuit board (PCB) hosting 110 identical AER chips, each communicating with its four neighbors through bidirectional bit-serial AER asynchronous links, for event-driven (spiking) communications. Each chip contains an AER processor, which in general is any array of neural processing units. This processor would receive events asynchronously, which shall be processed "as they flow," generating asynchronous output events. The chip also contains a block for programming and configuration of parameters, and a router block. The events interchanged between chips contain not only the standard Address Event (x,y,p) (where (x,y) is event coordinate and p its polarity), but also a header (a,b) indicating the chip address in the PCB. This chip address can indicate either the chip where the event was originated (source coding) or the destination chip (destination coding) [54]. The router block in each chip looks at the event header of each traveling event and decides whether to send it to its local event processor, or to one of its output ports. Similarly, for each event generated by the local event processor, the router adds a header and sends the event out through one of its output ports. The router takes all these decisions based on a local programmable routing table. The set of all routing tables in the array of chips in the PCB defines the architectural topology of the overall neural network.

In order to implement the STDP learning mechanism with memristors and spiking neurons as described throughout this chapter, each AER processor in a chip may contain an array of CMOS cells, each containing an input and an output neuron. The input neuron sends out a spike of programmable shape whenever it is stimulated by an incoming event with its (x,y) coordinate. The output neuron receives and integrates incoming currents, and when it reaches its threshold sends out an Address Event with its (x,y) coordinate and polarity p, while at the same time sends a backward spike of programmable shape through its current summing input terminal.

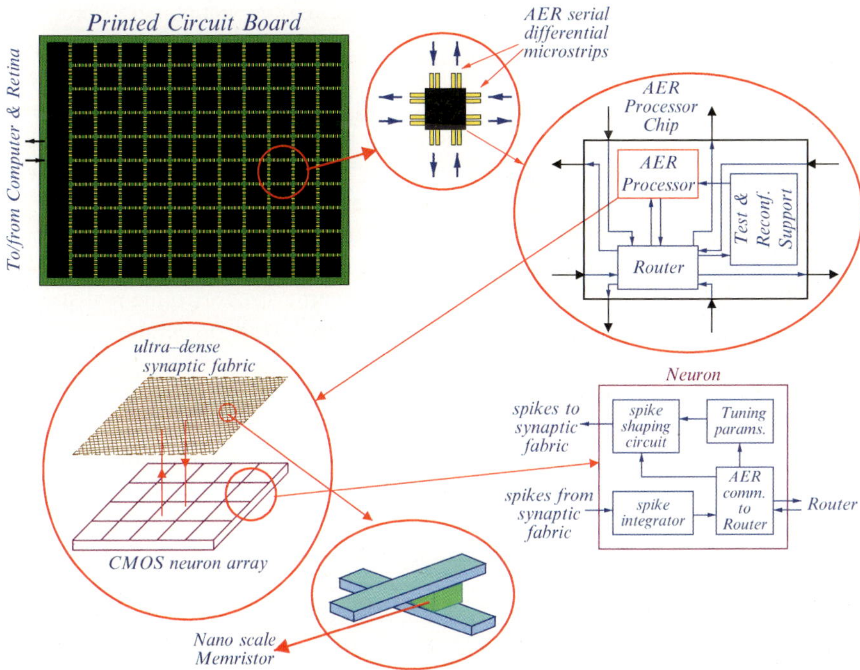

Fig. 12.9 Top down description of hybrid CMOS/nano multi-chip STDP event-driven system. A PCB holds a large number of chips arranged in a 2D grid communicating events serially through AER differential microstrip lines. Each chip contains an event-driven processor, a router, and test and configuration circuitry. The event-driven processor is made of an array of cells each containing one pre- and one post-synaptic neuron. On top of this array there would be two layers of perpendicular nanowires. At the crossing of two perpendicular nanowires there is memristive material implementing physically one synapse

This current summing input terminal of the output neuron and the output terminal of the input neuron of each cell connect to a crossbar of nano-wires assembled on top of the CMOS chip using the connection arrangement known as CMOL [95]. At the crossing of each nano-wire pair there is memristive material implementing physically one memristor synapse. This arrangement would implement the scheme shown in Fig. 12.8.

12.7 Practical Limitations, Realistic Sizes, Pitches, Density, Crosstalk and Power Considerations

Nanoscale memristor technology is still quite incipient and no realistic large-scale systems have been reported at the time of writing (as far as we know). However, we can estimate an orientative scale and density of what may realistically be achieved in the near future, and the main limitations which may be encountered in a real physical implementation.

Regarding the wiring density of synaptic memristors, a pitch of 100 nm is conservatively realistic for present day technologies [96, 97], while the near future might bring us closer to 10 nm [98]. Assuming technologies of 100 nm pitch 2D memristor arrays capable of interfacing reliably with lower CMOS become available some time soon, this would result in a synaptic density of 10^{10} synapses per cm^2.

In the brain, the number of synapses per neuron is about 10^3 to 10^4. If we want to maintain the 10^4 ratio, we would need to fabricate CMOS neurons with a pitch of 10 μm, resulting in 10^6 neurons per cm^2. Such neuron sizes are quite realistic for present day nanometer scale CMOS (45 nm or 32 nm), given the complexity of the neurons needed.

Another problem is that of resistance value ranges of the memristors' R_{min} (synapse ON) and R_{max} (synapse OFF). Reported memristors present resistance values from the $k\Omega$ range up to the $M\Omega$ range [4–6]. The memristor resistance value range affects the performance, reliability, crosstalk, and power dissipation of a full large-scale system. For example, it affects the driving capability of the neurons and their power consumption. If one neuron needs to drive 10^4 synapses of average value $1\,M\Omega$ to an average $1\,V$ level, it has to be able to provide an average current of 10 mA during a spike (of say 20 ms), delivering 10 mW per spike. If there are 10^6 neurons per cm^2 each firing at an average of 10 Hz (which is similar to biological neurons), the synapses would dissipate a power of about 2 kW. The neurons would need at least the same power, presumably more. It is obvious that such a structure would melt quickly. The resistance range needs to be increased by a minimum factor of 100, so that minimum resistances are at least $100\,M\Omega$, or even larger. As pitches are lowered, resistances would need to increase quadratically with pitch decrease, to maintain the power limitation. Another option would be to scale down voltage, but there is not much range. Even our $1\,V$ maximum voltage assumption is quite optimistic for available present day memristors, which tend to operate between $2-10sV$ [4–6]. Also, we have always assumed so far that voltage sources driving memristor terminals behave as ideal voltage sources, or at least, that the output resistance of such voltage sources is negligible compared to the total resistance they have to drive. Again, this will be achieved more easily if memristors present rather high resistance values. If driving voltage sources are no longer so ideal, then there will be crosstalk between lines. For example, if a spike is sent to a column, then the voltage on all rows would change slightly. The consequence of this is that part of the charge provided by the incoming spike will be lost through non-desired synapses and the impact of the spike on the target neurons will be weaker. During learning, the situation is less severe because for STDP update the memristor voltage has to exceed the learning threshold (v_{th} in (12.5)). The effect of having non-ideal voltage sources is that the terminal voltage difference on the memristors needing synaptic update would be slightly less than in the ideal situation and learning would be weaker than expected ideally. However, having non-ideal voltage sources would not induce STDP update in undesired synapses. Another parasitic issue related to crosstalk is parasitic capacitive crosstalk between lines, which can be more pronounced as pitch and line distances decrease.

Also, one highly critical aspect which needs to be evaluated is the influence of component mismatches. Nano scale devices suffer from high mismatch in general. Consequently, we should expect nano scale memristors too to suffer from great parameter variations from one to another. It is true that they will operate as adaptive devices that will learn their functionality hopefully compensating for (some) mismatches. However, their learning and adaptation rules will also suffer from mismatch, making some synapses learn faster than others, or in slightly different fashions. In any case, the main sources of mismatch in memristor devices still need to be identified, and then their influence in the overall system learning behavior evaluated. However, to undertake such an initiative, we first need ready access to large arrays of reliable memristors fabricated in a stable and repeatable manner.

In general, an important issue is precise memristor modeling. Throughout this chapter we have assumed an idealized voltage-driven memristor ideal model. This is useful to devise possible system architectures to achieve a desired functionality, such as STDP learning. However, to estimate realistic performance figures of resulting systems, it will be necessary to include non-ideal effects, both of the memristors and companion CMOS circuits. In this chapter, no high order effects have been modeled, such as those related to noise, mismatch, and other memristor non-idealities not yet reported.

12.8 Conclusions

In this chapter we have shown that STDP learning can be induced by the voltage/flux-driven formulation of a memristor device. We have used this formulation to develop fully asynchronous circuit architectures capable of performing STDP, by having neurons send their spikes not only forward but also backwards. We have seen that depending on the memristive mechanism taking place, the resulting STDP behavior can be of additive or quadratic type. We have shown how the shape of spikes is critical to achieve and modulate a specific STDP learning function. At the end we have also discussed possible limitations of present day memristors.

The presented results are ideal extrapolations based on behavioral simulations. As memristor devices are further developed and non-ideal effects become known, the impact of non-idealities in the presented architectures and methods can be further assessed. Future work has to evolve towards more realistic memristor models and improved memristor devices, specially devices with much higher resistivities. One critical property that memristors need to provide for efficient STDP and non-volatility is the central dead-zone in Fig. 12.9b, which the already reported memristor from Michigan University [6] seems to present. Another issue relates to the quadratic type of multiplicative STDP followed by the presented devices and architectures. This is a quite unusual form of STDP, which needs to be further investigated from a theoretical point of view. In general, there might be stability issues with generic STDP when used in complex biological models [99, 100]. Similarly, since the presented approach allows the shape of the neural spikes, and therefore the shape of the STDP learning curves to be changed in time, further theoretical studies are required to incorporate time-varying STDP learning functions for speeding up, stabilizing, or in general improving learning performance.

Acknowledgements This work was supported by EU grant 216777 (NABAB), Spanish grants (with support from the European Regional Development Fund) TEC2006-11730-C03-01 (SAMANTA2), TEC2009-10639-C04-01 (VULCANO), TEC2012-37868-C04-01 (BIOSENSE), PRI-PIMCHI-2011-0768 (PNEUMA) (granted by former Ministerio de Ciencia e Innovacin, and coordinated with the European CHISTERA program) and Andalusian grants P06TIC01417 (Brain System) and TIC-6091 (NANONEURO).

References

1. T. Serrano-Gotarredona, T. Prodromakis, B. Linares-Barranco, A proposal for hybrid memristor-CMOS spiking neuromorphic learning systems. IEEE Circuits Syst. Mag. **13**(2), 74–88 (2013)
2. C. Mead, *Analog VLSI and Neural Systems* (Addison Wesley, Reading, MA, 1989)
3. D.B. Strukov, G.S. Snider, D.R. Stewart, R.S. Williams, The missing memristor found. Nature **453**, 80–83 (1 May 2008)
4. J. Borghetti, Z. Li, J. Straznicky, X. Li, D.A.A. Ohlberg, W. Wu, D. R. Stewart, R.S. Williams, A hybrid nanomemristor/transistor logic circuit capable of self-programming. PNAS 106(6) 1699–1703 (10 February 2009)
5. S.H. Jo, T. Chang, I. Ebong, B.B. Bhadviya, P. Mazumder, W. Lu, Nanoscale memristor device as synapse in neuromorphic systems. Nano Lett. **10**(4), 1297–1301 (2010)
6. S.H. Jo, K.-H. Kim, W. Lu, High-density crossbar arrays based on a Si memristive system. NANO Lett. **9**(2), 870–874 (2009)
7. L.O. Chua, Memristor: the missing circuit element. IEEE Trans. Circuit Theory **18**, 507–519 (1971)
8. L.O. Chua, S.M. Kang, Memristive devices and systems. Proc. IEEE **64**(2), 209–223 (February 1976)
9. L.O. Chua, C.A. Desoer, E.S. Kuh, *Linear and Nonlinear Circuits* (McGraw-Hill, New York, 1987)
10. W. Gerstner, R. Ritz, J. L. Hemmen, Why spikes? Hebbian learning and retrieval of time-resolved excitation patterns. Biol. Cybern. **69**, 503–515 (1993)
11. W. Gerstner, R. Kempter, J. Leo van Hemmen, H. Wagner, A neuronal learning rule for sub-millisecond temporal coding. Lett. Nat. **383**, 76–78 (5 September 1996)
12. J. Sjöström, W. Gerstner, Spike-timing dependent plasticity, in *Scholarpedia, the peer-reviewed open-access encyclopedia* **5**(2), 1362 (2010). (Available from http://www.scholarpedia.org/article/STDP) (10 February 2010)
13. R.P.N. Rao, T.J. Sejnowski, Spike-time-dependent Hebbian plasticity as temporal difference learning. Neural Comp. **13**, 2221–2237 (2001)
14. B. Porr, F. Wörgötter, How the shape of pre- and postsynaptic signals can influence STDP: A biophysical model. Neural Comp. **16**, 595–625 (2004)
15. A. Delorme, L. Perrinet, S.J. Thorpe, Networks of integrate-and-fire neurons using Rank Order Coding B: Spike timing dependent plasticity and emergence of orientation selectivity. Neurocomputing **38–40**, 539–545 (2001)

16. R. Guyonneau, R. VanRullen, S.J. Thorpe, Temporal codes and sparse representations: a key to understanding rapid processing in the visual system. J. Physiol. Paris **98**(4–6), 487–497 (2004)
17. T. Masquelier, S.J. Thorpe, Unsupervised learning of visual features through spike timing dependent plasticity. PLoS Comput. Biol. **3**(2), e31 (2007)
18. U. Weidenbacher, H. Neumann, Unsupervised learning of head pose through spike-timing dependent plasticity, in *Perception in Multimodal Dialogue Systems*. Lecture Notes in Computer Science (Springer, Berlin/Heidelberg, 2008), pp. 123–131
19. T. Masquelier, S.J. Thorpe, Learning to recognize objects using waves of spikes and Spike Timing-Dependent Plasticity. *Proceedings of the 2010 IEEE International Joint Conference on Neural Networks*, Barcelona, 18–23 July 2010. doi:10.1109/IJCNN.2010.5596934
20. T. Masquelier, R. Guyonneau, S.J. Thorpe, Spike timing dependent plasticity finds the start of repeating patterns in continuous spike trains. PLoS One **3**(1), e1377 (2008)
21. T. Masquelier, R. Guyonneau, S.J. Thorpe, Competitive STDP-based spike pattern learning. Neural Comp. **21**(5), 1259–1276 (2009). doi:10.1162/neco.2008.06-08-804
22. T. Masquelier, E. Hugues, G. Deco, S.J. Thorpe, Oscillations, phase-of-firing coding and spike timing-dependent plasticity: an efficient learning scheme. J. Neurosci. **29**(43), 13484–13493 (2009)
23. J.M. Young, Cortical reorganization consistent with spike timing-but not correlation-dependent plasticity. Nat. Neurosci. **10**(7), 887–895 (2007)
24. L.A. Finelli, S. Haney, M. Bazhenov, M. Stopfer, T.J. Sejnowski, Synaptic learning rules and sparse coding in a model sensory system. PLoS Comput. Biol. **4**(4), e1000062 (2008)
25. D.O. Hebb, *The Organization of Behavior: A Neuropsychological Study* (Wiley, New York, 1949)
26. G.S. Snider, Self-organized computation with unreliable, memristive nanodevices. Nanotechnology **18**, 365202 (2007)
27. G.S. Snider, Spike-timing-dependent learning in memristive nanodevices. *IEEE International Symposium on Nano Architectures*, pp. 85–92, Anaheim, CA, 12–13 June 2008
28. B. Linares-Barranco, T. Serrano-Gotarredona, Memristance can explain Spike-Timc-Dependent-Plasticity in Neural Synapses. Available from *Nature Precedings*, March 2009, http://hdl.handle.net/10101/npre.2009.3010.1
29. C. Zamarreño-Ramos, L.A. Camuñas-Mesa, J.A. Pérez-Carrasco , T. Masquelier, T. Serrano-Gotarredona, B. Linares-Barranco, On spike-timing-dependent-plasticity, memristive devices, and building a self-learning visual cortex. Front. Neurosci. **5**, 26 (2011). doi:10.3389/fnins.2011.00026
30. T. Serrano-Gotarredona, T. Prodromakis, T. Masquelier, G. Indiveri, B. Linares-Barranco, STDP and STDP variations with memristors for spiking neuromorphic learning systems. Front. Neurosci. **7**, 2 (2013). doi:10.3389/fnins.2013.00002
31. H. Markram, J. Lübke, M. Frotscher, B. Sakmann, Regulation of synaptic efficacy by coincidence of postsynaptic APS and EPSPS. Science **275**(5297), 213–215 (1997)
32. G. Bi, M. Poo, Synaptic modifications in cultured hippocampal neurons: dependence on spike timing, synaptic strength, and postsynaptic cell type. J. Neurosci. **18**(24), 10464–10472 (1998)
33. G. Bi, M.M. Poo, Synaptic modification by correlated activity: Hebb's postulate revisited. Ann. Rev. Neurosci. **24**, 139–166 (2001)
34. L. Zhang, H. Tao, C. Holt, W. Harris, M. Poo, A critical window for cooperation and competition among developing retinotectal synapses. Nature **395**(6697), 37–44 (1998)
35. D. Feldman, Timing-based LTP and LTD at vertical inputs to layer II/III pyramidal cells in rat barrel cortex. Neuron **27**(1), 45–56 (2000)
36. Y. Mu, M.M. Poo, Spike timing-dependent LTP/LTD mediates visual experience-dependent plasticity in a developing retinotectal system. Neuron **50**(1), 115–125 (2006)
37. S. Cassenaer, G. Laurent, Hebbian STDP in mushroom bodies facilitates the synchronous flow of olfactory information in locusts. Nature **448**(7154), 709–713 (2007)
38. V. Jacob et al., Spike-timing-dependent synaptic depression in the in vivo barrel cortex of the rat. J. Neurosci. **27**(6), 1271–1284 (2007)

39. S. Thorpe, et al., Speed of processing in the human visual system. Nature **381**, 520–522 (1996)
40. J. Costas-Santos, T. Serrano-Gotarredona, R. Serrano-Gotarredona, B. Linares-Barranco, A Spatial Contrast Retina with On-chip Calibration for Neuromorphic Spike-Based AER Vision Systems. IEEE Trans. Circuits Syst. I **54**(7), 1444–1458 (July 2007)
41. J.A. Leñero-Bardallo, T. Serrano-Gotarredona, B. Linares-Barranco, A five-decade dynamic-range ambient-light-independent calibrated signed-spatial-contrast AER retina with 0.1-ms latency and optional time-to-first-spike mode. IEEE Trans. Circuits Syst. I **57**(10), 2632–2643 (October 2010)
42. J.A. Leñero-Bardallo, T. Serrano-Gotarredona, B. Linares-Barranco, A 3.6 s latency asynchronous frame-free event-driven dynamic-vision-sensor. IEEE J. Solid-State Circuits **46**(6), 1443–1455 (June 2011)
43. T. Serrano-Gotarredona, B. Linares-Barranco, A 128x128 1.5 % contrast sensitivity 0.9 % FPN 3us latency 4mW asynchronous frame-free dynamic vision sensor using transimpedance preamplifiers. IEEE J. Solid-State Circuits **48**(3), (March 2013)
44. P. Lichtsteiner, C. Posch, T. Delbrück, An 128x128 120dB 15 s-latency temporal contrast vision sensor. IEEE J. Solid State Circuits **43**(2), 566–576 (2008)
45. C. Posch, D. Matolin, R. Wohlgenannt, A QVGA 143 dB dynamic range frame-free PWM image sensor with lossless pixel-level video compression and time-domain CDS. IEEE J. Solid-State Circuits **46**(1), 259–275 (January 2011)
46. E. Culurciello, R. Etienne-Cummings, K. Boahen, A biomorphic digital image sensor. IEEE J. Solid State Circuits, Part 2 **38**, 281–294 (2003)
47. P.-F. Ruedi, P. Heim, F. Kaess, E. Grenet, F. Heitger, P.-Y. Burgi, S. Gyger, P. Nussbaum, A 128x128 pixel 120-dB dynamic-range vision-sensor chip for image contrast and orientation extraction. IEEE J. Solid-State Circuits **38**(12), 2325–2333 (December 2003)
48. K.A. Zaghloul, K. Boahen, A silicon retina that reproduces signals in the optic nerve. J. Neural Eng. **3**, 257–267 (December 2006)
49. T. Serrano-Gotarredona, R. Serrano-Gotarredona, A. Acosta-Jiménez, B. Linares-Barranco, A neuromorphic cortical-layer microchip for spike-based event processing vision systems. IEEE Trans. Circuits Syst. I **53**(12), 2548–2566 (December 2006)
50. L. Camuñas-Mesa, A. Acosta-Jiménez, C. Zamarreño-Ramos, T. Serrano-Gotarredona, B. Linares-Barranco, A 32x32 pixel convolution processor chip for address event vision sensors with 155 ns event latency and 20 Meps throughput. IEEE Trans. Circuits. Syst. I **58**(4), 777–790 (April 2011)
51. L. Camuñas-Mesa, C. Zamarreño-Ramos, A. Linares-Barranco, A.J. Acosta-Jiménez, T. Serrano-Gotarredona, B. Linares-Barranco, An event-driven multi-kernel convolution processor module for event-driven vision sensors. IEEE J. Solid-State Circuits **47**(2), 504–517 (February 2012)
52. T. Choi, P.A. Merolla, J.V. Arthur, K.A. Boahen, B.E. Shi, Neuromorphic implementation of orientation hypercolumns. IEEE Trans. Circuits Syst. I **52**(6), 1049–60 (2005)
53. R. Serrano-Gotarredona, M. Öster, P. Lichtsteiner, A. Linares-Barranco, R. Paz-Vicente, F. Gómez-Rodríguez, L. Camuñas-Mesa, R. Berner, M. Rivas-Pérez, T. Delbrück, S.-C. Liu, R. Douglas, P. Häfliger, G. Jiménez-Moreno, A. Civit-Ballcels, T. Serrano-Gotarredona, A.J. Acosta-Jiménez, B. Linares-Barranco, CAVIAR: A 45k Neuron, 5M Synapse, 12G Connects/s AER Hardware Sensory-Processing-Learning-Actuating System for High-Speed Visual Object Recognition and Tracking. IEEE Trans. Neural Netw. 20(9), 1417–1438 (September 2009)
54. C. Zamarreño-Ramos, A. Linares-Barranco, T. Serrano-Gotarredona, B. Linares-Barranco, Multi-casting mesh AER: A scalable assembly approach for reconfigurable neuromorphic structured AER systems. Application to ConvNets. IEEE Trans. Biomed. Circuits Syst. 2013. In Press. 7(1), 82–102 (Febraury 2013)
55. J.A. Pérez-Carrasco, B. Zhao, C. Serrano, B. Acha, T. Serrano-Gotarredona, S. Chen, B. Linares-Barranco, Mapping from frame-driven to frame-free event-driven vision systems by low-rate rate-coding. Application to Feed Forward ConvNets. IEEE Trans. Pattern Anal. Mach. Intell. in Press 35(11), 2706–2719 (November 2013)

56. C. Shoushun, P. Akselrod, Bo Zhao, J.A. Pérez-Carrasco, B. Linares-Barranco, E. Culurciello, Efficient feedforward categorization of objects and human postures with address-event image sensors. IEEE Trans. Pattern Anal. Mach. Intell. 34(2), 302–314 (February 2012)
57. T. Serrano-Gotarredona, J. Park, A. Linares-Barranco, R. Benosman, B. Linares-Barranco, Improved contrast sensitivity DVS and its application to event-driven stereo vision. *IEEE International Symposium on Circuits and Systems*, Beijing, 19–23 May 2013
58. Z. Ni, C. Pacoret, R. Benosman, S. Ieng, S. Régnier, Asynchronous event-based high speed vision for microparticle tracking. J. Microsc. **245**(3), 236–244 (March 2012)
59. P. Rogister, R. Benosman, S.-H. Ieng, P. Lichtsteiner, T. Delbrück, Asynchronous event-based binocular stereo matching. IEEE Trans. Neural Netw. Learn. Syst. **23**(2), 347–353 (February 2012)
60. M. Hofstatter, M. Litzenberger, D. Matolin, C. Posch, Hardware-accelerated address-event processing for high-speed visual object recognition. *18th IEEE International Conference on Electronics, Circuits and Systems*, Beirut, 11–14 December 2011
61. Z. Ni, A. Bolopion, J. Agnus, R. Benosman, S. Regnier, Asynchronous event-based visual shape tracking for stable haptic feedback in microrobotics. IEEE Trans. Robot. **28**(5), 1081–1089 (2012)
62. M. Litzenberger, C. Posch, D. Bauer, A. N. Belbachir, P. Schon, B. Kohn, H. Garn, Embedded vision system for real-time object tracking using an asynchronous transient vision sensor. *12th Signal Processing Education Workshop, 4th Digital Signal Processing Workshop*, pp. 173–178, Teton National Park, WY, 24–27 September 2006
63. R. Benosman, S.H. Ieng, P. Rogister, C. Posch, Asynchronous event-based hebbian epipolar geometry. IEEE Trans. Neural Netw. **22**, 1723–1734 (2011)
64. R. Benosman, S.-H. Ieng, C. Clercq, C. Bartolozzi, M. Srinivasan, Asynchronous frameless event-based optical flow. Neural Netw. **27**, 32–37 (March 2012)
65. J. Conradt, M. Cook, R. Berner, P. Lichtsteiner, R. J. Douglas, T. Delbrück, A pencil balancing robot using a pair of AER dynamic vision sensors. *International Symposium on Circuits and Systems*, ISCAS, 2009, pp. 781–784, Taipei, 24–27 May 2009
66. Z. Fu, T. Delbrück, P. Lichtsteiner, E. Culurciello, An address-event fall detector for assisted living applications. IEEE Trans. Biomed. Circuits Syst. **2**, 88–96 (2008)
67. D. Drazen, P. Lichtsteiner, P. Häfliger, T. Delbrück, A. Jensen, Toward real-time particle tracking using an event-based dynamic vision sensor. Exp. Fluids **51**(5), 1465–1469 (September 2011)
68. T. Delbrück, P. Lichtsteiner, Fast sensory motor control based on event-based hybrid neuromorphic-procedural system. *International Symposium on Circuits and Systems*, ISCAS, 2007, pp. 845–848, New Orleans, LA, 27–30 May 2007
69. S. Schraml, A.N. Belbachir, A spatio-temporal clustering method using real-time motion analysis on event-based 3D vision. *IEEE Computer Vision and Pattern Recognition Workshop* (CVPRW), 2010, pp. 57–63, San Francisco, CA, 13–18 June 2010
70. O. Bichler, D. Querlioz, S.J. Thorpe, J.P. Bourgoin, C. Gamrat, Extraction of temporally correlated features from dynamic vision sensors with spike-timing-dependent plasticity. Neural Netw. **32**, 339–348 (2012)
71. C. Zamarreño-Ramos, T. Serrano-Gotarredona, B. Linares-Barranco, An Instant-Startup Jitter-Tolerant Manchester-Encoding Serializer/Deserializar Scheme for Event-Driven Bit-Serial LVDS Inter-Chip AER Links. IEEE Trans. Circuits Syst. I **58**(11), 2647–2660 (November 2011)
72. C. Zamarreño-Ramos, T. Serrano-Gotarredona, B. Linares-Barranco, A $0.35\mu m$ sub-ns wake-up time ON-OFF switchable LVDS driver-receiver chip I/O pad pair for rate-dpendent power saving in AER bit-serial links. IEEE Trans. Biomed. Circuits Syst. **6**(5), 486–497 (October 2012)
73. C. Zamarreño-Ramos, R. Kulkarni, J. Silva-Martínez, T. Serrano-Gotarredona, B. Linares-Barranco, A 1.5 ns OFF/ON switching-time voltage-mode LVDS driver/receiver pair for asynchronous AER bit-serial chip grid links with up to 40 times event-rate dependent power savings. IEEE Trans. Biomed. Circuits Syst. in Press.

74. Y. LeCun, B. Boser, J.S. Denker, D. Henderson, R.E. Howard, W. Hubbard, L.D. Jackel, Backpropagation applied to handwritten zip code recognition. Neural Comput. **1**(4), 541–551 (December 1989)
75. C. Farabet, et al., Large-scale FPGA-based convolutional networks, in *Machine Learning on Very Large Data Sets*, ed. by R. Bekkerman, M. Bilenko, J. Langford (Cambridge University Press, Cambridge, 2011)
76. J.E. Rubin, R.C. Gerkin, G.-Q. Bi, C.C. Chow, Calcium time course as a signal for spike-timing-dependent plasticity. J. Neurophysiol. **93**, 2600–2613 (2005)
77. E.V. Lubenov, A.G. Siapas, Decoupling through synchrony in neuronal circuits with propagation delays. Neuron **58**, 118–131 (10 April 2008)
78. M.C.W. van Rossum, G.Q. Bi, G.G. Turrigiano, Stable hebbian learning from spike timing-dependent plasticity. The Journal of Neuroscience **20**(23), 8812–8821 (1 December 2000)
79. J. Rubin, D.D. Lee, H. Sompolinsky, Equilibrium properties of temporally asymmetric hebbian plasticity. Phys. Rev. Lett. **86**(2), 364–367 (8 January 2001)
80. R. Gütig, R. Aharonov, S. Rotter, H. Sompolinsky, Learning input correlations through non-linear temporally asymmetric hebbian plasticity. J. Neurosci. **23**(9), 3697–3714 (1 May 2003)
81. T. Prodromakis, I. Salaoru, A. Khiat, C. Toumazou, Concurrent resistive and capacitive switching of nanoscale TiO_2 memristors, in *Nature Conference on Frontiers in Electronic Materials: Correlation Effects and Memristive Phenomena*, Aachen, Germany, 2012
82. T. Prodromakis, B.P. Peh, C. Papavassiliou, C. Toumazou, A versatile memristor model with nonlinear dopant kinetics. IEEE Trans. Electron Devices **58**(9), 3099–3105 (2011)
83. F. Argall, Switching phenomena in titanium oxide thin films. Solid State Electron. Pergamon Press **11**, 535–541 (1968)
84. B. Swaroop, W.C. West, G. Martinez, M.N. Kozicki, L.A. Akers, Programmable current mode hebbian learning neural network using programmable metallization cell. *Proceedings of the IEEE International Symposium on Circuits and Systems* (ISCAS1998), vol. 3, pp. 33–36, Monterey, CA, 31 May–3 June 1998
85. T. Prodromakis, C. Toumazou, L.O. Chua, Two centuries of memristors. Nat. Mater. **11**, 478–481 (22nd May 2012). doi:10.1038/nmat3338
86. Y. Nian, J. Strozier, N. Wu, X. Chen, A. Ignatiev, Evidence for an oxygen diffusion model for the electric pulse induced resistance change effect in transition-metal oxides. Phys. Rev. Lett. **98**(14), 146–403 (April 2007)
87. J.J. Yang, M.D. Pickett, X. Li, D.A.A. Ohlberg, D.R. Stewart, R.S. Williams, Memristive switching mechanism for metal/oxide/metal nanodevices. Nat. Nanotech. **3**(7), 429–433 (June 2008)
88. J. Hur, M.-J. Lee, C. Lee, Y.-B. Kim, C. J. Kim, Modeling for bipolar resistive memory switching in transition-metal oxides. Phys. Rev. B **82**(15), 155321 (October 2010)
89. M. Wuttig, N. Yamada, Phase-change materials for rewriteable data storage. Nat. Mater. 6(11), 824–832 (2007)
90. D.H. Kwon et al., Atomic structure of conducting nanofilaments in TiO_2 resistive switching memory. Nat. Nanotech. **5**, 148–153 (2010)
91. Y. Yang, P. Gao, S. Gaba, T. Chang, X. Pan, W. Lu, Observation of conducting filament growth in nanoscale resistive memories. Nat. Commun. **3**, 732 (2012). doi:10.1038/ncomms1737
92. A. Chanthbouala, V. Garcia, R.O. Cherifi, K. Bouzehouane, S. Fusil, X. Moya, S. Xavier, H. Yamada, C. Deranlot, N.D. Mathur, M. Bibes, A. Bartélémy, J. Grollier, A ferroelectric memristor. Nat. Mater. **11**, 860–864 (October 2012)
93. B. Linares Barranco, T. Serrano Gotarredona, Exploiting memristance in adaptive asynchronous spiking neuromorphic nanotechnology systems. *9th IEEE Conference on Nanotechnology* (IEEE-NANO 2009), pp. 601–604, Genoa, 26–30 July 2009
94. O. Bichler, M. Suri, D. Querlioz, D. Vuillaume, B. DeSalvo, C. Gamrat, Visual pattern extraction using energy-efficient "2-PCM Synapse" Neuromorphic Architecture. IEEE Trans. Electron Devices **59**(8), 2206–2214 (August 2012)
95. D.B. Strukov, K.K. Likharev, CMOL FPGA: a reconfigurable architecture for hybrid digital circuits with two-terminal nanodevices. Nanotechnology **16**, 888–900 (2005)

96. J.E. Green, J.W. Choi, A. Boukai, Y. Bunimovich, E. Johnston-Halperin, E. DeIonno, Y. Luo, B.A. Sheriff, K. Xu, Y.S. Shin, H.-R. Tseng, J.F. Stoddart, J.R. Heath, A 160-kilobit molecular electronic memory patterned at 10^{11} bits per square centimetre. Nature **445**, 414–417 (25 January 2007)
97. G.-Y. Jung, E. Johnston-Halperin, W. Wu, Z. Yu, S.-Y. Wang, W.M. Tong, Z. Li, J.E. Green, B.A. Sheriff, A. Boukai, Y. Bunimovich, J.R. Heath, R.S. Williams, Circuit fabrication at 17 nm half-pitch by nanoimprint lithography. Nano Lett. **6**(3), 351–354 (February 2006)
98. H.-J. Jeon, K.H. Kim, Y.-K. Baek, D.W. Kim, H.-T. Jung, New top-down approach for fabricating high-aspect-ratio complex nanostructures with 10 nm scale features. Nano Lett. **10**(9), 3604–3610 (2010)
99. E.M. Izhikevich, N.S. Desai, Relating STDP to BCM. Neural Comput. **15**, 1511–1523 (2003)
100. A.J. Watt, N.S. Desai, Homeostatic plasticity and STDP: keeping a neuron's cool in a fluctuating world. Front. Synaptic. Neurosci. **2**, 5 (2010). doi:10.3389/fnsyn.2010.00005

Chapter 13
Memristor for Neuromorphic Applications: Models and Circuit Implementations

Alon Ascoli, Fernando Corinto, Marco Gilli, and Ronald Tetzlaff

13.1 Introduction

The current-controlled *ideal memristor* is a passive bipole linking charge $q(t)$ and flux $\varphi(t)$ through a nonlinear relation, i.e. $\varphi(t) = \varphi(q(t))$. From application of Faraday's Law and of the chain rule it follows that voltage $v(t)$ depends upon current $i(t)$ through

$$v(t) = \frac{d\varphi(t)}{dt} = M(q(t))\,i(t), \tag{13.1}$$

where $M(q) = \frac{d\varphi(q)}{dq}$ is the memristance (i.e. memory-resistance) of the bipole. Since $q(t) = \int_{-\infty}^{t} i(t')dt'$, then $M(q) = M(\int_{-\infty}^{t} i(t')dt')$. In other words the resistance of the memristor depends upon the time history of the current flowed through it. This explains the memory capability of the memristor, theoretically envisioned by Chua in 1971 [1] and later classified by Chua and Kang in 1976 as the simplest element from a large class of nonlinear dynamical systems endowed with memristance, the so-called memristive systems [2].

In [2] a memristive system (or memristor system[1]) is a nonlinear dynamical circuit element defined by the following differential-algebraic system of equations:

[1] In the following memristive systems are referred to as memristor systems, whereas the term ideal memristor is used for systems described by (13.1).

A. Ascoli (✉) • R. Tetzlaff
Technische Universität Dresden, Mommsenstraße 12, 01062 Dresden, Germany
e-mail: alon.ascoli@tu-dresden; ronald.tetzlaff@tu-dresden

F. Corinto • M. Gilli
Politecnico di Torino, Corso Duca degli Abruzzi 24, 10129 Torino, Italy
e-mail: fernando.corinto@polito.it; marco.gilli@polito.it

R. Tetzlaff (ed.), *Memristors and Memristive Systems*,
DOI 10.1007/978-1-4614-9068-5_13,

$$\frac{d\mathbf{x}(t)}{dt} = \mathbf{f}(\mathbf{x}(t), u(t)), \tag{13.2}$$

$$y(t) = \mathbf{g}(\mathbf{x}(t), u(t))u(t), \tag{13.3}$$

where[2] $\mathbf{x} \in \mathbb{R}^n$ is the state, $u \in \mathbb{R}$ refers to the input, $y \in \mathbb{R}$ describes the output, $\mathbf{f}(\mathbf{x}, u) : \mathbb{R}^n \times \mathbb{R} \to \mathbb{R}^n$ stands for the *state evolution function*, while $\mathbf{g}(\mathbf{x}, u) : \mathbb{R}^n \times \mathbb{R} \to \mathbb{R}$ denotes the *memductance* (*memristance*) if input u is in voltage (current) form.

Since 2008, when its existence at the nano-scale was certified at Hewlett-Packard (HP) Labs [3], the memristor has attracted a strong interest from both industry and academia for its central role in the setup of novel integrated circuit (IC) architectures, especially in the design of high-density nonvolatile memories [4], programmable analog circuitry [5], neuromorphic systems [6], and logic gates [7,8].

The development of innovative strategies for the design of memristor-based electronic systems requires the availability of mathematical models [3, 9–14, to name but a few] for the memristor nano-structures under study. A good model should be as general as possible, i.e. it should be able to capture the memristor dynamics of a large number of nano-films. In this respect the Boundary Condition Memristor (BCM) model, recently introduced in [12], was developed so as to meet this generality requirement. In fact the distinctive feature of the BCM model is the adaptability of the nano-device behavior at boundaries. In particular, the model makes use of adaptable[3] threshold voltages v_{th0} and v_{th1}, respectively, defining[4] the magnitude of the limit value the input voltage (i.e., the voltage drop across the memristance) needs to cross after its negative-to-positive and positive-to-negative sign reversal before the memristor state may be released from its lower and upper bound. It is straightforward to establish an optimization procedure, which, on the basis of observed data, sets the most suitable values for the threshold voltages, i.e. those values, let us identify them as v^*_{th0} and v^*_{th1}, minimizing the mean squared error between observed and modeled data. This enables the BCM model to stand out over other models available in the literature for the larger number of detectable dynamics, despite the extreme simplicity of the window function embedded into the state equation (when the state variable lies within its two bounds its time evolution is governed by the basic linear dopant drift model [3]). It is noteworthy that the class of detectable dynamics include not only all the behaviors observed in the HP memristor [3], but also phenomena exhibited by various other nano-structures

---

[2]For the sake of brevity the explicit time dependency is dropped where it is not strictly necessary.

[3]Note that by defining a time evolution rule for the threshold voltages, it was recently demonstrated [15] that an adaptable threshold voltage-based version of the memristor model from [6] may explain the Suppression Principle [16] of the Spike-Timing-Dependent-Plasticity (STDP) Rule [6], which may occur in the case of triplet spikes.

[4]Throughout the paper, unless stated otherwise and without loss of generality, we assume that the doped layer is spatially located to the left of the un-doped layer along the horizontal extension of the nano-film [12], and in this case we assign a value of $+1$ to the memristor polarity coefficient η (see (13.6)).

where memristor behavior arises from distinct physical mechanisms [17–20]. In order to enable the BCM model to support various neural learning rules, we recently developed a generalized version [21], in which the activation threshold property characterizing the boundary behavior in the original BCM model [12] is extended to the whole admissible range of the state variable, thus allowing the modeling of the degree of non-volatility of the nano-device.

Another necessary requirement for the investigation of potential applications of memristor devices is the implementation of the mathematical models into a software package for computer-aided integrated circuit design. In this chapter we shall first present the PSpice [22] implementation of the generalized BCM model. In the PSpice realization the voltage drop across a linear capacitor models the memristor state. Further the memristance is determined by the series combination of a linear resistance and a nonlinear resistance depending upon the capacitor voltage. The current through the memristance is first nonlinearly filtered so as to model the degree of non-volatility of the nano-structure. In other words, this current is multiplied by a nonlinear function which is responsible for the activation of the state dynamics as the control voltage crosses a tunable positive (negative) threshold v_{t0} $(-v_{t1})$ in its ascent (descent). The filtered current drives a current source, which, under positive (negative) input voltage polarity, charges (discharges) the capacitor. For each of the two lower and upper bounds of the memristor state, flexible boundary conditions are implemented in PSpice by means of a reference voltage source with value equal to that bound and by a pair of voltage-controlled voltage switches, one controlled by the voltage across the capacitor and responsible for clipping the memristor state at the lower (upper) limit under negative (positive) input voltage, the other calling for the release of the state from its lower (upper) bound as the input voltage cuts through yet another tunable positive (negative) threshold v_{th0} $(-v_{th1})$ in its ascent (descent).

The PSpice circuit of the generalized BCM model may be used to model dynamics typical of biological synapses. It is in fact capable to support various rules governing the way neurons learn from each other. As an example, this chapter demonstrates how the PSpice circuit favors associative learning based on the Hebbian rule, one of the most important adaptation rules in neural learning [23].

The last part of this chapter proposes a novel class of memristor emulators. Each element from the class is an electronic circuit comprising standard passive electrical components from circuit theory, namely static nonlinear devices such as diodes and linear dynamical elements such as resistors, inductors, and capacitors.

The structure of the manuscript is organized as follows. Section 13.2 reviews the most noteworthy memristor circuit models available in the literature. Section 13.3 briefly reviews the generalized BCM model and describes its PSpice implementation. Section 13.4 illustrates the ability of the PSpice circuit model to support the Hebbian neural learning rule. Section 13.5 introduces a novel class of memristor emulators. Finally Sect. 13.6 outlines the conclusions.

13.2 Brief Review of Memristor Models

Various memristor circuit models have been proposed in the literature. A large number of models assume that the control waveform is in current form (the voltage v-current i relationship is expressed by (13.1)), views the memristance as the series between two variable resistances, associated with the insulating and conductive layers of the nano-film, and sets the width w of the conductive layer, normalized with respect to the entire length D of the device, as the state $x = \frac{w}{D} \in [0,1]$ of the system. The linear drift model from Williams [3], where the time derivative of the state is proportional to the input waveform in current form, is valid under the assumption that the state is confined within its two bounds, since it does not take into account the boundary behavior.

In the nonlinear drift models from [9, 10] and [24] the rate of change of the state is proportional to the product between the input waveform in current form and a window function accounting for nonlinear dynamical behavior and imposing suitable boundary conditions.

In Joglekar's model [9] the window function is defined as $f_J(x) = 1 - (2x - 1)^{2p}$ ($p \in \mathbb{Z}_+$). Such window describes the suppression of dopant drift close to the extremities, but is not vertically scalable (i.e. its maximum value may not be up- or down-shifted) and introduces the so-called terminal-state problem [24], since if the state is at either of its two bounds it may not leave it for any subsequent time instant. Note that for $p = 1$ Joglekar's window is a scaled (by a factor of 4) version of yet another window previously derived by Strukov in [3], i.e. $f_S(x) = x(1-x)$. Benderli [25] presented a circuit realization of Strukov's model [3], where the use of comparators and logic gates allowed the emulation of the state clipping at or release from either bound.

In Biolek's model the window function depends on both state x and input current i, being defined as $f_B(x,i) = 1 - (x - stp(-i))^{2p}$, where $stp(x) = 1$ for $x \geq 0$ and 0 otherwise ($p \in \mathbb{Z}_+$). Such window resolves the "terminal-state problem," but has limited scalability (in particular, its maximum value may not exceed +1 [24]). PSpice implementations of Joglekar's and Biolek's models are reported in [10].

In the versatile model proposed by Prodomakis [24] the window function $f_P(x) = j(1 - ((x-0.5)^2 - 0.75)^p)$ has two control parameters j and p lying in $\mathbb{R}_+$ and is vertically scalable, i.e. $0 \leq max\{f_P(x)\} \gtreqless 1$. A PSpice version of such model may be easily derived by modifying the PSpice .circ [22] file available in [10].

Another model endowed with a PSpice circuit implementation was developed by Cserey [26]. In this model the state evolution function in Strukov's model [3] was augmented with an additive state-dependent linear term to resolve the "terminal-state problem."

One of the finest circuit emulators of memristor behavior is credited to Shin and Kang [11], which proposed a general model where the control waveform may be in either current or voltage form and the state is defined as the memristance. Their model, from which the charge-flux relationship of the memristor under modeling may be easily extracted, may be suitably tuned through the introduction of a window function depending on the memristor charge.

Kavehei [27] proposed a memristor model based upon the specification of a piecewise-linear charge q-flux φ relationship. In such model the state and output equations are not specified. Its PSpice implementation is based on Chua's [1] first circuit realization of a memristor through a type-1 memristor-resistor mutator.

An interesting model was presented in [28] to explain the memristor behavior of nanoparticle assemblies.

The nonlinear dependence of the time derivative of the state on the input signal is taken into account in Lehtonen's model [29], inspired by the experimental work from [30], where the current is related to the voltage by means of a rectifying exponential function in the off state (as in a diode) and of a sinh function in the on state (typical of electron tunneling). This model, where the control waveform is in voltage form, was implemented in PSpice to describe the neighborhood connections among cellular neural networks (CNNs) [31, 32].

An even more nonlinear function of the input governs the state equation in the voltage-controlled model from Poikonen [33], which studied the transition between non-programming and programming phases in memristor devices.

In the memristor emulator circuit from [34], used as basic building block of a 4-memristor bridge synapse for neuromorphic applications, the memristance, modeled by the input impedance of an active circuit, is made proportional to the time integral of the memristor current by constraining the voltage at one of the input terminals of an operational amplifier to be the analogue multiplication between the voltage across a resistor, proportional to the memristor current, and the voltage across a capacitor, proportional to the time integral of the memristor current.

In [35] Strukov and Williams demonstrated the exponential relationship between drift velocity and local electric field. Since this discovery a number of models have been introduced to support threshold-activated state dynamics.

Among them, one which merits mention, is the physics-based Pickett's model from [13], in which the dependency of the rate of change of the state on the current-form input is strongly nonlinear. In such model the memristor is seen as the series between a low resistance associated with the conductive layer of the nano-film and Simmons' electron tunneling barrier [36], whose width is chosen as the system state. A PSpice version of Pickett's model was presented in [37].

More recently Kvatinski developed a simplified version of the Pickett's model [13] and named it as ThrEshold Adaptive Memristor (TEAM) model [14]. In such model for input current magnitude below a certain adaptable threshold no state change occurs, otherwise the state evolution rule may be tuned to the memristor element under modeling through specification of an appropriate set of control parameters and of suitable window and memristance functions. The PSpice architecture of the TEAM model is similar to the one originally presented in [11].

Another activation-type state model, where the state variable expresses the memristance and the control signal is in voltage form, embedded in the PSpice software program [38], enabled to capture the adaptive behavior of a unicellular organism named amoeba through a simple memristor-based oscillator [39].

Another interesting model with threshold-activated state dynamics was proposed in [40] to explain Spike-Timing-Dependent-Plasticity (STDP) in neural synapses.

Most of these PSpice models have been classified in [41]. Another insightful discussion on the models available in the literature was recently published in [42], where a novel model inspired from Simmons' electron tunneling theory [36], endowed with programming threshold capability and PSpice circuit implementation, was also proposed.

The Boundary Condition Memristor (BCM) model is a simple yet accurate boundary condition-based mathematical model for memristor nano-structures made up of two layers with different conductivity levels, whose longitudinal extensions depend on the time history of the input. In comparison with the classical BCM model [12], the generalized version [21] is augmented with programming threshold capability [42], i.e. with tunable nonvolatile behavior.

Recently, in [43], assuming Pickett's model [13] as reference for comparison, various memristor models, including Biolek's, the TEAM and the BCM models, were first compared on the basis of the ability to reproduce (after an optimization process) the dynamics of the reference model in a particular simulation scenario, and secondly employed in a couple of memristor-based circuits to investigate the variance in the nonlinear dynamical behaviors they give rise to. The latter study revealed the model-dependency of the dynamics of memristor-based circuits, and thus raised a warning against a blind faith in the memristor models and pointed out the necessity to develop a universal mathematical model for exploring the full potential of the memristor and unfolding its unique properties.

Section 13.3 describes the recently proposed generalized BCM model and its PSpice-based circuit [21] (the PSpice emulator of the classical BCM model is reported in [44]).

13.3 Generalized BCM model and Its Circuit Implementation

Let R_{on} and R_{off} stand for the on and off resistances of a memristor nano-film. The memristor state variable x is chosen as the length $w(t)$ of the conductive layer of the nano-film normalized with respect to the entire longitudinal extension D of the nano-film (i.e. $x = \frac{w(t)}{D} \in [0,1]$). Denoting memristor current and voltage as i and v, respectively, the state-dependent input–output algebraic relationship of the generalized BCM model is expressed by

$$i(t) = W(x(t))\, v(t), \tag{13.4}$$

where $W(x(t))$ describes the state-dependent memductance, expressed by

$$W(x(t)) = \frac{G_{on}G_{off}}{G_{on} - \Delta G x(t)}, \tag{13.5}$$

with $G_{on} = R_{on}^{-1}$, $G_{off} = R_{off}^{-1}$, while $\Delta G = G_{on} - G_{off}$.

The state equation of the generalized BCM model is defined as

$$\frac{dx(t)}{dt} = \eta\, k\, W(x(t))\, v(t)\, f\,(x(t), \eta\, v(t), v_{th0}, v_{th1}, v_{t1}, v_{t2}, a, b)\,, \tag{13.6}$$

where $k \in \mathbb{R}$ is a constant depending on physical properties of the memristor (its dimensions are C^{-1}), $\eta \in \{-1, +1\}$ is a coefficient denoting the polarity of the nano-device, while $f\,(x(t), \eta\, v(t), v_{th0}, v_{th1}, a, b) \in \{0, a, b\}$, a switching window function defining not only the boundary behavior but also the degree of non-volatility [42], is expressed as

$$f\,(x, \eta\, v, v_{th0}, v_{th1}, v_{t0}, v_{t1}, a, b) = \begin{cases} b \texttt{ if } C_1 \texttt{ or } C_2 \texttt{ holds,} \\ 0 \texttt{ if } C_3 \texttt{ or } C_4 \texttt{ holds,} \\ a \texttt{ if } C_5 \texttt{ holds,} \end{cases} \tag{13.7}$$

where tunable conditions C_n ($n = 1, 2, 3, 4, 5$) are mathematically described by

$$C_1 = \{\ (x(t) \in (0,1) \texttt{ and } ((\eta\, v(t) > v_{t0}) \texttt{ or } (\eta\, v(t) < -v_{t1})))\}, \tag{13.8}$$

$$C_2 = \{\ (x(t) = 0 \texttt{ and } \eta\, v(t) > v_{th0}) \texttt{ or } (x(t) = 1 \texttt{ and } \eta\, v(t) < -v_{th1})\}, \tag{13.9}$$

$$C_3 = \{\ x(t) = 0 \texttt{ and } \eta\, v(t) \leq v_{th0}\}, \tag{13.10}$$

$$C_4 = \{\ x(t) = 1 \texttt{ and } \eta\, v(t) \geq -v_{th1}\}, \tag{13.11}$$

$$C_5 = \{\ (x(t) = \bar{x} \in (0,1) \texttt{ and } ((\eta\, v(t) \leq v_{t0}) \texttt{ and } (\eta\, v(t) \geq -v_{t1})))\}, \tag{13.12}$$

where $v_{th0} \in \mathbb{R}_+$, $v_{th1} \in \mathbb{R}_+$ represent the input thresholds at boundaries, $v_{t0} \in \mathbb{R}_+$, $v_{t1} \in \mathbb{R}_+$ define the programmability thresholds, ($v_{t0} \leq v_{th0}$ and $v_{t1} \leq v_{th1}$), while a and b are constants modulating the degree of non-volatility of the memristor ($b \in \mathbb{R}_+$, $a \in \mathbb{R}_{0,+}$, $a < b$).

The PSpice implementation of the generalized BCM model is depicted in Fig. 13.1. The source code is reported in Table 13.1.

In the circuit of Fig. 13.1 voltages at nodes y and z, the two terminals of the bipole, are, respectively, denoted as v_y and v_z, while $v = v_y - v_z$ and i, respectively, stand for voltage across and current through the memristor. The architecture of this circuit realization takes inspiration from the design of Batas and Fiedler [45], which, however, was lacking the adaptability of the boundary behavior and the tunability of the degree of non-volatility.

The memristor state x is modeled by the voltage v_θ across capacitance C_x. The series between linear resistor R_{off} and nonlinear voltage-controlled resistor $R(v_\theta) = -\Delta R v_\theta$, where $\Delta R = R_{off} - R_{on}$, implements the input–output equation (13.4).

If the value of window function (13.7) were unitary at all times, as in the original model from Williams [3], state equation (13.6) would be simply implemented by letting memristor current i flow through linear capacitor C_x (in any case a tiny conductance g is placed in parallel to the capacitor so as to prevent node z from floating). However, $f\,(v_\theta(t), \eta\, v(t), v_{th0}, v_{th1}, a, b) \in \{0, a, b\}$ and its behavior is regulated by conditions C_1 and C_5, governing the degree of non-volatility, and by conditions C_2-C_4, determining the boundary behavior.

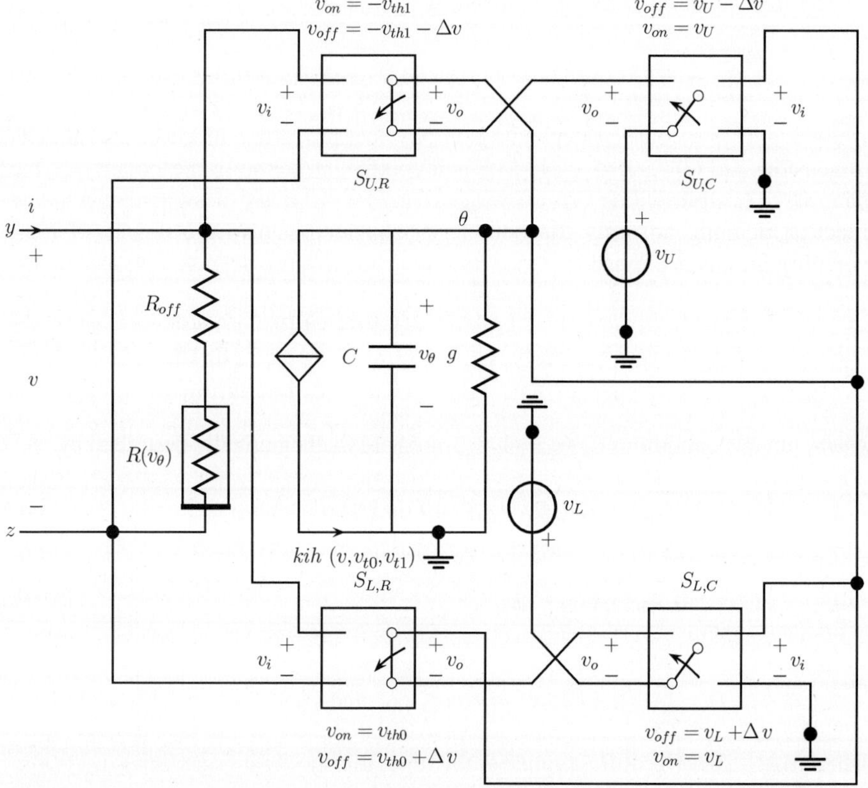

Fig. 13.1 PSpice implementation of the generalized BCM model. Note that Δv denotes for each switch the width of the transition region between on and off states

Conditions (13.8) and (13.12) are implemented by nonlinearly filtering memristor current i before letting it flow through capacitor C_x. This filtering consists of performing the multiplication between a k-scaled version of memristor current i and a nonlinear function $h(v, v_{t0}, v_{t1}, a, b)$, which, under $x \in (0,1)$, is responsible for the modulation of the evolution rate of the state. In particular, under positive (negative) input larger (smaller) than a suitable threshold v_{t0} ($-v_{t1}$) the right-hand-side of state equation (13.6) is multiplied by a factor (b) larger than the factor (a) by which it is multiplied in the sub-threshold input case. Nonlinear function $h(v, v_{t0}, v_{t1}, a, b)$ is mathematically expressed by

$$h(v, v_{t0}, v_{t1}) = b + \frac{a-b}{2}(sign(v + v_{t1}) - sign(v - v_{t0})), \tag{13.13}$$

Note that the multiplication between current ki and function (13.13) may be easily implemented by letting flow through capacitor C_x one of the currents of two complementary-activated parallel branches. One of these branches is activated

Table 13.1 Netlist of the PSpice implementation of the generalized BCM model in Fig. 13.1

```
* Local parameters:
* Ron, Roff: on and off resistances
* delta_R: difference between Roff and Ron
* x0: initial state
* vth0: activation threshold for x=0
* vth1: activation threshold for x=1
* vt0, vt1: activation thresholds for 0<x<1
* Cx: capacitance value
* k: memristor charge scaling factor
* a, b: constants modulating the degree of non-volatility
* delta_v: width of the transition region of the switches
.SUBCKT BCM_MEMRISTOR 1 2 3
* node 1: node y in Fig. 13.1
* node 2: node z in Fig. 13.1
* node 3: node theta in Fig. 13.1
R1 1 8 {Roff}
Vsense1 8 7 0
E1 7 2 VALUE={{-delta_R}*I(Vsense1)*V(3)}
C1 3 0 {Cx} IC={x0}
R3 3 0 1G
Vsense2 10 0 1
G1 0 3 VALUE={k*Cx*I(Vsense1)*V(10)*h(V(1)-V(2),vt0,vt1,a,b)}
S1 3 9 1 2 SMODRH
S2 9 4 3 0 SMODCH
S3 3 6 1 2 SMODRL
S4 6 5 3 0 SMODCL
V1 5 0 0
Vu 4 0 1
.MODEL SMODRH VSWITCH Roff=1G Von={-vth1} Voff={-vth1-delta_v}
.MODEL SMODCH VSWITCH Roff=1G Voff={1-delta_v} Von={1}
.MODEL SMODRL VSWITCH Roff=1G Von={vth0} Voff={vth0+delta_v}
.MODEL SMODCL VSWITCH Roff=1G Voff=delta_v Von=0
.func h(v,v0,v1,a,b)={b+(a-b)/2*SGN(v+v1)-(a-b)/2*SGN(v-v0)}
.ENDS
```

through a voltage-controlled voltage-switch for $v > v_{t0}$ or $v < -v_{t1}$ and carries a current equal to kbi. The other branch is activated through another voltage-controlled voltage-switch for $v \leq v_{t0}$ and $v \geq -v_{t1}$ and carries a current equal to kai.

Boundary conditions (13.9)–(13.11) are modeled by two reference voltage sources, i.e. $v_L = 0$ and $v_U = 1$, respectively, denoting the lower and upper limits of capacitor voltage v_θ (hence the use of letter L or U as subscript of symbol v for the reference voltage source), and by two pairs of voltage-controlled voltage switches, one pair for each of the two memristor state bounds $v_\theta = 0V$ and $v_\theta = 1V$ (the first subscript of symbol S for a switch indicates whether it refers to the lower or upper state bound, hence letter L or U is chosen). Within each pair of switches, the clipping switch is controlled by capacitor voltage v_θ, while the release switch is controlled by input voltage v (the second subscript of symbol S for a switch hints at whether it models the exit from or the entrance into condition C_2 expressed by

(13.9), i.e. the clipping or release event, hence letter C or R is chosen). Basically, for each state bound, node θ is connected to a reference voltage source through the series between the output resistances of the corresponding pair of clipping and release switches. With regard to the upper (lower) state bound, the relative clipping switch remains open, i.e. in the off state, as long as the memristor state keeps below the unitary (above the zero) value. In this case, due to the large output resistance of the clipping switch, reference voltage source v_U (v_L) is unable to constrain the voltage at node θ, irrespective of the behavior of the release switch. However, the clipping switch turns into on state in case v_θ approaches its upper (lower) bound in its ascent (descent). When this occurs, the associated release switch is always closed, i.e. in the on state, thus allowing the memristor state to be clipped at the upper (lower) bound. Only with memristor state v_θ clipped to $+1V$ ($0V$), do the dynamics of the release switch become relevant: this switch turns into off state in case the input voltage v goes below (above) a certain adaptable negative (positive) threshold voltage $-v_{th1}$ (v_{th0}), thus enabling the memristor state to be released from the upper (lower) bound.

Note that it is possible to develop a more realistic implementation of the PSpice circuit of Fig. 13.1 by replacing the voltage-controlled voltage switches with suitable combinations of Complementary-Metal-Oxide-Semiconductor (CMOS) transistors.

13.4 Case Study: Neuromorphic Applications

This section uses the PSpice circuit of the generalized BCM model to model dynamics typical of biological neural networks. One of the most natural ways in which neurons strengthen their synaptic connections is by sending signals to each other at the same time. This primitive form of neural learning is named Hebbian rule [23]. In order to demonstrate that the circuit of Fig. 13.1 does indeed favor Hebbian-based associative learning, we set up a transient simulation (with time step equal to $0.1\,ms$, initial and final time, respectively, fixed to $0\,s$ and $1.4\,s$) in which we excite nodes y and z with pulses of magnitude, let us call it v_p, equal to $-1\,V$ and $+1\,V$, respectively, width, let us name it Δt_p, of value $10\,ms$, rise and fall time $1\,ms$ and period $10\,s$ (i.e. larger than the simulation final time). The time delay of the pulse exciting node y (i.e. the post-synaptic signal), let us name it $t_{d,pos}$, was swept in steps of $0.1\,ms$ from $0.975\,s$ to $1.025\,s$, while that of the pulse exciting node z (i.e. the pre-synaptic signal), let us name it $t_{d,pre}$, was chosen as $1\,s$.

The memristor under modeling is a nano-structure of the kind discussed in Sect. 13.3, therefore $k = \frac{\mu R_{on}}{D^2}$. The BCM parameters were specified as follows: $R_{on} = 526.3158\,\Omega$, $R_{off} = 18182\,\Omega$, $v_\theta(0) = 1\,V$, $D = 10\,nm$, $\mu = 1e-10^{-14}m^2V^{-1}s^{-1}$ (therefore $k = 52631.58C^{-1}$), and $C_x = 50\,\mu F$. The activation threshold voltages at the boundaries (used in conditions (13.9–13.11)) and those within the boundaries (used in conditions (13.8) and (13.12)) are set to $v_{th0} = v_{th1} =$

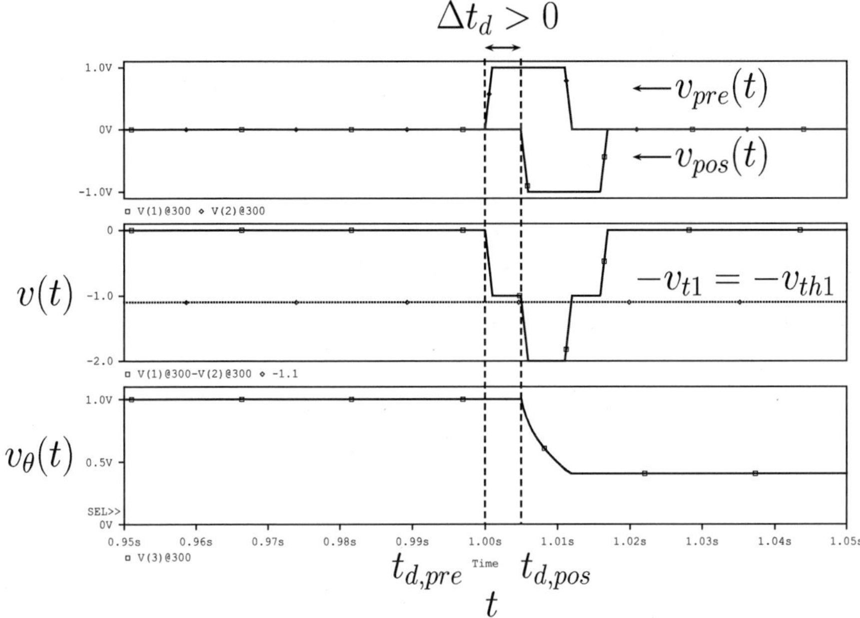

Fig. 13.2 Demonstration of Hebbian-based associative learning under partial temporal overlap between pre- and post-synaptic signals ($\Delta t_d = 0.0049\,s$). *Top plot*: Pre- and post-synaptic pulses. *Middle plot*: Memristor voltage (negative activation thresholds are shown with *dotted lines*). *Bottom plot*: Memristor state

$1.1\,V$ and to $v_{t0} = v_{t1} = 1.1\,V$, respectively. The parameters modulating the degree of non-volatility are set to $a = 0$ and $b = 5$. The width of the transition region of the switches is set to $\Delta v = 0.1\,V$.

Figure 13.2 shows for $t_{d,pos} = 1.0049\,s$ the pulse waveforms at nodes y and z, i.e. $v_y = v_{pos}$ and $v_z = v_{pre}$, the voltage across the memristor, i.e. $v = v_y - v_z$, and the memristor state, modeled by capacitor voltage v_θ in the PSpice circuit of Fig. 13.1. In this case the post- and pre-synaptic pulses overlap in time. The difference between the time delays of such pulses, defined as $\Delta t_d = t_{d,pos} - t_{d,pre}$, is $0.0049\,s$. Only within the overlapping time window is the memristor voltage below the negative activation threshold referring to upper boundary $v_\theta = 1V$, i.e. $-v_{th1}$ (and, since $v_{th1} \geq v_{t1}$, also below the negative activation threshold within the boundaries, i.e. $-v_{t1}$) and, as a result, does the memristor state decrease from its initial unitary value. As Fig. 13.3 demonstrates, the change in memristor state $\Delta v_{theta} = v_{theta} - v_{theta}(0)$ is more significant as the overlapping time window gets larger, i.e. as the magnitude of Δt_d gets smaller. The maximum of the absolute value of Δv_{theta} occurs in fact when the two pulses completely overlap in time, i.e. when $t_{pos} = 1\,s$, implying $\Delta t_d = 0\,s$.

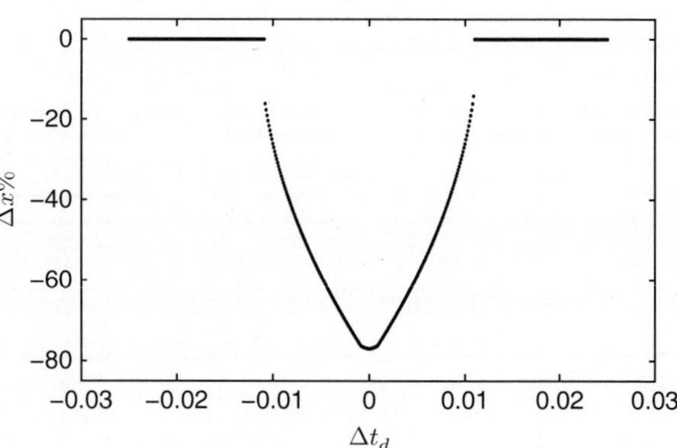

Fig. 13.3 Change in memristor state (recall that $x = v_\theta$, therefore $\Delta x = x - x(0) = v_\theta - v_\theta(0) = \Delta v_\theta$) versus difference of time delays of post- and pre-synaptic signals. The more simultaneous are the pulses, the more pronounced is the change in synaptic strength

13.5 A Novel Class of Passive Memristor Circuits

This section shall introduce a novel class of memristor systems. Each element from the class is an electrical circuit employing only purely passive components from circuit theory (diodes and linear capacitors, inductors and resistors).

Each of the circuits from the class to be presented shall be characterized by a system of differential-algebraic equations of the kind given in (13.2)–(13.3). Section 13.5.1 is devoted to the presentation of the core block of each element from the novel class of memristor systems, i.e. *a switching two-port based upon the Graëtz diode bridge*.

13.5.1 The Graëtz Diode Bridge

Let us consider the full-wave rectifier shown in Fig. 13.4. It is a two-port where v_i and i_i, respectively, denote input voltage and current, while v_o and i_o, respectively, refer to output voltage and current.

The voltage across and the current through diode D_j are, respectively, expressed as v_j and i_j, where $j = \{1,2,3,4\}$. Let us identify the constraints upon voltages and currents of the two-port. These constraints shall play a key role in the emergence of memristor behavior in the circuits from the class to be presented. Application of Kirchhoff's Current Law (KCL) to the input and output port, respectively, yields

$$i_i = i_1 - i_4, \tag{13.14}$$

$$i_i = i_3 - i_2, \tag{13.15}$$

$$i_o = i_1 + i_2. \tag{13.16}$$

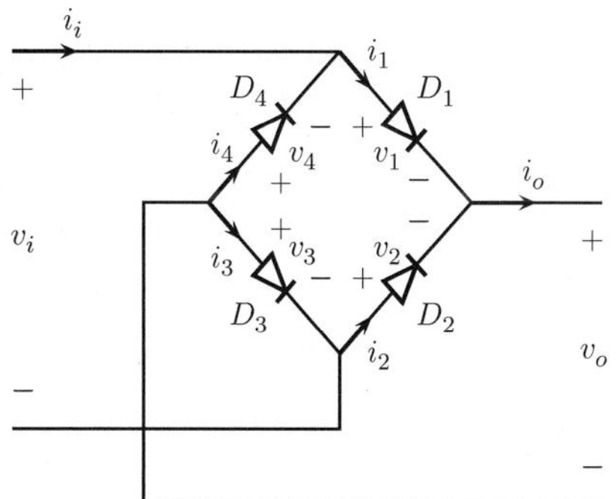

Fig. 13.4 The Graëtz diode bridge

Combining (13.14) and (13.15) yields

$$i_1 + i_2 = i_3 + i_4. \tag{13.17}$$

Applying Kirchhoff's Voltage Law (KVL) to the input and output port gives

$$v_i = v_1 - v_2, \tag{13.18}$$

$$v_i = v_3 - v_4, \tag{13.19}$$

$$v_o = -v_1 - v_4. \tag{13.20}$$

Combination of (13.18)–(13.19) results into

$$v_1 + v_4 = v_2 + v_3. \tag{13.21}$$

Assuming perfectly matched diodes, we express $i_j = i_j(v_j)$, where $j = \{1,2,3,4\}$, as $i_j = I_S\left(\exp\left(v_j n^{-1} V_T^{-1}\right) - 1\right)$, where I_S symbolizes the saturation current, $V_T = KTq^{-1}$ stands for the thermal voltage and n is the emission coefficient, where $K = 1.38 \cdot 10^{-23} JK^{-1}$ is the Boltzmann's constant, T represents the absolute temperature, and $q = 1.6 \cdot 10^{-19} C$ refers to the elementary electronic charge.

Defining $y_j = \exp\left(v_j n^{-1} V_T^{-1}\right)$, (13.17) and (13.21) may be recast as

$$y_1 + y_2 = y_3 + y_4, \tag{13.22}$$

$$y_1 y_4 = y_2 y_3. \tag{13.23}$$

Solving (13.22) for y_1 and inserting the resulting expression into (13.23) gives:

$$y_4^2 + (y_3 - y_2)y_4 - y_2 y_3 = 0,$$

from which, given the sign of y_4, the only acceptable solution is $y_4 = y_2$. Using (13.22), we also have $y_1 = y_3$. Recalling the definition of y_j, we then have $v_4 = v_2$ and $v_1 = v_3$. Note that these two voltage constraints, each involving one pair of parallel diodes, represent *the key mechanism at the origin of the memristor behavior of the circuits to be proposed*. Recalling the current-voltage relationship for a diode it follows that $i_4 = i_2$ and $i_1 = i_3$.

Equations (13.14) and (13.18) for input port current and voltage and (13.16) and (13.20) for output port current and voltage may thus be recasted as

$$i_i = i_1 - i_2, \tag{13.24}$$

$$v_i = v_1 - v_2, \tag{13.25}$$

$$i_o = i_1 + i_2, \tag{13.26}$$

$$v_o = -v_1 - v_2. \tag{13.27}$$

Equations (13.24)–(13.27) represent the four bridge constraints. Let us present the novel class of memristor electronic systems.

13.5.2 Classification and Properties

Each element from the proposed class is characterized by the following properties:

1. The switching two-port of Sect. 13.5.1 is cascaded with a suitable n^{th}-order dynamical one-port employing n linear dynamical elements (capacitors or inductors) and, not necessarily though, some linear resistor.
2. The input voltage v_i and current i_i of the bridge, taken in any prescribed order, denotes input and output of the memristor element.
3. Either the output voltage v_o or the output current i_o of the bridge denotes one of the n state variables of the memristor element. In the first (latter) case the linear dynamical one-port contains a capacitor (an inductor) with voltage v_o across it (current i_o through it).

The first and third properties constrain the set of one-port topologies which may be chosen as load to the Graëtz diode bridge.

Remark 1. The elements from the novel class, one of which was recently presented in [46], represent the first-ever circuit implementations of memristor systems employing only diodes and linear inductors, capacitors and resistors. This discovery contradicts common expectations according to which memristor behavior may not arise out of elementary circuits comprising solely purely passive components known in circuit theory before the advent of the memristor.

The novel class of memristor electronic systems may be split into two sub-classes, respectively, comprising *voltage-controlled* and *current-controlled* systems, i.e. systems where the input, respectively, is voltage v_i and current i_i (and thus the output, respectively, is i_i and v_i). The first sub-class is dealt with in Sect. 13.5.3, while the reader may derive the second class by duality. Within each of such sub-classes, two further sub-classes shall be identified, respectively comprising *voltage-state* and *current-state* systems, i.e. systems where one of the states respectively is voltage v_o and current i_o. Such systems shall be presented in Sects. 13.5.3.1 and 13.5.3.2, respectively.

13.5.3 Voltage-Controlled Systems

The input and output to each of these systems, respectively, are v_i and i_i. Let us present the two sub-classes a circuit of this kind may belong to.

13.5.3.1 Voltage-Controlled Voltage-State Systems

For these systems one of the states is v_o. The most appropriate representation of the two-port of Fig. 13.4 for the synthesis of such systems is the current-voltage form. Let us derive it. Solving (13.25)–(13.27) for v_1 and v_2 yields

$$v_1 = \frac{v_i - v_o}{2}, \tag{13.28}$$

$$v_2 = -\frac{v_i + v_o}{2}. \tag{13.29}$$

Recalling the current-voltage relationship for a diode and using (13.28)–(13.29) into (13.24) and (13.26), the current-voltage representation of the two-port of Fig. 13.4 is found to be:

$$i_i = 2I_S \exp\left(-\frac{v_o}{2nV_T}\right) \sinh\left(\frac{v_i}{2nV_T}\right), \tag{13.30}$$

$$i_o = 2I_S \exp\left(-\frac{v_o}{2nV_T}\right) \cosh\left(\frac{v_i}{2nV_T}\right) - 2I_S. \tag{13.31}$$

Equation (13.30) may be recast as

$$i_i = g(v_o, v_i) v_i, \tag{13.32}$$

with $g(\cdot,\cdot)$ expressed by

$$g(v_o, v_i) = \frac{I_S}{nV_T} \exp\left(-\frac{v_o}{2nV_T}\right) \sum_{k=0}^{\infty} \frac{\left(\frac{v_i}{2nV_T}\right)^{2k}}{(2k+1)!}, \tag{13.33}$$

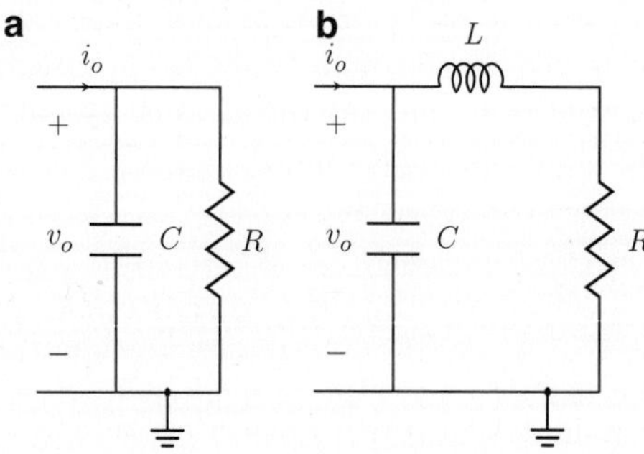

Fig. 13.5 First-order (**a**) and second-order (**b**) linear dynamic one-ports for voltage-controlled voltage-state circuits

where we used the Taylor series expansion of the hyperbolic sine [47]. From (13.32) it follows that any time $v_i = 0$, then $i_i = 0$ and vice versa. This is the so-called *zero crossing property*, typical of a memristor system [2]. Equation (13.32) models the input–output relation of the voltage-controlled voltage-state circuits, whose memductance function is expressed by (13.33).

The state equation of the elements from this class depends on the particular linear dynamic one-port chosen as load to the full-wave rectifier. After choosing a particular one-port topology (making sure it contains a capacitor with voltage v_o across it), the constitutive equations of the dynamical elements within the one-port are then written down. Then, inserting (13.31) into these constitutive equations yields the state equations of a voltage-controlled voltage-state system. Let us present examples of first- and second-order circuits of this kind, deriving their state equations.

- **First-order circuit**

With regard to a first-order case, let us close the output port of the diode bridge onto the parallel combination of a capacitor of value C and of a resistor of value R (see Fig. 13.5a). Inserting (13.31) into the constitutive equation of the capacitor, i.e. $i_o - \frac{v_o}{R} = C\frac{dv_o}{dt}$, the state equation of the resulting system is found to be

$$\frac{dv_o}{dt} = \frac{2I_S}{C}\exp\left(-\frac{v_o}{2nV_T}\right)\cosh\left(\frac{v_i}{2nV_T}\right) - \frac{2I_S}{C} - \frac{v_o}{RC}, \tag{13.34}$$

where v_o denotes the state of the system. This first-order voltage-controlled voltage-state memristor circuit is modeled by (13.32) and (13.34).

- **Second-order circuit**

Let us introduce a second-order example. Let the two-port be cascaded with the second-order one-port of Fig. 13.5b, which is an inductor L-capacitor C parallel circuit augmented with the series resistance R of the inductor and characterized by a resonance frequency expressed by $\omega_o = \sqrt{\frac{1}{LC} - \left(\frac{R}{L}\right)^2}$. Choosing v_o and i_L, the current through the inductor, as the states of the system, writing down the constitutive equations of the dynamical elements of the one-port, and using (13.31) into them, the following state equations are finally obtained:

$$\frac{d}{dt}\begin{bmatrix} v_o \\ i_L \end{bmatrix} = \begin{bmatrix} -\frac{1}{C}i_L + \frac{2I_S}{C}\exp\left(-\frac{v_o}{2nV_T}\right)\cosh\left(\frac{v_i}{2nV_T}\right) - \frac{2I_S}{C} \\ \frac{1}{L}\left(v_o - Ri_L\right) \end{bmatrix}. \tag{13.35}$$

In conclusion, (13.32) and (13.35) define this second-order voltage-controlled voltage-state memristor circuit.

13.5.3.2 Voltage-Controlled Current-State Systems

For these systems one of the states is i_o. The use of the inverse hybrid representation of the two-port of Fig. 13.4 is the most appropriate for the synthesis of these elements. Let us derive such representation. Rearranging (13.31), we have:

$$2I_S\exp\left(-\frac{v_o}{2nV_T}\right) = \frac{(i_o + 2I_S)}{\cosh\left(\frac{v_i}{2nV_T}\right)}. \tag{13.36}$$

Using (13.36) into (13.30) and extracting from (13.36) an expression for v_o as function of v_i and i_i, the inverse hybrid representation of the two-port turns out to be

$$i_i = (i_o + 2I_S)\frac{\sinh\left(\frac{v_i}{2nV_T}\right)}{\cosh\left(\frac{v_i}{2nV_T}\right)}, \tag{13.37}$$

$$v_o = -2nV_T\ln\left(\frac{i_o + 2I_S}{2I_S\cosh\left(\frac{v_i}{2nV_T}\right)}\right). \tag{13.38}$$

Equation (13.37) may be recast as

$$i_i = g(i_o, v_i)v_i, \tag{13.39}$$

with $g(\cdot,\cdot)$ given by

$$g(i_o, v_i) = \frac{(i_o + 2I_S)}{2nV_T}\frac{\sum_{k=0}^{\infty}\frac{\left(\frac{v_i}{2nV_T}\right)^{2k}}{(2k+1)!}}{\sum_{k=0}^{\infty}\frac{\left(\frac{v_i}{2nV_T}\right)^{2k}}{(2k)!}}, \tag{13.40}$$

where we used the Taylor series expansions of the hyperbolic sine and cosine [47].

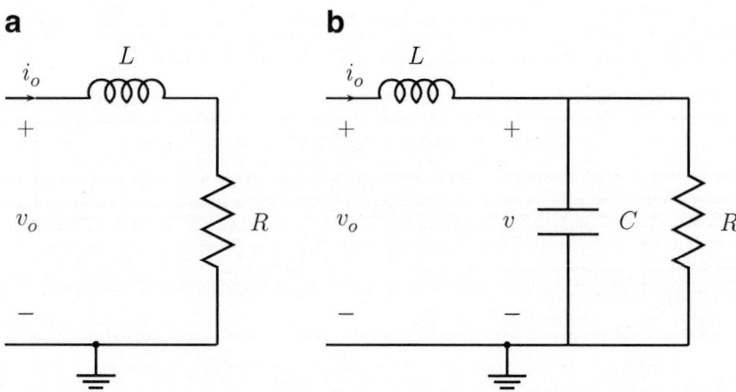

Fig. 13.6 First-order (**a**) and second-order (**b**) linear dynamic one-ports for voltage-controlled current-state circuits

From (13.39) we deduce that $v_i = 0$ implies $i_i = 0$ and viceversa. Equation (13.39) defines the input–output relation of the voltage-controlled current-state circuits, whose memductance function is modeled by (13.40).

The state equation of the elements from this class depends on the particular linear dynamic one-port chosen as load to the full-wave rectifier. After choosing a particular one-port topology (making sure it contains an inductor with current i_o through it), the state equations of a voltage-controlled current-state circuit are obtained by inserting (13.38) into the constitutive equations of capacitors and inductors of the one-port. Let us describe examples of first- and second-order circuits of this kind and determine their state equations.

- **First-order circuit**

With regard to a first-order case study, the series combination between an inductor L and a resistor R, as given in Fig. 13.6a, is taken as the load of the switching network of Fig. 13.4. Inserting (13.38) into the constitutive equation of the inductor, i.e. $v_o - Ri_o = L\frac{di_o}{dt}$, yields the following state equation:

$$\frac{di_o}{dt} = -\frac{2nV_T}{L}\ln\left(\frac{i_o + 2I_S}{2I_S\cosh\left(\frac{v_i}{2nV_T}\right)}\right) - \frac{R}{L}i_o, \tag{13.41}$$

where i_o denotes the state of the system. In conclusion, (13.39) and (13.41) define this first-order voltage-controlled current-state memristor circuit.

- **Second-order circuit**

With regard to a second-order example, let us close the output port of the full-wave rectifier of Fig. 13.4 onto the inductor L-capacitor C series circuit augmented with the parallel resistance R of the capacitor. The resonance frequency of such second-order one-port, shown in Fig. 13.6b, is expressed by $\omega_o = \sqrt{\frac{1}{LC} - \left(\frac{1}{RC}\right)^2}$. Writing

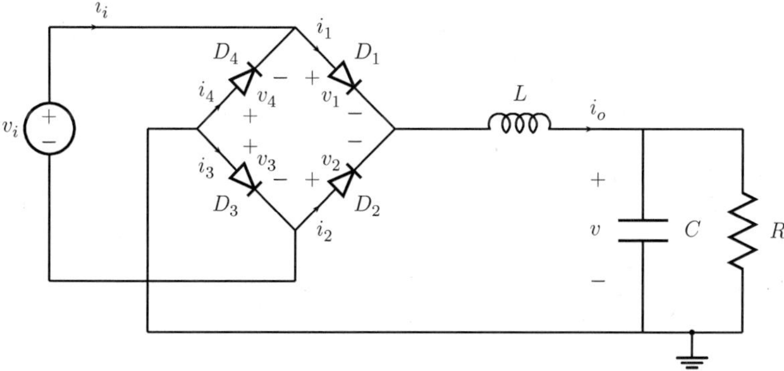

Fig. 13.7 A second-order voltage-controlled current-state memristor element from the proposed class. The element is driven by input voltage source v_i

down the constitutive equations of the dynamic elements of the one-port and the inserting (13.38) into them, the state equations are found to be:

$$\frac{d}{dt}\begin{bmatrix} v \\ i_o \end{bmatrix} = \begin{bmatrix} \frac{1}{C}\left(i_o - \frac{v}{R}\right) \\ -\frac{1}{L}v - \frac{2nV_T}{L}\ln\left(\frac{i_o+2I_S}{2I_S\cosh\left(\frac{v_i}{2nV_T}\right)}\right) \end{bmatrix}, \tag{13.42}$$

where v, the voltage across the capacitor, and i_o denote the states of the system.

In conclusion, the defining equations of this second-order voltage-controlled current-state memristor circuit are (13.39) and (13.42).

13.5.4 Simulation Results

With reference to the voltage-controlled current-state second-order memristor circuit of Fig. 13.7 [46], making use of the diode bridge of Fig. 13.4 loaded by the second-order one-port shown in Fig. 13.6b and discussed in Sect. 13.5.3.2, the system state is expressed as $\mathbf{x} = [x_1\, x_2]'$, where state variables are defined as $x_1 = v\,(V_T)^{-1}$ and $x_2 = i_o\,(I_S)^{-1}$. Further system input and output are chosen as $u = v_i\,(V_T)^{-1}$ and $y = i_i\,(I_S)^{-1}$ respectively, and dimensionless time variable is taken as $\tau = t\,(t_0)^{-1}$, where $t_0 = 2\pi\,(\omega_0)^{-1}$ stands for the time normalization factor and ω_0 is the resonant frequency of the one-port of Fig. 13.6b, which we previously defined. After some algebraic manipulation we get:

$$\frac{d\mathbf{x}}{d\tau} = \begin{bmatrix} \beta(x_2 - \alpha x_1) \\ \gamma\left(-x_1 - 2n\ln\left(\frac{x_2+2}{2\cosh\left(\frac{u}{2n}\right)}\right)\right) \end{bmatrix} \tag{13.43}$$

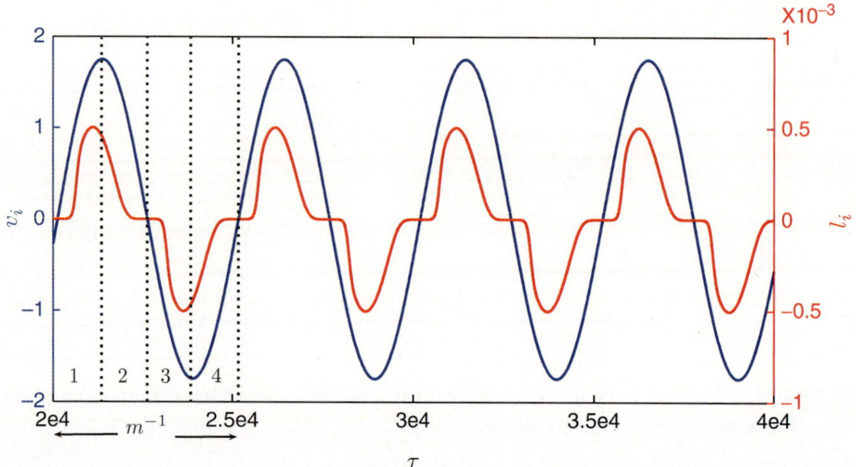

Fig. 13.8 Time waveforms of current i_i (*red signal*) and voltage v_i (*blue signal*) under sinusoidal excitation with $v_{io} = 1.75\,V$ and $f_i = 10\,Hz$. The dimensionless input period m^{-1} is divided into 4 intervals, numbered from 1 to 4, separating zeros, minimum and maximum of the voltage waveform

$$y = (x_2 + 2)\frac{\sinh\left(\frac{u}{2n}\right)}{\cosh\left(\frac{u}{2n}\right)} \tag{13.44}$$

where $\alpha = \frac{V_T}{RI_S}$, $\beta = \frac{t_0 I_S}{CV_T}$ and $\gamma = \frac{t_0 V_T}{LI_S}$ are dimensionless parameters. The Matlab software environment [48] was used for the numerical integration of the *mathematical model* of the memristor circuit of Fig. 13.7, i.e. (13.43)–(13.44), for a sine-wave input source with amplitude $v_{io} = 1.75\,V$ and varying frequency f_i, expressed as $v_i = v_{i0}\sin(2\pi f_i t)$, which yields $u = u_{i0}\sin(2\pi m\tau)$, where $u_{io} = v_{io}\,(V_T)^{-1}$ and $m = f_i\,t_0$ denotes the dimensionless input frequency. The values of the circuit components were set to $R = 1.5\,k\Omega$, $C = 4\,\mu F$, and $L = 2.5\,\mu H$. The values for saturation current I_S and emission coefficient n of the four matched diodes were respectively taken as $2.682 \cdot 10^{-9}$ and 1.836, i.e. as in the case of standard diode $D1N4148$. The initial conditions of the voltage across the capacitor and of the current through the inductor are respectively chosen as $v(0) = 0.01\,V$ and $i_L(0) = 0.01\,A$. Ordinary differential equation solver ode15*s* [48] was employed to integrate (13.43)–(13.44) from $\tau = 0$ up to τ equal to 10 times the dimensionless input period $m^{-1} = f_i^{-1}\,t_0^{-1}$. Under such parameter setting, letting the input frequency $f_i = 10\,Hz$, the time evolutions of voltage v_i and current i_i are depicted in Fig. 13.8, from which it is evident that voltage and current exhibit zeros at the same instants but have misaligned maxima and minima. As a result, the circuit of Fig. 13.7 manifests the typical pinched hysteretic current-voltage loop characterizing memristor systems, as it is shown in Fig. 13.9 (black bow-tie). With reference to Fig. 13.8, note that over each normalized period m^{-1} the maximum and minimum of the current always occur before the maximum and minimum of the voltage. As a result, following the path

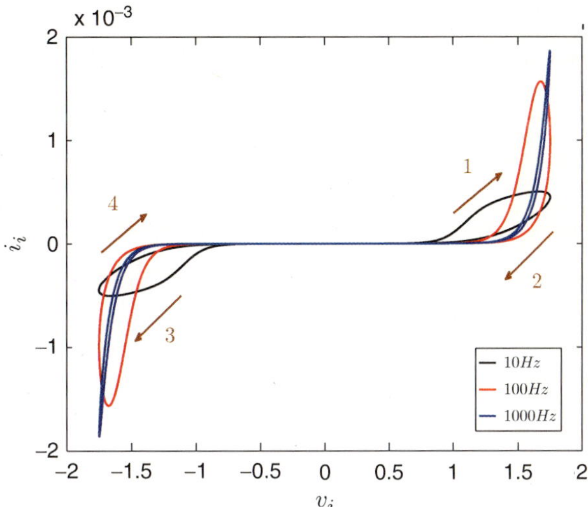

Fig. 13.9 Current i_i-voltage v_i bow-ties under sinusoidal excitation with $v_{io} = 1.75V$ and f_i, respectively, equal to $10Hz$ (*black loop*), $100Hz$ (*red loop*), and $1000Hz$ (*blue loop*). *Brown arrows*, mapping one-to-one with time intervals 1-4 in Fig. 13.8, show the non-self-crossing property of the i_i-v_i loop for $f_i = 10Hz$ (note that this property is exhibited by the other loops as well)

drawn by the trajectory point on the i_i-v_i plane in one period, as indicated by the four consecutively numbered brown arrows in Fig. 13.9 (corresponding to the four intervals in which the period is divided in Fig. 13.8), it may be realized that the loop is non-self-crossing, i.e. it is of type II, according to the definition given by Biolek in [49]. With reference to Fig. 13.7, the voltages across the bridge diodes may be expressed as

$$v_1 = v_3 = nV_T \ln\left(\frac{x_2 + 2}{2\cosh\left(-\frac{u}{2n}\right)\exp\left(-\frac{u}{2n}\right)}\right),$$
$$v_2 = v_4 = v_1 - v_i. \tag{13.45}$$

Figure 13.10 shows the time dependence of v_1 and v_2 in the simulation of Fig. 13.8.

Sweeping frequency above $10Hz$, the lobes of the loop get increasingly squeezed (while stretching along the i_i axis), as it is demonstrated in Fig. 13.9, where the red and blue bow-ties respectively refer to an input frequency f_i set to $100Hz$ and $1000Hz$. Note that these other two loops also are of type II.

It is worth pointing out that at infinite frequency, when the inductor and the capacitor respectively are an open and a short circuit, the electronic system of Fig. 13.7 behaves as a nonlinear resistor. Furthermore, sweeping frequency below $10Hz$ also yields a gradual flattening of the loop lobes. Finally, bear in mind that nonlinearly resistive behavior also arises at direct current (dc), when the inductor and the capacitor respectively are a short and an open circuit.

An experimental proof for the occurrence of memristor behavior in the circuit of Fig. 13.7 is reported in [50].

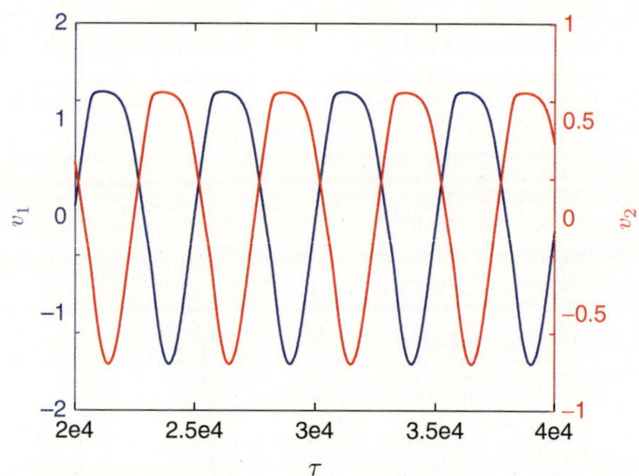

Fig. 13.10 Voltage drops across the bridge diodes for the sinusoidal excitation at frequency $f_i = 10Hz$

13.6 Conclusions

After a brief review of the memristor models available in the literature, this paper describes the PSpice-based implementation of the generalized Boundary Condition Memristor (BCM) model, which stands out over the other models thanks to the adaptability of the boundary behavior and to the tunability of the non-volatility degree. The first part of the paper ends with a case study where the use of the PSpice emulator sheds light into the synapse-like behavior of the memristor. Such circuit implementation of the generalized BCM model may be of great help to researchers willing to investigate in the user-friendly PSpice environment the extraordinary opportunities memristors offer in integrated circuit design.

The second part of the paper introduces a class of purely passive circuits, each made up of a nonlinear static two-port (a full-wave rectifier employing a four diode bridge) cascaded with a linear dynamic one-port (employing standard linear components from circuit theory, namely resistors, inductors and capacitors). The state equations of these circuits fall into the class of memristor systems, as originally formulated by Chua and Kang in 1976 [2]. This manuscript presents voltage-controlled elements from the proposed class. Dual memristor emulators with current-control may be derived in a similar manner [50]. These novel circuits may be used to introduce the undergraduate students to the concept of memory systems [51, 52]. In conclusion, it is important to note that all the novel memristor circuits proposed in this manuscript are *volatile*. However, we conjecture that non-volatility could be attained by inserting active elements into the one-port loading the diode bridge. This shall be the topic of future studies, where we aim at increasing the complexity of the circuits presented in this manuscript so as to model memristor systems within the Hodgkin-Huxley neuron [53].

Acknowledgments This work was partially supported by the CRT Foundation, under the project no. 2012.1121 and by the Ministry of Foreign Affairs "*Con il contributo del Ministero degli Affari Esteri, Direzione Generale per la Promozione del Sistema Paese*."

References

1. L.O. Chua, Memristor: the missing circuit element. IEEE Trans. Circuit Theory **18**(5), 507–519 (1971)
2. L.O. Chua, S.M. Kang, Memristive devices and systems. Proc. IEEE **64**(2), 209–223 (1976)
3. D.B. Strukov, G.S. Snider, D.R. Stewart, R.S. Williams, The missing memristor found. Nature **453**, 80–83 (2008)
4. P.O. Vontobel, W. Robinett, P.J. Kuekes, D.R. Stewart, J. Straznicky, R.S. Williams, Writing to and reading from a nano-scale crossbar memory based on memristors. Nanotechnology **20**(42), 425204(1)-(21) (2009)
5. Y.V. Pershin, M. Di Ventra, Practical approach to programmable analog circuits with memristors. IEEE Trans. Circuits Syst. I **57**(8), 1857–1864 (2010)
6. C. Zamarreño-Ramos, L.A. Camuñas-Mesa, J.A. Pérez-Carrasco, T. Masquelier, T. Serrano-Gotarredona, B. Linares-Barranco, On spike-timing-dependent-plasticity, memristive devices, and building a self-learning visual cortex. Frontiers in Neuromorphic Engineering. Front. Neurosci. **5**(26), 1–22 (2011)
7. G.S. Borghetti, P.J. Snider, J.J. Kuekes, D.R. Yang, R.S. Stewart, R.S. Williams, Memristive switches enable stateful logic operations via material implication. Nat. Lett. **464**(7290), 873–876 (2010)
8. Q. Xia, W. Robinett, M.W. Cumbie, N. Banerjee, T.J. Cardinali, J.J. Yang, W. Wu, X. Li, W.M. Tong, D.B. Strukov, G.S. Snider, G. Medeiros-Ribeiro, R.S. Williams, Memristor-CMOS hybrid integrated circuits for reconfigurable logic. Nano Lett. **9**(10), 3640–3645 (2009)
9. Y.N. Joglekar, S.T. Wolf, The elusive memristive element: properties of basic electrical circuits. Eur. J. Phys. **30**, 661–675 (2009)
10. Z. Biolek, D. Biolek, V. Biolková, Spice model of memristor with nonlinear dopant drift. Radioengineering **18**(2), 210–214 (2009)
11. S. Shin, K. Kim, S.O. Kang, Compact models for memristors based on charge-flux constitutive relationships. IEEE Trans. Comput. Aided Des. Integr. Circuits Syst. **29**(4), 590–598 (2010)
12. F. Corinto, A. Ascoli, A boundary condition-based approach to the modeling of memristor nanostructures. IEEE Trans. Circuits Syst. I **59**(11), 2713–2726 (2012)
13. M.D. Pickett, D.B. Strukov, J.L. Borghetti, J.J. Yang, G.S. Snider, D.R. Stewart, R.S. Williams, Switching dynamics in titanium dioxide memristive devices. J. Appl. Phys. **106**(7), 074508(1)-(6) (2009)
14. S. Kvatinski, E.G. Friedman, A. Kolodny, U.C. Weiser, TEAM: threshold adaptive memristor model. IEEE Trans. Circuits Syst. I **60**(1), 211–221 (2013)
15. W. Cai, R. Tetzlaff, Advanced memristive model of synapses with adaptive thresholds. Proceedings of IEEE International Workshop on Cellular Nanoscale Networks and Their Applications, Turin, 29–31 August 2012
16. R.C. Froemke, Y. Dan, Spike-timing-dependent synaptic modification induced by natural spike trains. Nature **416**(6879), 433–438 (2002)
17. T. Oka, N. Nagaosa, Interfaces of correlated electron systems: proposed mechanism for colossal electroresistance. Phys. Rev. Lett. **95**(26), 64031–64034 (2005)
18. A. Beck, J.G. Bednorz, Ch. Gerber, C. Rossel, D. Widmer, Reproducible switching effect in thin oxide films for memory applications. Appl. Phys. Lett. **77**(1), 140 (2000)
19. E. Linn, R. Rosezin, C. Kügeler, R. Waser, Complementary resistive switches for passive nanocrossbar memories. Nat. Mater. **9**, 403–406 (2010)

20. L.O. Chua, Resistance switching memories are memristors. Appl. Phys. A **102**(4), 765–783 (2011)
21. A. Ascoli, F. Corinto, R. Tetzlaff, Generalized Boundary Condition Memristor Model. IEEE Trans. Circuits Syst. I, under revision (2013)
22. Cadence Design Systems, in *OrCad PSpice User's Guide*, OrCAD, Inc., USA. Available online as PSpice.pdf, 1998, http://www.electronics-lab.com/downloads/schematic/013/
23. D.O. Hebb, *The Organization of Behavior: A Neuropsychological Theory* (Wiley, New York, 1949)
24. T. Prodromakis, B.P. Peh, C. Papavassiliou, C. Toumazou, A versatile memristor model with nonlinear dopant kinetics. IEEE Trans. Electron Devices **58**(9), 3099–3105 (2011)
25. S. Benderli, T.A. Wey, On Spice macromodelling of TiO_2 memristors. Electron. Lett. **45**(7), 377–379 (2009)
26. *Á*. R*á*k, G. Cserey, Macromodeling of the Memristor in SPICE. IEEE Trans. Comput. Aided Des. Integr. Circuits Syst. **29**(4), 632–636 (2010)
27. O. Kavehei, A. Iqbal, Y.S. Kim, K. Eshraghian, S.F. Al-Sarawi, D. Abbott, The fourth element: Characteristics, modelling, and electromagnetic theory of the memristor. Proc. Roy. Soc. A, Math. Phys. Eng. Sci. **466**(2120), 2175–2202 (2010)
28. T. Hee Kim, E.Y. Jang, N.J. Lee, D.J. Choi, K.-J. Lee, J.-T. Jang, J.-S. Choi, S.H. Moon, J. Cheon, Nanoparticle assemblies as memristors. Nano Lett. **9**(6), 2229–2233 (2009)
29. E. Lehtonen, M. Laiho, CNN using memristors for neighborhood connections. IEEE International Workshop on Cellular Nanoscale Networks and their Applications, pp. 1–4, Berkeley, CA, 3–5 February 2010
30. J.J. Yang, M.D. Pickett, X. Li, D.A.A. Ohlberg, D.R. Stewart, R.S. Williams, Memristive switching mechanism for metal/oxide/metal nanodevices. Nat. Nanotechnol. **3**(7), 429–433 (2008)
31. L.O. Chua, L. Yang, Cellular neural networks: theory. IEEE Trans. Circuits Syst. **35**(10), 1257–1272 (1988)
32. T. Roska, L.O. Chua, The CNN universal machine: an analogic array computer. IEEE Trans. Circuits Syst. II: Analog Digit. Signal Process. **40**(3), 163–173 (1993)
33. E. Lehtonen, J. Poikonen, M. Laiho, W. Lu, Time-dependence of the threshold voltage in memristive devices. IEEE International Symposium on Circuits and Systems, pp. 2245–2248, Rio de Janeiro, 15–18 May 2011
34. M.Pd. Sah, C. Yang, H. Kim, L.O. Chua, A voltage-mode memristor bridge synaptic circuit with memristor emulators. Sensors **12**(3), 3587–3604 (2012)
35. D. Strukov, R.S. Williams, Exponential ionic drift: Fast switching and low volatility of thin-film memristors. Appl. Phys. A Mater. Sci. Process. **94**(3), 515–519 (2009)
36. J.G. Simmons, Generalized formula for the electric tunnel effect between similar electrodes separated by a thin insulating film. J. Appl. Phys. **34**(6), 1793–1803 (1963)
37. H. Abdalla, M.D. Pickett, Spice modeling of memristors. IEEE International Symposium on Circuits and Systems, pp. 1832–1835, Rio de Janeiro, 15–18 May 2011
38. Y.V. Pershin, M. Di Ventra, Spice model of memristive devices with threshold. Available online at arXiv, 2012, http://lanl.arxiv.org/abs/1204.2600v4
39. Y.V. Pershin, S. La Fontaine, M. Di Ventra, Memristive model of amoeba learning. Phys. Rev. E **80**(2), 021926(1)–021926(6) (2009)
40. B. Linares-Barranco, T. Serrano-Gotarredona, Memristance can explain spike-time-dependent-plasticity in neural synapses. Available on-line at Nature Precedings, http://hdl.handle.net/10101/npre.2009.3010.1
41. G.E. Pazienza, J. Albo-Canals, Teaching memristors to EE undergraduate students. IEEE Circuits Syst. Mag. **11**(4), 36–44 (2011)
42. K. Eshraghian, O. Kavehei, K.-R. Cho, J.M. Chappell, A. Iqbal, S.F. Al-Sarawi, D. Abbott, Memristive device fundamentals and modeling: Applications to circuits and systems simulation. Proc. IEEE **100**(6), 1991–2007 (2012)
43. A. Ascoli, F. Corinto, V. Senger, R. Tetzlaff, Memristor model comparison. IEEE Circuits Syst. Mag. **13**(2), 89–105 (2013)

44. A. Ascoli, R. Tetzlaff, F. Corinto, M. Gilli, PSpice switch-based versatile memristor model. Proceedings of International Symposium on Circuits and Systems, Beijing, 19–23 May (2013)
45. D. Batas, H. Fiedler, A memristor SPICE implementation and a new approach for magnetic flux-controlled memristor modeling. IEEE Trans. Nanotechnol. **10**(2), 250–255 (2011)
46. F. Corinto, A. Ascoli, Memristive diode bridge with LCR filter. Electron. Lett. **48**(14), 824–825 (2012)
47. M. Abramowitz, I.A. Stegun, *Handbook of Mathematical Functions, with Formulas, Graphs and Mathematical Tables* (Dover Publications, Inc., New York, 1972)
48. B.R. Hunt, R.L. Lipsman, J.M. Rosenberg, K.R. Coombes, J.E. Osborn, G.J. Stuck, *A Guide to MATLAB: For Beginners and Experienced Users* (Cambridge University Press, New York, 2006)
49. D. Biolek, Z. Biolek, V. Biolkova, Pinched hysteretic loops of ideal memristors, memcapacitors and meminductors must be self-crossing. Electron. Lett. **47**(25), 1385–1387 (2011)
50. F. Corinto, A. Ascoli, The simplest class of passive Memristor Emulators. submitted to IEEE Trans. Circuits Syst. I, (2013)
51. L.O. Chua, The fourth element. Proc. IEEE **100**(6), 1920–1927 (2012)
52. Y.V. Pershin, M. Di Ventra, Memory effects in complex materials and nanoscale systems. Adv. Phys. **60**(2), 145–227 (2011)
53. F. Corinto, S.-M. Kang, A. Ascoli, Memristor based neural circuits. International Symposium on Circuits and Systems, Beijing , 19–23 May 2013

Index

R. Tetzlaff (ed.), *Memristors and Memristive Systems*,
DOI 10.1007/978-1-4614-9068-5,